# Fuel Cells

Supramaniam Srinivasan

# Fuel Cells

## From Fundamentals to Applications

With 50 Figures

 Springer

Supramaniam Srinivasan
(Deceased)

ISBN   978-1-4419-3772-8        e-ISBN   978-0-387-35402-6

Printed on acid-free paper.

Printed in the United States of America.     (MVY)

9  8  7  6  5  4  3  2  1

springer.com

# FOREWORD

Dr. Supramaniam Srinivasan (1932-2004) obtained his Bachelor of Science with Honors in Chemistry from the University of Ceylon in 1955 and his PhD in Physical Chemistry from the University of Pennsylvania in 1963. The title of his PhD is Mechanism of Electrolyte Hydrogen Evolution: An Isotope Effect Study, which he did under the direction of Professor John O'Mara Bockris. Srini, as he liked to be called, was very productive throughout his career. He spent several years as a post-doc with Dr. Bockris and published his first book on fuel cells with Dr. Bockris in 1969. The title of their book is *Fuel Cells: Their Electrochemistry*, which was published by McGraw-Hill. In 1966, Srini began serving as an Assistant Professor of Electrochemistry in Surgical Research at the Downstate Medical Center, State University of New York (DMC-SUNY) in Brooklyn, New York. While he was there, Srini published several articles on the use of electrochemical methods in medical applications and later wrote a chapter on the subject entitled "An Electrochemical Approach to the Solution of Cardiovascular Problems," which appeared as Chapter 8 in Volume 10 of the *Comprehensive Treatise of Electrochemistry, Biochemistry*. Srini left DMC-SUNY in 1973 and served at several institutions including the Los Alamos National Laboratory (LANL) where he returned to more energy related electrochemical topics. Srini left LANL in 1988 and began serving as the Deputy Director of the Center for Electrochemical Systems and Hydrogen Research at Texas A&M University where I was at that time. Srini and I worked closely with Dr. B. Dave who did some seminal work on the molten carbonate fuel cell system. I learned during that time how dedicated Srini was to excellence.

Srini's quest for knowledge was much more than a job to him; he felt it was his moral obligation to seek new information and share it with others. In 1996, he received the Energy Technology Division Research Award from the

Electrochemical Society and in 2001 he became an ECS Fellow because of his dedication to producing scholarly work and practical devices. Srini left TAMU in 1997 and spent the end of his career serving as a Visiting Research Collaborator at the Center for Energy and Environmental Studies at Princeton University. He will be remembered for his hard work and dedication. It was a pleasure for me and many others to have known him and his lovely wife, Mangai. I am sure that you will enjoy this last contribution from Dr. Supramaniam Srinivasan.

Srini's book has four parts and contains 12 chapters. Part 1 reviews the history of electrochemistry, electrode/electrolyte interfaces, structure and kinetics of charge transfer, and electrochemical technologies and applications. Part II covers fuel cell principles, electrode kinetics and electrocatalysis of fuel cell reactions, and experimental techniques. Part III covers modeling studies of fuel cell characteristics, and fuels, fuel processing and fuel storage/transmission/distribution. Part IV presents the status of fuel cell technologies, applications and economics of fuel cell power plants/power sources, competing technologies, and conclusions and prognosis.

In chapter 1, Srini includes a review of the report made by Volta at the Royal Society in London about chemical to electrical energy conversion. Srini next moves to the reverse process found by Nicholson and Carlisle reflecting conversion of electricity to chemicals. He then gives the definition of Faraday's two laws and moves to the electrochemical cell and the Volta battery. Srini gives various definitions and shows that the scope of electrochemistry is multi-disciplinary in nature. Srini discusses corrosion briefly and ends this chapter with a discussion of Nernst potentials for energy storage systems.

Srini begins chapter 2 with structure, charge, and capacitance characteristics of the electrode/electrolyte interface. He also discusses single and multi-step reactions at the electrode/electrolyte interface. He follows with a discussion of the concept of a rate-determining step and the dependence of current density on potential for activation controlled reactions. He also presents other types of rate limitations and overpotential and their effects on current density potential behavior. Srini finishes this chapter with a discussion on electrocatalysis and pseudocapacitors.

Chapter 3 begins with a discussion of the role of electrochemical technologies in the chemical industry. Aluminum production is reviewed followed by a discussion of chlorine and caustic soda. In the next two sections of this chapter Srini reviews electro-organic synthesis and electrowinning and refining of metals. Next, Srini presents a discussion of corrosion inhibition/passivation followed by a section on electrochemical energy storage of fuel cells and batteries. The last section, section 8, in this chapter includes a review of the development of electrochemical sensors.

Chapter 4 discusses fuel cell principles while chapter 5 reviews the development of electrode kinetics and electrocatalysis of fuel cell reactions. Chapter 6 presents experimental methods for studying fuel cells.

Chapter 7 begins with a general overview of simulation of fuel cells and the associated benefits to research and development in fuel cells. Modeling of half-cell reactions is presented in section 2 of this chapter. Section 3 includes a discussion of electrolyte overpotentials and Ohm's Law followed by a discussion of additional efficiency losses at the single cell level. Section 5 further discusses the efficiency lost in cell stacks. This chapter ends with a discussion of the modeling of fuel cell power plants.

Chapter 8 contains a discussion of fuels, fuel processing and fuel storage/transmission/distribution followed by a discussion in Chapter 9 of the status of fuel cell technologies. Chapter 10 discusses the application and economics of fuel cell power plants and power sources.

Chapter 11 follows the development of competing technologies with conclusions and prognoses in Chapter 12.

I wish to extend my thanks to those who directly and indirectly assisted in the completion of this book by Dr. Supramaniam Srinivasan.

Ralph E. White

# *Publisher's Preface*

I am greatly honored to have worked with Srini on this textbook, and pleased that the book is being carried to completion despite Srini's absence. I regret that it could not have been finished in time for the symposium in his honor at the fall 2005 meeting of the Electrochemical Society in Los Angeles, but I am proud to be one of the many people who have worked to fulfill Srini's vision for this volume.

While Srini is the author of this book, I would like to take this opportunity to record my gratitude to the many people without whose selfless help this project could not have been completed. Naming those people whom I know have helped courts the danger of omitting others, and I must make plain at the outset that I do not know everybody who has worked on this book. However, I must acknowledge the following: Tilak Bommaraju, Lakshmi Krishnan, Carlos Marozzi, F. J. Luczak (Glastonbury, CT), S. Sarangapani (ICET, Inc., Norwood, MA), Paola Costamanga, Joan Ogden, Brent Kirby, Edward Miller, and Christopher Yang.

I thank Andrew Bocarsly, whose group at Princeton reviewed and improved many of the chapters. Jonathan Mann devoted precious time to this book. James McBreen also deserves recognition, along with Hans Maru, who supplied most of the photographs for the cover. Protonex Technology Corporation also supplied one of the cover figures. Ralph White wrote the Foreword.

Dr. A. B. LaConti, T. F. O'Brien, Prof. T. Fuller, Dr. H. Maru, and Prof. D. Scherson all supplied invaluable help.

Endorsements were provided by Ken-ichiro Ota, President, Hydrogen Energy Systems Society of Japan (HESS) and Professor, Yokohama National University; and by Prof. Ernesto Rafael Gonzalez, IQSC-USP- Brazil.

Finally, Mangai Srinivasan and Maria Gamboa-Aldeco both worked unstintingly to complete this manuscript.

Ken Howell
Springer
New York, NY

# CONTENTS

## *Chapter 2*

## ELECTRODE/ELECTROLYTE INTERFACES: STRUCTURE AND KINETICS OF CHARGE TRANSFER

## Chapter 3

## ELECTROCHEMICAL TECHNOLOGIES AND APPLICATIONS

# Part II: FUNDAMENTAL ASPECTS FOR RESEARCH AND DEVELOPMENT OF FUEL CELLS

## Chapter 4

## FUEL CELL PRINCIPLES

*Chapter 6*

## EXPERIMENTAL METHODS IN LOW TEMPERATURE FUEL CELLS

## Part III:   ENGINEERING AND TECHNOLOGY DEVELOPMENT ASPECTS OF FUEL CELLS

### Chapter 7

### MODELING ANALYSES FROM HALF-CELL TO SYSTEMS

## Chapter 8

## FUELS: PROCESSING, STORAGE, TRANSMISSION, DISTRIBUTION, AND SAFETY

## Part IV: APPLICATIONS, TECHNO-ECONOMIC ASSESSMENTS, AND PROGNOSIS OF FUEL CELLS

### Chapter 9

### STATUS OF FUEL CELL TECHNOLOGIES

## Chapter 10

## APPLICATIONS AND ECONOMICS OF FUEL-CELL POWER PLANTS/POWER SOURCES

## Chapter 11

## COMPETING TECHNOLOGIES

## Chapter 12

# A TRIBUTE

The words of this final chapter are penned in great sadness. The author of this book, Dr. Supramaniam Srinivasan, wrote the book while valiantly battling progressive heart disease and passed away just prior to writing this concluding chapter. Although the disease severely diminished his strength, Dr. Srinivasan always maintained a positive, even energetic attitude toward science, the research students he mentored, and toward life as a whole. To those he knew, Dr. Srinivasan preferred to be called *Srini*, a simple gesture indicating his desire to be both a good friend and a teacher. To run into Srini while strolling through the lab guaranteed a bright smile, a good conversation, and a new idea to try out.

Although most people would either choose or be forced by health conditions to slow down with age and retirement, Srini did not know this concept. Upon retiring as Deputy Director of the Center for Electrochemical Systems and Hydrogen Research at Texas A&M University, Srini moved to Princeton and joined the Center for Energy and the Environment's fuel cell project before joining my research group in Princeton's Department of Chemistry. He introduced my group and two other engineering research groups to fuel-cell science, developing a strong laboratory and intellectual presence at Princeton in this endeavor, as well as serving as the primary mentor for a large group of graduate students and postdocs. He was tireless in doing his science; his energy was so great that it was often hard for his young students to keep up with his activities. His work ethic, enthusiasm for science, and expectation that all would rise to the top of their performance levels served as a model for all of us. He was passionate about his science and certain that fuel-cell technology was important to the future development of our society and the world. For me personally, it was a great blessing to have Srini as a colleague and a friend.

Although Srini was critical to the development of fuel-cell research in my laboratory, far more important to me was his interest in the lives and development of young scientists. It was that goal, the development of the next generation of scientists and engineers, which brought about this book. Srini desired a text that would teach the next generation about electrochemistry and fuel cells, so that they could carry on the work that he and others started. Thus, this work stands as a testament to Supramaniam Srinivasan's life and accomplishments. Perhaps more than any specific scientific breakthrough that the readers of this text might accomplish, he would take great pride in knowing that he had helped train the next generation of researchers through the words of this text, for above all, he was a friend and mentor of students.

Andrew B. Bocarsly
Chemistry Department
Princeton University

# PART I:

# BASIC AND APPLIED ELECTROCHEMISTRY AS RELEVANT TO FUEL CELLS

# CHAPTER 1

# EVOLUTION OF ELECTROCHEMISTRY

## 1.1. ELECTROCHEMISTRY: A FIELD BORN OUT OF CHEMISTRY AND ELECTRICITY

The field of electrochemistry was discovered in 1791 when Luigi Galvani was dissecting a frog. One of his coworkers touched its internal crucial nerves with the tip of a scalpel causing the muscles and nerves of the frog to contract. Nine years later, Volta reported to the Royal Society in London that by placing a membrane in contact with silver and zinc plates, on either side, and wetting it in salt water, an electric current would flow in the external circuit connecting the silver and zinc plates (chemical to electrical energy conversion). This discovery was soon followed by that of the *reverse process* when Nicholson and Carlisle demonstrated in the same year that by connecting two wires of platinum, immersed in a dilute acid, to a battery, bubbles of hydrogen and oxygen evolved on the two electrodes (electricity to chemicals). The pioneering researcher in the field of electrochemistry, Michael Faraday, started his work in 1832, and proposed the two quantitative laws of electrolysis:

- *Faraday's Law I.* The amount of chemical decomposition achieved is proportional to the quantity of electricity passed; and
- *Faraday's Law II.* The amount of different substances deposited or dissolved by the same quantity of electricity is proportional to their equivalent weights.

This chapter was written by S. Srinivasan.

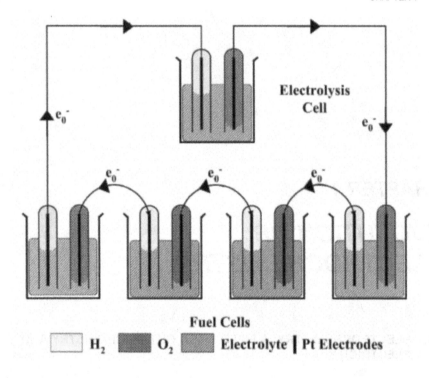

**Figure 1.1.** First demonstration of a fuel cell by Grove in 1839; four cells were connected in series and the electricity generated was used to decompose water.

According to Faraday, the flow of electrons in the external circuit was coupled with the flow of charged species in the electrolyte—i.e., the positively charged ions (cations) flow towards the cathode and the negatively charged ions (anions) flow towards the anode. An electrochemical cell (the fuel cell), akin to the Volta cell (a battery), was discovered in 1839 by Sir William Grove. He demonstrated that when the products of water electrolysis (hydrogen and oxygen) were fed to two platinum electrodes separated in individual cells and connected externally in series (Figure 1.1), an electric current was generated. This spontaneous reaction, which is the reverse of water electrolysis (a fuel cell), is equivalent to the chemical combustion of hydrogen to produce water, but with the added and attractive feature of electricity generation.

## 1.2. SCOPE OF ELECTROCHEMISTRY

All the above mentioned discoveries, made in the 19th century, illustrate that electrochemistry is a field of science which encompasses inter-relations of electrical and chemical phenomena. This field of science is thus a very broad one because practically all chemical reactions are fundamentally electrical in origin. According to Bockris and Reddy, one can sub-divide electrochemistry into two types:[1]

- *Ionics*. This field deals with ions in solution (aqueous, non-aqueous), molten salts, and solid electrolytes; and
- *Electrodics*. This field deals with an electrode/electrolyte interface, across which electron transfer can occur.

The field of *ionics* was considered as the central theme of electrochemistry until about the first half of the 20th century. Important contributions led to the fundamental understanding of topics such as (i) interactions of ions with other ions and solvents; (ii) ion transport in solutions; (iii) unique characteristics of the proton, as compared with other ions with respect to its solvation and transport (quantum mechanical proton-hopping and mobility, i.e., Grotthus mechanism) and homogeneous proton-transfer reactions; and (iv) ionic liquids - pure liquid electrolytes, hole models, transport phenomena, and molten electrolytes.

However, starting in the early 1920s but intensified from the 1950s, there was a transformation in the central theme of electrochemistry from ionics to *electrodics*. The focusing aspects in this field are the structure of interfaces (mainly solid/liquid but could also be solid/solid, solid/gas, liquid/liquid, liquid/gas) and the charge transfer processes that can occur across such interfaces.

An attempt is made in Figure 1.2 to demonstrate the quantum rate of growth of the field of electrochemistry and solid-state science and technology in the 20th century. Two societies (The Electrochemical Society and International Society of Electrochemistry), very active over the last 100 years, were founded in this century. They have been responsible for nurturing multidisciplinary areas of science and engineering, involving electrical/chemical phenomena into symposia presented at their annual meetings. It is not possible to cover, in much detail, the topics covered in the eight major areas of electrochemistry (see Figure 1.2) in this short chapter. However, it is worthwhile to highlight the major advances made in these areas as well as their significant relevance as tools for the pursuit of R&D on fuel cells, other electrochemical technologies, and electrochemistry in the biomedical sciences.

### 1.2.1. Electrochemistry and Surface Science (Wet Electrochemistry)

Even though bioelectrochemistry was the first invention in electrochemistry, the bulk of electrochemical research and development in the 19th century and in the first half of the 20th century was in wet-inorganic and -organic electrochemistry. As stated in the previous section, the field of ionics was foremost in the minds of

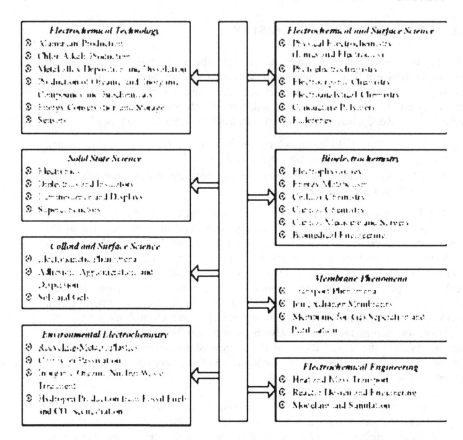

**Figure 1.2.** Scope of electrochemistry and its multi-disciplinary nature.

the early scientists; thereafter, the field of electrodics emerged and to this day much attention in electrochemical and surface science is focused on unraveling the mysteries of charge transfer reactions at electrode/electrolyte interfaces. An offshoot of the research activities since the late 1920s is identifying the role of the electrode materials, apart from their being a donor or acceptor of electrons for the charge transfer reaction. Significant accomplishments were the discoveries of the roles of electronic and geometric properties of the electrodes on the adsorption and charge transfer characteristics of intermediates formed during the reaction; just as in the case of heterogeneous catalysis, these properties influence the electrocatalytic activity. The Nobel Prize in Chemistry was awarded to Rudolf Marcus in 1992 for providing a quantum mechanical description of electron transfer at electrode/electrolyte interfaces, redox reactions, and biological reactions.

*Electroorganic chemistry* was also born in the 19th century, when Kolbe discovered that hydrocarbons could be synthesized by the electro-dimerization of carboxylic acids. Since then this field has grown by leaps and bounds via electrooxidation or reduction reactions, which involve hydrogenation, selective reduction of carboxylic groups, carbonyl groups, preparation of optically active compounds (in the biochemical area), etc. A revolutionary discovery was the conversion of adiponitrile to adipic acid, by Bazier at Monsanto; this reaction is an intermediate step in nylon synthesis.

*Electroanalytical chemistry* is also a reliably old topic in electrochemistry. The major invention here was polarography, for which Heyrovsky won the Noble Prize in chemistry (1924). Several new areas of research have evolved since in wet electrochemistry.

Recent interests have been in *photoelectrochemistry* involving light induced reactions at interfaces of electrodes with electrolytes. The electrode material is a semiconductor material (e.g. Si, Ga, As). By absorbing light, electrons are promoted from the lower, occupied valence band to the upper, unoccupied conduction band. The resulting electron-hole pair in the valence band can react with an electron acceptor or donor in the electrolyte to convert light energy to chemical energy, e.g., photo-electrolysis of water or of HBr to produce hydrogen.

A revolutionary evolution in the 1980s was the discovery of *electrochemically conducting organic polymers*. These were initially prepared by polymerization and ion doping of 5 organic compounds, i.e., acetylene, pyrrole, aniline, phenylone, and thiophene. These have oxidized and reduced forms. The oxidized form has an electronic conductivity very close to that of metals, while the reduced form is similar to a semiconductor. This field of science has grown by quantum jumps and Alan Heeger, Alan MacDiarmid, and Hideki Shirakawa, the discoverers, were awarded the Nobel Prize.

Another enlightening discovery, in recent times, was that of *fullerenes*, spherically caged molecules with carbon atoms located at the corners of polyhedral structures, consisting of pentagons and hexagons. For this discovery, Nobel Prizes were awarded in 1996, the recipients being R. F. Curl, H. W. Kroto, and R. F. Smalley. Fullerenes are also commonly referred to as bucky-balls or buckminister fullerenes. There is increasing interest in the applications of fullerenes in microelectronics, photo-electrochemistry, and in the area of superconductivity.

## 1.2.2. Solid-State Science (Dry Electrochemistry)

During the 20th century, *dry electrochemistry* reached the stage of being as extensive as wet electrochemistry. The main reason for this major advance was because of the ability to control electronic materials and materials-processing techniques for semiconductors, dielectric, magnetic, piezoelectric, opto-electronic, and optical fiber applications. Solid-state electrochemistry has also led to super-conducting materials.

Electronic materials and processes have been in the forefront of solid-state science and technology. The main reason for this impetus is the computer and communications era. Electronic and hole conduction in semiconductors gave birth to electronic and logic circuits and microprocessors. Interactions of photons with solids led to optical devices such as lasers and optical fiber networks. Highly sophisticated techniques have been developed for material processing from crystal growth to circuit fabrication, with the desired microstructure and electronic properties. Active areas of research and development in solid-state science are (a) silicon integrated circuits, (b) nano-electronics, and (c) meteorology and materials characterization. The transistor was a major discovery in electronics, for which the Nobel Prize was awarded to Bardeen, Brattain, and Schottky in the 1950s.

The area of *dielectric science and technology* deals with electrical, chemical, and mechanical properties of insulator materials. More recently, the interest has been in thin *dielectric films*, as applied to microelectronics. Faraday first used the term dielectric when he found that current, introduced at one plate, flows through the insulator to charge the other plate. The positively and negatively charged species are thus polarized. This phenomenon is analogous to that in a solid-state capacitor. Traditionally, insulating oxides and nitrides have been used for dielectric materials. However, in the recent past, polymeric materials have entered the arena, for example (a) plastic fibers for short optical data links, (b) polymeric films for non-linear optics, (c) radiation sensitive polymers for high-speed high-density integrated circuits, and (d) passivation by polymeric films for integrated circuits and packages in computers.

*Ceramic materials* are also used in packaging of semiconductor integrated circuits, automobiles, composites for space vehicles, and high efficiency power generation. The revolutionary discovery of super-conductivity in ceramic materials of the Ba-La-Cu-O class of oxides led to Bednorz and Muller being awarded the Nobel Prize in 1987. The goals of the technologies are to attain very high speed, high-density devices and interconnection systems (electrical and optical) for optical and telecommunication systems.

*Luminescence and display materials* are similar to electronic materials. Practical applications are envisioned for photoluminescence, electroluminescence, and cathode luminescence in the area of display technologies. More recently, interest has been in non-linear optics as applied to lasers, optical energy storage and holarography, femto second spectroscopy, infrared light sources and detectors, x-ray phosphors, and imaging.

*High temperature materials* encompass a wide area. They embody the physico-chemical characterization of materials, elucidation of reaction kinetics and of thermodynamics and phase equilibria, and investigations of use of such materials at high temperatures. These materials include metals, alloys, ceramics, and composites. The goal of research on these materials at high temperatures is to overcome the challenges in respect to chemical and mechanical instabilities which arise in the development of (a) clean energy technologies, (b) reliable electronic, optical, magnetic, and mechanical devices, (c) chemical sensors, (d) corrosion resistant and structural materials, and (e) cost-effective methods for recycling and safe disposal of waste materials. Of relevance to fuel cells, is the selection and evaluation of materials for solid oxide fuel cells. The research objectives are to: (a) find thermally compatible materials for the anode, cathode, electrolyte, and interconnect, (b) investigate the role of defects in the conduction mechanism of the oxide ions, (c) inhibit the interdiffusion of free elements in the component materials, and (d) develop thin film fabrication techniques for single and multi-cell stacks in fuel cells. Apart from this area, research on high temperature materials has been and continues to be vital in areas such as high temperature super-conducting materials, silicon based ceramics (SiC, $Si_3N_4$), borides, carbides, silicides, and diamond-like materials. These have wide applications in microelectronics and lightweight structural components at high temperatures. High temperature research on new alloys and composite materials has also been most beneficial in the development of tungsten, tantalum, nickel, and Ni-Co-Al based alloys for a multitude of applications. One example is the use of such materials in high temperature gas turbines.

### 1.2.3. Colloid and Surface Science

*Colloid* and *surface science* is another major field in electrochemistry and deals with thermodynamics, kinetics, and electrochemical phenomena at insulator/electrolyte interfaces. Thus, it is unlike the topics dealt with in the preceding sub-section, where the solid phase is a metal, alloy, or semiconductor.

An area of major interest, starting in the 19th century, is the topic of *electrokinetic phenomena*. It deals with charge separation at the insulator material (colloid, smooth and rough surface, porous plug)/electrolyte interface. The charge separation occurs because of the presence of charged ionic species on the surface of these materials. There are basically four types of electrokinetic phenomena:

- *Electroosmosis.* Flow of electrolyte in an insulating or porous plug when an electric potential is applied at opposite ends.
- *Streaming potential.* It is the opposite of electroosmosis, i.e., the setting up of a potential when there is flow from one end of the tube or porous plug to the other end.
- *Electrophoresis.* The mobility of colloidal particles in an electric field.
- *Sedimentation potentials.* The setting up of an electric field when colloidal particles are in motion in a stationary electrolyte.

All these phenomena are connected with the surface charge characteristics of these particles. The pH of the electrolyte has a significant influence on electrolyte phenomena. Other factors, which affect the surface charge characteristics of insulator materials, are the adsorption of ions and of neutral molecules. Investigating these electrokinetic effects has proven to be vital to solving the problems of adhesion and agglomeration, as well as methods to keep nano- to micro-sized particles dispersed.

In the colloidal state, double layer interactions are as important as bulk interactions. A colloidal suspension of discrete micron sized particles is a *sol*. Conversely, one could have a continuous matrix with pores of very fine dimensions, constituting a porous mass or membrane, i.e., a *gel*. Double layer interactions occur within the gel. Sols and gels are frequently encountered in biological processes, for example blood clotting and thrombosis.

### 1.2.4. Membrane Phenomena

*Membrane phenomena* are very extensive in the chemical industry, mainly for separation processes. Some notable examples are filtration, reverse osmosis, gas permeation, dialysis and electrodialysis, and electrofiltration. Membranes are, in general, permeable or semipermeable. The main characteristics of membranes are: (a) flux or permeation rate (i.e., the flow rate of fluid passing through membrane per unit area of membrane per unit time), and (b) selectivity of different constituents in a mixture. The driving force for the permeation may be hydrostatic pressure, concentration gradient, electric potential, or temperature.

Of significant interest to fuel cell technologies is the *proton-conducting membrane*, used in proton-exchange fuel cells and ceramic-oxide membranes such as the zirconia-yttria membrane, which is the selective oxide ion conductor in solid oxide fuel cells. Other electrochemical applications of such membranes are in gas separation or purification, as for example in concentration cells to separate hydrogen from reformed gas or oxygen from air. Membranes are also used in water purification systems and in separations of ions and solutes from aqueous systems (electrodialysis). A major application of ion exchange membranes is in chlor-alkali plants for the production of chlorine, caustic, and hydrogen. The membrane used is similar to the perfluorosulfonic acid membrane in the proton-exchange membrane fuel cells. One difference is that it is a laminated membrane, the other component being a carboxylic membrane that inhibits the transport of the hydroxyl ion produced at the cathode to the anode in a chlor-alkali cell.

### 1.2.5. Bioelectrochemistry/Biomedical Sciences

*Bioelectrochemistry* is another area of electrochemistry that has attracted the attention of biochemists and physiologists. Because of the strong orientation towards biological sciences, there has been little interaction until the 1960s between the physical and natural scientists. But with the growing interest in the molecular biological/physico-chemical approach, collaborative efforts among biochemists, physiologists, clinicians, surgeons, physicists, chemists, material scientists, and engineers to investigate these areas are increasing at a rapid rate. Figure 1.3 illustrates the active areas of research, which have led to unravel the mysteries of biological systems and to use the electrochemical approach to solve complex biomedical problems.

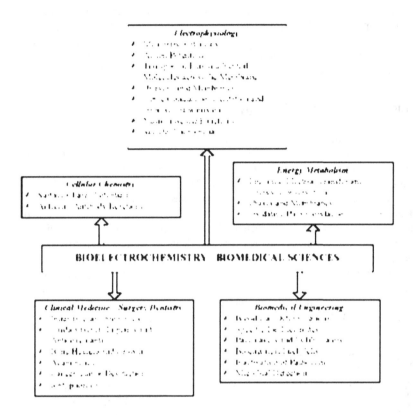

**Figure 1.3.** Active areas of bioelectrochemical research and engineering to unravel mysteries of biological systems and biomedical problems.

## 1.2.6. Electrochemical Engineering

*Electrochemical engineering* is a field in electrochemistry, which has advanced by leaps and bounds, mainly due to the group set up at the University of California, Berkeley, by Charles W. Tobias. In fact, Tobias and John Newman trained several of the faculty members at other universities. Electrochemical engineering deals with the fundamental aspects of the approaches in the development, design, and operation of electrochemical reactors (electrosynthesis, electrochemical energy conversion, and storage). The processes in such reactors involve concepts of heat and mass transfer, energy balance, current and potential distribution, thermodynamics, kinetics, scale-up, sensing, control, and optimization. The computer age has been of great value to electrochemical engineers, who are quite active in modeling processes from the half-cell to the stack-level. Sophisticated software programs are available for such modeling studies, which have been proven to be valuable in predicting and optimizing the performance of reactors. There is an increasing exchange of ideas among electrochemists, electrochemical engineers, and technologists for this purpose.

### 1.2.7. Electrochemical Technology

About 5% of the electricity generated in the USA is utilized by the electrochemical industry. Approximately 65% of the electricity consumption by the electrochemical industry is used for aluminium production and about 30% for the production of chlorine and sodium hydroxide. The remaining 5% is used in: (a) metal winning and refining, (b) membrane processes, and (c) electrosynthesis of organic and inorganic compounds and biochemicals. In the above mentioned processes, electrical energy is used for the production of chemicals. Electrochemistry also involves the use of chemicals to generate electricity via batteries (primary and secondary), fuel cells, and supercapacitors. Since the 1950s, there has been a rapid growth in these areas. Another emerging area is that of the *electrochemical sensors*. There are different types of sensors: solid-state, electrochemical, and optical. The topic of electrochemical technologies is presented in detail in Chapter 3. Thus, only a brief description is presented in this section.

### 1.2.8. Environmental Electrochemistry

Though *environmental electrochemistry* has been in existence for more than five decades, it is only since the mid eighties that there has been an exponential increase in the number of emerging areas where challenging problems are being encountered. This is because of strict environmental legislation in many states to (a) reduce air, land, and water pollution, and (b) recycle materials such as metals, plastics, paper, etc. Within the last decade, the problem of global-warming has attracted worldwide attention and the Kyoto agreement stipulated a 10% reduction by the year 2010 in the carbon dioxide level from what it was in 1991. Thus, there are several ongoing projects to sequester $CO_2$ and store it under pressure in aquifers, salt beds, and the ocean. An electrochemical solution to the global-warming problem is a solar-hydrogen economy. The use of hydrogen powered cars will also reduce the emission of pollutants such as carbon monoxide, sulfur dioxide, nitric oxide, and volatile organic compounds to zero levels. Wastewater treatment by electrochemical methods to remove organic impurities via peroxide species produced in-situ and lower levels of metallic impurities (including nuclear materials) is also gaining momentum. Methods have been developed for the removal of $H_2S$ from sour gas (natural gas which contains impurities to the level of about 10–12%) and electrochemically decompose it to pure hydrogen and sulfur. Another area of investigation is the conversion of nitrates in radioactive waste materials to ammonia. Recycling of materials is a growing field. Well-established technologies are available for the complete recovery of lead from lead acid batteries and platinum in catalytic converters. Of future interest will be the recovery of platinum from low to intermediate temperature fuel cells, after they have served their lifetime periods of operation. In many of the above-mentioned clean-up or recycling processes, at least one intermediate step is electrochemical.

## 1.3. TYPES OF ELECTROCHEMICAL REACTIONS

There are a wide variety of *charge-transfer reactions* (electron transfer) that occur at electrode/electrolyte interfaces. In general, the electrode material is metallic but there are several cases of charge transfer reactions that occur at interfaces of semiconductor electrodes with electrolytes. There are a wide variety of electrolytes– i.e., aqueous, non-aqueous, molten salt, and solid electrolytes. The bulk of electrochemical processes (chemical to electrical energy and electrical energy to chemical) are in aqueous electrolytes. A synopsis of the types of electrochemical reactions is presented in the following subsections.

### 1.3.1. Gas Evolution and Gas Consumption

A good example of gas evolution and gas consumption is the electrolysis of water in an acid or alkaline electrolyte; the half-cell reactions are represented in the acid medium by:

Cathodic:               $2H^+ + 2e_0^- \Leftrightarrow H_2$                          (1.1)

Anodic:               $H_2O \Leftrightarrow \frac{1}{2}O_2 + 2H^+ + 2e_0^-$                          (1.2)

Overall:               $H_2O \Leftrightarrow H_2 + \frac{1}{2}O_2$                          (1.3)

The reversible signs ($\Leftrightarrow$) are shown to illustrate that the electrochemical reactions can occur in both directions. The forward ones, as shown in Eqs. (1.1) and (1.2), are for gas evolution (water electrolysis), and the reverse ones are for gas consumption (fuel cell). It must be noted that the electrodes are inert, meaning that they are neither consumed, nor is there any deposition on them. However, for these types of reactions, it is necessary for them to provide electrocatalytic activities.

### 1.3.2.  Metal Deposition and Dissolution

This type of reaction occurs in the field of metallurgy and is extensively used in industry for the production of metals such as aluminum, copper, zinc, and alkali and alkaline metals. The electrowinning of copper occurs via the reaction:

$$Cu^{2+} + 2e_0^- \Leftrightarrow Cu \tag{1.4}$$

The Cu electrode plays an active part in the reaction since the metal is deposited on it (*electrodeposition*), or, in the reverse direction, dissolved from it (*electrodissolution*).

### 1.3.3.  Redox Reactions

*Redox reactions* involve electron transfer and can occur in a homogeneous medium or at an electrode/electrolyte interface. In a homogeneous reaction, reduction of one chemical species and the simultaneous oxidation of another chemical species occur. At an electrode/electrolyte interface, the electrode is an inert conductor and serves as the donor or acceptor of electrons; this is similar to the role of the electrode in a gas evolution/gas consumption reaction. An example of this type of reaction is:

$$Fe^{3+} + e_0^- \Leftrightarrow Fe^{2+} \tag{1.5}$$

### 1.3.4.  Electrodes of the Second Kind

The electrodes, described in the above Sections may be referred to as *electrodes of the first kind*. An *electrode of the second kind* is one in which the metal is in contact with an insoluble salt of the metal, and this electrode is immersed in a solution containing the same anion, as that in the insoluble salt. The most common of these types of electrodes, used as reference electrodes in electrochemical research, are the standard calomel and standard silver/silver chloride electrode. The electrochemical reactions, which occur at these electrodes, are:

$$Hg_2Cl_2 + 2e_0^- \Leftrightarrow 2Hg_s + 2Cl_{aq}^- \tag{1.6}$$

and

$$AgCl_s + e_0^- \Leftrightarrow Ag + Cl_{aq}^- \tag{1.7}$$

The symbols $s$ and $aq$ designate the solid and liquid state. These electrode reactions are extremely fast and hence, they are ideally non-polarizable. Such electrodes are most commonly used as reference electrodes in electrode kinetics and electrocatalysis investigations.

## 1.3.5. Corrosion/Passivation

The *corrosion* of a metal or alloy is an electrochemical phenomenon. The main difference between a corrosion reaction and an electrochemical reaction that occurs in a cell with two electrodes separated by the electrolyte is that the anodic and cathodic reactions occur on the same material in the former, while these reactions occur individually on the two separate electrodes in the latter (Figure 1.4).

The partial reactions, which occur during the corrosion of iron, may be represented by:

$$Fe \Leftrightarrow Fe^{2+} + 2e_0^- \qquad\qquad (1.8)$$

$$2H^+ + 2e_0^- \Leftrightarrow H_2 \qquad\qquad (1.9a)$$

or

$$\frac{1}{2}O_2 + 2H^+ + 2e_0^- \Leftrightarrow H_2O \qquad\qquad (1.9b)$$

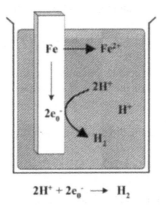

$$2H^+ + 2e_0^- \longrightarrow H_2$$

**Figure 1.4.** Anodic and cathodic reactions occurring during corrosion of iron. Note that these reactions occur at different sites on the iron sample. Electron transfer is through the iron from the anodic site to the cathodic site.

The electron transfer during this process does not occur in an external circuit, as in the case of the conventional electrochemical reactions in a cell with the anodes and cathodes separated by the electrolyte. The electron transfer is within the iron itself from the site where the anodic reaction occurs to the site where the cathodic reaction occurs.

*Passivation* is a process that is used to significantly inhibit the rate of a corrosion reaction. In order to passivate, for example, a metal like nickel, a small anodic current is passed through it in a cell with a counter electrode in an alkaline medium. The reaction is:

$$Ni + 2OH^- \Leftrightarrow Ni(OH)_2 + 2e_0^-$$                               (1.10)

Since the electronic conductivity of the film of nickel hydroxide is quite low, it retards the passage of electrons from the metal to the species in the electrolyte or vice versa.

### 1.3.6. Electroorganic Reactions

There has been interest in electroorganic synthesis since the latter part of the $19^{th}$ century. One of the early investigations was in the synthesis of the long chain hydrocarbon by the dimerization of semi-esters of dicarboxylic acid. Another reaction of interest has been the reduction of nitrobenzene to aniline in acidic media on a platinum amalgam electrode in an inert medium. The half-cell reaction is:

$$C_6H_5NO_2 + 6H^+ + 6e_0^- \rightarrow C_6H_5NH_2 + 2H_2O$$                    (1.11)

The standard thermodynamic reversible potential for this reaction is 0.87/NHE. According to the revised edition of the book *Modern Electrochemistry* by Bockris and Reddy, 110 chemicals were produced by electroorganic synthesis, at a rate of more than 10,000 tons/year.

Of relevance to fuel cells are the direct oxidation of organic fuels, e.g.,

$$CH_3OH + H_2O \rightarrow CO_2 + 6H^+ + 6e_0^-$$                              (1.12)

$$CH_4 + 2H_2O \rightarrow CO_2 + 8H^+ + 8e_0^-$$                               (1.13)

### 1.3.7. Photoelectrochemical Reactions

Photoelectrochemical reactions occur when light is shone on an electrode in contact with an electrolyte. For such a reaction, it is necessary for the electrode to be a semiconductor capable of absorbing the incoming light to generate excited electron-hole pairs in its valence band. In a photoelectrochemical reaction, p-doped semiconductors stimulate cathodic reactions when illuminated with light of the needed energy; conversely, n-doped semiconductors stimulate anodic reactions when illuminated. A reaction, which has been of great interest since the early 1970s, is the photoelectrochemical decomposition of water (Figure 1.5).[2] The main reason for this interest from a technological point of view, was the large-scale production of hydrogen fuel from water and solar energy. The reactions, which occur at the photoanode and photocathode, are exactly the same as an electrochemical cell for the electrolysis of water.

## 1.3.8. Bioelectrochemistry and Electrophysiology

Biochemical and physiological processes (e.g., ion transport, energy metabolism, nerve and muscle conduction, bone growth and healing, blood clotting, cancer, etc.) involve electrochemical mechanisms (ion fluxes, adsorption of solid/electrolyte interfaces, proton transfer, electron transfer, electrokinetic effects). Only since the 1950s, has there been an awakening to the electrochemical mechanisms in biological processes. Figure 1.3 depicts the active areas of research in bioelectrochemistry/electrophysiology. As illustrated in this figure, bioelectrochemistry, or better-stated *electrochemistry in the medical sciences*, covers a wide spectrum of electrochemical reactions in biological systems. Of most relevance to fuel cells is the nature of energy conversion in biological systems. As in a fuel cell, living systems consume organic food and oxygen (from the air) to initially produce electrical energy. The products of the fuel cell reaction are carbon dioxide and water. The overall efficiency for energy conversion is greater than 35%.

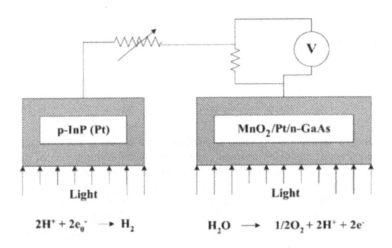

**Figure 1.5.** A self-driven photo-electrochemical cell for the decomposition of water to hydrogen and oxygen.

It is only because of electrochemical energy conversion that such a high efficiency can be obtained.[3] For example, on a simplistic basis, the half-cell reactions may be represented by the electro-oxidation of, for example, glucose and the electroreduction of oxygen:

$$C_6H_{12}O_6 + 6H_2O \rightarrow 6CO_2 + 24H^+ + 24e_0^- \qquad (1.14)$$

$$O_2 + 4H^+ + 4e_0^- \rightarrow 2H_2O \qquad (1.15)$$

Specific enzymes are the electrocatalysts for the reactions. It must be noted, however, that energy metabolism in living systems is considerably more complex and involve the conversion of adenosine diphosphate (ADP) to adenosine triphosphate (ATP) in the mitochondrion. The ATP is then transported from the mitochondrion to the nearest point at which the energy is needed–i.e., for muscle action. During this process, ATP is converted back to ADP and the energy production process is repeated.

In addition to the vital area of energy metabolism there is a host of electron/proton transfer reactions and other ion transport phenomena that occur in biological systems. Examples of such phenomena can be found in neurotransmission, membrane potentials, antigen-antibody reactions, use of electric currents in bone-healing and bone-growth, electrochemical reactions involving the blood coagulation factors that cause intravascular thrombosis, etc.

## 1.4. POTENTIALS AT ELECTRODE/ELECTROLYTE INTERFACES AND ELECTROMOTIVE SERIES

When an electrode (e.g., Cu) is immersed in an electrolyte (e.g., $CuSO_4$), there is, in general, an electrochemical reaction occurring at the interface, which is in equilibrium. For the example chosen, it is:

$$Cu^{2+} + 2e_0^- \Leftrightarrow Cu \qquad (1.16)$$

As a result of this, there is a charge separation which occurs at the interface, with a surface excess of positively charged copper ions on the solution side of the interface and negatively charged electrons on the solid side. This charge separation sets up a potential difference across the interface. However, this potential difference cannot be measured with a voltmeter because one of its terminals will be connected to the copper electrode and the latter will be hanging loose. Therefore, one has to connect the other terminal to another electrode in the same solution (Figure 1.6). The measured value of the potential will then be the difference in potential across one interface with respect to the other. If the second electrode is also the same as the

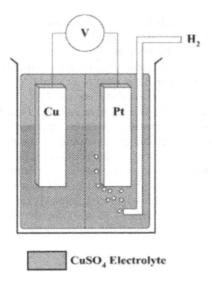

**Figure 1.6.** Electrical circuit for measurement of potential at an electrode (Cu)/electrolyte ($CuSO_4$) interface *vs.* the potential at the reference electrode (hydrogen) /electrolyte ($CuSO_4$) interface.

first one (i.e., Cu), the potential difference will be zero. To overcome this dilemma of determining the potential across an electrode/electrolyte interface, electrochemists proposed a standard electrode potential for the hydrogen evolution/ionization reaction. This reaction occurs, for example, at a platinum electrode, immersed in an acid electrolyte (e.g., $H_2SO_4$), which has an acid strength of unit activity (1N) and through which hydrogen is bubbled at 1 atm pressure (Figure 1.6).

The potential across this interface is assumed to have a zero value, and it is referred to as the potential of the *normal-hydrogen electrode* (NHE). The reactions which occur at this single electrode and which are in equilibrium are represented by:

$$2H^+ + 2e_0^- \Leftrightarrow H_2 \tag{1.17}$$

The above anodic (forward) and cathodic (backward) reactions are relatively fast and occur in fuel cells and water electrolysis reactions. Furthermore, because the reactions are relatively fast on the platinum electrode and their rates are not severely affected by impurities, the potential is quite stable. In addition, even if a small current (a few mA/cm$^2$) is drawn from the cell by connecting this electrode to another electrode via a load, there will be minimal variation in its potential. Such an electrode is referred to as an *ideally non-polarizable electrode*. This is not the case with the oxygen electrode (i.e., a Pt electrode, with oxygen bubbling through the solution). With such an electrode, the passage of even a small electric current will significantly change the potential across the electrode/electrolyte interface. Such an electrode is referred to as an *ideally polarizable electrode*.

Using the normal hydrogen electrode as represented in Figure 1.6, it is now possible to determine the potential at another electrode/electrolyte interface, where an electrochemical reaction is occurring. As an example, one could have a two compartment cell (Figure 1.7) containing 1N hydrochloric acid and two platinum electrodes. If hydrogen is bubbled through one compartment and chlorine through the other at 1 atm pressure and the cell maintained at 25 °C, the measured value of the potential would be 1.35 V. Since our reference potential is zero for the hydrogen electrode, the recorded potential value of 1.35 V is for the reversible chlorine electrode at which the reaction is:

$$2Cl^- \Leftrightarrow Cl_2 + 2e_0^- \tag{1.18}$$

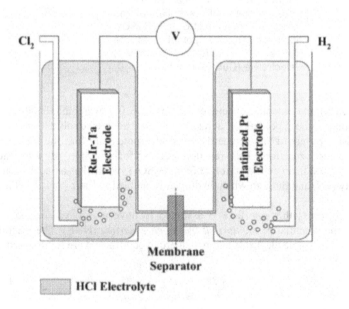

**Figure 1.7.** Schematic of the cell for measurement of the potential at a standard chlorine electrode with respect to hydrogen reference electrode.

By using the methodology represented in Figure 1.7, it is possible to measure the standard electrode potentials of other electrochemical redox couples. This led to the tabulation of the standard electromotive series that appears in classical books of electrochemistry and in the *Handbook of Physical and Chemical Constants*. The *thermodynamic-reversible potentials* ($E_r^0$) for some reactions of relevance to electrochemical energy conversion and storage are presented in Table 1.1. The metal with the most negative value of $E_r^0$, lithium is conventionally referred to as the most electropositive element (with the highest capability of donating electrons); conversely, fluorine is the most electronegative element (with the highest capability of accepting electrons). A lithium/fluorine battery will have a standard electrode potential of 5.915 V. From this table, it can also be observed that the hydrogen/oxygen fuel cell, under standard conditions, has a reversible potential of 1.229 V.

## 1.5.  FREE ENERGY CHANGES OF REACTIONS AND THEIR REVERSIBLE POTENTIALS (NERNST POTENTIALS)

The thermodynamic reversible cell potential ($E_{cell}^0$) is directly related to the free energy change ($\Delta G^0$) of an electrochemical reaction because the former represents a measure of the electrical work for the transfer of electrons from the anode to the cathode, as in the case of the fuel cell reaction:

$$H_2 + \frac{1}{2}O_2 \rightarrow H_2O \qquad (1.3)$$

The relation between $\Delta G^0$, and $E^0{}_{cell}$ for the reaction is:

$$\Delta G^0 = -nFE_{cell}^0 \qquad (1.19)$$

where $n$ represents the number of electrons transferred in the overall reaction and $F$ is Faraday's constant, i.e., 96,500 coulombs/equivalent. For this reaction, the value of $\Delta G^0$ is –237.09 kJ/mole, the value of 2 is $n$ and 96.40 kJ/V eq is $F$. $E_{cell}^0$ is defined as:

$$E_{cell}^0 = E_{r,c}^0 - E_{r,a}^0 \qquad (1.20)$$

where the subscripts $c$ and $a$ refer to the thermodynamic reversible potentials of the cathodic and anodic reactions, respectively. Thus, the $E_{cell}^0$ of 1.229 V for the reaction given in (1.3) represents the thermodynamic reversible

**TABLE 1.1**

**Thermodynamic Reversible Potentials for the Anodic and Cathodic Reactions, and for the Single Cells in Some Electrochemical Energy Conversion and Storage Systems**

| Electrochemical energy converter/storer | Electrolyte | Anodic reaction | $E_{r,a}^0$ (V) | Cathodic reaction | $E_{r,c}^0$ (V) | $E_{cell}^0$ (V) |
|---|---|---|---|---|---|---|
| **Fuel Cells** | | | | | | |
| Hydrogen/Oxygen | Acid | $H_2 \rightarrow 2\,H^+ + 2\,e_0^-$ | 0.00 | $O_2 + 4\,H^+ + 4\,e_0^- \rightarrow 2\,H_2O$ | 1.23 | 1.23 |
| Hydrogen/Oxygen | Alkaline | $H_2 + 2\,OH^- \rightarrow 2\,H_2O + 2\,e_0^-$ | −0.83 | $O_2 + 2\,H_2O + 4\,e_0^- \rightarrow 4\,OH^-$ | 0.40 | 1.23 |
| Methanol/Oxygen | Acid | $CH_3OH + H_2O \rightarrow CO_2 + 6\,H^+ + 6\,e_0^-$ | 0.01 | $O_2 + 4\,H^+ + 4\,e_0^- \rightarrow 2\,H_2O$ | 1.23 | 1.22 |
| Methane/Oxygen | Acid | $CH_4 + 2\,H_2O \rightarrow CO_2 + 8\,H^+ + 8\,e_0^-$ | 0.17 | $O_2 + 4\,H^+ + 4\,e_0^- \rightarrow 2\,H_2O$ | 1.23 | 1.06 |
| Carbon/Oxygen | Acid | $C + 2\,H_2O \rightarrow CO_2 + 4\,H^+ + 4\,e_0^-$ | 0.21 | $O_2 + 4\,H^+ + 4\,e_0^- \rightarrow 2\,H_2O$ | 1.23 | 1.01 |
| **Primary Batteries** | | | | | | |
| Zinc/MnO$_2$ | Alkaline | $Zn + 2\,OH^- \rightarrow Zn(OH)_2 + 2\,e_0^-$ | −1.25 | $MnO_2 + H_2O + e_0^- \rightarrow MnOOH + OH^-$ | 0.30 | 1.55 |
| Zinc/Air | Alkaline | $Zn + 2\,OH^- \rightarrow Zn(OH)_2 + 2\,e_0^-$ | −1.25 | $O_2 + 4\,H^+ + 4\,e_0^- \rightarrow 2\,H_2O$ | 0.40 | 1.65 |

## CHAPTER 2

# ELECTRODE/ELECTROLYTE INTERFACES: STRUCTURE AND KINETICS OF CHARGE TRANSFER

## 2.1. DOUBLE LAYER AT ELECTRODE/ELECTROLYTE INTERFACES

### 2.1.1. Structure, Charge, and Capacitance Characteristics

When a metal is partly immersed in an electrolyte, a potential is set up across the two phases, i.e., at the electrode/electrolyte interface. The phases may be solids (metals or alloys, semiconductors, insulators), liquids (ionic liquids, molten salts, neutral solutions), or gases (polar or non polar). The more common terminology in electrochemistry is that a *double layer* is set up at the interface. There are several reasons for a potential difference being set up across the interface of two phases, the most common one being the charge transfer occurring across the interface. During this process, a charge separation will occur because of electron transfer across the interface. Other reasons for the occurrence of potential differences are due to surface-active groups in the ionizable media (liquid, solid, or gas) and orientation of permanent or induced dipoles. The double layer at the interface between two phases has *electrical*, *compositional*, and *structural* characteristics. The electrical and compositional characteristics deal with the excess charge densities on each phase and the structural one with the distribution of the constituents (ions, electrons, dipoles, and neutral molecules) in the two phases, including the interfacial region. For the purposes of understanding and analyzing the electrical, compositional, and structural aspects relevant to the electrochemical reactions that occur in fuel cells, a

---

This chapter was written by S. Srinivasan.

brief description of the evolution of the theoretical aspects of the structure of the double layer, as applied to electrode/electrolyte interfaces, across which charge-transfer reactions occur, is presented in this section.

2.1.1.1. *Parallel-Plate Condenser Model: Helmholtz Model.*[1] This was the first type of model proposed for the structure of the double layer at an electrode/electrolyte interface. It is analogous to that in a solid-state capacitor—i.e., two layers of charge of opposite sign are separated by a fixed distance (Figure 2.1a). In this case, one may assume electrons in the metal and positive ions in solution. The potential drop across the interface will be linear and the capacitance ($C_{M-2}$) of the double layer is, as in the case of a parallel-plate condenser, given by:

$$C_{M-2} = \frac{\varepsilon}{4\pi d} \qquad (2.1)$$

where $\varepsilon$ is the dielectric constant in the medium between the plates and $d$ is the distance between them. Assuming $\varepsilon = 6$ and $d = 3$ Å, the value of $C_{M-2}$ can be expected to be about 18 $\mu$F cm$^{-2}$. For the Helmholtz model, the differential and integral capacity are equivalent and have a constant value, even when there is a change in the charge density on the two layers (electrode and electrolyte).

However, what is measured experimentally is the differential capacity and one finds that this capacity varies with potential (Figure 2.2). This poses a question as to whether the model of fixed charges on the two layers constituting the double layer is valid, and therefore, it leads to the second model.

2.1.1.2. *Diffuse-Layer Model: Gouy[2] and Chapman[3] Model.* According to Gouy and Chapman, ions in the electric double layer are subjected to electrical and thermal fields. This allows the Maxwell-Boltzmann statistics to be applied to the charge distribution of ions as a function of distance away from the metal surface akin to the distribution of negatively charged ions surrounding a positive ion (Figure 2.1b). The analysis in both cases is similar except that for in the ion-ion case there is spherical symmetry, while for the electrode/electrolyte layers, it is planar. For this model, the diffuse charge ($q_D$) for a 1-1 electrolyte is given by:

$$q_D = \left(\frac{2kTn_0\varepsilon}{\pi}\right)^{1/2} \sinh\frac{e_0V}{2kT} \qquad (2.2)$$

and the differential capacity, $C_{2-b}$, by:

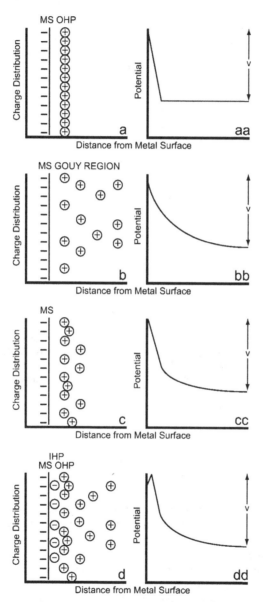

**Figure 2.1.** Evolution of the models of the double layer at electrode/electrolyte interface. Charge distribution vs. distance and potential variation vs. distance: (a) and (aa) Helmhotz model; (b) and (bb) Gouy-Chapman model; (c) and (cc) Stern model, and (d) and (dd) Esin and Markov, Grahame, and Devanathan model. Reprinted from Cited Reference 1 in Chapter 1.

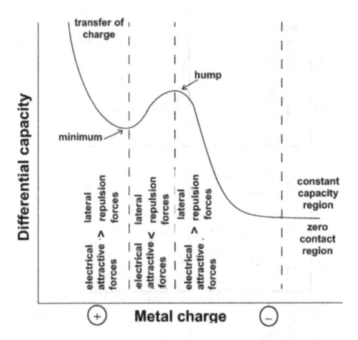

**Figure 2.2.** The lateral-repulsion model for the analysis of the capacity vs. potential plot. Reprinted from J. O'M. Bockris, A. K. N. Reddy, and Maria Gamboa-Aldeco, Modern Electrochemistry, Vol. 2A, 2nd edition. Copyright © 2000, with permission from Kluwer/Plenum Publishers.

$$C_{2-b} = \left( \frac{n_0 \varepsilon e_0^2}{2\pi kT} \right)^{1/2} \cosh \frac{e_0 V}{2kT} \tag{2.3}$$

where $n_0$ is the number of ions of positive and negative sign per unit volume in the bulk of the electrolyte and $V$ is the potential drop from the metal to the bulk of the electrolyte.

According to this model, the capacitance is a minimum at $V = 0$ and rises to very high values symmetrically and parabolically on either side of $V = 0$. A typical potential-distance (from the electrode surface) plot for this model is presented in Figure 2.1bb. This model also had deficiencies because: (a) the experimental capacity-potential relations did not behave in a symmetrical-parabolic manner, except at potentials close to the potential of zero charge of the metal and in very dilute solutions; (b) it neglected ion-ion interactions which become increasingly important at higher concentrations; and (c) it assumed a constant value of the dielectric constant in the region between the electrode and electrolyte.

2.1.1.3.  *Compact Diffuse-Layer Model: Stern[4] Model.* This model is a hybrid one, consisting of the above two models. First, ions are considered to have a finite size and are located at a finite distance from the electrode. Second, the charge distribution in the electrolyte is divided into two contributions: (i) as in the Helmholtz model immobilized close to the electrode, and (ii) as in the Gouy-Chapman model, diffusely spread out in solution (Figure 2.1c). Thus,

$$q_M = Q_s = q_H + q_G \qquad (2.4)$$

where $q_M$ is the charge on the metal. $Q_s$ is the total charge on the solution side (of opposite sign) comprising the Helmholtz fixed charge, $q_H$, and the Gouy-Chapman diffuse charge, $q_G$. The potential drops may be represented by:

$$V_M - V_{el} = (V_M - V_H) + (V_H - V_{el}) \qquad (2.5)$$

where the subscripts $M$, $H$, and $el$ denote the electrode, the Helmholtz layer on the solution side, and bulk electrolyte, respectively. As in the Helmholtz model, there is a linear variation of potential with distance across the Helmholtz component and a semi-exponential variation in the Gouy-Chapman component of the double layer (Figure 2.1cc). By differentiating Eq. (2.5) with respect to the charge, it can be shown that the double layer capacitance across this electrode/electrolyte interface is given by:

$$\frac{1}{C} = \frac{1}{C_H} + \frac{1}{C_G} \qquad (2.6)$$

where $C_H$ and $C_G$ are the contributions of the Helmholtz and Gouy-Chapman capacitances, which are in series. There are two implications of this model. One is that in concentrated electrolytes the value of $C_H^{-1}$ is considerably greater than that of $C_G^{-1}$. This model is very similar to that of Helmholtz (i.e., most of the charge is concentrated in the Helmholtz layer). The other implication is that in extremely dilute solution, $C_G^{-1} \gg C_H^{-1}$ and therefore, $C = C_G$. Thus, the double-layer structure approaches that of the Gouy-Chapman model. This model shows reasonable values of $C$ vs. $V$ relations for electrolytes with non-adsorbable ions such as $Na^+$ or $F^-$. However, it is not applicable for electrolytes with specifically adsorbable anions as it cannot describe the differential capacity data shown in Figure 2.2. Furthermore, this model does not take into account the role of the solvent as related to the hydration of the ions and its influence on the structure of the double layer.

2.1.1.4. *Triple-Layer Model: Esin and Markov,[5] Grahame,[6] and Devanathan[7] Model.* The subtle feature of this model (Figure 2.1d), proposed by the three groups, was to take into consideration that ions could be dehydrated in the direction of the metal and specifically adsorbed on the electrode. Thus, an *inner layer* between the electrode surface and the Helmholtz layer further modifies the structure of the double layer. This inner layer is the locus of centers of unhydrated ions strongly attached to the electrode. For this case, Devanathan derived the relation:

$$\frac{1}{C} = \frac{1}{C_{M-1}} + \left( \frac{1}{C_{M-2}} + \frac{1}{C_{2-b}} \right)\left( 1 - \frac{dq_1}{dq_M} \right) \tag{2.7}$$

where $C_{M-1}$ and $C_{M-2}$ are the integral capacities of the space between the electrode and the inner Helmholtz plane (IHP) and between the inner and outer Helmholtz planes (OHP), $C_{2-b}$ is the differential capacity of the diffuse double layer, and $(dq_1/dq_M)$ represents the rate of change of the specifically adsorbed charge with charge on the metal. Some interesting analyses may be obtained from Eq. (2.7):

- If $(dq_1/dq_M)$ is zero, the expression for the capacity ($C$) is equivalent to that for 3 capacitors in series, that is, inner Helmholtz, outer Helmholtz, and Gouy. Hence, this model is referred to as the *triple-layer model*.
- The capacity is a minimum when $dq_1/dq_M$ is zero, because the latter can have only positive values.
- If $dq_1/dq_M$ exceeds unity, the differential capacity attains large values. When $C$ tends to infinity, the electrode becomes non-polarizable.
- The minimum in the capacity is in the vicinity of the potential of zero charge.

Specific adsorption of ions occurs because of different types of electrical interactions between the electrodes and ions: electric field forces, image forces, dispersion forces, and electronic or repulsive forces. When the image and dispersion forces are larger than the electronic force, the specific adsorption of ions occurs (*physical adsorption*). However, a stronger bond could be formed by partial electron transfer between the ion and the electrode (*chemisorption*); small cations (e.g., $Na^+$) have a strong hydration sheath around them and are minimally adsorbed. On the other hand, large anions ($Cl^-$, $Br^-$) have only a few water molecules in the primary hydration sheath and since the ion-solvent interaction in this case is considerably less than the above mentioned ion-electrode interaction, specific adsorption of the ions occurs with some partial charge transfer of an electron. The variation of potential with distance, across the electrode/electrolyte interface (Figure 2.1dd), reveals a steep drop between the electrode and IHP and then a small rise between the IHP and OHP, and thereafter the variation is similar to that in the diffuse layer.

2.1.1.5. *Water-Dipole Model: Bockris-Devanathan-Muller[8] Model.* For the above described models, the structure of the double layer is based on the interfacial charge characteristics of the electrode and of the ionic species in the electrolyte. However, in electrochemistry, the bulk of the charge-transfer reactions occur in aqueous media. There are, of course, reactions in non-aqueous media such as organic solvents, molten salts, and solid electrolytes. The solvents like water or organic liquids (e.g., methanol, acetonitrile) are polar in character and contribute to the potential drop across the electrode/electrolyte interface. Thus, an innovative model for the structure of the double layer was proposed by Bockris, Devanathan, and Muller, which is schematically represented in Figure 2.3. The principal feature of this model is that, because of a strong interaction between the charged electrode and water dipoles, there is a strongly held, oriented layer of water molecules attached to the electrode. In this layer, because of competitive adsorption, there could also be some specifically adsorbed ions which are possible partially solvated. The locus of centers of these ions is the *inner Helmholtz plane* (IHP).

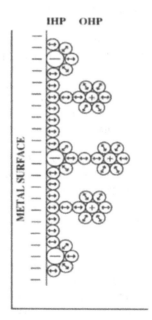

**Figure 2.3.** Water dipole model of the double layer at an electrode/electrolyte interface, (Bockris, Devanathan and Muller). Reprinted from Reference 8, Copyright © 1963, with permission from the Royal Society of London.

Adjacent to this layer is the layer of solvated ions, which is the locus of centers of the hydrated ions, i.e., the outer Helmholtz plane (OHP). Next to this layer is the diffuse layer, which is predominant in dilute electrolytes. Just as in the case of a primary hydration sheath surrounding an ion, the first layer of water molecules has a strong orientation (either parallel or anti-parallel to the electric field depending on the charge of the metal). Such a complete orientation yields a dielectric constant of about 6 for this layer. Next to this layer is a second layer of water molecules, somewhat disoriented due to electrical and thermal forces (this is similar to the secondary hydration sheath around an ion). This layer has a dielectric constant of about 30 to 40. The succeeding layers of water molecules behave like bulk water, which has a dielectric constant of *ca*. 80.

### 2.1.2. Effect of Specific Adsorption of Ions on the Double Layer Structure and their Adsorption Isotherms

Using the Bockris-Devanathan-Muller model (water-dipole model), the constant capacity on the negative side of the capacity-potential plot can be rationalized. The interpretation of the variation of capacity with potential at less negative and positive potentials is more complex and is dealt with in detail in the revised edition of the book *Modern Electrochemistry*. For the purpose of this chapter as well as its relevance to interfaces at which fuel cell reactions occur, a brief analysis of the regions of the capacitance-potential plot is presented next.

2.1.2.1. *The Region of Constant Capacity at Negative Potentials.* This is attributed to the Helmholtz type double layer and is not affected by the nature of ions in the double layer. The capacity thus attains a constant value of about 16 µF cm$^{-2}$ at the mercury electrode/electrolyte interface. It must be noted that higher values will be observed on metals, even if polished and smooth, because of some degree of roughness of these metals, as compared with mercury.

2.1.2.2. *Capacity Hump.* As illustrated in Figure 2.2, with an increase of potential in the positive direction, there is an increase of capacity leading to a maximum (*capacity hump*), followed by a decrease to a minimum, and then a sharp increase. For an explanation of this behavior, it is necessary to examine the general expression for the capacity at an electrode/electrolyte interface:

$$\frac{1}{C} = \left( \frac{1}{K_{M-OHP}} \right) - \left( \frac{1}{K_{M-OHP}} - \frac{1}{K_{M-IHP}} \right) \left( \frac{dq_{ca}}{dq_M} \right) \qquad (2.8)$$

where $K_{M-OHP}$ and $K_{M-IHP}$ refer to the integral capacity of the M-OHP and M-IHP regions, respectively.

If $(dq_{ca}/dq_M) \to 0$, the capacity will be a constant, but the limiting case will not be when there is no contact adsorption of ions. This equation also shows that $C$

increases when $q_{ca}$ increases with $q_M$, i.e., the capacity increases with potential difference across the double layer. The increase occurs until the electrode charge becomes positive, the extent of specific adsorption further increases and then the rate of its growth decreases. The question then is: Why is there this rate of decrease? The answer is that apart from the chemical and electrical forces that promote specific adsorption of ions, ion-ion lateral interaction forces play a role with increasing coverage of the electrode surface by the specifically adsorbed ions. Thus, there is the hump or a maximum in the differential capacity vs. the metal charge plot (Figure 2.2). After this decrease in capacity, there is again a sharp rise in the capacity, and this is due to the strong interaction between the highly, positively charged metal and the specifically adsorbed negatively charged ion. It has been demonstrated that the specific adsorption of the ions in this region of potential could also involve a partial charge transfer between the ion and the electrode. The above description shows that the specific adsorption of ions affects the structure of the double layer, which is reflected in the variation of the differential capacity with potential. It also has the effect on the variation of potential across the double layer with distance (Figure 2.1aa). A third effect is that specifically adsorbed ions can block sites for electrochemical reactions on the surfaces of the electrodes. Thus, knowledge of the adsorption behavior of these ions as a function of potential and/or charge on the metal is essential in elucidating the kinetics of electrode reactions. The adsorption behavior of ions on electrodes has been elucidated following an approach similar to that used in heterogeneous catalysis, and involves the adsorption of reactants, intermediates, and/or products on the surfaces of the catalysts. Common terminology is to express the adsorption behavior as *adsorption isotherms*, which are essentially equations of state relating physical quantities to the extent of adsorption. For example, if one considers the adsorption of an anion, A⁻, on an electrode, M, in the equilibrium state, as represented by

$$A_{sol}^- + M \overset{\rightarrow}{\underset{\leftarrow}{\phantom{x}}} MA_{sol}^- \tag{2.9}$$

the corresponding chemical potentials ($\mu$) can be written as,

$$\mu_{A_{sol}} = \mu^0_{A_{sol}} + RT \ln a \tag{2.10}$$

and

$$\mu_{MA_{ads}} = \mu^0_{MA_{ads}} + RT \ln f(\theta) \tag{2.11}$$

where $a$ is the activity of the ions A⁻ in solution and $f(\theta)$ is some function of the coverage of the electrode by the adsorbed species, $\theta$. In order to find an expression for $f(\theta)$, it is necessary to have knowledge of the type of adsorption and for this, several types of adsorption isotherms have been proposed. The oldest is the Langmuir isotherm,[9] and in this case it is assumed that for the reaction represented by Eq. (2.9), the *rate of adsorption* ($v_1$) is proportional to the free surface on the electrode. Thus,

$$v_1 = k_1(1-\theta)a_{A_{sol}} \tag{2.12}$$

where $k_1$ is the rate of the adsorption reaction. The *rate of the desorption* reaction ($v_{-1}$) is given by:

$$v_{-1} = k_{-1}\theta \tag{2.13}$$

where $k_{-1}$ is the rate of the reverse reaction. Under equilibrium conditions,

$$v_1 = v_{-1} \tag{2.14}$$

and

$$\frac{\theta}{1-\theta} = (\frac{k_1}{k_{-1}})a_{A^-} \tag{2.15}$$

Frumkin (1925) considered lateral interactions of the adsorbed species and the expression for the adsorption isotherm was modified to:[10]

$$\frac{\theta}{1-\theta}\exp(-2A\theta) = \frac{k_1}{k_{-1}}a_{A^-} \tag{2.16}$$

where

$$A = -\frac{N_a\varphi_a}{2kT} \tag{2.17}$$

and $\varphi_a$ is the interaction energy of one molecular pair and $N_a\varphi_a$ is the interaction energy of one molecule with $N_a$ nearest neighbors. A positive value of $A$ arises when there is attraction of the adsorbed particles and a negative value results for repulsion.

Temkin (1941) assumed heterogeneity of the surface but no molecular interactions, and for this case the adsorption isotherm is expressed by:[11]

$$\theta = \frac{1}{f}\ln\beta_0 a_{A^-} \tag{2.18}$$

where $f = K/kT$ with $K$ being a constant and $\beta_0 = W\exp(Q/kT)$ where $W$ is a parameter related to the distribution of molecules in solution and in the adsorbed state, and $Q$ to the enthalpy of adsorption.

In the Flory[12]-Huggins[13] type of isotherm, it was assumed that the adsorption process involves the displacement of water molecules from the surface, i.e.,

$$A_{sol}^- + nH_2O_{ads} \underset{\leftarrow}{\overset{\rightarrow}{\quad}} A_{ads} + nH_2O_{sol} \qquad (2.19)$$

The resulting isotherm has the form:

$$\frac{\theta}{(1-\theta)^n}e^{(1-n)} = \beta A_{sol}^- \qquad (2.20)$$

A more generalized form of the isotherm for the specific adsorption of ions, developed by Bockris, Gamboa-Aldeco, and Szklarczyk,[14] takes into account surface heterogeneity, solvent displacement, charge transfer, lateral interactions, and ion size. The expression for this isotherm is:

$$\theta = \frac{1}{f}\left(-\frac{\Delta G_{ch,i}}{kT} - \frac{\Delta G_{ch,w}}{kT} - \frac{\Delta G_E}{kT} - \frac{\Delta G_L}{kT} + \frac{f}{2} + \ln\frac{c}{c_w}\right) \qquad (2.21)$$

where $\Delta G_{ch,i}$ and $\Delta G_{ch,w}$ represent the chemical part, $\Delta G_E$, the potential dependent part, $\Delta G_L$ the lateral part of the free energy change for the adsorption of the ion, and $c$ and $c_w$ the bulk concentration of the ion and water in the solution, respectively. The term $f$ is equal to $-2v_0/kT$, where $v_0$ is an energy parameter of the lattice. This equation was tested with the experimental results for adsorption of chloride and bisulfate ions on platinum and the agreement was quite good (Figure 2.4). Knowledge of such type of adsorption behavior of ions is essential in an analysis of the kinetics of the electrode reactions because ionic adsorption can block sites for the reaction and can also affect the potential distribution across the electrode/electrolyte interface.

### 2.1.3. Effect of Adsorption of Neutral Molecules on the Structure of the Double Layer

An understanding of the adsorption behavior of neutral molecules (mainly organics) is also essential in elucidating the reaction kinetics across electrode/electrolyte interfaces, particularly in the low to intermediate temperature range in aqueous electrolytes. The significance of the effects of organic adsorption in electrochemical energy conversion (fuel cells) cannot be overstated because adsorption of species like carbon monoxide and of intermediates formed during the electrooxidation of organic fuels, such as methanol, causes significant overpotential

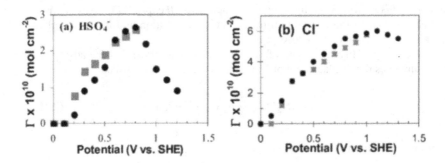

**Figure 2.4.** Adsorption of (a) chloride ions (b) bisulfate ions on platinum as a function of potential. Reprinted from Reference 14, Copyright © (1992), with permission from Elsevier.

(polarization) losses at the anodes in these fuel cells. Since the 1920s, it has been realized that the potential dependence of adsorption of organic species on an electrode is essentially parabolic in behavior (Figure 2.5).[15] The water-dipole model or the structure of the double layer provides an explanation for this behavior in the

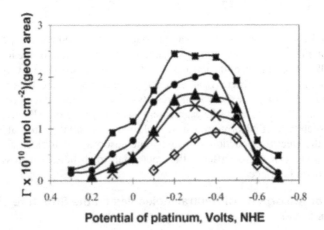

**Figure 2.5.** Typical coverage vs. potential plot for adsorption of an organic compound (naphthalene) on an electrode at an interface with electrolyte. Case illustrated is for adsorption of napthalene on platinum for different bulk concentrations of naphthalene in the electrolyte: (■) $1 \times 10^{-4}$ M, (●) $5 \times 10^{-5}$ M, (▲) $1 \times 10^{-5}$ M, (×) $5 \times 10^{-6}$ M, and (◊) $2.5 \times 10^{-6}$ M. Reprinted from Reference 15, Copyright © (1964), with permission from The Electrochemical Society, Inc.

following manner. The strong interaction between the water dipole and the electrode is mainly due to the electric field at the electrode/electrolyte interface, but there is also a small contribution to the electric field at the electrode/electrolyte interface due to the chemical interaction of the water molecule with the electrode. The electric field causes the orientation of the water molecules on either side of *the potential of zero charge* (pzc), i.e., if the electrode is negatively charged the positive end of the water dipole will be oriented towards the electrode and vice versa. The weakest interaction between the electrode and the water molecule is at a potential close to the pzc. It is for this reason that a pseudo-symmetrical inverse relation is observed for the variation of extent of adsorption with potential across the interface—i.e., the maximum adsorption is at a potential close to the pzc and there is a steady decrease in the extent of adsorption on either side of the pzc.[1]

Research studies on the behavior of organic adsorption in electrochemistry are quite extensive. Organic adsorption plays key roles not only in electrochemical energy conversion, but also in electroorganic synthesis, corrosion protection, electrodeposition, electrochemical sensors, etc. Apart from the adsorption behavior being controlled by the dependence of the orientation of the water molecules, organic adsorption depends on the chemical interaction between the organic molecules and the electrode surface (for example, the $\pi$ orbital interaction of aromatics with the electrode) and charge-transfer reactions, which could occur between the organic species and the electrode.

In general, organic adsorption may be represented by a displacement of adsorbed water molecules according to the reaction:

$$[organic]_{sol} + nH_2O_{ads} \underset{\leftarrow}{\rightarrow} [organic]_{ads} + nH_2O_{sol} \tag{2.22}$$

There could also be an adsorption via charge transfer and the adsorbed species could be without or with any breakdown in the chemical structure, the latter being an intermediate during an electron transfer reaction. Just as in the case of ionic adsorption, the adsorption behavior of organic species can be expressed in terms of adsorption isotherms (Langmuir, Frumkin, Temkin, Flory-Huggins, Bockris-Gamboa-Aldeco-Szklarczyk, etc.). A generalized isotherm was developed by Bockris and Jeng[16] by considering the adsorption process as a solvent substitution process, and the water molecules being adsorbed in three configurations as monomers in flipped-up and flopped-down positions and as dimers with no net dipole as described by

$$[organic]_{sol} + o[H_2O]\uparrow + p[H_2O]\downarrow + q[H_2O]\uparrow\downarrow$$
$$\underset{\leftarrow}{\rightarrow} [organic]_{ads} + (o + p + 2q)H_2O_{sol} \tag{2.23}$$

---

[1] The maximum does not occur exactly at the pzc because of the difference in chemical interactions between the electrode and the water molecules that depend on whether the H or the O atom is oriented towards the metal.

Comparison of the adsorption behavior of n-valeric acid (an aliphatic compound) and phenol on platinum with theory (Figure 2.6) reveals that the maxima in the coverage-potential curve are in the vicinity of the pzc. The factors, which affect adsorption of organic compounds/species on electrodes, may be summarized as follows:

- Aliphatic hydrocarbons, linear or branched, interact weakly with the electrode or water molecules, and thus the extent and strength of their adsorption is small. However, if the aliphatic molecules have functional groups (CO, CO-NH$_2$), they interact with water molecules and the surface of the electrode, and also have higher amounts of potential dependent adsorption.

- Aromatic compounds generally have $\pi$ electron interactions with neighboring atoms and with electrodes and these exhibit potential dependent adsorption. Further, unlike linear aliphatic molecules, which are vertically oriented when adsorbed on the electrode, aromatics at low coverage have a flat orientation on the electrode (see Figure 2.6 for comparison). Similarly, unsaturated linear compounds tend to have an orientation with the multiple bonds parallel to the surface. At higher coverages of the aromatics, there could be a reorientation in the vertical position to accommodate more molecules.

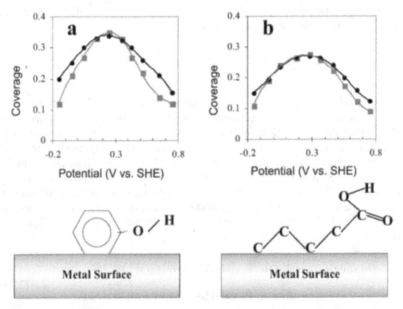

**Figure 2.6.** Comparison of coverage vs. potential for phenol and valeric acid on platinum. Reprinted from Reference 16, Copyright © (1992), with permission from Elsevier.

- The morphology of the electrode surface also affects the adsorption characteristics. The extent of adsorption appears to be less on rough surfaces than on smooth ones and on the former, there is less reorientation (horizontal to vertical) of adsorbed aromatic molecules.
- The electrolyte also affects organic adsorption. Organic molecules are, in general, considerably larger in size than the ions. Thus, more water molecules adsorbed on the electrode will have to be displaced. In addition, the organic molecules will have to be dissociated from any water molecules with which they are hydrated because they have polar groups. In general, the lower the solubility of the organic compound in the electrolyte, the higher the adsorbability on the electrode.

This short description of the extensive topic of the adsorption behavior of organic compounds on electrodes signifies its important role in affecting the characteristics of the double layer across electrode/electrolyte interfaces. Adsorption of organic compounds (i) affects the electric field across the interface; (ii) blocks sites for the desired electron transfer across the interface; and (iii) poisons electrode surfaces with strongly adsorbed species such as CO from reformed fuels or formed as intermediates during electrooxidation of organic fuels. For more details on this topic, the reader can refer to the revised edition of the book *Modern Electrochemistry,* Vol. 2A by Bockris, Reddy, and Gamboa-Aldeco.

## 2.1.4. Brief Analysis of Structures of Semiconductor/Electrolyte and Insulator/Electrolyte Interfaces

Semiconductor/electrolyte and insulator/electrolyte interfaces are also often encountered in electrochemical systems. They are of particular relevance to (i) photo-electrochemical reactions for use of solar energy to produce hydrogen and oxygen at semiconductor/electrolyte interfaces, and (ii) colloid and interfacial phenomena at insulator/electrolyte interfaces. Figure 2.7 schematically represents the potential distribution of a semiconductor electrode/electrolyte interface.

There are three characteristic regions. The regions across the interface (B-C) and that into the bulk (C-D) of the electrolyte are analogous to that for a metal/electrolyte interface. It is the region A-B, the *space-charge region,* which is different. It extends to a considerable distance within the semiconductor; for a carrier concentration of $10^{14}$ electrons $cm^{-3}$, and it is about $10^{-4}$ cm. The variation of potential with distance is similar to that in the diffuse-layer region. The charged species, which contributes to the potential distribution in the space charge region, are electrons, holes, or may be immobile impurity ions. Brattain and Garrett, who were the first to investigate the double layer characteristics of semi-conductor/electrolyte interfaces,[17] used an approach similar to that of Gouy and Chapman. A major part of the potential drop is within the semiconductor in region A-B, which is unlike the case of the metal/electrolyte interface, where it is in the

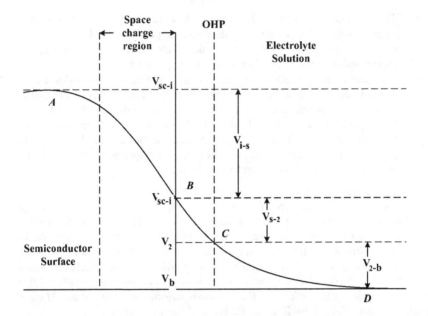

**Figure 2.7.** Potential distribution across semiconductor/electrolyte interface. Reprinted from J. O'M Bockris and S. Srinivasan, *Fuel Cells: Their Electrochemistry*, Copyright © 1969, with permission from McGraw Hill Book Company, with permission of The McGraw-Hill Companies.

Helmholtz layer. The equivalent circuit for the semiconductor/electrolyte circuit consists of three capacitors in series, and thus the overall capacity can be expressed by:

$$\frac{1}{C} = \frac{1}{C_{SC}} + \frac{1}{C_H} + \frac{1}{C_{dl}} \tag{2.24}$$

where the subscripts $SC$, $H$ and $dl$ represent the space charge layer, Helmholtz layer, and the Gouy layer, respectively. It must be noted that in this simple model, electrons become trapped in the surface, whereby quantum states for electrons on the surface differ from that in the bulk causing surface states. In the presence of surface states, the potential variation within the semiconductor resembles that of the double layer across a metal-electrolyte interface with the specific adsorption of ions.

When there is a movement of one phase relative to the other, electrokinetic phenomena arise because of the presence of a surface charge on the insulator material. Just as in the case of metals and semiconductors, there is some interaction of the surface charge species in the insulator material and ionic and dipolar species in the electrolyte. Specific adsorption of ions could also play a role. Studies have shown that the potential drop across the insulator material/electrolyte interface is significant in very dilute electrolytes. An important parameter in electrokinetic phenomena is the *zeta potential* ($\xi$) across the insulator/electrolyte interface, defined as the potential drop from the shear plane to the bulk of the electrolyte. It is somewhat less than the potential drop in the Gouy region at an interface between a metal and an electrolyte. The zeta potential is a valuable parameter in the sense that we can calculate free surface charge characteristics of the insulator material by using an equation similar to that in the Gouy-Chapman theory relating the diffuse-layer potential to the surface charge.

Zeta potentials can be measured in four different ways. If the insulator material is in the form of a narrow tube or porous plug, a flow of the electrolyte under pressure ($P$) gives rise to a *streaming potential* ($E$) or *streaming current* ($i$). The zeta potential ($\xi$) is calculated from the slope of the streaming potential-pressure plot according to the relation:

$$\xi = \frac{4\pi\eta\kappa}{\varepsilon}\left(\frac{dE}{dP}\right) \tag{2.25}$$

where $\eta$, $\kappa$ and $\varepsilon$ are the viscosity, specific conductance, and dielectric constant of the electrolyte.

The reverse of this electrokinetic phenomenon is the *electroosmostic flow*. In this case, a passage of current through the tube or porous plate the causes the flow of electrolyte from one end to the other. The zeta potential can be calculated from the expression:

$$\xi = \frac{4\pi\eta\kappa v}{\varepsilon i} \tag{2.26}$$

where $\eta, \kappa$, and $\varepsilon$ are as defined above, $v$ is the rate of flow, and $i$ is the current passing through the tube or porous plug.

Extensive studies related to electrokinetic phenomena have been made with colloidal particles (inorganic, organic, and biological). Two types of measurement are made, i.e., *electrophoresis* and *sedimentation potential*. Electrophoresis is the migration of colloidal particles in an electrolyte under the influence of an electric field. The zeta potential may be calculated from the *electrophoretic mobility* ($v$), according to the equation:

$$\xi = \frac{4\pi\eta v}{\varepsilon E} \tag{2.27}$$

where $E$ is the electric field in the electrolyte. This technique has been widely used to determine the surface charge characteristics of colloidal *isoelectric* points (i.e., the pH at which the $\xi$ potential is zero), and the effect of dispersing or agglomerating agents on the surface charge characteristics. The sedimentation potential method has been used only to a limited extent to determine surface charge characteristics. In this case, the colloidal particles are allowed to fall through a vertical column and the potential difference between two electrodes, vertically separated, is measured. The zeta potential across the colloid particle-electrolyte interface is given by the expression:

$$\xi = \frac{3\eta\kappa E}{\varepsilon\gamma^3(\rho - \rho')ngl} \tag{2.28}$$

where $\gamma$ is the radius of the colloidal particles, $\rho$ and $\rho'$ are the densities of the particles and of the solution respectively, $n$ is the number of particles in a unit volume, $g$ is the acceleration due to gravity, $l$ is the distance between the electrodes, and $E$ is the sedimentation potential.

## 2.2. VITAL NEED FOR MULTI-DISCIPLINARY APPROACH

A wide variety of charge-transfer reactions occur at the interfaces of electrodes and electrolytes. The electrode materials could be metals or alloys, semiconductors, or enzymes. For practically all the types of reactions mentioned earlier, the electrolyte is aqueous. However, there are several cases in which the electrolytes are non-aqueous (e.g., ionizable inorganic or organic compounds, molten or solid state ionic conductors). An attempt is made in Figure 2.8 to represent the needed multi-disciplinary approach for investigation of the mechanisms of charge-transfer reactions.

Thermodynamics lay the groundwork to determine whether a charge-transfer reaction can occur spontaneously or needs to be driven using electrical energy. A charge-transfer reaction involves either the donation or acceptance of electrons by the electrode to or from a species in the electrolyte or adsorbed on the surface of the electrode. Since the electron has a very low mass and the thickness of the double layer is of the order of a few angstroms, the transfer of electrons occurs by quantum mechanical tunneling. Physical chemistry plays a significant role in elucidating the kinetics and electrocatalysis of the charge-transfer reactions. It is for this reason that electrode kinetics has been conventionally treated as a topic in physical chemistry, particularly in European countries. But just like material science, electrode kinetics embodies a multitude of disciplines. Statistical mechanical treatments of reaction rates have been most helpful, particularly in studies of reactions involving chemisorption of reactants and intermediates, as well as isotopic reactions. Metallurgy and solid-state science are involved in investigations of effects of electronic and geometric factors (crystal structure, defects) on electrocatalysis and their role in nucleation and crystal growth. The link between electrochemistry and surface science has been growing by leaps and bounds,

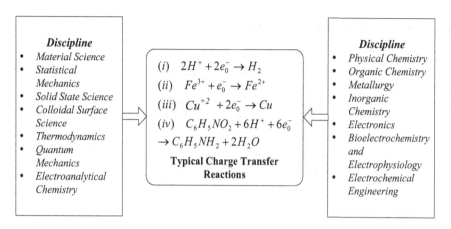

| Discipline | | Discipline |
|---|---|---|
| • Material Science <br> • Statistical <br>   Mechanics <br> • Solid State Science <br> • Colloidal Surface <br>   Science <br> • Thermodynamics <br> • Quantum <br>   Mechanics <br> • Electroanalytical <br>   Chemistry | $(i) \quad 2H^+ + 2e_0^- \rightarrow H_2$ <br> $(ii) \quad Fe^{3+} + e_0^- \rightarrow Fe^{2+}$ <br> $(iii) \quad Cu^{+2} + 2e_0^- \rightarrow Cu$ <br> $(iv) \quad C_6H_5NO_2 + 6H^+ + 6e_0^-$ <br> $\rightarrow C_6H_5NH_2 + 2H_2O$ <br><br> **Typical Charge Transfer <br> Reactions** | • Physical Chemistry <br> • Organic Chemistry <br> • Metallurgy <br> • Inorganic <br>   Chemistry <br> • Electronics <br> • Bioelectrochemistry <br>   and <br>   Electrophysiology <br> • Electrochemical <br>   Engineering |

**Figure 2.8.** Multidisciplinary approaches for investigations of mechanisms of charge-transfer reactions.

particularly by the use of highly sophisticated *in situ* electrochemical/spectroscopic techniques, specifically for the examination of reactants, intermediates, and products adsorbed on surfaces, and the formation of passive films. Other techniques that are useful are Scanning Tunneling Microscopy (STM) and Atomic Force Microscopy (AFM) to examine electrodeposition or electrodissolution processes. Electrochemistry and material science are often grouped together because the structure, composition, and characteristics of the electrode material and solid electrolytes play key roles during the course of the charge-transfer reactions. They also depend on the methods of preparation of these materials; further, in electrocatalysis, the application of nanostructured materials is gaining momentum. Electronics is governed by solid-state electrochemistry. Charge-transfer reactions at electrode/electrolyte interfaces involve the physics of current flow and electric fields.

The progress made in electronics has been beneficial in designing circuitry to control potentials across interfaces as well as to investigate transient behavior, employing techniques such as electrochemical impedance spectroscopy. From a technological point of view, electrochemical engineering plays a major role toward understanding mass transport (diffusion, convection), hydro-dynamics of flow of solutions, transport of ions to surfaces, process control, etc. Last, but not least, bioelectrochemical charge transfer processes involve the disciplines of biochemistry and electrophysiology. Charge-transfer reactions in these systems are fascinating and involve both electron and proton transfer. The electrochemical mechanisms, by which biological systems function with respect to energy metabolism and nerve transmission, obey electrochemical laws. Their high efficiency and high speed can hardly be matched by simple organic or inorganic charge-transfer reactions.

## 2.3.  SINGLE AND MULTI-STEP REACTIONS

Electrochemical reactions are similar to chemical reactions with one major difference: at least one step in the overall electrochemical reaction, the electron transfer reaction occurs across the electrode/electrolyte interface. In the case of chemical or biochemical reactions, there are three types of reactions: *single-step*, *consecutive-step*, and *parallel* reactions. Examples of these types of reactions are as follows:

- *Single-step reactions.* The electrodeposition/dissolution of a metal-like lithium occurs in a single step:

$$Li^+ + e_0^- \leftrightarrow Li \tag{2.29}$$

This is the reaction occurring in a secondary lithium ion battery. Even though the reaction is represented by Eq. (2.29) in a single step, the electrodeposition step   is followed by nucleation, surface diffusion of lithium, and crystal growth. In general, the kinetics of this reaction are relatively simple.

- *Consecutive-step reactions.* In a consecutive reaction, two or more intermediate steps occur in series, i.e., an intermediate produced in the first step is consumed in the second; if more than two intermediate steps are involved, the species produced in the second step takes part in the third step, etc. An example of this reaction of relevance to fuel cells is the four-electron transfer electroreduction of oxygen to water. A possible reaction pathway of this reaction on a platinum electrode in acid medium is:

$$O_2 + M + H_3O^+ + e_0^- \rightarrow MHO_2 + H_2O \tag{2.30}$$

$$MHO_2 + M \rightarrow MO + MOH \tag{2.31}$$

$$MO + H_3O^+ + e_0^- \rightarrow MOH + H_2O \tag{2.32}$$

$$2MOH + 2H_3O^+ + 2e_0^- \rightarrow 2M^+ + 4H_2O \tag{2.33}$$

The overall reaction is, thus,

$$O_2 + 4H_3O^+ + 4e_0^- \rightarrow 6H_2O \tag{2.34}$$

where $M$ represents the electronically conducting electrode material (say Pt) and is not involved in the overall reaction. It plays the role of an electrocatalyst for the reaction. It must be noted that the intermediate step represented by Eq. (2.33) occurs in two identical consecutive steps; the reason for this is that electron transfer occurs by quantum mechanical tunneling, which involves only one electron transfer at a time.

- *Parallel-step reactions.* When multistep reactions take place there is the possibility of parallel-intermediate steps. The parallel-step reactions could lead to the same final product or to different products. These types of reactions are more often encountered in electroorganic chemistry and bioelectrochemistry than in electrochemical reactions involving inorganic reactants and products. A fuel cell reaction, which sometimes exhibits this behavior, is the direct electrooxidation of organic fuels, such as hydrocarbons or alcohols. For instance, in the case of methanol, a six-electron transfer complete oxidation to carbon dioxide can occur consecutively in six or more consecutive steps; in addition, partially oxidized reaction products could arise producing formaldehyde and formic

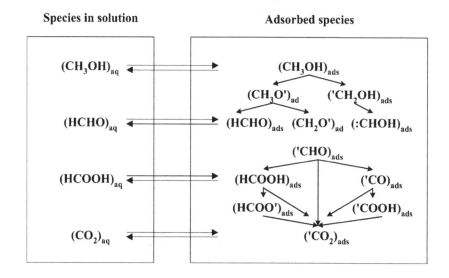

**Figure 2.9.** Various possible pathways for the electrooxidation of methanol.

acid in parallel reactions.[2] These, in turn, could then be oxidized to methanol. Such possible reaction pathways for methanol oxidation[18] are represented in Figure 2.9.

## 2.4.    CONCEPT OF RATE-DETERMINING STEP

The term *rate-determining step* (rds) is frequently referred to the step in reactions that proceed in two or more intermediate stages, either consecutively or in parallel. Most often, it is only one of these intermediate steps, which controls the rate of the overall reaction; this step is given the terminology *the rate-determining step* or *rds*. Several analogies have been proposed to visualize the concept of the rds in a consecutive reaction. One is that of an electrical circuit with a series of two or more resistances and a power source, as shown in Figure 2.10. This figure shows three resistors, $R_1$, $R_2$, and $R_3$ in series; in addition, the power source (a fuel cell or a battery) has an internal resistance, $R_i$. The current ($I$) through the electrical circuit is given by the expression:

$$I = \frac{E}{R_1 + R_2 + R_3 + R_i}$$

(2.35)

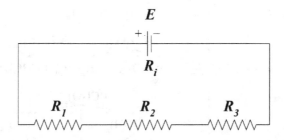

**Figure 2.10.** Electrical analogy for a rate-determining step in a consecutive reaction. Reprinted from J. O'M Bockris and S. Srinivasan, *Fuel Cells: Their Electrochemistry*, Copyright © 1969, with permission from McGraw-Hill Book Company.

---

[2] It must be noted that the proton in the intermediate steps of oxygen reduction (Eqs. 2.30 to 2.33) and in the overall reaction (Eq. 2.34) is designated as $H_3O^+$: the bare proton does not exist as such in an aqueous medium. Due to the charge on the proton, its size and the strong ion-dipole reaction with the $H_2O$ molecule, it forms the hydronium ion, $H_3O^+$, which is the discharging entity. The dissociation energy for breaking this bond is 183 kJ mol$^{-1}$.

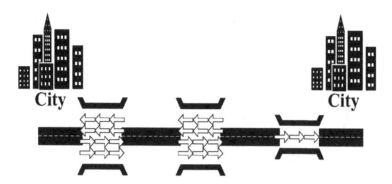

**Figure 2.11.** Roadblock analog for the rate-determining step in a consecutive reaction. Reprinted from J. O'M Bockris and S. Srinivasan, *Fuel Cells: Their Electrochemistry*, Copyright © 1969, with permission from McGraw-Hill Book Company.

where $E$ is the electric potential of the power source. Assuming that the resistances $R_1$, $R_3$, and $R_i$ are very small in comparison to the resistance $R_2$, the current in the electrical circuit may be expressed by the equation:

$$I = \frac{E}{R_2} \qquad (2.36)$$

which means that the resistor $R_2$ determines the current through the external circuit, and it is the *current-determining resistor*.

Another analogy is that of a roadblock between two cities, A & B, as represented by Figure 2.11. If one imagines several bridges between the cities, the flow of traffic in both directions will be quite fast but if there is one bridge which will let only one car travel on the bridge at a time, the speed of this car through the bridge will have a significant effect on the time for travel between the two cities.

Transforming these analogies to that of a consecutive reaction with about five intermediate steps, one can show from a plot of the free energy vs. distance along a reaction coordinate, that the step exhibiting the highest energy state with respect to the initial or final state controls the rate of the reaction. An analytical treatment of consecutive reactions carried out by Christiansen,[19] has shown that the rate-determining step controls the rate of the consecutive reaction in the forward and reverse direction, and that all other steps are virtually in equilibrium. Thus, for the chemical reaction represented in Figure 2.12, the rate of the forward reaction is given by:

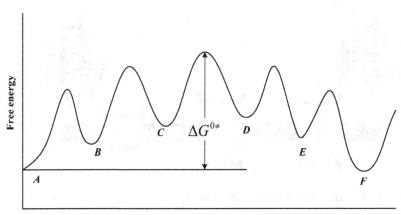

Distance along reaction coordinate

**Figure 2.12.** A typical free energy vs. distance along reaction coordinate plot for a consecutive reaction. Reprinted from J. O'M Bockris and S. Srinivasan, *Fuel Cells: Their Electrochemistry*, Copyright © 1969, with permission from McGraw-Hill Book Company.

$$\vec{v} = k_{C \to D}[C] \tag{2.37}$$

The rate-determining step has the highest barrier with respect to the initial or final state. It must be pointed out that according to the classical mechanical

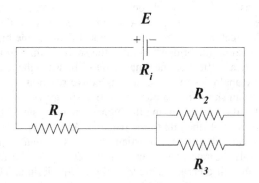

**Figure 2.13.** Electrical analogue for rate-determining step in a parallel reaction. Reprinted from J. O'M Bockris and S. Srinivasan, *Fuel Cells: Their Electrochemistry*, Copyright © 1969, with permission from McGraw-Hill Book Company.

treatment of reaction rates, it is only the reactant particles with sufficient energy to surmount the barriers from A to D for the forward reaction and F to C for the reverse reaction that are effective for the occurrences of the forward and reverse reactions. By using a Maxwell-Boltzmann statistical analysis, the rate of the reaction will be the same as expressed by Eq. (2.37) for the forward reaction.

In the case of a parallel reaction too, the electrical and roadblock analogs are helpful at understanding the rate-determining step. From Figure 2.13 for the former case, the current ($I$) in the external circuit is given by:

$$I = \frac{E}{R_1 + \dfrac{R_2 R_3}{R_2 + R_3} + R_i} \tag{2.38}$$

Assuming that $R_i$ and $R_1$ are much less than $R_2$ or $R_3$ and that $R_2 \ll R_3$, $I$ will approximate to

$$I \approx \frac{E}{R_2} \tag{2.39}$$

Thus, in a parallel circuit, the smaller resistor controls the current.

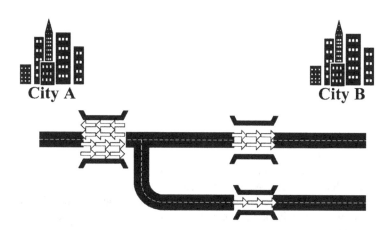

**Figure 2.14.** Roadblock analog for rate-determining step in a parallel reaction. Reprinted from J. O'M Bockris and S. Srinivasan, *Fuel Cells: Their Electrochemistry,* Copyright © 1969, with permission from McGraw-Hill Book Company.

For the roadblock analogy (Figure 2.14), the parallel road, which is considerably wider, determines the rate of the cars travelling between the cities A and B.

In terms of a chemical reaction, one may consider the sequence:

$$A \underset{v_{-1}}{\overset{v_1}{\rightleftharpoons}} B \overset{v_2}{\underset{v_{-2}}{\rightleftarrows}} \begin{matrix} C_1 \\ C_2 \end{matrix} \qquad (2.40)$$

For simplicity, the rates ($v$) of the intermediate steps $C_1$ to B and $C_2$ to B may be considered negligible. Thus,

$$v_1 - v_{-1} = v_2 + v_3 \qquad (2.41)$$

If

$$v_2 \gg v_3 \qquad (2.42)$$

then

$$v_1 - v_{-1} \approx v_2 \qquad (2.43)$$

Further if

$$v_1 \gg v_2 \qquad (2.44)$$

then,

$$v_1 \approx v_{-1} \qquad (2.45)$$

Thus, the step A $\rightarrow$ B is virtually in equilibrium and the step B $\rightarrow$ C controls the rate of the overall reaction.

Another possible type of consecutive reaction is one with a *dual* or *coupled mechanism*. In such a case, the standard free energies of the activated complexes in two steps of a consecutive reaction could be nearly the same, and the forward velocities of the two steps will be identical. These two steps control the rate of the overall dual reaction. A consecutive reaction in which the velocities of the two reverse steps are negligible in comparison with the forward rates is referred to as a *coupled reaction*.

## 2.5. DEPENDENCE OF CURRENT DENSITY ON POTENTIAL FOR ACTIVATION-CONTROLLED REACTIONS: THEORETICAL ANALYSIS

### 2.5.1. Classical Treatment to Determine Electrode Kinetic Parameters

In electrochemical reactions, the potential across the interface affects the rate of the reaction. The reason is that in an electrochemical reaction an electron transfer reaction occurs across the interface, and the rate of this reaction could be significantly affected by the electric field across the double layer at the interface.

Let us first consider a chemical reaction such as:

$$A + B \rightarrow [AB]^{\ddagger} \rightarrow C + D \tag{2.46}$$

where A and B are the reactants, C and D the products, and $[AB]^{\ddagger}$ represents the activated state for the reaction. Assuming that this reaction occurs in a single step and that the potential energy for the reaction is represented as a function of the reaction coordinate (Figure 2.15), the velocity of the reaction ($v$) may be expressed by:

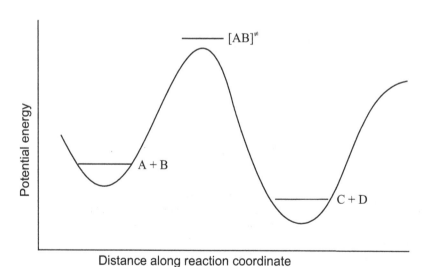

Distance along reaction coordinate

**Figure 2.15.** Potential energy profile along the reaction path for a single step reaction. Reprinted from J. O'M Bockris and S. Srinivasan, *Fuel Cells: Their Electrochemistry*, Copyright © 1969, with permission from McGraw-Hill Book Company.

$$v = \frac{kT}{h}[\text{A}][\text{B}]\exp\left(-\frac{\Delta H^{0\ddagger}}{RT}\right)\exp\left(\frac{\Delta S^{0\ddagger}}{R}\right) \qquad (2.47)$$

where $\Delta H^{0\ddagger}$ and $\Delta S^{0\ddagger}$ are the enthalpies and entropies for activation. The rate constant for this reaction, $k_0$, is thus:

$$k_0 = \frac{kT}{h}\exp\left(-\frac{\Delta H^{0\ddagger}}{RT}\right)\exp\left(\frac{\Delta S^{0\ddagger}}{R}\right) \qquad (2.48)$$

Only a simplified version of the chemical kinetics is presented above for the case of a single-step reaction. The calculations of potential energy versus reaction coordinate are much more complex and sophisticated for multi-step reactions because if more than three atoms are involved, there is an increase in the number of interaction energies to be considered in the potential energy calculations. Statistical mechanical treatments, involving calculations of the translational, rotational, and vibrational partition functions for the activated and initial states, have been made.

However, conceptually it is possible to arrive at a general expression, such as Eq. (2.47) for the rate of a reaction occurring in consecutive steps. The rate constants for the forward, $\vec{k}$, and reverse, $\overleftarrow{k}$, reactions can then be modified to:

$$\vec{k} = \frac{kT}{h}\exp\left(-\frac{\Delta G^{0\ddagger}_{i\to g}}{RT}\right) \qquad (2.49)$$

and

$$\overleftarrow{k} = \frac{kT}{h}\exp\left(-\frac{\Delta G^{0\ddagger}_{n\to g}}{RT}\right) \qquad (2.50)$$

for a reaction which occurs in n steps with the $g^{th}$ step being the rate-controlling step. Similar treatments have been carried out for multistep reactions that occur by parallel-reaction paths or by dual or coupled mechanisms. For more details, the reader is referred to textbooks on the kinetics of chemical reactions.

The above description can now be extended to electrochemical reactions such as:

$$\text{M} + \text{AB}^+ + e_0^- \to \text{MA} + \text{B} \qquad (2.51)$$

An example of this reaction is proton discharge during hydrogen evolution, e.g.,

$$M + H_3O^+ + e_0^- \rightarrow MH + H_2O \qquad (2.52)$$

For this case, a potential energy profile (Figure 2.16) can be constructed by assuming that in the initial state a strong $H^+$-$H_2O$ bond stretches as this species comes towards the metal surface and an M-H bond formation (chemisorption of hydrogen) starts to occur, as represented by:

$$M + H^+ \text{---} H_2O + e_0^- \rightarrow [M \text{---} H^+ \text{---} H_2O] + e_0^- \rightarrow MH + H_2O \qquad (2.53)$$

Butler first proposed that the transfer of the electron from a stretched $H^+$-$H_2O$ to a stretched M–H bond occurs at the activated state via a tunnelling mechanism.[20] When an electric field is applied across the interface, it affects the potential energy profile for the stretching of the $H^+$-$H_2O$ bond plus the electron but not for the final state, $MH_x$. It must be noted that the electric energy across the interface varies linearly with distance. The net result is that the potential energy vs.

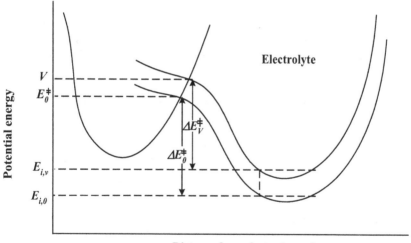

**Distance from electrode surface**

**Figure 2.16.** Potential energy vs. distance along reaction coordinate plot for transfer of proton from $H^+$–$H_2O$ to metal forming the adsorbed M–H bond as an intermediate during hydrogen evolution. $E_{i,0}$ and $E_{i,v}$ represent the zero point energy levels of the initial state $H^+$–$H_2O$ when the potentials across the metal/electrolyte interface are $V = 0$ and $V = V$, respectively; $\Delta E_0^{\ddagger}$ and $\Delta E_V^{\ddagger}$ and are the corresponding energies for the activated state; and $E_0^{\ddagger}$ and $E_V^{\ddagger}$ are the corresponding activation energies for the reaction. Reprinted from J. O'M Bockris and S. Srinivasan, *Fuel Cells: Their Electrochemistry*, Copyright © 1969, with permission from McGraw-Hill Book Company.

distance plot for the stretching of the initial state is raised effectively by a value of $VF$ for the initial state, where $V$ represents the potential across the interface. In the activated state, the effect of the field is only a fraction of that of the initial state. Thus, the rate constant for the proton discharge step ($k$) can be given by an equation of the form:

$$k = k_0 \exp\left( -\frac{\beta VF}{RT} \right)$$  (2.54)

where $k_0$ denotes the value of the rate constant when there is no electric field across the interface (i.e., $V = 0$); $\beta$ is referred to as the *symmetry factor*, and it represents a fraction of the field which changes the potential energy of the activated state when there is a potential of $V$ volts across the interface. It must be stressed that a simplified version is presented here, and for more rigorous treatments, the reader is referred to books on electrochemistry.

Just as in chemical kinetics, the next step is to obtain an expression for the *velocity* or *rate* of electrochemical reactions, which will depend on the kinetic parameters such as the concentration of the reactants, reaction order etc. The rate of an electrochemical reaction at an electrode/electrolyte interface is expressed as a current density (A or mA cm$^{-2}$) and is measured at constant temperature. Determination of the manner in which the current density is dependent on the potential is one of the most important diagnostic criteria in elucidating the mechanism of an electrochemical reaction, i.e., the reaction path, intermediate steps, and rate-determining step. In this chapter we present only the essential theoretical analysis for determining the mechanism of the electrochemical reactions. Chapters 5 and 6 will discuss the electrocatalytic factors involved in fuel cell reactions, and illustrate the experimental methods to (i) determine the mechanisms of the electrochemical reactions of fuel cells; and (ii) elucidate the intermediate steps and the rate-determining step, which leads to an evaluation of the electrode kinetic parameters for the reaction.

In the remainder of this section, we shall deal with the expressions for current density as a function of potential, reaction orders, and exchange current densities. A short description will also be made about stoichiometric number, which is often encountered in investigations of the kinetics of reactions via consecutive or parallel-reaction pathways.

2.5.1.1. *Expression for Current Density as a Function of Potential.* Just as in the case of a chemical reaction, the rate constant for an electrochemical reaction is given by $k$ (Eq. 2.54) and the rate of the reaction (v) depends on the activities of the reactants for the forward reaction and of the products for the reverse reaction. The net rate of the reaction expressed by Eq. (2.52) is given by:

$$v = \overrightarrow{v} - \overleftarrow{v}$$  (2.55)

where $\vec{v}$ and $\overleftarrow{v}$ are the rates of the forward and reverse reactions, respectively. They are expressed as

$$\vec{v} = \vec{k}_0 c_{H_3O^+} (1-\theta) \exp\left(-\frac{\beta VF}{RT}\right) \tag{2.56}$$

$$\overleftarrow{v} = \overleftarrow{k}_0 \theta \exp\left[\frac{(1-\beta)VF}{RT}\right] \tag{2.57}$$

where $c_{H_3O^+}$ is the concentration of $H_3O^+$ and $\theta$ is the fractional degree of coverage of the species MH at the surface of the electrode. In these equations, it is assumed that the reaction is first order with respect to the concentrations of reactants and products. The terms $\theta$ and $(1 - \theta)$ arose in the above two equations because we assumed a Langmuir adsorption for the chemisorption of MH on the electrode. The velocities of the forward and reverse reaction are in mol s$^{-1}$ for unit area of the electrode. To convert the velocities to current densities, one has to multiply the velocity by $nF$, which represents the number of coulombs involved during the charge transfer. In the chosen reaction (Eq. 2.57), $n$ is equal to unity. For the transformation of the proton plus one electron to an adsorbed hydrogen atom (MH), the current density ($i$) can then be expressed by:

$$i = \vec{i} - \overleftarrow{i} = F\left[\vec{k}_0 c_{H_3O^+}(1-\theta)\exp\left(-\frac{\beta VF}{RT}\right) - \overleftarrow{k}_0 \theta \exp\left(\frac{(1-\beta)VF}{RT}\right)\right] \tag{2.58}$$

The above expression is for the net current density for the discharge of the proton on the metal to form an adsorbed hydrogen atom on the metal. The species MH is an intermediate and not a final product. This step is then followed by one of the following intermediate steps for the overall electrolytic evolution of hydrogen:

$$2MH \underset{k_{-2}}{\overset{k_2}{\longrightleftharpoons}} 2M + H_2 \tag{2.59}$$

$$MH + H_3O^+ + e\bar{0} \underset{k_{-3}}{\overset{k_3}{\longrightleftharpoons}} M + H_2 + H_2O \tag{2.60}$$

for which the current density-potential relations are given by:

$$i = 2F\left[k_2\theta^2 - k_{-2}(1-\theta^2)P_{H_2}\right] \tag{2.61}$$

or

$$i = F\left[k_3 c_{H_3O^+} \theta \exp\left(-\frac{\beta VF}{RT}\right) - k_{-3}(1-\theta)P_{H_2} \exp\left(\frac{(1-\beta)VF}{RT}\right)\right] \qquad (2.62)$$

The intermediate step, as represented by Eq. (2.59), is referred to as the *recombination step,* and by Eq. (2.60) as the *electrochemical-desorption step.* It must be noted that there is no apparent potential dependence for the former but there is one for the latter. However, in the former case, there is the indirect potential dependence from the first step and a modified one for the second step because of the dependence of $\theta$ (concentration of the $MH_{ads}$ species) on the potential.

In order to arrive at an expression of $i = f(V)$ for the overall reaction, it is necessary for one to assume (i) a reaction pathway (i.e., discharge-recombination or discharge-electrochemical desorption) and (ii) whether the first or the second step is the rate-determining step in the two consecutive pathways. These expressions are presented in Table 2.1 for both pathways and the two possible rate-determining steps for each pathway. One can also make further approximations, particularly for the Langmuir conditions of adsorption, for $\theta$ tends to zero or unity and simplify the expressions as shown in Table 2.1. Another possible pathway frequently encountered is the coupled reaction slow discharge/slow electrochemical desorption. In this case, both the discharge and the electrochemical desorption intermediate steps have equal rates.

### 2.5.1.2. Reaction Orders, Transfer Coefficients, and Stoichiometric Numbers.

Just as in the case of rates of chemical reactions, the rates of electrochemical reactions depend on the activities (concentrations) of reactants and products. One frequently encounters the term *reaction order* with respect to a particular reactant. Depending on the simplicity or complexity of a reaction which may occur in a single step or multi-step and on the rate-determining step, the reaction order could be unity, zero, or greater than or less than one (whole or fractional). In general terms, for an overall electrochemical reaction of the type

$$aA + bB + \ldots\ldots + ne_0^- \rightarrow xX + yY + \ldots\ldots \qquad (2.63)$$

If the rate-determining step is

$$lL + mM + \ldots\ldots + n'e_0^- \rightarrow pP + qQ + \ldots\ldots \qquad (2.64)$$

and its velocity is given by

$$v = ka_L^l a_M^m \ldots\ldots \exp\left(\frac{\alpha VF}{RT}\right) \qquad (2.65)$$

The reaction order, $l$, for the reactant, L, is given by:

$$l = \left( \frac{\partial \ln v}{\partial \ln a_L} \right)_{a_M \dots V, T} \tag{2.66}$$

In Eq. (2.65), $\alpha$, the transfer coefficient, is related to the symmetry factor $\beta$; it may or may not be equal to $\beta$. The numerical value of $\alpha$ would depend on how many of the preceding or succeeding steps involve an electron transfer. Table 2.1 shows the reaction orders and the transfer coefficients for the different possible rate-determining steps according to the two reaction pathways for the hydrogen evolution reaction.

The *stoichiometric number*, $v$, is a term used in one of the columns in Table 2.1. This term represents the number of times the rate-determining step has to take place for one act of the overall reaction. For instance, for a mechanism involving slow discharge followed by a fast recombination step, the stoichiometric number is two, but for a slow discharge followed by an electrochemical desorption step, the stoichiometric number is unity. One of the fundamental aspects of electrode kinetics is that the parameters, Tafel slopes, symmetry factors, transfer coefficients, reaction orders, stoichiometric numbers, and separation factors[3] are diagnostic criteria for determining the mechanisms of electrochemical reactions, i.e., the reaction path and the rate-determining step.

2.5.1.3. *Exchange-Current Density and Reversible Potential.* To explain the concept of exchange-current density ($i_0$), let us consider a single step reaction:

$$O + ne_0^- \rightarrow R \tag{2.67}$$

where O and R are the oxidized and reduced species, respectively, in the reaction. Thus, for this reaction, the net current density, $i$, may be expressed by:

$$i = F \left[ \overrightarrow{k}^0 c_O \exp\left( -\frac{\beta V_r F}{RT} \right) - \overleftarrow{k}^0 c_R \exp\left( \frac{(1-\beta)V_r F}{RT} \right) \right] \tag{2.68}$$

As represented above, the net current density is for the cathodic reduction of species O to R. Under equilibrium conditions, i.e., at the reversible potential, the net current

---

[3] In Table 2.1 the parameter $S_T$ represents the hydrogen-tritium *separation factor*, defined as the ratio of hydrogen/tritium in the gas phase to the electrolyte. This parameter is also a diagnostic criteria in elucidating the mechanism of electrolytic-hydrogen evolution.

## TABLE 2.1
### Kinetic Parameters for the Most Probable Mechanisms of the Hydrogen Evolution Reaction[a]

| Mechanism | Condition θ → 0 | | | | Condition θ → 1 | | | | $\nu$ | $S_T$ |
|---|---|---|---|---|---|---|---|---|---|---|
| | $\dfrac{\partial\eta}{\partial\ln i}$ | $\dfrac{\partial\eta}{\partial\ln a_{H_3O^+}}$ | $\dfrac{\partial\eta}{\partial\ln p_{H_2}}$ | $\dfrac{\partial\eta}{\partial\ln\theta}$ | $\dfrac{\partial\eta}{\partial\ln i}$ | $\dfrac{\partial\eta}{\partial\ln a_{H_3O^+}}$ | $\dfrac{\partial\eta}{\partial\ln p_{H_2}}$ | $\dfrac{\partial\eta}{\partial\ln\theta}$ | | |
| Slow discharge-fast recombination | $-\dfrac{RT}{\beta F}$ | $\dfrac{RT}{F}$ | $\dfrac{RT}{2F}$ | $0$ | $-\dfrac{RT}{\beta F}$ | $\dfrac{RT}{F}$ | $-\dfrac{RT}{2F}\dfrac{(1-\beta)}{\beta}$ | $0$ | 2 | 5 |
| Fast discharge-slow recombination | $-\dfrac{RT}{2F}$ | $0$ | $\dfrac{RT}{2F}$ | $-\dfrac{RT}{F}$ | $\infty$ | $0$ | | $0$ | 1 | 11 |
| Coupled discharge recombination | $-\dfrac{RT}{\beta F}$ | $\dfrac{RT}{F}$ | $\dfrac{RT}{2F}$ | $\dfrac{2RT}{\beta F}$ | $-\dfrac{RT}{\beta F}$ | $\dfrac{RT}{F}$ | $-\dfrac{RT}{2F}\dfrac{(1-\beta)}{\beta}$ | $-\infty$ | | 5 |
| Slow discharge-fast electrochemical desorption | $-\dfrac{RT}{\beta F}$ | $\dfrac{RT}{F}$ | $\dfrac{RT}{2F}$ | $\dfrac{RT}{F}$ | $-\dfrac{RT}{(1+\beta)F}$ | $\dfrac{RT}{F}\dfrac{(1-\beta)}{(1+\beta)}$ | $-\dfrac{RT}{2F}\dfrac{(1-\beta)}{(1+\beta)}$ | $0$ | 1 | 6 |
| Fast discharge-slow electrochemical desorption | $-\dfrac{RT}{(1+\beta)F}$ | $\dfrac{RT}{F}\dfrac{(1-\beta)}{(1+\beta)}$ | $\dfrac{RT}{2F}$ | $-\dfrac{RT}{F}$ | $-\dfrac{RT}{\beta F}$ | $\dfrac{RT}{F}$ | $\dfrac{RT}{2F}$ | $0$ | 1 | 23 |
| Coupled discharge-electrochemical desorption | $-\dfrac{RT}{\beta F}$ | $\dfrac{RT}{F}$ | $\dfrac{RT}{2F}$ | $0$ | $-\dfrac{RT}{\beta F}$ | $\dfrac{RT}{F}$ | $\dfrac{RT}{2F}$ | $0$ | 1 | 7 |
| Slow molecular hydrogen-ion discharge | $-\dfrac{RT}{(1+\beta)F}$ | $\dfrac{RT}{F}\dfrac{(1-\beta)}{(1+\beta)}$ | $\dfrac{RT}{2F}$ | $\dfrac{RT}{F}$ | $-\dfrac{RT}{\beta F}$ | $\dfrac{RT}{F}$ | $\dfrac{RT}{2F}$ | $0$ | 1 | 6 |
| Slow molecular hydrogen diffusion | $-\dfrac{RT}{2F}$ | $0$ | | $-\dfrac{RT}{F}$ | $\infty$ | $0$ | | $0$ | 1 | 8 |

[a] From Suggested Reading 1.

density is zero. The *exchange-current density* ($i_0$) may be defined as the rate of the forward or reverse reaction under equilibrium conditions. Thus,

$$i_0 = \vec{i} - \overleftarrow{i} = 0 \tag{2.69}$$

and

$$i_0 \equiv \vec{i} = \overleftarrow{i} = F\vec{k}^0 c_O \exp \frac{-\beta V_r F}{RT} = F\overleftarrow{k}^0 c_R \exp \frac{(1-\beta)V_r F}{RT} \tag{2.70}$$

From Eq. (2.70), it follows that an expression for $V_r$, the *reversible potential*, is:

$$V_r = \frac{RT}{F} \ln \frac{\vec{k}^0}{\overleftarrow{k}^0} + \frac{RT}{F} \ln \frac{c_O}{c_R} \tag{2.71}$$

This equation for the half-cell reaction is exactly the same as for the Nernst reversible potential, which was derived in Section 1.5. The first term on the right hand side represents the standard reversible potential for conditions of unit activities of reactants and products and the temperature is assumed to be 25 $^0$C. The second term reflects the change in the reversible potential with the change in concentrations of reactants and products. The expressions are similar for more complex reactions involving more than one reactant and one product and for multi-electron transfer reactions, which occur in consecutive or parallel steps but the formats for $i_0$ and $V_r$ are similar. In general terms, for the reaction expressed by Eq. (2.64), these could be expressed by:

$$i_0 = \vec{i} = \overleftarrow{i} = F\vec{k}^0 c_O \exp\left(-\frac{\beta V_r F}{RT}\right) = F\overleftarrow{k}^0 c_R \exp\left(\frac{(1-\beta)V_r F}{RT}\right) \tag{2.72}$$

From Eq. (2.72), the expression for $V_r$, the reversible potential, is:

$$V_r = \frac{RT}{F} \ln\left(\frac{\vec{k}^0}{\overleftarrow{k}^0}\right) + \frac{RT}{F} \ln\left(\frac{c_O}{c_R}\right) \tag{2.73}$$

## 2.5.2. Quantum Mechanical Treatment

In the preceding section, a classical kinetic treatment was followed to derive the expression for the rate of a charge-transfer reaction, i.e., the first step of proton discharge on the electrode surface to form an adsorbed hydrogen atom on the metal;

this step is then followed by the recombination or electrochemical desorption step in the overall hydrogen evolution reaction (see Eqs. 2.52, 2.59, and 2.60). In such a treatment, it is assumed that the reactants, which have sufficient kinetic energy (i.e., greater than the potential energy barrier), can proceed at a rate dependent on (i) $kT/h$, (ii) the product of reactant concentrations, and (iii) the free energy of activation. It was also assumed that the proton discharge step involves the stretching of the $H^+$-$H_2O$ bond and that it occurs, only when the bonds are stretched to the intersection of the two Morse curves for the $H^+$-$H_2O$ and M-H bonds. Another assumption was that the electron transfer occurs when the proton is stretched to the intersection point of the two Morse curves (Figure 2.16). Though the latter assumption is correct, it is only a quantum-mechanical treatment of electron transfer that can provide an explanation for the considerably higher rates for the proton discharge step than that predicted according to the classical treatment. It is also necessary to present another plot (Figure 2.17) for this derivation, i.e., the potential energy vs. distance plot for the transfer of an electron from the metal to the solvated proton.

According to Gurney,[21] the energy barrier for electron transfer is obtained by taking into consideration: (i) the *image interaction* given by $e^2/4x$, where $e$ is the

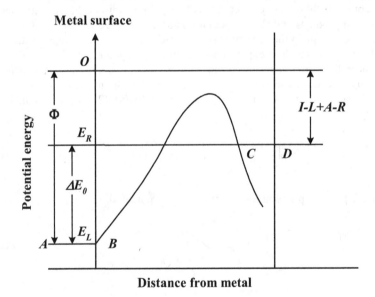

**Figure 2.17.** Potential energy vs. distance from the surface of the metal electrode for transfer of electron from $H_3O^+$ to metal. Reprinted from J. O'M Bockris and S. Srinivasan, *Fuel Cells: Their Electrochemistry*, Copyright © 1969, with permission from McGraw-Hill Book Company.

electronic charge and $x$ is the distance of the electron from the metal surface and (ii) *the coulombic interaction* between the electron and the solvated proton. The level AB (cf. Figure 2.17) in the potential energy plot is determined by the *work function* of the metal, $\varphi$, defined as the work done in bringing an electron from an infinite distance, in vacuum, to the metal. The level CD is the work done in bringing an electron from infinity to the solvated proton. This process involves (i) the desolvation of the ion $H_3O^+$, (ii) the electron acceptance, (iii) formation of the M-H bond, and (iv) a repulsion between the adsorbed hydrogen atom and water molecules. The energy levels at AB ($E_L$) and CD ($E_R$) may be expressed by:

$$E_L = -\varphi \tag{2.74}$$

$$E_R = -\left(I - L + A - R\right) \tag{2.75}$$

where $I$ is the ionization energy of the hydrogen atom, $L$ is the interaction energy between the protons and solvent molecules with the proton at the assumed distance from the metal, $A$ is the adsorption energy between the hydrogen atom and the metal for the specified distance, and $R$ is the repulsion energy between the hydrogen atom and water molecule to which the proton was attached as $H_3O^+$, prior to charge transfer.

Numerical calculations show that according to the classical treatment, the activation energy for the transfer of electrons from the metal to the protons in solution is too high and the observed current densities cannot be explained. The only other possibility is the tunneling of electrons from the metal to the protons in solution. One of the conditions for the tunneling of the electrons from the Fermi level of the metal to the protons in solution is that there must be vacant levels in the solvated protons with energy equal to that of the former. $E_L$ and $E_R$ (horizontal lines) in Figure 2.17 represent the *ground states* (initial states) of the electron in the metal and the protons in the OHP. Tunneling of electrons can occur only when:

$$E_L \geq E_R \tag{2.76}$$

Using Eqs. (2.74) and (2.75) this condition becomes:

$$\varphi \leq I - L + A - R \tag{2.77}$$

The parameter $I$ is a constant but the parameters $R$, $A$, and $L$ are not. They depend on the distances of the hydrogen atom and ion from the metal. The potential energy curves for the stretching of the proton from its equilibrium position in the OHP and the hydrogen atom from the metal are shown in Figure 2.16. As first proposed by Butler, the only way of reducing the energy gap for electron transfer to occur is by stretching the $H^+$–$OH_2$ bond. Figure 2.16 shows that this gap is reduced to zero at the intersection point of the potential energy distance plots for the stretching of the

$H^+$–$OH_2$ bond and M–H bonds. It also shows that only a fraction of the energy (represented by LN) is required to bring the proton up to the intersection point. It must be noted that *electron tunneling* can occur left of the intersection point for stretching of the $H^+$–$OH_2$ bond but not to its right. The *discharge step* may then be represented by:

$$H_3O^+ + M(e\bar{0}) \rightarrow \left[H_2O - H^+ + M(e\bar{0})\right]^{\neq} \underset{\leftarrow}{\rightarrow} [H_2O + H - M]^{\neq} \rightarrow H_2O + H - M$$

$$(2.78)$$

The potential energy plot in Figure 2.17 is illustrated for the case when the potential drop across the electrode/electrolyte interface is zero. When the potential drop is $V$, as applied in the cathodic direction for speeding up the reaction,

$$\Delta E_V = \Delta E_0 + VF \qquad (2.79)$$

where $\Delta E_V$ and $\Delta E_0$ are the energies at the potentials $V$ and zero, respectively. It must be noted that $V$ has a negative value for the cathodic direction. Since

$$\Delta E_0{}^{\neq} = \beta\Delta E_0 \quad \text{and} \quad \Delta E_V{}^{\neq} = \beta\Delta E_V \qquad (2.80)$$

it also follows from the two preceding equations that

$$\Delta E_V{}^{\neq} = \beta(\Delta E_0 + VF) \qquad (2.81)$$

where $\beta$, the symmetry factor, represents the fraction of the energy gap between AB and CD to close the gap for electron tunneling.

The electron transfer rapidly occurs when the proton is stretched up to the intersection point of the two Morse curves (Figure 2.16). When the electrode/electrolyte potential is zero, the rate of the overall hydrogen-evolution reaction may be expressed by

$$i_V = ek_{ef}W(E_f)(W\Delta E^{\neq})n_{H_3O^+} \qquad (2.82)$$

where $e$ is the electronic charge, $k_{ef}$ is the *frequency factor* (number of electrons with Fermi energy, colliding with the unit area of the surface at the electrode/electrolyte interface, $W(E_f)$ is the probability of the electron being able to tunnel through the barrier, $(W\Delta E^{\neq})$ is the probability that an $H_3O^+$ is in a suitably stretched state for electron transfer to occur, and $n_{H_3O^+}$ is the number of $H_3O^+$ ions populating the unit area of the OHP. Approximate expressions for $k_{ef}$ and $W(E_f)$ are:

$$k_{ef} = \frac{4\pi m (kT)^2}{h^3} \qquad (2.83)$$

and

$$W_{ef} = \exp - \frac{4\pi l}{h} \left[ 2m \left( E_x - E_f \right)^{1/2} \right] \qquad (2.84)$$

where $m$ is the mass of the electron, $E_x$ is the energy at the top of the electron transfer barrier, and $l$ is the width of the barrier (the barrier is assumed to be rectangular). For $(W \Delta E_0^{\neq})$, a Boltzman expression may be used, i.e.,

$$(W \Delta E_0^{\neq}) = \exp \frac{-\Delta E_0^{\neq}}{kT} \qquad (2.85)$$

Substituting Eqs. (2.83) to (2.85) in Eq. (2.82),

$$i_{V=0} = 4\pi e m \frac{(kT)^2}{h^3} \exp \left[ -\frac{4\pi l (2m)}{h} (E_x - E_F)^{1/2} \right] \exp \left( \frac{-\Delta E_0^{\neq}}{kT} \right) n_{H_3O^+} \qquad (2.86)$$

The main difference in the above expression for the current density at a potential $V$ will be that $\Delta E_0$ in the last term is replaced by $\Delta E_V$. From Eqs. (2.80) and (2.81), it follows that,

$$\Delta E_V^{\neq} - \Delta E_0^{\neq} = \beta V F \qquad (2.87)$$

Thus, at a potential $V$, Eq. (2.86) is transformed to:

$$i_{V=V} = i_{V=0} \exp(-\frac{\beta V F}{RT}) \qquad (2.88)$$

By using $V = V_r + \eta$, where $\eta$ is the *activation overpotential* and $V = V_r$ for the reversible potential at which $i - i_0$, one arrives at the equation:

$$i = i_0 \exp(-\frac{\beta \eta F}{RT}) \qquad (2.89)$$

which is the empirical Tafel equation that was also arrived at following the classical treatment. The main difference between the classical and quantum mechanical treatments is that the calculations yield reasonable values of the exchange current density, consistent with the experimental ones in the former case but would yield considerably lower values in the latter case.

Section 1.3 dealt with several types of electrochemical reactions. The charge transfer steps in practically all these types of reactions play key roles in determining the rates of the overall reactions. In the case of electrocatalytic reactions such as in fuel cells, water electrolysis, electroorganic oxidation and reactions, intermediate chemical steps of adsorption/desorption can also be slow enough to contribute to activation overpotentials.

## 2.6. CONCEPT OF ACTIVATION OVERPOTENTIAL, EXPRESSION FOR CURRENT DENSITY AS A FUNCTION OF ACTIVATION OVERPOTENTIAL, AND CHARGE TRANSFER RESISTANCE

To express the current density-potential relations in simpler terms, another term deserves a definition, i.e., the *overpotential,* $\eta$, for the reaction. The *activation overpotential* ($\eta$), for a reaction controlled by charge transfer in an electrochemical reaction, is the extent of departure of the potential, across the electrode/electrolyte interface (at which this reaction occurs), from the reversible potential, when the reaction occurs at a net current density of $i$ A cm$^{-2}$. The *overpotential*, like *overheat* for a chemical reaction, is the driving force for the electrochemical reaction. For a cathodic reaction (as the one represented by Eq. 2.67) the overpotential is negative, and for an anodic reaction, it is positive. For the reaction expressed by Eq. (2.67), one can transform Eq. (2.68) into the form:

$$i = i_0 \left[ \exp\left( -\frac{\beta \eta F}{RT} \right) - \exp\left( \frac{(1-\beta)\eta F}{RT} \right) \right] \tag{2.90}$$

using the equation

$$V = V_r + \eta \tag{2.91}$$

and the expressions for $i_0$ are given by Eqs. (2.69) and (2.70).

In more general terms, the familiar form of the equation for a multistep reaction involves $\alpha$, the *transfer coefficient*, instead of $\beta$, the symmetry factor, which is valid for a single-step reaction. Thus, in general terms,

$$i = i_0 \left\{ \exp\left[ -\frac{\alpha \eta F}{RT} \right] - \exp\left[ \frac{(1-\alpha)\eta F}{RT} \right] \right\} \tag{2.92}$$

Two limiting cases can now be considered. One is at a low overpotential, when the exponential terms in Equation (2.92) can be linearized. This applies at less than 20 mV at room temperature. For such a case, this equation becomes:

$$i = \frac{i_0 \eta F}{RT} \tag{2.93}$$

Thus, at potentials close to the reversible potential for the reaction, there is a linear dependence of current density on overpotential. Considering an electrical analog, one may introduce the term *charge transfer resistance* $(R_{ct})$ for the electrochemical reaction in this linear region, which is defined as:

$$R_{ct} = \frac{d\eta}{di} = \frac{RT}{i_0 F} \tag{2.94}$$

From this equation, it is clear that if the exchange current density, $i_0$, has low values for the charge-transfer reaction, the charge transfer resistance is high and vice versa. The term *highly polarizable* and *pseudo-nonpolarizable* reactions are often used for such cases. The former is for irreversible reactions and the latter is for pseudo-reversible reactions. For highly reversible reactions, $i_0 \geq 10^{-3}$ A cm$^{-2}$ (e.g., hydrogen evolution/ionization on platinum electrodes in acid media, copper deposition/dissolution) and for highly irreversible reactions, generally $i_0 \leq 10^{-6}$ A cm$^{-2}$ (e.g., oxygen evolution/reduction, methanol oxidation).

The other limiting case occurs when $\eta$ exceeds about 20 or more mV. Under these conditions, the current density for the reverse reaction is small in comparison with the forward reaction. Thus, the expression for the current density as a function of overpotential reduces from Equation (2.92) to

$$i = i_0 \exp\left(-\frac{\alpha \eta F}{RT}\right) \tag{2.95}$$

which may be written in the form

$$\eta = \frac{RT}{\alpha F} \ln i_0 - \frac{RT}{\alpha F} \ln i \tag{2.96}$$

This is identical with the equation:

$$\eta = a + b \ln i \tag{2.97}$$

empirically proposed by Tafel in 1905. The parameters $a$ and $b$ are the Tafel parameters. According to this equation, there is a semi-logarithmic dependence of the overpotential on the current density. In the Tafel region, the charge transfer resistance $(R_{ct})$ is given by

$$R_{ct} = \frac{d\eta}{di} = \frac{b}{i} \qquad (2.98)$$

Thus, in this region $R_{ct}$ decreases inversely with $i$. Alternatively stated, there is a semi-exponential increase of $\eta$ with an increase of $i$ for a cathodic reaction.

Quite often in electrochemical reactions, the first electron transfer step is rate determining and the value of $\alpha$ is then equal to the value of $\beta$ (about 0.5). Thus, a Tafel slope of 120 mV decade$^{-1}$ is frequently encountered. However, there are several cases where the second, third, and in some cases, subsequent steps are rate determining and these will generally have higher values of the transfer coefficients and correspondingly lower Tafel slopes, i.e., 60 mV, 30 mV, and 15 mV decade$^{-1}$. More detailed interpretations of the Tafel slopes for multi-step fuel cell reactions are presented in the Chapter 5 on *Electrocatalysis of Fuel Cell Reactions*, as well as in several books on electrode kinetics.

## 2.7. OTHER TYPES OF RATE LIMITATIONS AND OVERPOTENTIALS AND THEIR EFFECTS ON CURRENT DENSITY POTENTIAL BEHAVIOR

### 2.7.1. Mass-Transport Overpotential

So far we have focused on charge-transfer reactions that occur at interfaces of electrodes with electrolytes. The assumption was made that the transport of the reactant species to and from the OHP has no hindrances. This assumption is valid for several electrochemical reactions and for several fuel cell reactions, particularly at low to intermediate current densities. However, when the concentrations of reactants are low, particularly for gases with very low solubility in the electrolyte (for example hydrogen or oxygen), limitations occur due to the slowness of transport of these species from the bulk to the OHP where the charge transfer occurs. One type of transport limitation for the example chosen is the *diffusion* of the reactant species to the electrode surface, if ionic species are involved. There is an effect of the electric field on the rate of transport of the ionic species toward or away from the electrode. Additionally, limitations due to *convective* transport could be caused by differences in densities as a result of temperature or concentration. One can overcome most limitations caused by *migration* or convection by using supporting electrolytes in the former case and by working in still solutions in the latter. Under such conditions, diffusion alone governs mass transport. Diffusion in electrochemistry is analogous to heat transfer in solid, liquid, or gas media.

The current density–overpotential relation for an electrochemical reaction, which is controlled by the rate of diffusion of a reactant species from the bulk electrolyte to the interface, is shown below. We shall consider a reaction:

$$M^{n+} + ne_0^- \rightarrow M \qquad (2.99)$$

which occurs at a planar electrode. We shall also assume that the kinetics of the reaction is determined by the rate of one-dimensional diffusion of the reactant $M^{n+}$ to the electrode, as represented by the equation:

$$M_b^{n+} \rightarrow M_e^{n+} \qquad (2.100)$$

The suffixes $b$ and $e$ denote the bulk and electrode-electrolyte interface where the charge-transfer reaction occurs. According to the theory of mass transport, Fick's first law expresses the *diffusion flux, Q,* as

$$Q = -D\frac{dc}{dx} \qquad (2.101)$$

In order to evaluate the concentration gradient, Nernst and Merriam[22] introduced the concept of a *diffusion layer* near the electrode across which the concentration of the reactant species varies linearly with distance. It was further assumed that the concentration maintains the bulk value from the diffusion layer to the bulk of the electrolyte. There is sufficient experimental evidence to show that in unstirred electrolytes the thickness of the diffusion layer is about $5 \times 10^{-2}$ cm. Using the Nernst-Merrium model, Eq. (2.101) for the diffusional flux will be modified as:

$$Q = -D\left(\frac{dc}{dx}\right)_{x=0} = D\frac{c_b - c_e}{\delta} \qquad (2.102)$$

Under steady state conditions, the diffusional flux will be equal to the rate of the charge transfer of the reaction. Since $Q$ is in mol cm$^{-2}$ and $n$ electrons are consumed in the charge-transfer reaction, the current density for the reaction under diffusion control becomes:

$$i = DnF\frac{c_b - c_e}{\delta} \qquad (2.103)$$

It must be noted that the unit for $D$ is cm$^2$ s$^{-1}$. With an increase of current density, the concentration gradient across the diffusion layer becomes steeper and the maximum gradient occurs when the steady state concentration at the electrode reaches zero. Under such conditions, all the $M^{n+}$ ions reaching the electrode undergo the charge-transfer reaction. The expression for the *limiting current density, $i_L$,* is, then,

$$i_L = \frac{DnFc_b}{\delta} \qquad (2.104)$$

Using Eqs. (2.103) and (2.104), one may express the current density as a function of $i_L$, $c_e$, and $c_b$ as follows:

$$i = i_L \left( 1 - \frac{c_e}{c_b} \right) \tag{2.105}$$

To express the current density as a function of potential for the diffusion-controlled reaction, one must make the assumption that the charge-transfer reaction is fast, i.e., quasi-reversible or virtually in equilibrium. Under these conditions, one can use the Nernst equation for the potential at the electrode/electrolyte interface ($V$). When $i = 0$, the concentration of the reactant $M^{n+}$ at the electrode is the same as in the bulk ($c_b$). Thus,

$$V_{i=0} = V_r^0 + \frac{RT}{nF} \ln c_b \tag{2.106}$$

where $V_r^0$ is the reversible potential of the electrode. At a current density of $i$, the concentration at the electrode is $c_e$. Thus,

$$V_{i=i} = V_r^0 + \frac{RT}{nF} \ln c_e \tag{2.107}$$

The suffixes $i = 0$ and $i = i$ for $V$ denote the potential of the electrode during equilibrium conditions and the passage of current density $i$, for each electrochemical reaction, respectively. We may now introduce the concept of *diffusion overpotential*, $\eta_D$, as given by:

$$\eta_D = V_{i=i} - V_{i=0} = \frac{RT}{nF} \ln \frac{c_e}{c_b} \tag{2.108}$$

Substituting Eqs. (2.106) to (2.108) into Eq. (2.105), we arrive at the current density–overpotential relation for the diffusion-controlled reaction:

$$i = i_L \left( 1 - \exp \frac{n\eta_D F}{RT} \right) \tag{2.109}$$

When $\eta_D$ is small ($< 20$ mV), the exponential term can be linearized and the expression for $i$ as a function of $\eta_D$ reduces to:

$$i = -\left( \frac{i_L \eta_D F}{RT} \right) \tag{2.110}$$

The reason there is a negative sign on the right hand side of Eq. (2.110) is that the electrochemical reaction is a cathodic one and $\eta_D$ has negative values.

A typical plot of potential vs. current density for a diffusion-controlled reaction is represented in Figure 2.18. AB corresponds to activation overpotentials; BC is the linear region representing ohmic overpotentials; CD is the region of limiting current density. At D, other electrochemical reactions with more negative reversible potentials take place as shown by region DE. The values of the diffusion coefficients for most species undergoing electrochemical reactions in aqueous media are of the order of $10^{-5}$ cm$^2$ s$^{-1}$ at room temperature.

The range of values of the solubilities of the reactant species is very high. However, for fuel cell reactants such as hydrogen, oxygen, and several other gases, the solubility at room temperature is only of the order of $10^{-4}$ moles dm$^{-3}$. Using these values in Eq. (2.105), the limiting current densities for the electrooxidation of hydrogen or for the electroreduction of oxygen (fuel cell reactions in aqueous media), at planar electrodes, will be only about $10^{-4}$ A cm$^{-2}$. For this reason, three-dimensional porous gas diffusion electrodes are used in fuel cells to enhance the three-dimensional reaction zone and the diffusion of the reactant species to the electroactive sites by radial diffusion. In this case, for diffusion in fine pores or to spherical particles, $i_L$ is expressed by $DnFc/r$ where $r$ is the radius of the fine pore or the particle. In micropores or with nanoparticles, $i_L$ can be significantly increased.

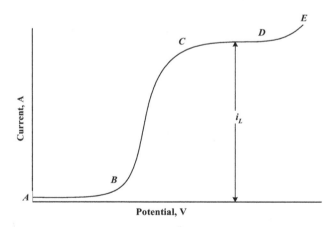

**Figure 2.18.** A typical plot of potential vs. current density for a diffusion-controlled reaction. Reprinted from J. O'M Bockris and S. Srinivasan, *Fuel Cells: Their Electrochemistry*, Copyright © 1969, with permission from McGraw-Hill Book Company.

## 2.7.2. Case of Activation plus Diffusion-Controlled Reactions

In Section 2.5, we dealt with only *activation-controlled reactions* and the preceding section dealt with a solely *diffusion-controlled reaction*. However, in real life situations, there are regions of current density where both activation and diffusion play roles, and thus, the current density-potential relation needs modification. In the case of the reaction expressed by Eq. (2.67) under conditions of solely activation control, the current density-overpotential relation is as expressed by Eq. (2.92). In arriving at this expression, one assumes that the concentrations of the reactants and products are the same at the interface and in the bulk. But at higher current densities and particularly with reactants and/or products having low solubilities, this assumption may not be valid because of the slow rate of mass transport. Under such conditions, the equation may be modified to

$$i = i_0 \left[ \frac{c_{O,e}}{c_{O,b}} \exp \frac{-\beta\eta F}{RT} - \frac{c_{R,e}}{c_{R,b}} \exp \frac{(1-\beta)\eta F}{RT} \right] \tag{2.111}$$

where $\eta$ is still the activation overpotential; the diffusion overpotentials are absorbed within the terms $(c_{O,e}/c_{O,b})$ and $(c_{R,e}/c_{R,b})$ (the suffixes $b$ and $e$ represent concentrations in the bulk and at the electrode). By taking into consideration that mixed control occurs at higher values of activation overpotentials, one can arrive at an expression without the unknown $c_e$ terms. Since the rate of the reverse reaction is negligible, Eq. (2.111) reduces to

$$i = i_0 \frac{c_{O,e}}{c_{0,b}} \exp\left( \frac{-\beta\eta F}{RT} \right) \tag{2.112}$$

Introducing the expression for $(c_{O,e}/c_{O,b})$ from Eq. (2.106) into the above equation, one arrives at:

$$i = i_0 \left( 1 - \frac{i}{i_L} \right) \exp \frac{-\beta\eta F}{RT} \tag{2.113}$$

and thus,

$$\eta = -\frac{RT}{\beta F} \ln \frac{i i_L}{i_0 (i_L - i)} \tag{2.114}$$

Another way of visualizing Eq. (2.113) is by considering the term

$$i_F = i_0 \exp\frac{-\beta\eta F}{RT} \tag{2.115}$$

where $i_F$ is the *activation-controlled current* with no diffusional limitations. Thus, Eq. (2.114) becomes

$$\frac{1}{i} = \frac{1}{i_F} + \frac{1}{i_L} \tag{2.116}$$

The reciprocal relation provides the interpretation of an activation-diffusion controlled reaction as occurring in series and the reciprocal terms represent the charge transfer and diffusion resistances.

### 2.7.3. Ohmic Overpotential

Ohmic overpotential arises predominantly during the passage of an electric current and it is due to electrical resistances for the transport of ions from one electrode to the other in an electrochemical cell. The electrochemical cell is akin to an electrical circuit with a power supply and resistors in series or in parallel. In the latter case, the electrical resistance is due to transfer of electrons via the resistors. In both cases, the potential drop across the resistance (ionic in the former case and electronic in the latter) varies linearly with the amount of current passing through it. In the electrochemical case, the tip of a reference electrode (or a Luggin capillary, c.f. Chapter 6) can be placed very close to the cathode or the anode, (where the charge-transfer reaction is occurring) to record the potential with respect to the reference electrode. This half-cell potential will still include the ohmic potential drop between the test electrode and the reference electrode due to the passage of current from cathode to anode, but not the total electrolyte resistance between the anode and cathode. It was stated earlier that ohmic overpotential is predominantly due to ionic resistance. Contributions to ohmic overpotential could also be from small values of electronic resistances of the electrodes and current collectors. In the case of metallic electrodes, their electrical resistances are negligible. However, if there are passive films formed on the electrodes or if the electrodes are semiconducting materials, these could have higher electronic resistances. In order to measure the activation overpotential accurately, the ohmic resistance is measured using a *transient method* (current interrupter or ac impedance, c.f. Chapter 6). An $iR$ corrected Tafel equation is of the form:

$$i = i_0 \exp\frac{\alpha F(\eta - \eta_{ohm})}{RT} \tag{2.117}$$

where $\eta$ and $\eta_{ohm}$ are the measured *total* and *ohmic overpotentials*, respectively. It is assumed that there is no mass transport overpotential. The *ohmic resistance* of the electrolyte, $R_{el}$, can be expressed by:

$$R_{el} = \rho \frac{l}{A} \tag{2.118}$$

where $\rho$ is the specific resistance of the electrolyte, $l$ is the distance between the tip of the Luggin capillary (or reference) and the electrode, and $A$ is its cross-sectional area. It must be noted that

$$\rho = \frac{1}{\kappa} \tag{2.119}$$

where $\kappa$ is the specific conductivity of the electrolyte.

## 2.8.  ELECTROCATALYSIS

### 2.8.1.  Electrocatalysis Vital Role in Electrosynthesis and Electrochemical Energy Conversion and Storage

Electrocatalysis is an important topic in electrochemical reactions; the electrode plays a catalytic role, being the donor of electrons for the cathodic reaction or the acceptor of electrons for the anodic reaction. However, just as in heterogeneously catalyzed chemical reactions at gas/solid or at solid/liquid interfaces, the catalysts enhance steps of adsorption/desorption of intermediates. Electrocatalysis plays a significant role in electrochemical gas-evolution/gas-consumption reactions, electroorganic reactions, and bioelectrochemical reactions. Beginning in the 1920s, but more so since the 1950s, tens of thousands of publications on catalysis and electrocatalysis have appeared in the literature. Probably, the most extensively investigated electrochemical reactions are those of hydrogen and oxygen evolution (water electrolysis) and their reverse reactions (fuel cells). Electroorganic synthesis and electrooxidation of organic fuels (mostly methanol and ethanol) have also been researched but to a lesser scale. Enzymatic reactions almost always occur via electron transfer intermediate steps with the enzymes serving as electrocatalysts.

The topic of electrocatalysis is of vital importance in the investigations of fuel cell reactions, particularly the low temperature—alkaline fuel cell (AFC), proton exchange membrane fuel cell (PEMFC), direct methanol fuel cell (DMFC)—and intermediate temperature—phosphoric acid fuel cell (PAFC)—fuel cells. Even for the high temperature molten carbonate fuel cells (MCFC) and solid oxide fuel cells (SOFC), electrocatalysis has some influence on the rates of the anodic and cathodic reactions but to a considerably lesser extent than those due to ohmic and mass transport resistances. Since the early 1960s, a high percentage of the published literature is on the electrocatalysis of fuel cell reactions. In this book, Chapter 5 is devoted to the electrode kinetics and electrocatalysis of fuel cell reactions. In this chapter only summarizing remarks will be made on the topics of the distinctive and similar characteristics in the two types of catalysis, electrocatalysis and heterogeneous catalysis.

## 2.8.2. Distinctive Features of Electrocatalysis

*2.8.2.1. Net-electron Transfer in Overall-Electrocatalytic Reaction.* Electrocatalysis is a field akin to heterogeneous catalysis but with one major difference: in one or more of the intermediate steps in the overall reaction, there is a net-electron transfer across the interface in electrocatalysis but not in heterogeneous catalysis. It must be noted, however, that in several cases of heterogeneously catalyzed reactions, electron transfer mechanisms are involved but there is no net electron transfer between the two phases. Thus, the potential across the interface is a variable in the expression for the rate constant for the electrocatalytic reaction. The electrocatalyst has a positive or negative effect on enhancing the chemical rate constant and often on altering the reaction path and the intermediate and rate-determining steps for the reaction.

*2.8.2.2. Wide Range of Reaction Rates Attained by Altering the Potential across the Interface at Constant Temperature.* Because of the fact that for most charge-transfer reactions there is an exponential dependence of the current density on potential, it is possible for several electrocatalytic reactions to attain reaction rates covering more than two orders of magnitude on the same electrocatalyst at constant temperature. In the case of heterogeneous catalytic reactions, the only variables for enhancing the rates on the same catalysts are temperature and the concentrations of the reactants. These also influence the rates of electrocatalytic reactions.

*2.8.2.3. An Electrochemical Pathway for Chemical Reactions.* There are a wide variety of chemical reactions, which can also occur via electrochemical pathways. An example of relevance to fuel cells, the combustion of methane to produce $CO_2$ and water can be carried out at a relatively high temperature catalytically or at a relatively low temperature electrochemically:

Chemical: 
$$CH_4 + 2O_2 \rightarrow CO_2 + 2H_2O \tag{2.120}$$

Anode: 
$$CH_4 + 2H_2O \rightarrow CO_2 + 8H^+ + 8e_0^- \tag{2.121}$$

Cathode: 
$$2O_2 + 8H^+ + 8e_0^- \rightarrow 4H_2O \tag{2.122}$$

The overall reaction that occurs electrochemically is the same as that occurring chemically, as described by Eq. (2.121). The advantage of the electrochemical route is that in addition to the chemical products, electricity is directly generated, while in the case of the combustion route, the conversion of the heat energy (released by the reaction expressed by Eq. 2.121) to electrical energy has to occur in a subsequent step in a thermal power plant. It is noteworthy to mention that the German expression for fuel cells is *Kalte Verbrennung* (i.e., cold combustion) because fuel cells can directly convert chemical to electrical energy at low temperatures, while high temperatures are necessary for thermal power plants.

2.8.2.4. *Different Products in Different Ranges of Potential.* In electrocatalysis, it is possible to carry out selective oxidation or reduction reactions in different ranges of potential across the electrode/electrolyte interface on the same electrode material under isothermal conditions. For such types of reactions to take place in heterogeneously catalyzed reactions, the only variables are the operating conditions (temperature and pressure) or concentration of reactants. The simplest example of the former is the reactions occurring on a platinum electrode in an aqueous medium. Under these conditions hydrogen evolution occurs at a potential below 0.0 V/RHE, an oxide starts forming at about 0.8 V/RHE, and oxygen evolution occurs above 1.5 V (note that oxygen is not significantly evolved in the potential range 1.23–1.5 V because of the high irreversibility of the reaction). Another good example is the progressive reduction of nitrobenzene on a platinum electrode according to the following reaction:

$$C_6H_5NO_2 + 2e_0^- \rightarrow C_6H_5NO + 2e_0^- \rightarrow C_6H_5NHOH + 2e_0^- \rightarrow C_6H_5NH_2$$

$$(2.123)$$

Nitrobenzene $\rightarrow$ Nitroso-benzene $\rightarrow$ Hydroxylamine $\rightarrow$ Aniline

2.8.2.5. *Change in Path of Reactions Using Redox Systems.* The electrooxidation of methanol has a high overpotential on even the most active electrocatalyst (Pt-Ru) with an overpotential of about 400 mV at a current density of 300 mA cm$^{-2}$ at 80 $^0$C in a DMFC. However, the reaction pathway for the electrooxidation of methanol could be altered with a significant acceleration of the reaction rate by using a redox system. An example is the electrooxidation of the species $Mo^{5+}$ (present in the electrolyte), at the electrode/electrolyte interface to $Mo^{6+}$, according to the following reaction:

$$Mo^{5+} \rightarrow Mo^{6+} + e_0^-$$

$$(2.124)$$

and the chemical oxidation of methanol by the $Mo^{6+}$ near the electrode as:

$$CH_3OH + H_2O + 6Mo^{6+} \rightarrow CO_2 + 6H^+ + 6Mo^{5+}$$

$$(2.125)$$

2.8.2.6. *In-Situ Reactivation of Electrocatalysts.* One of the problems encountered in heterogeneous catalysis is the poisoning of catalyst surfaces by intermediate species and/or byproducts and/or impurities. Poisoning problems arise in electrocatalytic reactions, especially those involving organic reactants and impurities such as CO. In the case of heterogeneous catalytic chemical reactions, for the regeneration of active sites, it is often possible to stop the reaction periodically and give the catalyst high-temperature heat treatments. In an electrochemical cell, it is possible to regenerate the electrocatalytic sites periodically by the application of electric pulses to either oxidize some organic impurities or to reduce the films (i.e., oxide films) formed on the surfaces.

2.8.2.7. *Electrochemical Nature of Biological Reactions.* In heterogeneous biochemical reactions occurring in living systems, enzymes serve as the catalysts. Most enzyme reactions occur via electron transfer reactions and a more appropriate term for these catalysts is *bioelectrocatalysts*. In several biocatalyzed reactions, proton transfer also occurs. The turnover rates for enzymatic reactions (i.e., the number of individually catalyzed events that occur per active site per sec) are several orders of magnitude (may be as high as six) higher than for heterogeneously catalyzed chemical reactions. Further, their specificities for reactions can be very high, for instance, the enzyme glucose oxidase oxidizes $\alpha$-glucosidic but not $\beta$-glucoside bonds. Enzymes are giant molecules compared to inorganic or organic species involved in electrochemical reactions. For example, cytochrome-C, a relatively small molecule, has a molecular weight of 12,400. In spite of its relatively large size, electron transfer occurs in this enzyme through the modifier to the heme group at a rate about as fast as in a redox reaction involving oxidation of a metallic ion in solution from a lower to a higher valence state.

### 2.8.3. Similarities Between Electrocatalysis and Heterogeneous Catalysis

2.8.3.1. *Effect of Geometric Factors.* The term *geometric factors* of catalysts refers to their structural and morphological aspects, which include the heterogeneity of surfaces. Also included in the geometric factors are crystal orientation and lattice spacing, crystal defects such as edges, steps and kinks, particle size, and amorphous nature (amorphicity). An original hypothesis for the importance of geometric factors of the catalysts on the rates of reactions was proposed by Balandin.[23] It indicates that reactants and/or intermediates must be sufficiently strongly adsorbed on the surface to accelerate the reaction, but not be too strongly adsorbed to hinder the rate of the subsequent intermediate step. Interatomic distances and crystal structure in the catalyst are critical parameters for the adsorption of a reactant. According to Balandin, bonds between atom in the reactant molecules are weakened, distorted, and in the limiting case may undergo rupture, as illustrated in Figure 2.19, for a reaction of the type:

$$A B + C D \rightarrow A D + BC \qquad (2.126)$$

The catalytic activity depends on lattice spacing and structure. The first supporting evidence for the Balandin hypothesis was found that when the

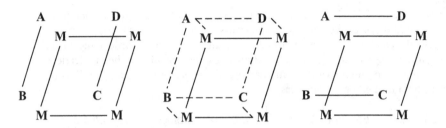

**Figure 2.19.** Reaction sequence for heterogeneously catalyzed reaction AB + CD and AD+ BC. M's represent metal atoms. Reprinted from J. O'M Bockris and S. Srinivasan, *Fuel Cells: Their Electrochemistry*, Copyright © 1969, with permission from McGraw-Hill Book Company.

hydrogenation benzene and its reverse reaction were considerably more active on catalysts with face-centered cubic or with close-packed hexagonal than with body centered cubic structures. In electrocatalysis, it has been shown that hydrogen evolution has a higher exchange current density on the (111) crystal plane than on its (100) or (111) crystal planes. The reasons why defects in crystals—such as edges, kinks, and steps—play a role in heterogeneous catalysis or electrocatalysis are that the free energies of adsorption reactants/intermediates are higher at these sites because of their intrinsic nature, as well as their having higher coordination numbers of surrounding atoms than those for a planar site. Particle sizes and intersite distances also play significant roles in heterogeneous and electrocatalysis. For example, the rate of oxygen reduction in PAFCs has a practically constant value for Pt particles above 20 to 30 Å, but there is a decrease in rate for smaller particles. Point defects (vacancies and impurities) affect catalytic activities. An example is that the rate of electrolytic hydrogen evolution is higher on an iron electrode, containing 0.2 % carbon than on zone-refined iron with less than 0.01 % impurities. Line defects (dislocations) do not seem to affect rates of fuel cell reactions.

2.8.3.2. *Electronic Factors.* Heterogeneous catalytic and electrocatalytic reactions invariably involve the intermediate steps of adsorption/desorption in the overall reaction. Since the bonds formed between the catalysts and the reactants/intermediates are covalent, these bonds are relatively strong (unlike physical adsorption of molecular hydrogen or oxygen on surfaces). It has for a long time been recognized, first in heterogeneous catalysis and considerably later in electrocatalysis, that transition metals and their alloys are the most active catalysts. In the short space available for this section, it is impossible to summarize the voluminous, innovative research findings to unravel the role of electronic factors of the materials in catalysis and electrocatalysis. One statement, which can support all these findings, is that relative strong covalent-chemical bonds (covalent) are formed between atoms in the catalyst, or in the electrocatalyst, and those in the adsorbed species. As stated in the previous Section, these bonds should be strong enough to

| Metal | Electronic Structure in | |
| --- | --- | --- |
| | 3d orbital | 4s orbital |

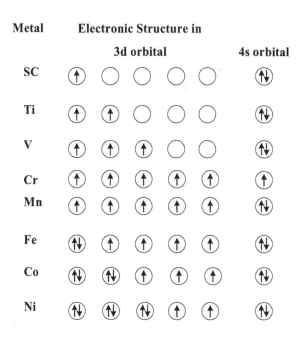

**Figure 2.20.** The electronic structures for the 3d and 4s orbitals for the transition metals in the first long period of the periodic table. Reprinted from J. O'M Bockris and S. Srinivasan, *Fuel Cells: Their Electrochemistry,* Copyright © 1969, with permission from McGraw-Hill Book Company.

accelerate the formation of these intermediates in the overall reaction but not too strong to have the inhibiting effect of decelerating the subsequent desorption steps of the reaction.

There are two electronic properties that affect the rates of catalytic/electrocatalytic reactions: *d-band vacancies* (a related parameter is the percentage d-band character), and the *work function* of the metal. The transition metals have unpaired d-electrons and their number could vary from one to ten (note that there are 5d orbitals in the two long periods of the periodic table). The electronic structure of the 3d and 4s orbitals for the transition metals in the first long period are shown in Figure 2.20.

The metals Fe, Co, and Ni have 4, 3, and 2 unpaired d-electrons, respectively, in the gaseous phase. However, the situation is different in the metallic state. In the latter state, according to the electron-band theory of metals, there is an electron overlap of the d-levels with the immediately higher s-level (in the case of the above three metals, these are the 3d and 4s levels). For Ni, with 12 spaces available for electron occupancy in the 3d and 4s levels, 0.54 rather than 2 electrons enter the 4s level. This value is close to the measured value of the saturation magnetic moment of the metal for Ni (0.61 Bohr magnetons).

A parameter related to the d-band vacancy is the *percentage d-band character*. This arises out of Pauling's valence band theory of metals.[24] According to this theory, promotion of electrons to higher orbitals (e.g., 3d to 4s) plays a role in the bonding of metals. In turn, for the metals in the first long period, 4s levels could be promoted to the 4p states and the electronic structure in the solid state involve the 3d, 4s, and 4p states. This process is referred to as *dsp hybridization*. The percentage d-band character is defined as the extent to which dsp hybridization occurs. Table 2.2 illustrates the percentage d-band character of the transition metals in the first, second, and third long periods of the periodic table. Metals, having more unpaired electrons in the d-band, have a lower percentage d-band character.

The question that one might now ask is what does heterogeneous catalysis/electrocatalysis have to do with this. The answer is apparent from the preceding section: catalytic reactions involve intermediate steps of adsorption/desorption via covalent bonds. For such type of bond formation, the catalyst plays the role of being a donor or acceptor of electrons to or from the adsorbed atoms or molecules. For an adsorption reaction, a metal with a lower percentage d-band character (i.e., more unpaired electrons) is more favorable; for a desorption reaction, it is the reverse case.

Another electronic factor, which has proven to be very beneficial in a fundamental understanding of heterogeneous catalysis/electrocatalysis and tailor-making novel catalytic materials, is the *work function* ($\varphi$) of the metal. The work function of a metal is defined as the energy required to remove an electron from its bulk to a point well outside it. Even though the work function is a bulk property of the metal, it is strongly affected by the surface because the transfer of the electron, from the bulk of the metal to a point well outside it, involves its passage across the metal/vacuum interface. For the transition metals, it has been found that the work function increases with the d-band character because with increasing d-band character, the number of unpaired electrons decreases, making it more difficult to extract an electron out of the metal. From a catalytic point of view, it would then

## TABLE 2.2

### Percentage *d*-Band Character in the Metallic Bond of Transition Elements[a]

| Sc | Ti | V  | Cr | Mn   | Fe   | Co   | Ni | Cu |
|----|----|----|----|------|------|------|----|----|
| 20 | 27 | 35 | 39 | 40.1 | 39.5 | 39.7 | 40 | 36 |
| Y  | Zr | Nb | Mo | Tc   | Ru   | Rh   | Pd | Ag |
| 19 | 31 | 39 | 43 | 46   | 50   | 50   | 46 | 36 |
| La | Hf | Ta | W  | Rh   | Os   | Ir   | Pt | Au |
| 19 | 29 | 39 | 43 | 46   | 49   | 49   | 44 |    |

[a]From Suggested Reading 1.

follow that the enthalpy of adsorption of a species will decrease with an increase of the percentage of the d-band character and an increase of the work function of the metal.

A third electronic factor, which is useful in understanding its effect in heterogeneous catalysis/electrocatalysis is the *electronegativity* of an element ($\chi$). Pauling proposed an empirical equation for the bond energy of say $D_{M-H}$, a hydrogen atom (H) adsorbed on a metal (M) by the equation:

$$D_{M-H} = 0.5\left(D_{M-M} + D_{M-H}\right) + 23.06\left(\chi_M - \chi_H\right)^2 \qquad (2.127)$$

where $D_{M-M}$ and $D_{M-H}$ are the bond dissociation energies of two neighboring metal atoms and of the hydrogen molecule, respectively. The parameters $\chi_M$ and $\chi_H$ are the electronegativities of the metal and hydrogen atom, respectively.

In Chapter 6, the roles of the aforementioned electronic factors influencing electrocatalysis of fuel cell reactions are addressed. The reader is referred to the voluminous literature for analyses of heterogeneously-catalyzed reactions.

### 2.8.4. At What Potentials Should One Compare Reaction Rates to Elucidate the Roles of Electronic and Geometric Factors?

This topic has been discussed over the last 40 years. Should we compare the reaction rates at the reversible potential or at the potential of zero charge of the metal? If the transfer coefficients ($\alpha$) are the same for the electrochemical reaction on different electrocatalysts, the accepted view is that the exchange current density is a reasonably good measure of inherent electrocatalytic activity; however, if the transfer coefficients are different, causing a change in Tafel slope with increasing current density, it is more desirable to make comparisons at desired overpotentials. The current density at the potential of zero charge, reveals an inherent activity of the metal, devoid of any electrical aspects of the double layer. Knowledge of the influence of the potential across the interface on the rate of an electrochemical reaction is a valuable and additional diagnostic criterion for determining the mechanism of an electrochemical reaction.

## 2.9. ADSORPTION ISOTHERMS AND PSEUDOCAPACITANCE

### 2.9.1. Types of Adsorption Isotherms and Their Influence on Electrode Kinetics and Electrocatalysis

2.9.1.1. *Langmuir Isotherm.* In heterogeneous catalysis involving multi-step reactions at solid/gas or solid/liquid interfaces, as well as in multi-step charge-transfer reactions, the adsorption and desorption of reactants and intermediate species play a dominant role. This is more so for reactions in which the solid phase serves a catalytic/electrocatalytic role. The kinetics of the adsorption/desorption steps depends on the physicochemical characteristics of the solid phase (e.g., geometric and electronic factor, and the availability of free sites on the surface.) In the case of electrocatalytic reactions, the intermediate adsorption/desorption steps may or may not involve electron transfer. Chemical steps may precede or succeed the charge transfer step. The kinetics of the adsorption/desorption steps strongly depends on the types of isotherms governing the processes. There are also cases where diffusion of the species plays a role. Knowledge of the type of isotherm and the kinetics of the adsorption/desorption step is vital in elucidating the mechanism of several types of electrochemical reactions, particularly gas-evolution/gas-consumption reactions (as in water electrolysis and fuel cells) and electroorganic oxidation and reduction reactions.

Sections 2.1.2. and 2.1.3. provided a brief description of types of adsorption isotherms and of the effects of the adsorption of ions and of neutral molecules on the structure of the double layer. Of the different types of isotherms considered in these sections, the two most common ones are the Langmuir and the Temkin adsorption isotherms. In this section, an analysis will be made of the kinetics of a two-step electrochemical reaction, which involves an adsorption/desorption step of an intermediate governed by the Langmuir isotherm.

Let us consider an electrochemical reaction of the type:

Step1 $$M + A^- \rightleftharpoons MA + e_0^-$$ (2.128)

Step 2 $$2MA \rightleftharpoons 2M + A_2$$ (2.129)

Overall Anodic $$2A^- \rightarrow A_2 + 2e_0^-$$ (2.130)

where M is the electrocatalyst. $A^-$ is the ion in solution (e.g., a chloride ion), MA is the adsorbed intermediate and $A_2$ is the product (e.g., $Cl_2$). According to the Langmuir-isotherm kinetics, the free sites on the surface of the electrocatalyst determine the rate of the first step of the adsorption $v_1$, while the rate of the reverse reaction, $v_{-1,}$ is dependent on the coverage of this species. Thus,

$$v_1 = k_1 c_A (1-\theta) \exp\frac{\beta VF}{RT}$$ (2.131)

and

$$v_{-1} = k_{-1} \theta \exp\frac{-(1-\beta)VF}{RT}$$ (2.132)

In the above equations, $k_1$ and $k_{-1}$ are rate constants for the forward and reverse steps, respectively, $\theta$ is the fractional coverage of the surface by the intermediate MA, $V$ is the potential across the electrode/electrolyte interface, and $\beta$ is the symmetry factor. The rates of the second step in the forward and reverse directions, respectively, $v_2$ and $v_{-2}$, are given by:

$$v_2 = k_2\theta^2 \tag{2.133}$$

and

$$v_{-2} = k_{-2}(1-\theta)^2 P_{A_2} \tag{2.134}$$

where $k_2$ and $k_{-2}$ are the respective rate constants and $P_{A_2}$ is the pressure of the product gas. It must be noted that since this second step does not involve electron transfer, it is not directly dependent on the potential across the interface. However, since the fractional coverage, $\theta$, is dependent on potential, there is an indirect dependence of the rates (forward and backwards) on potential, as will be seen by the following analysis. Suppose one assumes that the second step (Eq. 2.129) is rate determining in the anodic-gas evolution reaction (Eq. 2.130). Then, one may consider that the first intermediate step is in equilibrium and obtain an expression for $\theta$ as:

$$\theta = \frac{K_1 c_{A^-} \exp\dfrac{VF}{RT}}{1 + K_1 c_{A^-} \exp\dfrac{VF}{RT}} \tag{2.135}$$

where $K_1 = k_1/k_{-1}$.

The Langmuir-adsorption isotherm is applicable at very low or very high coverage of the adsorbed species. If we assume the case of $\theta \to 0$, then:

$$\theta = K_1 c_{A^-} \exp\frac{VF}{RT} \tag{2.136}$$

When the second forward step is the rate-determining step, its rate is:

$$v_2 = k_2 K_1^2 \exp\frac{2VF}{RT} \tag{2.137}$$

and since $(1 - \theta) \approx 1$, the rate of the reverse step is:

$$v_{-2} = k_{-2} P_{A_2} \tag{2.138}$$

From this overall rate, it is clear that with the assumptions made, the Tafel slope for the gas evolution reaction is $RT/2F$, while the reverse reaction rate is independent of potential and depends only on pressure of the gas $A_2$. In Chapter 5, a more detailed analysis of the electrode kinetics of electrolytic hydrogen evolution/ionization reactions will be made. It will be based on the Langmuir adsorption isotherm of the intermediate. The hydrogen electrode reaction is analogous to the above type of reaction but it is in the reverse electrochemical direction (i.e., gas evolution is the cathodic reaction and gas consumption is the anodic one).

**2.9.1.2. *Temkin Isotherm*.** It was stated in Sections 2.1.2. and 2.1.3. that the Langmuir isotherm is based on the availability of free sites on the electrode. This type of behavior is mostly applicable at low and high coverage ($\theta \rightarrow 0$) and ($\theta \rightarrow 1$). However, in the case of several electrochemical reactions, particularly electroorganic oxidation (as in fuel cells using organic fuels directly) and reduction reactions (electrosynthesis), as well as the oxygen-electrode reaction (as in water electrolysis or fuel cells), intermediate species are adsorbed to a relatively high extent. The availability of free sites is not the only factor that governs the adsorption of the reactant or intermediate species. As stated in Section 2.1.2., the Temkin isotherm represents such types of behavior. The free energy of adsorption ($\Delta G_{ads}$) is then dependent on lateral interaction between the adsorbed species and the heterogeneity of the surface. In most cases, it was found that there is a linear relation between $\Delta G$ and $\theta$. Thus, one may write:

$$\Delta G_{\theta}{}^0 = \Delta G_0{}^0 + r\theta \tag{2.139}$$

where $\Delta G_{\theta}{}^0$ and $\Delta G_0{}^0$ are the standard free energies of adsorption at coverage values of $\theta = \theta$ and $\theta = 0$, and $r$ is an interaction-energy parameter.

We shall now examine how the adsorption behavior of a Temkin-isotherm type of the intermediate, MA, affects its electrode kinetics. Taking into consideration the variation of the standard free energy of adsorption with coverage, as expressed by Eq. (2.139), the rates of the forward and backward intermediate steps will be given by:

$$v_1 = k_1 (1 - \theta) c_{A^-} \exp\left(\frac{\beta VF}{RT}\right) \exp\left(\frac{-\gamma r\theta}{RT}\right) \tag{2.140}$$

and

$$v_{-1} = k_{-1} \theta \exp\left[-(1 - \beta)\frac{VF}{RT}\right] \exp\left[(1 - \gamma)\frac{r\theta}{RT}\right] \tag{2.141}$$

The second exponential term in each of the above equations represent the changes in the activation energies for these reactants as a result of the change in adsorption energy of the species MA with coverage. The term $\gamma$ is analogous to the symmetry factor. An approximation is made that in the range of $0.2 < \theta < 0.8$, the variations in the linear terms of $\theta$ are considerably less than those in the exponential terms of $\theta$. Thus, Eqs. (2.140) and (2.141) may be reduced to:

$$v_1 = k_1 c_A \exp\left(\frac{\beta VF}{RT}\right) \exp\left(\frac{-\gamma r\theta}{RT}\right) \tag{2.142}$$

$$v_{-1} = k_{-1} \exp\left[(1-\beta)\frac{VF}{RT}\right] \exp\left[(1-\gamma)\frac{r\theta}{RT}\right] \tag{2.143}$$

The rates of the second forward and reverse steps are:

$$v_2 = k_2 \theta^2 \exp\left(\frac{2\gamma r\theta}{RT}\right) \cong k_2 \exp\frac{2\gamma r\theta}{RT} \tag{2.144}$$

and

$$v_{-2} = k_{-2}(1-\theta)^2 \exp\left[-2(1-\gamma)\frac{r\theta}{RT}\right] \cong k_{-2} \exp\left[-2(1-\gamma)\frac{r\theta}{RT}\right] \tag{2.145}$$

If we assume that the second step is rate determining and the first step is virtually in equilibrium, the rate of the overall forward and backward directions will be given by

$$v_2 = k_2 \left(k_1 c_{A^-}\right)^2 \exp\frac{2\gamma rF}{RT} \tag{2.146}$$

and

$$v_{-2} = k_{-2}\left(\frac{1}{K_1 c_A}\right)^{-2(1-\gamma)} \exp\left[-2(1-\gamma)\frac{VF}{RT}\right] \tag{2.147}$$

Since the currents are proportional to the velocities:

$$\frac{\partial V}{\partial \ln i_2} = \frac{RT}{2\gamma F} \tag{2.148}$$

and

$$\frac{\partial V}{\partial \ln i_{-2}} = \frac{RT}{2(1-\gamma)F} \qquad (2.149)$$

An examination of our analyses of the same reaction, governed by the two types of adsorption isotherms, reveals their influence on the electrode kinetics, including the dependence of the reaction rates on potential.

The mechanism of the oxygen-electrode reaction (evolution, as during water electrolysis, and consumption, as in fuel cells) well exemplifies the influence of the Langmuir type and Temkin type of adsorption of the intermediate M–O or M–OH (where M is the electrocatalyst on the electrode kinetics). This aspect is dealt with in detail in Chapter 5.

### 2.9.2. Adsorption Pseudocapacitance

2.9.2.1. *What is Pseudocapacitance?*  The topics of the structure of the double layer and the variation of its capacity with potential as well as the effects of adsorption of ions and neutral molecules were dealt with in Section 2.1. In the preceding Section, we analyzed effects of the types of isotherms on the electrode kinetics of the reactions. Another important characteristic of certain types of electrochemical reactions, including fuel cell reactions, is their exhibition of another type of capacitance, namely *adsorption pseudocapacitance*. Adsorption pseudocapacitance is defined as the differential capacitance of an electrode/electrolyte interface caused by the change in coverage of an electroactive species with potential across the interface. It is usually observed in an electrochemical reaction in which a charge transfer step precedes the rate-determining step. One of the best examples is the electrolytic hydrogen evolution reaction in which the first electron transfer step to form the MH species is fast, and the second step involving the combination of two hydrogen atoms to form the hydrogen molecule is slow (see Eqs. 2.52 and 2.59). In such a case, pseudocapacitance is observed during transient studies, as in cyclic voltammetry or during switching on and off the current. The pseudocapacitance is a measure of the change in charge with potential. The charge is related to the coverage $\theta$ by the expression:

$$q = q_0\theta \qquad (2.150)$$

where $q_0$ is the charge at $\theta = 1$. The differential capacitance, i.e., the pseudocapacitance, $C_{ps}$, is expressed by:

$$C_{ps} = \left(\frac{\partial q}{\partial V}\right) = q_0\left(\frac{\partial \theta}{\partial V}\right) \qquad (2.151)$$

2.9.2.2.  *Theoretical Analysis of Dependence of Pseudocapacitance on the Type of Adsorption Isotherm.*  The Langmuir type adsorption of an intermediate in a reaction is represented by Eq. (2.136). Using Eq. (2.135) in Eq. (2.150) one can show that the expression for $C_{ps}$ is:

$$C_{ps} = \frac{q_0 F}{RT} \theta(1 - \theta)$$
(2.152)

It can also be shown by analysis of this expression that $C_{ps}$ has a maximum at $\theta = 0.5$, which is given by the equation:

$$C_{ps} = \frac{q_0 F}{4RT}$$
(2.153)

A plot of $C_{ps}$ vs. $\theta$ will reveal a parabolic behavior symmetrical around $\theta = 0.5$. The expression for $C_{ps}$ as a function of potential for this case is

$$C_{ps} = \frac{q_0 F}{RT} \frac{c_A - K_1 \exp\dfrac{VF}{RT}}{\left(1 + c_A - K_1 \exp\dfrac{VF}{RT}\right)^2}$$
(2.154)

A plot of $C_{ps}$ vs. potential for hydrogen adsorption on platinum is shown in Figure 2.21. For this case, with $10^5$ atomic sites of Pt cm$^{-2}$, the maximum value of $C_{ps}$ (i.e., at $\theta = 0.5$) is $1.6 \times 10^3$ $\mu$F cm$^{-2}$. This value is considerably higher (by a factor of 100) than the double layer capacity at an ideally polarizable interface, which behaves very much like a solid-state capacitor. It is for this reason that electrochemical capacitors are referred to as supercapacitors and are gaining interest for applications such as high power density energy storage devices (c.f. Chapter 3).

We shall now analyze the pseudocapacitance-potential behavior for the case of adsorption of the intermediate species governed by the Temkin isotherm. From the derivation of the expression for $\theta$ for this isotherm, and Eq. (2.151), it follows that:

$$\frac{q_0}{C_{ps}} = \frac{dV}{d\theta} = \frac{RT}{F} d\ln\frac{\theta}{1-\theta} + \frac{r}{F}$$
(2.155)

From the above equation and Eq. (2.153), one may arrive at the expression:

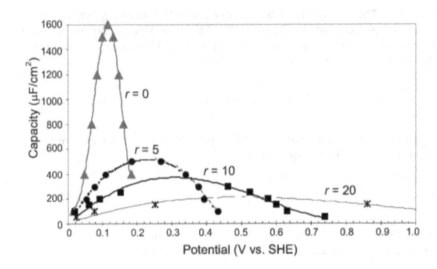

**Figure 2.21.** Plots of pseudocapacitance versus potential for hydrogen adsorption/desorption on platinum for different values of the heterogeneity parameter, $r$. The case $r = 0$ is for Langmuir conditions of adsorption/ desorption and the cases $r > 0$ are for Temkin conditions. Reprinted from B. E. Conway and E. Gileadi, Transactions of the Faraday Society **58**, 2493, Copyright © 1962, with permission from The Royal Society of Chemistry.

$$\frac{1}{C_{ps}} = \frac{1}{C_L} + \frac{1}{C_T} \qquad (2.156)$$

This equation shows that the pseudocapacitance is a series combination of two capacitors: one as in Langmuir case, $C_L$, and the other involving the heterogeneity parameter. Plots for the variation of $C_{ps}$ with potential $V$ are also shown in Figure 2.21 for different $r$ values. The case $r = 0$ represents Langmuir adsorption behavior. With increasing values of $r$, there is a decrease in peak height as well as a broadening of the parabolic curves. For $r \neq 0$, the capacitor maximum is given by:

$$C_{ps,M} = \frac{q_0 F}{4RT + r} \qquad (2.157)$$

Knowledge of the dependence of the pseudocapacitance on potential is a useful diagnostic criterion for the determination of the mechanism of an electrochemical reaction (i.e., whether Langmuir or Temkin type adsorption prevails). This is quite applicable to the electrooxidation of hydrogen and of organic fuels as well as electroreduction of oxygen in fuel cells.

2.9.2.3. *Electrical Equivalent Circuits for Reactions Exhibiting Pseudocapacitances.* An experimental technique that has been gained momentum since the 1970s to elucidate the mechanism of several types of electrochemical reactions is Electrochemical Impedance Spectroscopy (EIS). In order to analyze the results of the EIS experiments, the most commonly used method is that based on an equivalent circuit for the reaction being investigated. The rationale for bringing up this topic is that EIS is one of the most valuable transient methods for investigating mechanisms of reactions, which exhibit pseudocapacitance. For the reaction sequence considered in the preceeding Sections, the equivalent electrical circuit generally used is shown in Figure 2.22, where $C_{DL}$ represents the double-layer capacity, $C_{ps}$ the pseudocapacity for the charge transfer step, $R_{CT}$ the charge transfer resistance of this step, and $R_{rec}$ the reaction resistance for the recombination step. By an analysis of the EIS experimental results for this reaction using this type of equivalent circuit, it is possible to determine the relevant electrode kinetic parameters.

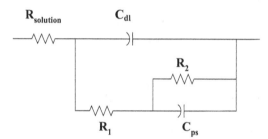

**Figure 2.22.** Equivalent circuit for the hydrogen evolution reaction according to the (fast discharge)-(slow recombination) mechanism. $C_{dl}$ is the double layer capacity at the electrode/electrolyte interface, $C_{ps}$ is the pseudocapacitance for the hydrogen adsorption/desorption step, $R_1$ and $R_2$ are the charge transfer resistance for the fast discharge step and recombination steps, and $R$ is the electrolyte resistance. Reprinted from J. O'M Bockris and S. Srinivasan, *Fuel Cells: Their Electrochemistry*, Copyright © 1969, with permission from McGraw-Hill Book Company.

## Suggested Reading

1. J. O'M Bockris and S. Srinivasan, *Fuel Cells: Their Electrochemistry* (McGraw-Hill Book Company, New York, 1969).
2. A. J. Bard and L. Faulkner, *Electrochemical Methods* (Wiley, New York, 1980).

3.   C. H. Hamman, A. Hammnett, and W. Vielstich, *Electrochemistry* (Wiley-VCH Verlag GmbH, Weinheim, Germany, 1998).
4.   P. M. Natishan (Ed.), *What is Electrochemistry?*, 4th edition (The Electrochemical Society, Pennington, N. J., 1997).

## Cited References

1.   H. Helmholtz and W. Abhandl, *Physik. Tech. Reichsanst Alt.* **1,** 925 (1879).
2.   A. Gouy, *Ann, Chim. Phys.* **29,** 145 (1903); *Compt. ReNd.* **149,** 654 (1909).
3.   D. L. Chapman, *Phil. Mag.* **25,** 475 (1913).
4.   O. Stern, *Z. Electrochem.* **30,** 508 (1924).
5.   O. A. Esin and B. F. Markov, *Zh. Fiz. Khim.* **13,** 318 (1939).
6.   D. C. Grahame, *J. Electrochem Soc.* **98,** 313 (1951).
7.   M. A. V. Devanathan, *Trans. Faraday Soc.*, **50,** 373 (1954).
8.   J. O'M Bockris, M. A. V Devanthan, and K. Mueller, *Proc. Roy. Soc., Ser. A.* **274,** 55 (1963).
9.   I. Langmuir, *J. Amer. Chem. Soc.* **40,** 1361 (1918).
10.  A. N. Frumkin, *Z. Physik.* **35,** 792 (1926).
11.  M. I. Temkin, *Zh. Fiz. Khim.* **15,** 296 (1941).
12.  P. J. Flory, *J. Chem. Phys.* **10,** 51 (1942).
13.  M. L. Huggins, *Ann. NY Acad. Sci.* **43,** 6 (1942).
14.  J. O'M Bockris, M. Gamboa, and M. Szklarczyk, *J. Electroanal. Chem.* **339,** 355 (1992).
15.  J. O'M Bockris. M. Green, and D. A. J. Swinkels, *J. Electrochem. Soc.* **11,** 743 (1966).
16.  J. O'M Bockris and K. T. Jeng, *J. Electroanal. Chem.* **330,** 541 (1992).
17.  W. H. Brattain and C. G. B. Garrett, *Ann. NY Acad. Sci.* **58,** 951 (1954).
18.  V. S. Bagotzky and Y. B. Vassilyev, *Electrochim. Acta* **12,** 1323 (1967).
19.  J. Christiansen, *Z. Physik. Chem. Ser. B* **33** (1936) 145; **37,** 374 (1937).
20.  J. A. V. Butler, *Proc. Roy. Soc. Ser. A* **157,** 423 (1936).
21.  R. W. Gurney, *Proc. Roy. Soc. Ser. A* **134,** 137 (1931).
22.  W. Nernst and E. S. Merriam, *Z. Physik. Chem.* **53,** 235 (1905).
23.  A. A. Balandin, *Z. Physik. Chem.* **B2,** 28 (1929); **B3,** 167 (1929).
24.  L. Pauling, *Phys. Rev.* **54,** 899 (1938); *Proc. Roy. Soc. Ser. A* **196,** 343 (1949).

## PROBLEMS

1.   There has been a progressive advance in elucidating the structure of the electric double layer at an electrode/electrolyte interface starting from the Helmholtz model to the water dipole model. Identify the distinctive features of these Double Layer models.

2.   What is the main difference between the structure of the double layer at (i) a metal/electrolyte interface, and (ii) a semiconductor/electrolyte interface?

What is the common feature in the space charge region of the semiconductor and in the diffuse-layer region in the electrolyte?

3. What are adsorption isotherms? How do these isotherms affect the structure of the double layer in the case of (i) specific adsorption of ions, and (ii) adsorption of neutral organic molecules (in this case consider both polar and non-polar molecules)?

4. Express the overall electrochemical reactions, as equations, for the following reference electrodes:
   (a) Reversible hydrogen electrode
   (b) Calomel electrode
   (c) Silver/silver chloride electrode
   (d) Mercury/mercuric oxide electrode
   Reference electrodes are also referred to as *ideally non-polarizable electrodes.* What is the reason for this terminology from an electrode kinetic point of view? (Hint: What will be the order of magnitude values of the exchange current densities for these reference electrodes?).

5. The cathodic reduction of nitrobenzene to aniline follows a consecutive reaction path (see Eq. 2.123). It is also possible to have intermediate reaction products by partial reduction of $C_6H_5NO$ and $C_6H_5NHOH$. Express the equations leading to these products by complete or partial reduction. What is the reason for referring to the electrode at which these reactions occur as a non-polarizable electrode? Explain it in terms of probable values for their exchange current densities.

6. Examine Figure 2.12, *Typical plot of free energy* vs. *distance along reaction path for a consecutive reaction.* What is the reason for the intermediate step from C to D for the forward reaction and from D to C for the reverse reaction being the rate-determining step (rds)? With these being their rds, derive the expressions for the velocities of the forward reactions, assuming that the steps from A to C and F to D are in equilibrium.

7. In a reaction occurring by the consecutive pathway, the rate-determining step (rds) is the slowest one in the sequence of intermediate steps. What is the rds in a reaction that occurs by the parallel reaction pathway? Give an example of a reaction that occurs by this route. What type of reaction is referred to as a coupled reaction?

8. Take into consideration electrolytic hydrogen evolution. It occurs via the reaction path, as expressed by the proton discharge step according to Eq. (2.52), and is followed by the recombination step according to Eq. (2.59). Assuming that the proton discharge step is the rds step, what are the physico-chemical characteristics that have to be taken into account in constructing the *potential energy* vs. *distance from electrode surface* plots?

9. Assuming the expression represented by Eq. (2.92) for the dependence of overpotential on current density, construct the Tafel plots for the following cases:

   a) Slow discharge–fast recombination: for the high overpotential metal electrodes (e.g., mercury, lead, thallium), $i_o$ is about $10^{-10}$ A cm$^{-2}$, and the transfer coefficient is 0.5. Construct the overpotential vs. current density plot for values of $\eta$ in the range 10 to 500 mV.

   b) For the medium overpotential metals (e.g., Fe, Ni, Cu), the electrochemical desorption step, $i_o$, is about $10^{-6}$ A cm$^{-2}$, and $\alpha = 1.5$ up to a current density of 100 mA cm$^{-2}$. Above this current density, $\alpha = 0.5$ and $i_0 = 10^{-4}$ A cm$^{-2}$.

   c) Fast discharge–slow recombination: for the low overpotential metals (e.g., Pt, Rh, Ir), assume $i_o = 10^{-3}$ A cm$^{-2}$ and $\alpha = 2$. Make all of these plots on one sheet of paper and propose some comments as to which type of electrode material will be best for a fuel cell application. Also, at what values of overpotential can the current density for the reverse reaction be less than 5% of that for the forward direction so that the former can be neglected in the expression for $i$ as a function of $\eta$ (Eq. 2.92)? What will then be the equation for i as a function of current density? What is this plot referred to as and what is the name of the scientist who used it empirically in 1905?

10. For what reason is the quantum mechanical treatment better than the classical for electron transfer reactions? (Hint: Read Section 2.5.2. before you answer this question).

11. At what values of overpotential can the exponential terms in Eq. (2.92) be linearized for the three cases in Problem 6? For these cases, what does the expression for $i$ as a function of $\eta$ reduce to?

12. What are the expressions for the charge transfer resistance as derived from the expressions for $i$ as a function of current density using (i) Eq. (2.93) and (ii) Eq. (2.95)? For the three cases considered in Problem 6, plot the charge transfer resistance as a function of current density. What is the significant difference in the characteristics considering cases a and b on one hand and case c on the other for the charge transfer resistance as a function of current density?

13. What is meant by the term *mass transport overpotential*? Using values of $i_L = 10^{-4}$ A cm$^{-2}$, $n = 2$, 100 mV $\leq \eta_D \leq$ 500 mV, and $T = 25$ °C in Eq. (2.109), plot $i$ as a function of $\eta_D$ on a semi-logarithmic scale ($i$ vs. log $\eta$) and linear scale ($i$ vs. $\eta$). Then, plot mass transport resistance ($d\eta_D/di$) vs. $i$. What specific features do you visualize at low and high current density? At what values of $\eta_D$ can Eq. (2.109) be linearized?

14. What is *ohmic overpotential*? What are the main contributions to ohmic overpotential?

# CHAPTER 3

# *ELECTROCHEMICAL TECHNOLOGIES AND APPLICATIONS*

## 3.1. ROLE OF ELECTROCHEMICAL TECHNOLOGIES IN CHEMICAL INDUSTRY

### 3.1.1. Background

The consumption of electricity in the U.S. in the year 2000 was 3,613 billion kWh, of which 890 billion kWh was used by the chemical industry to manufacture a wide variety of chemicals.[1,2] A breakdown of the energy usage by application in the chemical industry (Table 3.1) indicates that ~130 billion kWh was utilized by the electrochemical industry to generate commodity and specialty chemicals.

### 3.1.2. Principles of Technologies

Electrochemical processing is an acknowledged means to achieve inter-conversion of electricity and chemicals via reactions at electrode/electrolyte interfaces. Electrons in an electrochemical process are the reagents like the chemicals in chemical reactions. The rate of a chemical reaction, $k_c$, is dependent on its free energy of activation, $\Delta G^{\neq}$, the reactant concentration, $c_r$, and temperature, $T$, as expressed by:

---

This chapter was written by S. Srinivasan and T. Bommaraju.

**TABLE 3.1**
**Breakdown Of Electricity Usage In Electrochemical Operations**

| Product | Production in 2000 million metric tons | Electric energy consumption, billion kWh |
|---|---|---|
| Aluminum | 3.78 | 53.00 |
| Chlorine | 14.06 | 40.00 |
| Sodium chlorate | 0.74 | 4.46 |
| Zinc | 0.23 | 0.75 |
| Copper | 1.59 | 2.00 |
| Manganese | 0.76 | 6.00 |
| Others | | 24.00 |

$$k_c = k_c^0 \; c_r \; \exp(-\Delta G^{\neq} / RT) \tag{3.1}$$

where $k^o{}_c$ is the concentration-independent-rate constant. Thus, increasing the concentration of the reactant and temperature can increase the rate of a chemical reaction. On the other hand, the rate of an electrochemical reaction, $k_e$, can be described as:

$$k_e = k_e^0 \; c_r \; \exp\left(nF\Delta E^{\neq} / RT\right) \tag{3.2}$$

where $\Delta E^{\neq}$ is the potential difference at the electrode/electrolyte interface and $nF\Delta E$ is equivalent to $-\Delta G^{\neq}$. Hence, the rates of electrochemical reactions can be altered not only by the concentration of the reactants and temperature, but also by changing the potential difference across it by the passage of current. Thus, there is a parallelism between consumption of free energy in a chemical reaction and the electrochemical free energy in an electrochemical process utilizing the passage of electric current. The electrochemical processes have several advantages over the conventional chemical routes, which include:

- control of the rate and selectivity by manipulating the electrode potential;
- high thermodynamic efficiency allowing reactions to proceed at ambient temperatures and pressures;
- pollution-free operation;
- use of less expensive starting materials; and
- reduction in the number of steps involved to make the final product and avoiding the down stream purification steps.

It is because of these benefits that electrochemical technologies have gained commercial status over the last several decades.

Let us now examine the free energy of the reaction $H_2O \rightarrow H_2 + O_2$. The change in free energy for this reaction $\Delta G^0$, is 237 kJ/mole. The positive $\Delta G^0$ signifies that the reaction, as written, is not spontaneous. However, by applying a potential above that corresponding to the $\Delta G^0$ of 237 kJ/mole, i.e., 1.23 V, the water decomposition reaction can be driven in the desired direction. This is the basis for electrochemical processes to produce chemicals. For the reaction of hydrogen and oxygen to form water, $\Delta G^0$ is –237 kJ/mole. Since $\Delta G^0$ for this reaction is negative, the reaction occurs spontaneously. Therefore, one can theoretically extract electrical energy equivalent to that of $\Delta G^0$ per mole of hydrogen. This is the basis for producing electrical energy, be it a fuel cell reaction or other reactions in batteries. It is this capability of inter-conversion of electricity and chemicals that makes the electrochemical processing elegant and attractive. The various technologies that are possible and have been commercialized are presented in Table 3.2.

Electricity is, of course, the primary source of energy supplied to an electrochemical reactor. It drives the reaction and heats the electrolyzer. In practice, the cells sometimes require more electric current than that calculated by the Faraday's law because a part of the electricity is consumed by the side reactions, resulting in coulombic losses. Also, the cell potential is higher than the thermodynamic value due to the overpotentials at the two electrodes and the ohmic overpotentials in the electrolyte and the metallic components (current collectors, electrodes, cables).

The thermodynamic-decomposition potential, $E^0_r$, is the minimum potential required for a given electrochemical process to proceed in a given direction and is based on the standard free energy change, $\Delta G,^0$ of the overall reaction as:

$$E_r^o = -\frac{\Delta G^0}{nF}$$

(3.3)

## Table 3.2
## Leading Electrochemical Technologies

| Electrochemistry | |
| --- | --- |
| Electrochemical processing | Energy conversion and storage |
| • Electrowinning of metals<br>  e.g., Al, Na, Mg, Cu, Ni, Zn | • Primary batteries<br>  e.g., Lechlanche cell |
| • Electrorefining of metals<br>  e.g., Al, Au, Cu, Ni, Zn | • Secondary batteries<br>  e.g., lead-acid, Ni-Cd, Ag-Zn |
| • Production of inorganic chemicals<br>  e.g., chlorine/caustic, chlorates,<br>  perchlorates | • Fuel cells<br>  e.g., hydrogen-oxygen |
| • Production of organic chemicals<br>  e.g., adiponitile | • Supercapacitors<br>  e.g., carbon and noble-metal oxides |
| • Corrosion and corrosion prevention | |
| • Electroforming and electromachining | |

The cell potential corresponding to the heat of the reaction, is:

$$E_t^0 = -\frac{\Delta H^0}{nF}$$

(3.4)

and is referred to as the thermoneutral potential, $E_t^0$, representing the potential at which neither heat is lost to the surroundings nor required by the system. However, commercial electrolytic cells operate at higher than the thermodynamic decomposition potential to allow the reaction to proceed at the desired rate in the forward direction to generate the desired products. The excess potential, constituting the overpotentials leads to the generation of heat, which should be taken out to ensure isothermal operating conditions of the electrolytic cells.

Let us now exemplify the importance of thermoneutral potential by comparing the $E_r^0$ and $E_t^0$ for the electrolysis of HCl solutions. The thermodynamic decomposition potential for the reaction: $HCl \rightarrow 0.5\ H_2 + 0.5\ Cl_2$ is 1.36 V at 25 $^0$C, whereas the thermoneutral potential is 1.74 V. The value of $E_t^0$ is greater than $E_r^0$, since the $E_t^0$ term contains the heat associated with the entropy change for the reaction. If the cell operates at potentials less than $E_t^0$ (and above $E_r^0$), the cell will cool as the cell dissipates the heat corresponding to the entropy change, irreversibly. On the other hand, at voltages higher than $E_t^0$, the cell is heated by the excess energy generated by the joule heat caused by the overpotentials and it has to be cooled to operate the cells at a given temperature. The amount of heat generated or absorbed by the system can be calculated as follows. The amount of heat released, $Q_{rev,}$ is given as:

$$Q_{rev} = -T\Delta S = \Delta G^0 - \Delta H^0 = nFE_r^0 - \Delta H^0$$

(3.5)

The cell potential ($E$) can be written in terms of its constituents as:

$$E = E_r^0 + \Sigma iR + \Sigma \eta_{act}$$

(3.6)

where $\Sigma\ \eta_{act}$ are the activation overpotentials at the anode and cathode.

The irreversible heat generated, $Q_{irr}$, is given by:

$$Q_{irr} = (E - E_r^0)nF$$

(3.7)

Since the heat generated ($Q$) is equal to $Q_{irr} + Q_{rev}$, $Q$ can be expressed as:

$$Q = nFE - \Delta H \tag{3.8}$$

When $Q$ is positive, heat is released by the system and when $Q$ is negative, heat is absorbed by the system.

The thermoneutral voltage defined by Eq. (3.4) does not reflect the inevitable heat losses from the convection and radiation losses. This was addressed by defining a practical thermoneutral potential, called the *thermobalance voltage*, $E_t$, by LeRoy et al.[3,4] for water-electrolysis cells. The thermobalance potential is defined as:

$$E_{tb} = E_{tn} + E_{rad} + E_{conv} \tag{3.9}$$

where $E_{rad}$ and $E_{conv}$ refer to the potential corresponding to the energy losses via radiation and convection, respectively. The $E_{rad}$ and $E_{conv}$ can be estimated using the following equations:

$$E_{rad} = A\varepsilon\sigma(T^4 - T_a^4)10^{-3} / I \tag{3.10}$$

$$E_{conv} = 1.77 A(T - T_a)^{1.25} 10^{-3} / I \tag{3.11}$$

where $A$ is the radiating area (in m$^2$), $\varepsilon$ its emissivity, $\sigma$ is the Stefan-Boltzmann constant (5.67 x 10$^{-8}$ W/m$^2$ degree$^4$), $T_a$ is the ambient temperature (K), $T$ is the operating cell temperature, and $I$ is the load in kA. Figure 3.1 illustrates the results of the calculations described above for water-electrolysis cells.[4]

The broken line shows the cell voltage as a function of temperature, which crosses with the curve of thermal balance potential at about 90 °C. At temperatures lower than 90 °C, the cell voltage is higher than the thermal balance voltage or the voltage corresponding to the overall requirement of heat, and hence the cell dissipates excess heat, which must be removed. On the other hand, at temperatures higher than 90 °C, the heat generation is insufficient to compensate for the total heat required for the reaction so that the cell has to be heated. Thus, the cell can be operated at about 90 °C without any heating and/or cooling. The discussion, presented above, emphasizes the need to seek the optimal conditions to realize effective utilization of the energy.

Another important consideration towards achieving energy savings in an electrolytic process is an understanding of the energy needed for the electrolytic cell, which is directly proportional to the cell potential and inversely proportional to the cell efficiency. Components, constituting the cell potential and cell efficiency, should be clearly understood to pursue options to realize energy savings. These aspects will be discussed in the following Sections in the context of the technologies presented in this Chapter. The intent of this Chapter is to (i) provide a brief

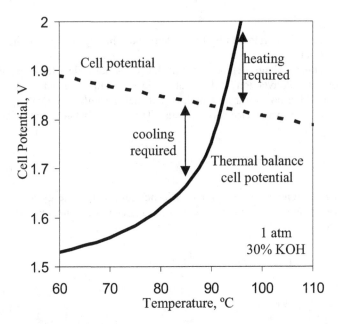

**Figure 3.1.** Isothermal and thermal-balance cell potentials ($E$) of a water electrolyzer as a function of temperature at 1 atm total pressure. Reproduced from Reference 4, Copyright (1983), by permission of The Electrochemical Society, Inc.

description of some important electrochemical technologies and (ii) demonstrate benefits to or from fuel cell technologies. Two of the well-developed hydrogen energy technologies, i.e., electrolytic-hydrogen production (which has also been commercialized) and fuel cells (which are in an era of the beginning of commercialization) are not included in this Chapter. The former is presented in Chapter 8 and the latter in Chapters 4, 9, and 10.

## 3.2.    ALUMINUM PRODUCTION

### 3.2.1.  Background and Applications

The world production of aluminum was 23 million metric tons in the year 2001 and about 11 % of it was produced in the U.S.A. Aluminum is a lightweight and strong material that is used in a wide variety of engineering and construction applications. It has also been used to replace copper as the conductor of electricity.

Domestic consumption of aluminum for transportation is 35 %, while 25 % is for packaging, 8 % for consumer durables, 7 % for electrical, and 11 % for others.[6-8]

## 3.2.2. Principles of Technology

Aluminum is produced by the molten salt electrolysis of $Al_2O_3$ dissolved in $Na_3AlF_6$ by the Bayer-Hall-Hèrault process. This process is over 100 years old and is the only technology used in the world to produce this metal. The overall reaction to produce aluminum is:

$$2Al_2O_3 + 3C \rightarrow 4Al + 3 CO_2 \qquad (3.12)$$

the cathodic reaction being the reduction of $Al^{3+}$ to Al and the anodic reaction being the oxidation of C to $CO_2$. The changes in the enthalpy and the free energy of this reaction at 1250 K (977 °C) are −547.6 kJ/mol and −338.6 kJ/mol, respectively, and the entropy change is −168 J/mol K. The free energy change of the reaction corresponds to a reversible cell voltage of 1.18 V. The stoichiometric consumption of carbon is 0.33 tons/ton of Al, which may be compared to the actual consumption of 0.4–0.5 tons/ton of Al in commercial cells.

Aluminum can also be produced by the electrolysis of $Al_2O_3$ as:

$$2Al_2O_3 \rightarrow 4Al + 3 O_2 \qquad (3.13)$$

However, the reversible voltage for this reaction is 2.21 V vs. 1.18 V for the reaction (3.12). Thus, while the use of carbon anodes results in the consumption of carbon and generation of $CO_2$, reaction (3.12) has the advantage of allowing the operation of the aluminum electrolysis cells at a lower voltage.

Alumina required for the Hall-Hèrault process is produced from bauxite ores by the Bayer process. Bauxite contains 55–60 % $Al_2O_3$, 5–10 % $Fe_2O_3$, 3–7 % $SiO_2$ and a small amount of $TiO_2$ along with water, depending on the source. Since $SiO_2$ forms an insoluble $Na_2O$. x $Al_2O_3$. y $SiO_2$, ores with low silica are preferred for the Bayer process. Briefly, the Bayer process consists of calcining the ore with caustic at 350–500 °C to remove organics and water, followed by grinding it to 20 mesh size or smaller in ball mills. It is then extracted with caustic at 150–170 °C to form sodium aluminate, as expressed by the equation:

$$Al_2O_3 + 2NaOH \rightarrow 2 NaAlO_2 + H_2O \qquad (3.14)$$

When the NaOH concentration in the reaction mixture is ~40–50 % by weight and the $Al_2O_3$: NaOH concentration ratio is 0.55–0.65, the solution is filtered to remove "red mud" which is mostly an oxyhydroxide of Fe. The filtrate is then transferred to a hydrolysis tank, where it is seeded at 50 to 70 °C to facilitate the formation of

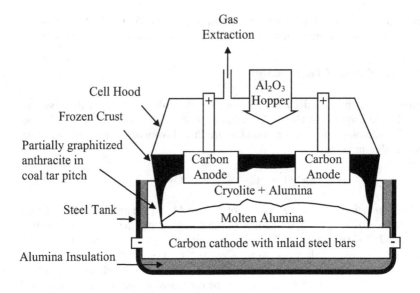

**Figure 3.2.** Electrochemical cell for the Hall-Heroult process for aluminum production. Reprinted from Reference 5.

aluminum hydroxide precipitates. The aluminum hydroxide is then filtered and calcined in a rotary kiln at 1200–1300 $^0$C to form $\alpha$-$Al_2O_3$. Electrolysis is conducted at 975 $^0$C in cells consisting of a steel shell, lined with alumina that acts as a refractory, thermal insulator, and then with carbon. The cells are typically 9–16 m x 3–4 m x 1.0–1.3 m in size. The base of the tank is lined with pre-baked carbon blocks with inlaid steel bars to lower electrical resistance. The sides are lined with partially graphitized anthracite in coal tar pitch. During electrolysis, a layer of solid cryolite and alumina forms at the sides of the cell and a solid crust on the surface, which acts as a barrier to corrosion and heat losses from the cell. The cell has provision for periodic addition of alumina through the crust and for removal of aluminum by suction. Figure 3.2 depicts a typical Hall-Hèrault cell.

Although not completely understood, the reactions participating in the cathodic deposition of aluminum[6] and anodic oxidation of carbon are generally believed to follow the scheme noted below. Cryolite ionizes in the following:

$$Na_3AlF_6 \rightarrow 3\ Na^+ + AlF_6^{3-} \tag{3.15}$$

$$AlF_6^{3-} \rightarrow AlF_4^- + 2F^- \tag{3.16}$$

Alumina forms complex oxyfluoride ions depending on the concentration of alumina. At low concentrations of alumina, $Al_2OF_6^{2-}$ is formed according to Eq. (3.17), and at high concentrations, $Al_2O_2F_4^{2-}$ is formed according to Eq. (3.18):

$$2Al_2O_3 + 4AlF_6^{3-} \rightarrow 3Al_2OF_6^{2-} + 6F^- \qquad (3.17)$$

$$2Al_2O_3 + 2AlF_6^{3-} \rightarrow 3Al_2O_2F_4^{2-} \qquad (3.18)$$

The cathodic reaction is generally assumed to be:

$$12Na^+ + 4\ AlF_6^{3-} + 12e_o^- \rightarrow 12\ (Na^+\text{-}F^-) + 4Al + 12F^- \qquad (3.19)$$

The anodic reaction involves the discharge of oxyfluoride ions to form $CO_2$ and $AlF_3$:

$$2Al_2O_2F_4^{2-} + C \rightarrow CO_2 + 2Al_2OF_4 + 4e_o^- \qquad (3.20)$$

$$Al_2OF_4 + Al_2OF_6^{2-} \leftrightarrow Al_2O_2F_4^{2-} + 2AlF_3 : \qquad (3.21)$$

Combining Eqs. (3.17) and (3.20) results in the overall reaction (3.12). Aluminum produced by the Hall-Hèrault process is 99.9 % pure, but still contains small amounts of impurities such as Si (0.05 %) and Fe (0.05 %). This purity is generally sufficient for most end uses. However, some markets require high quality aluminum. Hence, it is further purified in a specially designed three-phase electrolyzer, to produce 99.995 % aluminum.

## 3.2.3. Economics

The capital cost involved in manufacturing aluminum varies with the size of the plant and its location. Capital costs for a new plant for producing aluminum are $5000/ton of installed annual capacity, whereas the capital costs for producing alumina are $1000/ton of annual capacity. The average operating costs vary widely depending on the source and location, as shown in Table 3.3. On an average, the power costs vary from > 30 mils/kWh for high cost smelters in Asia to as low as 9.7 mils/kWh in Canada, and 18.2 mils/kWh in the western world.

## TABLE 3.3
### Operating Costs ($/Ton Of Aluminum) in 1998

|  | Western World | Russia | China |
|---|---|---|---|
| Power | 283 | 230 | 586 |
| Alumina | 351 | 491 | 469 |
| Other materials | 238 | 302 | 267 |
| Labor | 118 | 40 | 25 |
| Other costs | 164 | 174 | 190 |
| Total | 1154 | 1236 | 1536 |

**TABLE 3.4**
**Consumption of Raw Materials for Producing 1 Ton of Aluminum**

| Raw materials | Weight, kg | Electricity, kWh | |
|---|---|---|---|
| Alumina | 1948 | DC power for | 15,338 |
| Cryolite | 29 | electrolysis | |
| Recovered cryolite | 21 | Losses in bus bar | 413 |
| Aluminum fluoride | 32 | and rectifier | |
| Anode paste | 585 | For melt | 68 |
| Coke | 420 | | |
| Pitch | 188 | Labor (man h) | |
| Cathode carbon | 15 | Direct | 2.03 |
| | | Indirect | 1.07 |

Table 3.4 shows the consumption of raw material and energy requirements for the production of aluminum. This data shows that the major component of the energy usage in aluminum production is the electricity for operating the cells, which is ~ 13–15 kWh/kg of aluminum.[9-11]

### 3.2.4. Energy Conservation Measures

The theoretical energy requirement for producing Al is $60 \times 10^6$ kJ/ton of Al (assuming 30 % energy efficiency), whereas the process needs are five times higher, electrolysis alone consuming three times the theoretical energy needs. As a result, there has been a major focus towards lowering the energy consumption and significant progress was made to reduce the energy consumption from ~ 28 kWh/kg of Al in the 1920s to 13–15 kWh/kg of Al. The breakdown of the energy consumption in terms of the components of cell voltage shows (see Table 3.5) that the ohmic drop between the anode and cathode is one of the major contributors to the cell voltage.

The need for the large anode-cathode gap of 5 cm in the cells arises as a consequence of the instability of the aluminum pool from the intense magnetic fields. The large anode–cathode gap minimizes the shorting of the electrodes and

**TABLE 3.5**
**Components of Cell Voltage of Aluminum Electrolyzer (with Pre-Baked Anodes)**

| | |
|---|---|
| Decomposition voltage, $E^0_{th}$ | 1.60 V |
| IR ohmic drop in electrolyte, $IR_e$ | 1.45 V |
| Anode overpotential, $\eta_a$ | 0.30 V |
| Cathode overpotential, $\eta_c$ | 0.40 V |
| IR ohmic drop in hardware, $IR_{hw}$ | 0.10 V |
| Anode effect | 0.15 V |
| Total cell voltage | 4.00 V |

the chemical reaction of the products of the reaction, which results in lower current efficiency. It is interesting to note that in the aluminum cell the joule heat is effectively used to maintain the temperature at ~1000 °C. The use of carbon anodes in the Hall-Hèrault cells causes several problems, which include: $CO_2$ emissions, generation of perfluorinated carbon compounds at the anode, disposal of cyanide contaminated spent pot-lining, and unpleasant working conditions. These factors coupled with the high-energy needs forced the aluminum manufacturers to examine measures to avoid these problems. Two types of carbon anodes used are the prebaked and self-baking ones (also called Soderberg anodes). The Soderberg anodes are fed at the top with the ground carbon and a pitch binder, which bakes in situ as the anode is lowered into the molten electrolyte to form a hard, dense anode material. The Soderberg anodes are used as they eliminated the anode-manufacturing step. However, the in situ baking process created fumes that are difficult to control. Therefore, the industry reverted to the use of prebaked anodes. Except for the design of anodes, the industry has changed little since the last century.

Three approaches that were pursued to achieve energy conservation and environmental benignity are (i) electrolysis of $AlCl_3$, (ii) development of non-consumable $TiB_2$ anodes, and (iii) refractory hard metal (RHM) composite cathodes.

(a) *Electrolysis of AlCl₃.* Alcoa examined the electrolysis of $AlCl_3$ to produce Al and $Cl_2$. The chlorine generated in this process can be used to prepare $AlCl_3$ as shown in the equation:

$$2\ Al_2O_3 + 3C + 6Cl_2 \rightarrow 4\ AlCl_3 + 3CO_2 \tag{3.22}$$

The advantages of this route include lower overall consumption of carbon and 10% better energy efficiency than the Hall-Hèrault process.

(b) *TiB₂ anodes.* The use of titanium diboride anodes will eliminate the need to use the carbon anodes and thereby avoid the environmental problems noted earlier. However, the thermodynamic decomposition voltage will be higher since the anode reaction is the discharge of oxygen. This approach was investigated extensively and only recently has it been brought to a commercial stage. DeNora North America developed the "Veronica Anode", which is a $TiB_2$-based anode with long life and low solubility in cryolite, and a Moltech Tinor 2000 cathode that is a Ni-Fe alloy with special additives such as Cu, Al, Ti, Y, Mn, and Si to improve corrosion resistance. Pilot tests[8] have shown a wear rate of ~3 mm/yr at current densities greater than 1.1 A/cm$^2$. Preliminary estimates show a savings of 20 % of the production cost and 2.5 % increase in cell life.

(c) *RHM cathodes.* It is necessary for the cathode to be wetted with a thin film of aluminum, which would drain to a sump and provide a stable cathode surface. Presently, the aluminum deposited on the cathode stays on the cathode as a pool. This pool sloshes due to the electrohydrodynamic effect and as a result, the anode is kept away to prevent shorting and to prevent the recombination reaction.

### 3.2.5. Benefits to or from Fuel Cell Technologies

Though there is no apparent benefit to or from fuel cell technologies, aluminum will be a useful material for fabrication of some fuel cell components. As an example, thin aluminum sheets coated with graphitic carbon or gold are being considered as candidate materials for the construction of bipolar plates.

## 3.3.    CHLOR-ALKALI TECHNOLOGY

### 3.3.1.  Background and Applications

Chlor-alkali technology is one of the largest electrochemical industries in the world, the main products being chlorine, sodium hydroxide (also called as caustic soda), and hydrogen generated simultaneously by the electrolysis of sodium chloride. It is an energy intensive process. The chlor-alkali industry is the second largest consumer of electricity (next to aluminum) among the electrochemical industries, accounting for about 40 billion kWh of electrical energy consumed for the production of 14.06 million short tons of chlorine in the year 2000. Chlorine and caustic soda are indispensable intermediates for the chemical industry. Chlorine is a strong oxidizer and has the largest volume use in the production of vinyl chloride monomer, which, in turn, is polymerized to polyvinyl chloride (PVC). Chlorine is also widely used as a bleaching agent, especially in the pulp and paper industry, and as a disinfectant, as for example in swimming pools. Chlorinated organic compounds, such as chlorinated ethanes and fluorocarbons are used as intermediates in the manufacture of polymers, like polyesters and urethanes. Caustic soda, on the other hand, has wide industrial applications in mineral processing, the pulp and paper industry, and the textile and glass manufacturing operations.

### 3.3.2.  Principles of Technology

Electrolysis of aqueous solutions of sodium chloride (NaCl), commonly referred to as brine, simultaneously generates chlorine, caustic soda (or potash), and hydrogen according to the overall chemical reaction:

$$2\ NaCl + 2\ H_2O \xrightarrow{\text{Electrical Energy}} 2NaOH + H_2 + Cl_2 \quad (3.23)$$

Reaction (3.23) has a positive free energy change of 422.2 kJ/mol of chlorine at 25 °C. Therefore, dc-electrical energy has to be provided to force the reaction to proceed in the forward direction. The amount of electrical energy required depends on the amount of the product needed and the electrolytic cell parameters, current density, and cell potential—the latter being dictated by the nature of the anode and cathode material, the separator, the inter-electrode gap, and the cell design.[12-15]

Production of chlorine, caustic, and hydrogen from brine is accomplished in three types of electrolytic cells: the mercury cell, the diaphragm cell, and the membrane cell. The distinguishing characteristic of these cells is the manner in which the electrolytic products are prevented from mixing with each other. In Figure 3.3 a comparison is made of the three types of cells and Figures 3.4 to 3.6 show schematics of the process flow diagrams for these technologies.

The primary electrochemical reactions, which occur during the electrolysis of brine, are the discharge of the chloride ion at the anode to form chlorine:

$$2\ Cl^- \rightarrow Cl_2 + 2e_o^- \tag{3.24}$$

and the generation of hydrogen and hydroxide ions, $OH^-$, at the cathode:

$$2H_2O + 2\ e_o^- \rightarrow H_2 + 2\ OH^- \tag{3.25}$$

Chlorine is generated at the anode in all the three types of electrolytic cells. The cathodic reaction in diaphragm and membrane cells is the electrolysis of water to generate hydrogen as shown by Eq. (3.25), whereas the cathodic process in mercury cells is the discharge of the sodium ion, $Na^+$, to form sodium amalgam, containing 0.2 to 0.3 % sodium:

$$Na^+ + Hg + e_o^- \rightarrow Na(Hg) \tag{3.26}$$

This amalgam subsequently reacts with water in denuders, or decomposers, to generate hydrogen and caustic:

$$2Na(Hg) + 2\ H_2O \rightarrow 2\ NaOH + 2Hg + H_2 \tag{3.27}$$

Separation of the anode and cathode products in diaphragm cells is realized using an asbestos polymer-modified asbestos composite, or a non-asbestos material deposited on a mesh cathode. In membrane cells, the separator is a cation-exchange membrane. Mercury cells require no diaphragm or membrane, because the mercury itself acts as a separator.

The catholyte from diaphragm cells typically contains 9–12 % caustic soda and 14–16 % NaCl. This cell liquor is concentrated to 50 % NaOH in a series of evaporation steps involving three or four stages. Membrane cells produce 30–35 % NaOH, which is evaporated in two or three stages to produce 50 % caustic soda.

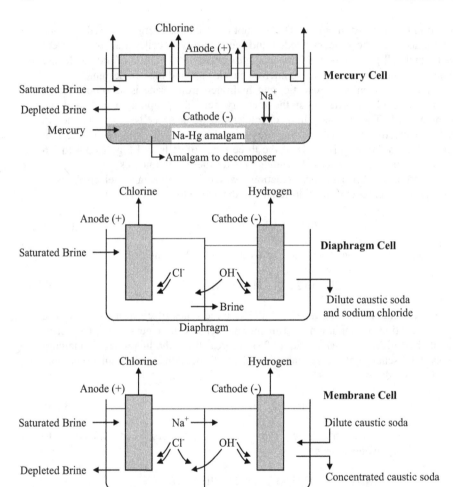

**Figure 3.3.** Schematics of the three types of chlor-alkali cells.

The 50 % caustic soda, containing very little salt, is made directly in the mercury cell process by reacting the sodium amalgam with water in the decomposers.

The anodes, used in all these three cell technologies, are ruthenium oxide and titanium oxide coated titanium, which operate at a low chlorine overpotential with excellent dimensional stability and long life. The cathode material in diaphragm cells is made of carbon steel and exhibits an overpotential of 300–400 mV for the

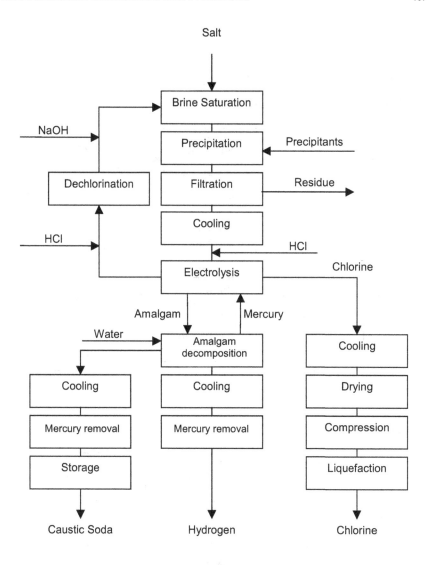

**Figure 3.4.** Flow diagram of the Mercury cell process for chlor-alkali production. Reproduced from Reference 12, Copyright (1985), with permission from John Wiley and Sons, Inc.

hydrogen evaluation reaction. Energy savings, by reducing the overpotential as much as 200–280 mV, are achievable, in principle, by using nickel cathodes with a catalytic coating. The coatings that are commercially employed in membrane cells

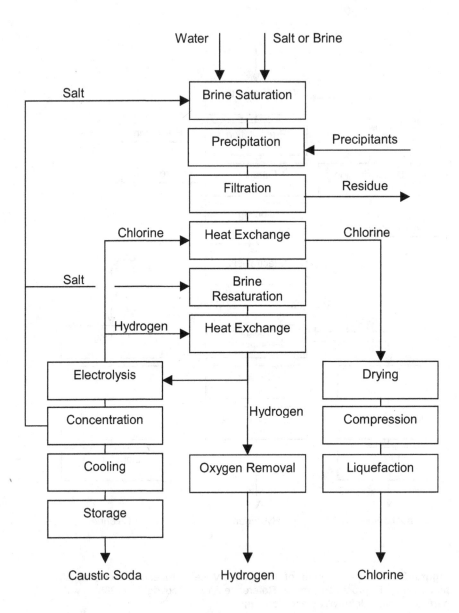

**Figure 3.5.** Flow diagram of the Diaphragm cell process for chlor-alkali process. Reproduced from Reference 12, Copyright (1985), with permission from John Wiley & Sons, Inc.

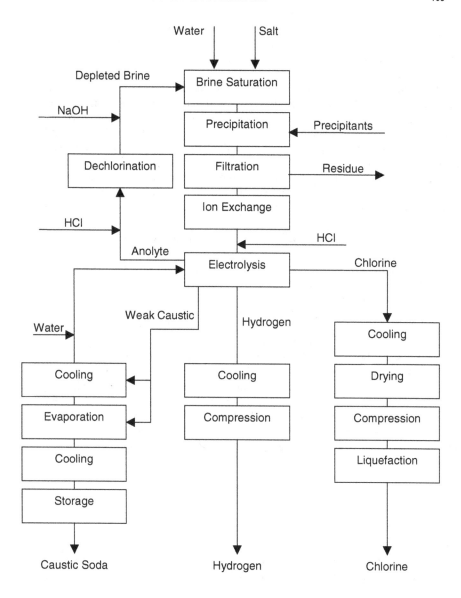

**Figure 3.6.** Flow diagram of the Membrane process for chlor-alkali production. Reproduced from Reference 12, Copyright (1985), with permission from John Wiley & Sons, Inc.

are nickel-sulfur, nickel-aluminium, nickel-nickel oxide mixtures, and nickel coatings containing the platinum group metals. Although electrocatalytic cathode technology is utilized in membrane cells, commercialization is still awaited in diaphragm cells.

The membrane is a critical component of the membrane-cell technology as it determines the current efficiency, cell voltage, and hence the energy consumption for the production of chlorine and caustic soda. The membranes currently used are composite membranes with perfluorosulfonic acid layers facing the anolyte side and perfluorocarboxylic acid layers facing the catholyte side, with an intermediate reinforcing fabric between them. These composite membranes provide high current efficiency and low cell voltage at current densities above 5 kA/m$^2$. These composite membranes are prepared by lamination of the perfluorocarboxylate and perfluorosulfonate films, by the chemical conversion of the perfluorosulfonic acid to realize a carboxylate layer thickness of 5–10 μm, or by co-extension of the two different polymer films, which will provide high efficiency with a low voltage penalty.

The minimum energy required to produce chlorine, hydrogen, and caustic from salt is the same (i.e., 1686.32 kWh/ton of chlorine) for all three cell technologies. However, the actual energy consumed is much higher than the minimum. The energy consumed in the mercury-cell process is the greatest because the combined voltages required by reactions (3.26) and (3.27) are higher than those encountered in the diaphragm or membrane-cell processes.

Electrolysis of brine (Eq. 3.23) is endothermic. The overall heat of the reaction is 446.68 kJ/mol (106.76 kcal/mol of chlorine) and hence, the thermoneutral voltage, i.e., the voltage at which heat is neither required by the system nor lost by the system to the surroundings, would therefore be 2.31 V. In practice, however, chlor-alkali cells operate in the range of 3.0 to 3.5 V, at an average coulombic efficiency for chlorine production (CE) of 95 %, resulting in heat generation to the extent of 3960 kJ/kg of Cl$_2$ for a cell potential (E) of 3.5 V as derived from the following equation:

$$Q = \left[\left[\frac{100}{CE}\right][46.05 \text{ V}]\right] - \Delta H \qquad (3.28)$$

Heat produced in these cells operating at voltages of above 2.31 V is generally removed by water evaporation and radiation losses. Figure 3.3 illustrates the basic principles of the three cell processes and Table 3.6 summarizes the differences in the cell technologies and their performances.

The values of energy consumption presented in Table 3.6 are not optimal for the indicated technology, since the actual value depends on the current density, cell voltage, and current efficiency, which are affected by the process variables. Similarly, the energy for evaporation varies with the type of the evaporator system used. In the case of membrane cells, the energy requirements would be 15–25 % lower than those for diaphragm cells—the major savings resulting from caustic

## TABLE 3.6
**Comparison of Components and of Performance Characteristics of Membrane, Diaphragm, and Mercury Cells in Chlor-Alkali Plants**

| Component | Mercury Cell | Diaphragm Cell | Membrane Cell |
|---|---|---|---|
| Anode | $RuO_2 + TiO_2$ coating on Ti substrate | $RuO_2 + TiO_2 + SnO_2$ on Ti substrate | $RuO_2 + IrO_2 + TiO_2$ coating on Ti substrate |
| Cathode | Mercury on steel | Steel (or steel coated with activated nickel) | Nickel coated with high area nickel based or noble metal based coatings |
| Separator | None | Asbestos, polymer-modified asbestos, or nonasbestos diaphragm | Ion-exchange membrane |
| Cathode product | Sodium amalgam | 10-12% NaOH + 15-17% NaCl + 0.04–0.05% $NaClO_3$, and $H_2$ | 30-33% NaOH + <0.01% NaOH and $H_2$ |
| Decomposer product | 50% NaOH and $H_2$ | None | None |
| Evaporator product | None | 50% NaOH with ~1.1% salt and 0.1-0.2% $NaClO_3$ | 50% NaOH with ~0.01% salt |
| Steam consumption | None | 1500-2300 kg/t NaOH | 450–550 kg/t NaOH |
| Cell voltage, V | 4–5 | 3.0–4 | 2.8–3.3 |
| Current density, $kA/m^2$ | 7–10 | 0.5–3 | 2.0–10. |
| **Energy consumption (kW h/ton of $Cl_2$)** | | | |
| Electricity for electrolysis | 3200-3600 | 2800-3000 | 1950–2220** |
| Steam for caustic evaporation* | 0 | 600–800 | 200–350*** |
| Total | 3200-3600 | 3400-3800 | 2150-2570 |

* 1 ton of steam = 400 kW h    ** Load: 3–6 $kA/m^2$    *** MP steam(10 bar a, 200 °C) double effect evaporator

evaporation. Thus, the energy needed to concentrate 33 % NaOH to 50 % caustic can vary from 720 kWh/ton of caustic to 314 kWh/ton, depending on whether a single effect or a triple effect is used for caustic concentration.

There are several inefficiencies arising from parasitic reactions occurring at the electrodes and in the bulk, which are described below. The two parasitic reactions offsetting anode efficiency are:

- co-generation of oxygen, from the anodic discharge of water:

$$2\, H_2O \rightarrow O_2 + 4\, H^+ + 4e^- \tag{3.29}$$

- electrochemical oxidation of hypochlorite ion, $OCl^-$, to chlorate, $ClO_3^-$:

$$6\, OCl^- + 3\, H_2O \rightarrow 2\, ClO_3^- + 4\, Cl^- + 6\, H^+ + 3/2\, O_2 + 6\, e^- \tag{3.30}$$

The oxygen generated from these reactions is dependent on the nature of the anode material and the pH of the medium. The current efficiency for oxygen evolution is generally 1–3% when using commercial metal anodes.

At the cathode, water molecules are discharged to form gaseous hydrogen and hydroxide ions, $OH^-$. Some of the caustic generated in the cathode compartment back migrates to the anode compartment and reacts with dissolved chlorine ($Cl_{2,aq}$) to form chlorate as follows:

$$Cl_{2(aq)} + OH^- \rightarrow HOCl + Cl^- \tag{3.31}$$

$$HOCl + OH^- \leftrightarrow H_2O + OCl^- \tag{3.32}$$

$$2\, HOCl + OCl^- \rightarrow ClO_3^- + 2\, H^+ + 2\, Cl^- \tag{3.33}$$

There are two reactions that influence the cathodic efficiency, namely the reduction of $OCl^-$ and $ClO_3^-$:

$$OCl^- + H_2O + 2\, e^- \rightarrow Cl^- + 2\, OH^- \tag{3.34}$$

$$ClO_3^- + 3H_2O + 6\, e^- \rightarrow Cl^- + 6\, OH^- \tag{3.35}$$

Although these reactions are thermodynamically possible, they are not kinetically significant under normal operating conditions. Hence, the cathodic efficiency is usually high (> 95 %) in diaphragm and membrane cells. In mercury cells, the cathodic inefficiency arises from the discharge of $H_2$ at the cathode as a result of the impurities in the brine. Reactions contributing to anodic inefficiency in mercury cells are the same as in diaphragm or membrane cells.

Of the three electrolytic technologies used to produce chlorine and caustic, diaphragm-cell technology is the source for generating the largest volume of chlorine, followed by the membrane-cell process, and then the mercury process.

However, the membrane process is the preferred technology for the future because of its ecological and economical benefits over the other two technologies. Over the past ten years, neither new diaphragm nor mercury plants have been built. Depending on operating parameters (i.e., current density, cell voltage, and current efficiency), the electrical energy consumption in membrane chlor-alkali electrolysis is between 1950 and 2300 kWh/t of chlorine. The amount of energy required to concentrate caustic soda from 32 wt% to 50 wt% varies between 700 and 800 kWh/t (basis: medium pressure steam) of chlorine for a double-effect evaporator. The electrical energy consumption in diaphragm chlor-alkali electrolysis is about 10 to 20 % higher than that for the membrane technology. In addition, diaphragm cells generate caustic soda at a low concentration of ~ 11 wt%, and hence, the energy requirements to achieve the commercial concentration of 50 wt% of caustic soda are much higher compared to the membrane process. The current membrane technology suppliers are: Asahi-Kasei, Chlorine Engineers, Uhde, ELTECH, and INEOS Chlor.

### 3.3.3. Economics

The choice of technology, the associated capital, and operating costs for a chlor-alkali plant are strongly dependent on the local energy and transportation costs, as well as environmental constraints. The primary difference in operating costs between membrane, diaphragm, and mercury plants results from variations in electricity and steam consumption for the three processes. The cost of constructing a grass root plant significantly depends on the actual plant configuration, procurement conditions, and of course, the production capacity. A breakdown for the total investment for a grass root membrane electrolysis plant of about 160,000 Mt/y of chlorine production capacity is presented in Table 3.7.

Conversions from mercury to membrane technology require new facilities to protect the membranes from traces of mercury and place restrictions on the use of existing equipment. Diaphragm plants are often more easily adapted to the needs of membrane technology.

### 3.3.4. Benefits to or from Fuel-Cell Technologies

Adopting fuel-cell based technologies can significantly lower the energy consumption involved in manufacturing chlorine. One approach, that is particularly applicable for diaphragm-cell operations, involves concentrating the weak diaphragm cell catholyte from ~10–12 % to 50 % NaOH. This scheme involves feeding the diaphragm caustic to the anode compartment of a membrane cell and using the hydrogen ionization reaction as the anodic process when the protons released by the anodic process will neutralize the hydroxyl ions in the anolyte. The cathodic reaction in this scheme is the discharge of water to form hydrogen and the

**TABLE 3.7**
**Typical Cost Breakdown For A Chlor-Alkali Plant**

| Item | Estimated investment in 1000 US$ |
|---|---|
| Cells | 27,200 |
| Brine purification | 14,000 |
| Chlorine processing | 16,000 |
| Waste gas treatment | 2,300 |
| Caustic evaporation | 6,900 |
| Utilities | 4,500 |
| Rectifiers | 10,000 |
| Engineering | 10,000 |
| Total | 90,900 |

hydroxyl ions in the catholyte. Occidental Chemical Corporation has examined this concept. However, it has not proceeded beyond the laboratory stage.

Another fuel cell based idea involves substituting an oxygen-reduction reaction for the hydrogen-evolution reaction at the cathode in chlor-alkali electrolysis. This will reduce the total cell potential by about one volt (theoretically 1.23 V), thereby realizing a substantial electrical energy savings of > 900 kWh/short ton of chlorine. The anode reaction is the same as in a conventional chlor-alkali cell, where the chloride ions are discharged to form a chlorine gas product and the sodium ions migrate to the cathode compartment through the ion-exchange membrane. At the cathode, oxygen is reduced to the $OH^-$ ions, which combine with the $Na^+$ ions to form sodium hydroxide. The oxygen consumed in this reaction enters the air cathode compartment either as water-saturated pure oxygen gas or as air and gets reduced at the porous air cathode (a fuel cell cathode). The additional advantages of using the oxygen-reduction reaction include avoidance of costly downstream-treatment of hydrogen and absence of the gas void fraction in the catholyte, resulting in a reduced ohmic drop in the cell. However, this scheme requires a high-performance air scrubbing system to remove all carbon dioxide from the air in order to protect the air cathode from the accumulation of sodium carbonate and a premature failure of the cells. As the anode side of the process does not differ from the conventional chlor-alkali cell, these components can be used without any modification.

However, the cathode side has several special needs, the central one being the adjustment of the local differential pressure between the caustic and the oxygen compartment on the other side of the electrode. Due to the porous nature of the gas-diffusion electrode (GDE), a pressure balance across the electrode has to be established in order to avoid the flow of the fluid from one side to the other. The local differential pressure is a function of height because of the different densities of the fluids, and it can be made small using two different approaches. The first one involves splitting the cathode compartment into several horizontal compartments, called gas pockets in which the height of each sub-compartment limits the hydrostatic pressure of the caustic to a tolerable value. The lean caustic flows through the pockets successively, by overflowing from one pocket to the next one below. This *gas-pocket principle* is patented and now being tested by the Bayer AG group. A second approach to the problem is the *falling-film principle*. The development of this type was initiated by the Hoechst group in the 1980's and is presently continued by Uhde. The idea here is to decrease the hydrostatic pressure of the caustic successively by establishing a falling film of caustic between the electrode and the membrane. This is realized by means of a layer of hydrophilic material, which is fixed between the anode and the cathode. This design ensures a constant gap between the GDE and the membrane itself. Because of the electroosmotic water transport from the anolyte to the catholyte, the caustic flow increases from the top to the bottom of the cell. A high flow in the hydrophilic layer will lead to a flooding of the GDE, and hence a breakthrough of caustic into the oxygen-compartment, caused by an increased differential pressure. This flooding can be avoided by a proper design of the hydrophilic layer. The falling-film technology shows some inherent advantages. Unlike the gas-pocket principle, the falling-film technology does not need an extensive gasket system throughout the surface of the GDE, allowing a simplified design. Furthermore, the GDE can be run close to the atmospheric pressure, so that it will always operate in the optimum operating pressure range. Both processes show comparable operating data. However, they are still at an experimental level. Nevertheless, the initial results with the GDE system are promising towards achieving an energy savings of ~30 %, compared to the conventional membrane process.[16,17]

An alternate approach to reduce energy consumption is to use the hydrogen generated in the chlor-alkali process in a fuel cell to extract the same energy benefits as retrofitting or developing a cell with oxygen cathodes.

## 3.4.  ELECTRO-ORGANIC SYNTHESIS

### 3.4.1.  Background and Applications

Electro-organic synthesis dates back to at least 1834 when Faraday described the anodic oxidation of acetate ions to carbon dioxide.[18] It was fifteen years later that Kolbe discovered the generation of n-butane by the electrolysis of aqueous valeric-acid solutions. Since then a large number of industrially important organic compounds were synthesized and several of them were commercialized. Table 3.8 describes some on-going commercial processes. There are many other fine chemicals that are produced electrochemically, but they remain in the proprietary domain.

**TABLE 3.8**
**Some Current Electroorganic Technologies**

| Product | Starting Material | Company |
|---|---|---|
| Acetoin | Butanone | BASF |
| Acetylenedicarboxylic Acid | 1,4-Butynediol | BASF |
| Adipoin Dimethyl Acetal | Cyclohexanone | BASF |
| Adiponitrile | Acrylonitrile | Monsanto (Solutia), BASF, Asahi Chemical |
| 4-Aminomethylpyridine | 4-Cyanopyridine | Reilly Tar |
| Anthraquinone | Anthracene | L. B. Holliday, ECRC |
| Bleached Montan Wax | Raw Montan Wax | Hoechst |
| Calcium Gluconate | Glucose | Sandoz, India |
| Calcium lactobionate | Lactose | Sandoz, India |
| S-Carbomethoxymethylcysteine | Cysteine and Chloroacetic | Spain |
| L-Cysteine | L-Cystine | Several |
| Diacetone-2-ketogluconic Acid | Diacetone-L-sorbose | Hoffman-LaRoche |
| Dialdehyde Starch | Starch | India, Others |
| 1,4-Dihydronaphthalene | Naphthalene | Hoechst |
| 2,5-Dimethoxy-2,5-dihydrofuran | Furan | BASF |
| 2,5-Dimethoxy-2,5-dihydrofuryl-1-ethanol | Furfuryl-1-ethanol | Otsuka |
| Dimethylsebacate | Monomethyladipate | Asahi Chemical |
| Gluconic Acid | Glucose | Sandoz, India |
| Hexafluoropropyleneoxide | Hexafluoropropylene | Hoechst |
| Mucic Acid | Galacturonic Acid | EDF |
| Perfluorinated hydrocarbons | Alkyl substrates | 3M, Bayer, Hoechst |
| Phthalide and t-Butylbenzaldehyde Acetal | Dimethyl Phthalate and t-Butyltoluene | BASF |
| p-Methoxybenzaldehyde | p-Methoxytoluene | BASF |
| Polysilanes | Chlorosilanes | Osaka Gas |
| p-t-Butylbenzaldehyde | p-t-Butyltoluene | BASF, Givaudan |
| Salicylic Aldehyde | o-Hydroxybenzoic Acid | India |
| Succinic Acid | Maleic Acid | CECRI, India |
| 3,4,5-Trimethoxybenzaldehyde | 3,4,5-Trimethoxytoluene | Otsuka Chemical |
| 3,4,5-Trimethoxytoluoyl Alcohol | 3,4,5-Trimethoxytoluene | Otsuka Chemical |

## 3.4.2. Principles of Technology

The initial step during the course of an electrochemical reaction involving an organic molecule is generally the formation of a reactive radical or radical ion as:

$$\text{Molecule} \xrightarrow{+e} \text{Radical ion} \xrightarrow{\text{reaction, fast}} \text{Intermediate} \xrightarrow{+e, \text{fast}} \text{Product}$$

$$(3.36)$$

**TABLE 3.9**
**Classification of Electroorganic Reactions**

| | |
|---|---|
| 1. Direct: Charge transfer is the primary act with the organic substrate of interest. | |
| a. Cation radical formation | $R \rightarrow R^+ + e^-$ |
| b. Anion radical formation | $RX \rightarrow RX^{\cdot-} - e^-$ <br> where X may be halogen, H or other functional group. |
| c. Carbonium ion formation | $RH \rightarrow R^+ + H^+ + 2e$ <br> e.g., $CH_3\text{-}COO^- \rightarrow CH_3^+ CO_2 + 2e^-$ |
| d. Carbanion formation | $RX + 2e^- \rightarrow R^- + X^-$ |
| e. Reduction of carbonium ions or oxidation of anions | 1. $R^- + e^- \rightarrow R^{\bullet}$ <br> e.g., $\varphi\text{-CH=N}^+(CH_2)_4 \rightarrow \varphi\text{-C}^+\text{H-N}(CH_2)_4 \rightarrow$ product <br> (dimerizes on reation) <br> 2. $R^- \rightarrow R^{\bullet} + e^-$ <br> e.g., $CH_3COO^- \rightarrow CH_3CO_2 + e\ CH_3 \rightarrow$ dimerizes |
| 2. Indirect: Charge transfer occurs with some other species, which then reacts with the substrate of interest. | |
| a. With electroregenerated redox species | e.g., $Cr^{6+}$, $Ce^{4+}$, $Ag^{2+}$, $Br_2$, etc. |
| b. With adsorbed intermediates generated during the course of a reaction | e.g., $Cl_{ads}$, $CO_2^-$, $HO_2^-{}_{(ads)}$, $H^+$ or $OH^-$ |

The manner by which the electron transfer can occur can be direct or indirect, as shown in Tables 3.9 and 3.10. The electrochemically generated radicals (radical anions or cations) react in a multitude of pathways, either chemically or electrochemically to form a wide variety of products. These reaction pathways could become more involved when adsorbed intermediates interact with these species. Figures 3.7 and 3.8 depict some of these reaction sequences that can occur with radical cations and anions from unsaturated hydrocarbons.

Thus, a wide variety of organic reactions can be performed electrochemically, the basic processes being anodic oxidation or cathodic reduction often without the use of any other chemical reagents. Generally, desirable concentrations of highly reactive cation or anion radicals can be easily generated electrochemically. Other reactive species that can be conveniently made include superoxide ions, hydroxyl radicals, peroxide, $CO_2$, anion radicals, hydrogen atoms, and halogens, including fluorine. Another advantage with electrochemical routes for organic synthesis is the role played by the electrode materials, which could alter the course of reactions. Some examples of the influence of electrode materials are shown in Figures 3.9 to 3.12 to illustrate the specificity that can be achieved electrochemically.

**TABLE 3.10.**
**Indirect Electrolysis for Electroorganic Syntheses.**

| Redox Couple | Electrochemical Conversion |
|---|---|
| $Ti^{4+}/Ti^{3+}$ | Nitroaromatics $\rightarrow$ anilines, |
| | Quinine $\rightarrow$ hydroquinone |
| $Fe^{3+}Fe^{2+}$ | Acrylonitrile polymerization |
| $Fe(CN)_6^{3-}/Fe(CN)_6^{4-}$ | Benzene oxidation |
| $MnO_4^-/MnO_4^{2-}$ | Oxidation of aromatics |
| $Ni^{3+}/NiF_6^{2-}$ | Electrofluorination |
| $Tl^{3+}/Tl^+$ | 1-Butene to methyl ethyl ketone |
| $Co^{3+}/Co^{2+}$ | Oxidation of aromatics |
| $Sn^{4+}/Sn^{2+}$ | Reduction of nitrocompounds |
| $Ce^{2+}/Ce^{3+}$ | Anthracene to anthraquinone |
| $Cu^{2+}/Cu^+$ | Hydroxylation of aromatics |
| $VO_3^-/VO^{2+}$ | Oxidation of aromatics |
| $HIO_4/HIO_3$ | Dialdehyde starch process |
| NaHg/Hg | Hydrodimerization |
| NaOCl/NaCl | Propylene oxide from propylene |
| or NaOBr/NaBr | or Oxidation of sugars |
| $OsO_4/[OsO_2(OH)_4]^{2-}$ | Olefins to glycols |
| $Br_2/Br^-$ | Alkoxylation of furans |
| $I_2/I^-$ | Halofunctionalization; Prevost reaction |

Electrofluorination of organic compounds occurs at only a few anode materials (Figure 3.9). These include nickel in HF solution, porous carbon in molten HF/KF, and Pt in organic solvents containing fluoride ion. The nickel anode leads to nonspecific fluorination apparently involving nickel hexafluoride ($NiF_6^{2-}$) ionic species as the fluorinating agent. This high-valent species is formed as an insoluble, continuously renewable anode surface coating that attacks the organic. The porous carbon anode employed in the Phillips electrofluorination process also leads to nonspecific fluorinations by the electrogenerated fluorine atoms. In contrast, platinum behaves differently because the organic is directly oxidized to a cationic species, which then undergoes nucleophilic attack by the fluoride ions.

Depending on the choice of the cathode material, acetone can be converted to isopropyl alcohol, pinacol, propane, or diisopropyl mercury (Figure 3.10). The mechanism involved in the formation of isopropyl alcohol is direct, but the formation of di-isopropyl mercury is a result of the interaction of electrogenerated radical species with the mercury cathode by metal-atom abstraction.

The diversity of products formed during the electroreduction of nitrobenzene depends largely on the nature of the cathode material, the solvent and the supporting electrolyte, and other factors such as temperature and reactant concentration (Figure 3.11). The selectivity depends on the nature of and stability of the intermediates formed and the rates at which the consecutive reactions proceed either in the bulk or at the interface. These factors dictate the appropriate choice of the hydrodynamic regimes in the cell that allow the reactions to proceed at significant rates in the

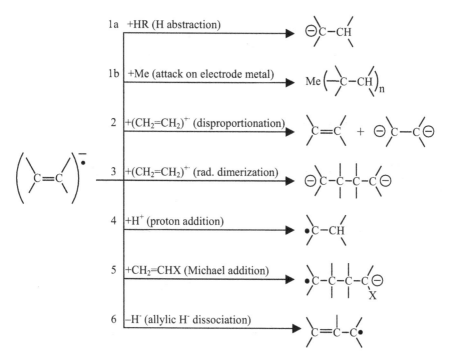

**Figure 3.7.** Reaction routes [6] for radical anions from unsaturated hydrocarbons (e.g., vinyl compounds): (1-3) radical reactions; (4-6) anionic reactions. Reproduced from Reference 19.

desired direction. It must be emphasized that the penalties associated with poor selectivity and low yields are enormous, as the overall process economics would be adversely affected by the high-energy requirements and the costs involved in the separation and purification of the products.

Hydrodimerization of acrylonitrile provides a further example of the role of the substrate in organic synthesis (Figure 3.12). Thus, acrylonitrile is reduced to propionitrile at Pt and Ni cathodes by adsorbed H atoms, whereas on Pb electrodes allylamine is formed. Sn electrodes yield the organometallic and Hg, Pb, and C cathodes lead to the formation of adiponitrile and propionitrile. The large-scale electro-organic industry, at present, is the Monsanto process for producing adiponitrile by the electrohydrodimerization of acrylonirile. The estimated total production of adiponitrile worldwide is ~340,000 tons/year, and this product is an intermediate for manufacturing Nylon-66. The chemical route for making adiponitrile is the catalytic reaction of adipic acid with ammonia, which is then converted to hexamethylenediamine and finally to Nylon-66 as shown below:

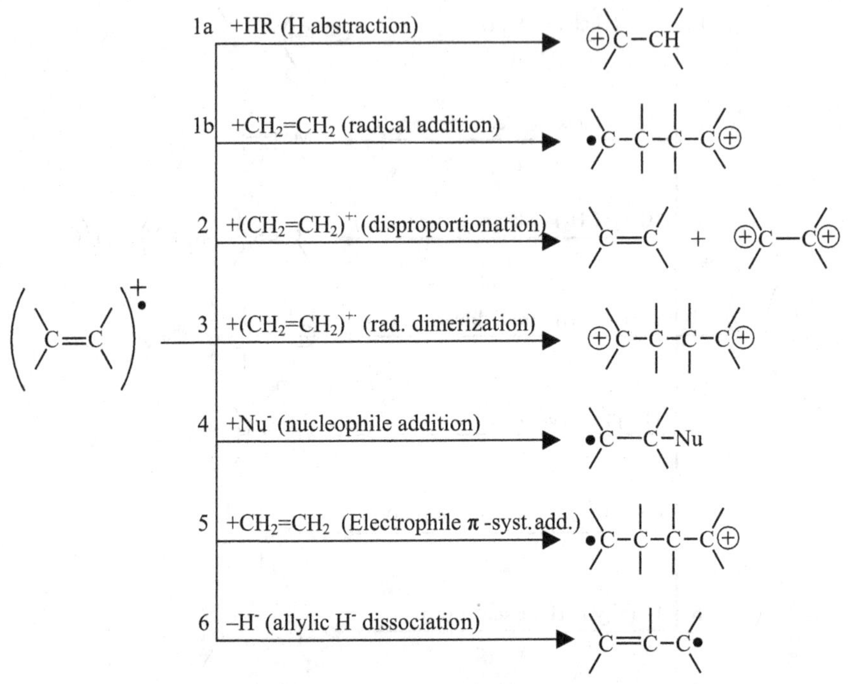

**Figure 3.8.** Reaction routes [6] for radical cations from unsaturated hydrocarbons: (1-3) radical reactions; (4-6) cationic reactions. Reproduce from Reference 19.

$$HOOC-(CH_2)_4-COOH + 2\,NH_3 \xrightarrow{\text{catalyst}} NC-(CH_2)_4-CN + 4\,H_2O \quad (3.37)$$

$$NC-(CH_2)_4-CN + 4\,H_2 \longrightarrow NH_2-CH_2-(CH_2)_4-CH_2-NH_2 \quad (3.38)$$

$$NH_2-CH_2-(CH_2)_4-CH_2-NH_2 + HOOC-(CH_2)_4-COOH \longrightarrow \text{Nylon 66}$$

$$(3.39)$$

The driving force for the electrochemical route to produce adiponitrile from acetonitrile is the high cost of adipic acid. This process was discovered by M. Baizer in 1959, and was commercialized by D. Danly at the Monsanto Corporation. Basically, the reaction involved is the electrohydrodimerization of acrylonitrile as follows:

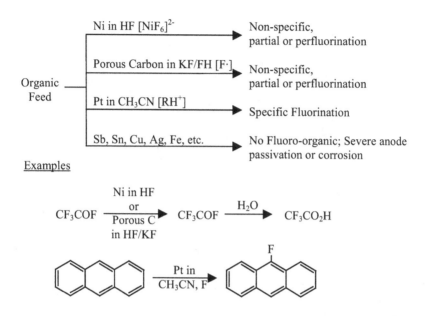

**Figure 3.9.** Influence of electrode material during electrofluorination. Reprinted with permission from Reference 20, Copyright (1979), American Chemical Society.

$$2\,CH_2 = CH - CN + 2\,H^+ + 2e^- \longrightarrow NC(CH_2)_4 CN \tag{3.40}$$

The early Monsanto process used a two-compartment cell with Pb cathodes and $PbO_2$ + AgO anodes, separated by a cation exchange membrane. The catholyte was 40 % tetraethyl ammonium sulfate containing 28 % water, 16 % acetonitrile and, 16 % acrylonitile, the anolyte being 5 % sulfuric acid. This process, although successful, suffered from several problems such as membrane fouling, low conversion rates, byproduct formation, high-energy consumption of 6700 kWh/ton of adiponitrile, and expensive product isolation processes. All these problems were addressed, and a new Monsanto process was developed in the late 1970's (see Figure 3.13).

This second generation process is based on an undivided cell and employs Cd cathodes and steel anodes in a bipolar cell configuration. The electrolyte is a two-phase recirculating aqueous emulsion of acrylonitrile, adiponitrile, a bisquarternary salt (hexamethylene (bisethyltributyl) ammonium phosphate), a phosphate buffer, and anode anti-corrosion additives, borax, and EDTA. Electrolysis is conducted at 55 °C at 2 kA/m². A fraction of the organic phase is continuously removed from the

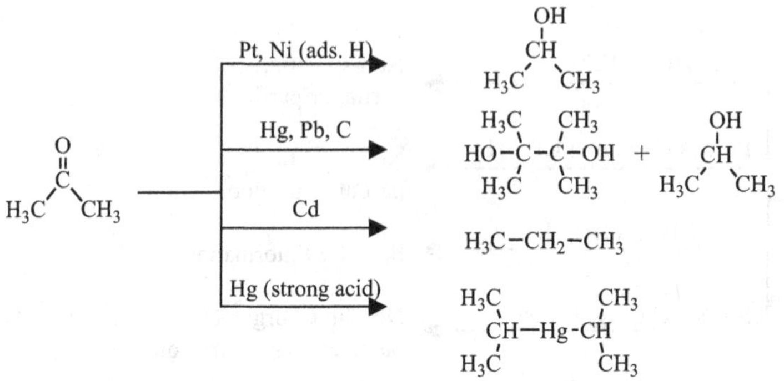

Some of the electrogenerated species involved:

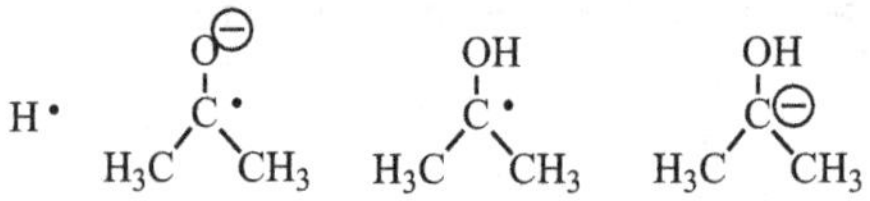

**Figure 3.10.** Influence of electrode material on the electrochemical reduction of acetone. Reprinted with permission from Reference 20, Copyright (1979), American Chemical Society.

emulsion reservoir for separation of the product. The aqueous phase is also reacted continuously to prevent accumulation of byproducts and metallic salts from electrode corrosion. Table 3.11 depicts a comparison of these processes.

**TABLE 3.11.**
**Comparison of the Monsanto 1965 Divided Cell Process With the Recent Monsanto Undivided Cell Process.**

|                                          | Divided cell | Undivided cell |
|------------------------------------------|--------------|----------------|
| Adiponitrile selectivity (%)             | 92           | 88             |
| Inter-electrode gap (cm)                 | 0.7          | 0.18           |
| Electrolyte resistivity ($\Omega$ cm)    | 38*          | 12             |
| Electrolyte flow velocity ($ms^{-1}$)    | 2            | 1-1.5          |
| Current density (A $cm^{-2}$)            | 0.45         | 0.20           |
| Voltage distribution (V)                 |              |                |
|    Estimated reversible cell voltage | 2.50 | 2.50 |
|    Overpotentials         | 1.22         | 0.87           |
|    Electrolyte $iR$       | 6.24         | 0.47           |
|    Membrane $iR$          | 1.69         | 0.87           |
|    Total                  | 11.65        | 3.84           |
| Energy consumption (kW h $t^{-1}$)       | 6700         | 2500           |

*catholyte

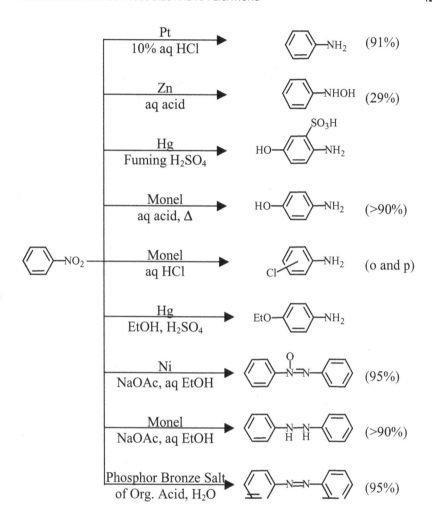

**Figure 3.11.** Influence of electrode material during electrochemical reduction of nitrobenzene. Reprinted with permission from Reference 20, Copyright (1979) American Chemical Society.

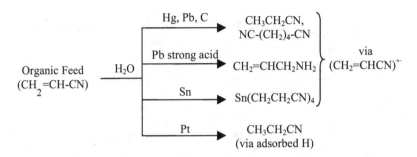

**Figure 3.12.** Influence of electrode material during hydrodimerization of acrylonitrile. Reprinted with permission from Reference 20, Copyright (1979), American Chemical Society.

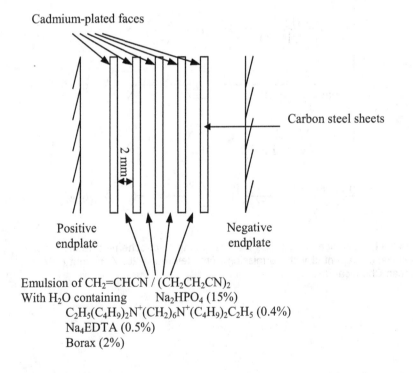

**Figure 3.13.** The new Monsanto process for the hydrodimerization of acrylonitrile.[20]

### 3.4.3. Economics

Generally, in electroorganic processes, 6–15 % of the operating costs are towards electricity and 20–70 % for the starting materials. The investment costs towards the electrochemical cell can vary between 5 to 50 % depending on the specific process. Adiponitrile (ADN) made by the Monsanto process is an intermediate for making hexamethylene diamine (HMDA), which is then converted to Nylon-66. The capital costs for producing 100,000 kg/year of HMDA were estimated to be $101MM in 1987,[21,22] of which about $61MM are for processing AN to HMDA. Since the yield of HMDA from ADN is 96.6 %, the capital costs for producing about 100,000 kg/year will be about $50MM, the cost of the undivided electrolytic cells being $7.5 MM and the power system being $0.9 MM. The total direct production costs for this capacity of HMDA are $0.26/kg of which the raw materials cost is $.49/kg and electricity cost is $0.03/kg at an electricity price of $0.05/kWh.

### 3.4.4. Benefits to or from Fuel Cell Technology

Depending on the type of organic reaction that is being conducted at the working electrode, the electrode reaction at the counter electrode is either the hydrogen evolution or the oxygen evolution reaction. These reactions can be supplanted by the oxygen reduction reaction or the hydrogen ionization reaction, as long as the organic species in the solution are not reacting at these interfaces, to achieve significant energy savings. The idea of using fuel-cell reactions in electroorganic syntheses was proposed in 1984.[23-25] The hydrogen ionization reaction was studied as the anodic reaction during the hydrodimerization of acetonitrile and a voltage savings of 400 mV at 100 mA/cm$^2$ was claimed with a gas diffusion electrode containing a platinum catalyst.[24] Another process that was examined using the hydrogen depolarized anode was the hydrodimerization of formaldehyde to ethylene glycol. Efficiencies as high as 90 % for ethylene glycol formation were obtained[23-25] with the hydrogen depolarized anode in an undivided cell. However, the low molecular weight compounds such as formaldehyde and methanol were oxidized at the anode, resulting in the generation of > 80 % $CO_2$ on the back side of the gas diffusion electrode. A third process examined with the anodic reaction as the hydrogen ionization reaction was the reduction of 3-hydroxybenzoic acid to the corresponding benzylic alcohol in a divided cell in aqueous sulfuric acid solutions. These investigations showed a 40% reduction in the cell potential, relative to a divided cell of comparable construction. Thus, while the investigations with the fuel cell based electroorganic reactions are limited, the results show significant potential towards achieving energy savings in these operations.

## 3.5.    ELECTROWINNING AND REFINING OF METALS

### 3.5.1.  Background and Applications

Almost all metals exist in their native state as oxides or sulfides. The sulfide ores are calcined, leached with sulfuric acid, and the resulting sulfate solution is electrolyzed to produce high quality metals. The large tonnage metals, produced electrolytically, are copper, nickel, and zinc; cobalt, chromium, manganese, cadmium, gallium, thallium, indium, silver, and gold are produced by electrolysis on a smaller scale. The annual production amounts of copper, nickel, and zinc were 13.2, 0.37 and 8.7MM tons, respectively, in 2001.[5]

Zinc is used for metallic coatings to protect iron and steel for corrosion protection by galvanizing, plating, and painting with zinc-bearing paints. Its use, as a structural material, is for pressure die-casting alloys for automotive and builders' hardware, foundry alloys, and slush alloys. Zinc dust and powder find extensive use in atomizing process, paint coatings, as a reducing agent to produce hydrosulfite compounds for the textile and paper industries and to enhance the physical properties of plastics and lubricants. Zinc powder is also used in primary batteries, frictional materials, spray metallizing, mechanical plating, and chemical formulations. 55% of the zinc produced is galvanizing, 17% for Zn-based alloys, 15% for making brass and bronze, and 15% for others.[7,26]

The largest single market for refined copper is for electrical uses because of its high electrical conductivity. 40 % of the total consumption of copper is for building construction (i.e., electrical products, plumbing goods, and roofing sheet). An average modern U.S. home contains about 200 kg of copper. It is also used for transportation and industrial machinery equipment. Other uses include coinage, agricultural fungicides, wood preservatives, food additives, utensils, and cutlery. About 39% of the copper is consumed in the building construction, 28 % for electric and electronic products, and 11% each for transportation equipment, industrial machinery and equipment, and consumer and general goods.[7,27]

Nickel is used to produce a wide range of corrosion- and heat-resistant materials by alloying with copper, chromium, iron, molybdenum, and others. Pure nickel powder is employed in the production of porous plates for batteries and powder metallurgy parts. 39% of the nickel is produced for stainless steel and alloy steel production, 38 % for non-ferrous and super alloys, 15% in electroplating, and 15% for others.[7,28]

### 3.5.2.  Principles of Technology

The processes of electrowinning and electrorefining generally employ a similar process scheme. In electrorefining, plates of crude metal are anodically dissolved in a suitable electrolyte, while pure metal is deposited on the cathode. Electrorefining is practiced to produce copper, lead, nickel, silver and other minor metals. The soluble anode, used in nickel refining, is nickel matte containing 20% sulfur. This process is similar to refining with a metallic anode, but it is not a true refining process and is called electrowinning with a soluble anode. A process that is related to electrorefining is electrowinning with insoluble anodes. In this process, the metal present in the ore is leached from the calcined ore and then the pure metal is "electrowon" with insoluble anodes. Thus in both processes, the cathodic reaction is the reduction of the cationic species, of interest, in the solution to the metallic state, whereas the anodic reaction is the anodic dissolution of the metal from the impure electrode and in some cases the oxygen evolution reaction (in electrowinning operations).

In an ideal electrorefining process, the metal deposition and dissolution reactions take place under nearly reversible conditions. Impurities, which are more electropositive to the electrorefined metal will not dissolve at this potential and will end up in the anode slime. Metals, which are electronegative to the refined metal, will dissolve anodically and accumulate in the electrolyte but will not deposit on the cathode. This simple description applies well to the metals with low dissolution and deposition potentials and with impurities for which the standard reversible electrode potentials are sufficiently different from that of the refined metal.

Selectivity[5, 29] for the anodic process is often helped by the refractory nature of the impurity-containing compounds in the anode metal. Thus, Ag, Se, and Te, present in the copper anodes, form refractory selenides and tellurides of Ag and Cu. Silver, present in the anodes, tends to plate on cathodes. The dissolution of silver is circumvented by maintaining a chloride level of 30 ppm in the copper refining electrolyte so that the silver ion concentration will not exceed the solubility product of AgCl. Nickel, present in the copper anodes, forms a refractory NiO, and ends up partially in the anode slime. In sulfate containing electrolytes, lead forms insoluble $PbSO_4$. When impurities from the anode become a problem, as in the case of nickel refining, a diaphragm separates the anode and cathode compartments. In electrowinning operations with insoluble anodes, it is not possible to eliminate noble metal impurities. Therefore, the electrolyte is treated to remove these impurities by cementation with active nickel powder in nickel electrowinning and with zinc powder in zinc electrowinning operations.

Selectivity of the cathodic process, with respect to the soluble impurities in electrorefining and electrowinning, is better with reversible metals such as Ag, Cu, Pb, and Zn than with nickel. Thus, electrorefined copper contains < 5 ppm Ni and < 1 ppm As and Sb, although they are present in relatively high concentrations in the refining electrolyte (~ 20 g/l Ni, ~ 1 g/l As, ~ 0.1 g/l Sb). Similarly, high purity silver can be deposited from solutions with ~1 g/l Cu and Pd, and high-purity zinc is produced from solutions containing ~ 1g/l Mn. However, in the case of nickel, soluble anode metal impurities (Cu, Fe, Co, Zn, Pb, As) are readily co-deposited. It should be noted that the most frequent cause of cathode contamination is the mechanical occlusion of anode slime, especially when the cathode is rough and porous. An example of this effect is the presence of > 20ppm Pb in electrorefined copper.

In electrowinning and electrorefining operations, thick metal deposits are grown over a period of several days (15 to 30, depending on the metal) and it is essential these be smoothed out without any dendritic growth or porosity. This is achieved by the use of addition agents. In copper refining, animal glue is used by most refineries, the typical addition rates being 3–100 g/ton of cathode weight. Other additives used in copper refining include thiourea, lignin related additives, avitone, safranin, and casein. Addition agents used in zinc electrowinning are: sodium silicate, gum Arabic, glue, cresylic acid, and a soya bean extract. Copper electrowinning operations often use guar gum—a natural colloid.

The relatively slow mass-transfer rates in the electrorefining and electrowinning operations present unique problems not encountered in other process industries.[29] For example, an average copper refinery, producing 500 tons Cu/day needs ~ 0.2 km$^2$ of electrode area. This corresponds to 50,000 anodes and 50,000 cathodes suspended in 1500 tanks, occupying a total floor area of 6000 m$^2$. A tank house is the heart of these operations regardless of the metal produced. The unique technological problems result from the fact that each of the many cells should be provided with a continuous supply of fresh electrolyte, continuous withdrawal of spent electrolyte, supply of electrical energy to the anodes and cathodes, periodic supply of new cathode blanks, removal of finished cathodes, and supply of soluble anodes, and periodic removal of anode slime.

The technological principles to achieve these needs are similar in the various refining and electrowinning operations as follows:

(a) *Anodes.* The soluble anodes cast from the impure metal are usually 80–100 cm wide, 90–110 cm long, and 3–6 cm thick, depending on the anode life cycle and operating current density. They are continuously cast using automated equipment developed by some industries, such as Mitsubishi, Outokumpu, and some other companies. The anodes used in electrowinning from sulfate solutions are usually made of lead alloys. Zinc electrowinning anodes are Pb-Ag alloys containing 0.5 % Ag and 0.5 % Pb. Anodes made of Sb are used for copper electrowinning. Coated titanium anodes are not widely used for electrowinning because of the high cost of the oxygen evolving dimensionally stable anodes (DSA's), although they are used in some small electrolytic cells for electrowinning of Co and Ni.

(b) *Cathodes.* Two types of cathodes are used for the electrodeposition of metals: reusable cathode blanks and sacrificial starter sheets. The reusable blanks, used in the zinc and cobalt electrowinning industry, are aluminum and stainless steel, respectively. Other electrowinning and refining operations use starter sheets, prepared by deposition on copper, stainless steel, or titanium blanks for ~ 24 hrs in a separate section of the tank house. Starter sheets are stripped from the blanks manually or by automatic stripping machines and prepared for use in the cells. Most of these operations are presently automated.

(c)  *Electrical circuits.* The most important aspect of the electrical circuitry in these operations is ensuring that all cathodes receive the same and uniform current density. This is achieved by using high quality anodes and cathodes, uniform electrode spacing, clean contacts, and suitable electrolyte composition having high throwing power.

(d)  *Electrolyte circulation.* In refining cells, electrolyte circulation is needed to provide a constant supply of addition agents, maintain a constant temperature and to ensure proper composition by removing the soluble impurities. Copper refining operations circulate the electrolyte through steam heat exchangers and addition agent feed tanks at 20–40 $l$/m per cell. A small part of this stream is treated to remove soluble impurities such a Ni, As, and Sb. Soluble impurities in nickel-refining are removed from the common anode compartment, purified and returned to the cathodes boxes of the Hybinette cells. In electrowinning operations, electrolyte circulation ensures a constant metal ion concentration in the cathode compartment. Transferring the spent electrolyte to the chemical replenishing operation and returning it as pregnant electrolyte achieves this requirement.

(e)  *Slime handling.* In copper and nickel-refining cells, the anode slimes from the dissolving anodes are accumulated at the cell bottom. They are removed either manually or by submerged "vacuum cleaners". The anode slimes in lead and nickel refining, with a sulfide anode, remain on the anode and are removed with a slime removal machine with water spray and rotating brushes.

(f)  *Material handling.* A 500 tons/day copper refinery typically handles about 2000 new cathodes, 2000 spent anodes, 4000 starter sheets, and 4000 finished cathodes. Because of the large number of electrodes involved in these operations, efficient material handling systems are required. These operations are presently automated.

(g)  *Electrolyte mist in electrowinning.* Oxygen generated at the anodes forms a fine mist of electrolyte and makes the tank house atmosphere unhealthy. The maximum mist content is set by some local authorities to ~ 1mg of major electrolyte component (e.g., $H_2SO_4$) per 1 $m^3$ of air above the cells. While some copper electrowinning operations use frothing agents such as Dowfax 2AO, some use several layers of plastic balls to suppress the mist. Nickel and zinc electrowinning circuits use forced ventilation systems or properly ventilated hoods.

Figure 3.14 shows a schematic of a typical electrorefining operation. Table 3.12 depicts the operating conditions[5] for the electrorefining of Cu, Ni, Pb, Co, and Sn, and Table 3.13 describes the components of cell voltage for the refining and electrowinning of Cu, Ni, and Zn.

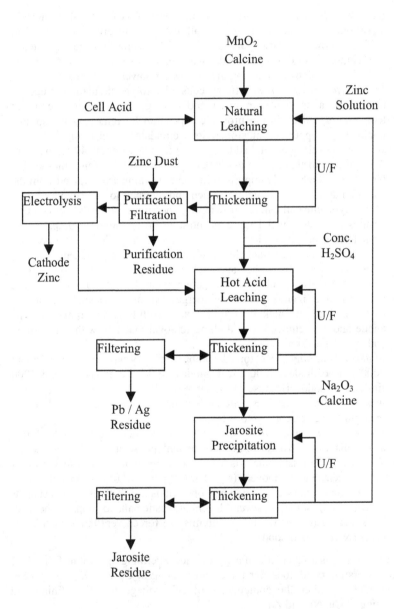

**Figure 3.14.** Typical leaching-jarosite precipitation flow sheet. Reprinted from Reference 29.

## TABLE 3.12
### Operating Parameters for Electrorefining of Metals

| Metal | Concentration $g/dm^3$ | Current density $mA/cm^2$ | Cell Voltage V | Temperature °C | Current efficiency % | Impurities | |
|---|---|---|---|---|---|---|---|
| | | | | | | slime | solution |
| Copper | $CuSO_4$ (100-140) $H_2SO_4$ (180-250) | 10-20 | 0.15-0.30 | 60 | 95 | Ag, Au, Ni, Pb, Sb | Ni, As, Fe, Co |
| Nickel | $NiSO_4$ (140-160) NaCl (100) $H_3BO_3$ (10-20) | 15-20 | 1.5-3.0 | 60 | 98 | Ag, Au, Pt | Cu, Co |
| Cobalt | $CoSO_4$ (150-160) $Na_2SO_4$ (120-140) NaCl (15-20) $H_3BO_3$ (10-20) | 15-20 | 1.5-3.0 | 60 | 75-85 | --- | Ni, Cu |
| Lead | $Pb^{2+}$ (60-80) $H_2SiF_6$ (50-100) | 15-25 | 0.3-0.6 | 30-50 | 95 | Bi, Ag, Au, Sb | --- |
| Tin | $Na_2SnO_3$ (40-80) NaOH (8-20) | 5-15 | 0.3-0.6 | 20-60 | 65 | Pb, Sb | --- |

**TABLE 3.13**
**Typical Energy Requirements in Electrorefining and Electrowinning**

| | Copper | | Nickel refining | | Zinc electrowinning |
|---|---|---|---|---|---|
| | Refining | Winning | Metal anode | Sulfide anode | |
| Current density, $kA/m^2$ | 2.1 | 3 | 2 | 2 | 5.7 |
| Current efficiency, % | 97 | 85 | 96 | 96 | 90 |
| Cell voltage, V | 0.28 | 2.0 | 1.9 | 3.7 | 3.5 |
| Reversible cell potential, V | 0 | 0.9 | 0 | 0.35 | 2.0 |
| Cathode overpotential, V | 0.08 | 0.05 | 0.25 | 1.5 | 0.15 |
| Anode overpotential, V | 0.03 | 0.6 | 0.3 | 0.25 | 0.6 |
| Ohmic drops, $IR$, V | 0.1 | 0.4 | 1.05* | 1.25* | 0.5 |
| Cell hardware, $IR$, V | 0.07 | 0.5 | 0.3 | 0.35 | 0.25 |
| Energy consumption, kW h/kg | 0.25 | 2 | 1.9 | 3.5 | 3.3 |

*with diaphragm

**TABLE 3.14**
**Breakdown of Capital Costs for Electrorefining**

| | $ per annual ton |
|---|---|
| Anode weighing, straightening, lug milling, sampling equipment | 50 |
| Electrolysis equipment, includes transformers, rectifiers, cells | 250 |
| Electrolyte purification and circulation equipment | 100 |
| Cathode preparation equipment | 25 |
| Materials handling and anode casting equipment | 75 |
| Total | 500 |

**TABLE 3.15**
**Breakdown of Capital Costs for Cu Electrowinning***

|  | $ per annual ton of Cu |
|---|---|
| Leach heap pad, piping, pumps, collection ponds, etc. | 270 |
| Solvent extraction plant | 310 |
| Electrowinning plant | 250 |
| Pb-Sn-Ca anodes | 50 |
| Stainless steel mother blank cathodes | 60 |
| Cranes and other moving equipment | 10 |
| Cathode handling equipment | 70 |
| Control system for solvent extraction and electrowinning plants | 50 |
| Engineering services, contingency, etc. | 370 |
| Total | 1440 |

*by the heap leach/solvent extraction/electrowinning route

## 3.5.3. Economics

The capital costs vary significantly for different mining operations because of differences in ore grades, mining methods, ground conditions, and mine sizes. The costs provided in this Chapter are approximate and are based on several assumptions. Typically, the capital cost[30] of a new copper refinery, in 1993 dollars, is about $500/annual ton of electrorefined cathodes, whereas the capital cost for a heap leach/solvent extraction and electrowinning plant is $1440/annual ton of copper. The breakdown of these costs is presented in Tables 3.14 and 3.15. The direct operating costs for producing electrorefined cathode are $0.10 per kg of copper, of which 20% is for electricity, 40% for manpower, 25% for maintenance materials, and the rest for overheads. The breakdown of the copper electrowinning costs is presented in Table 3.16.

The capital costs[31] for zinc production are $2500/annual ton for production levels of 100,00 tons/year, while the operating costs are $0.45–$0.70/kg of zinc. The electricity cost is about 34% of the operating cost, while labor accounts for 32% and others constitute the rest.

## TABLE 3.16
### Breakdown of Operating Costs for Cu electrowinning*

|  | $ per kg of Cu |
|---|---|
| Heap operation and maintenance | 0.09 |
| Sulfuric acid | 0.03 |
| Solvent-extraction plant operation and maintenance | 0.03 |
| Reagent make-up | 0.04 |
| Electrowinning-tankhouse operations, includes electrical energy and maintenance | 0.13 |
| Overheads | 0.05 |
| Total | 0.37 |

*by the heap leach/solvent extraction/electrowinning route

### 3.5.4.  Benefits to or from Fuel Cell Technology

Table 3.17 shows the process energy (PE) requirements to produce Al, Cu, Zn, Mg, and Ni. These PE values are expressed as total kJ equivalents of primary fuel requirements to make these metals and include the fuel requirements for mining, crushing, processing, production of reagents, and for the electricity consumed. Electricity is assumed to be produced in a fossil fuel-fired power plant at an average conversion efficiency of 32%, which corresponds to 11,100 kJ of primary fuel per kWh. It is interesting to note that the electric energy usage for electrolysis is significant for Al and Zn, whereas copper requires a small fraction of the PE. Therefore, the price of aluminum is more sensitive to the cost of energy when compared to the other metals. However, declining grades of available ores for Cu, Zn, and other metals will increase the overall PE, and thus the selling price will be dictated by the cost of energy. A comparison of the PE values[32] with the theoretically needed chemical energy to convert the ore into metal shows (see

### Table 3.17
#### Energy Use in Electrometallurgy

| Metal | Total PE MM kJ/ton | Energy use in electrolysis | |
|---|---|---|---|
|  |  | Electrolysis | Others |
| Al | 284 | 195 | 33 |
| Cu | 130 | 3 | 3 |
| Zn | 70 | 44 | 1 |
| Mg | 308 | 186 | 5 |
| Ni | 167 | 20 | 7 |

**TABLE 3.18**
**Comparison of Process Energy (PE) and Free Energy**

| Reaction | Energy (MM kJ/ton of metal) | |
|---|---|---|
| | $\Delta G^*$ at 25 $^0$C | PE |
| $2\ Al_2O_3{\cdot}H_2O + 3\ C \rightarrow 4\ Al + 3\ CO_2 + 2\ H_2O$ | 19 | 284 |
| $2\ CuFeS_2 + 5\ O_2 + SiO_2 \rightarrow 2\ Cu + 4\ SO_2 + 2\ FeO{\cdot}SiO_2$ | -11 | 130 |
| $ZnS + O_2 \rightarrow Zn + SO_2$ | - 1.5 | 70 |

*$\Delta G$ refers to the free energy change of the reaction.

Table 3.18) that the chemical energy is less than 10% of PE for Al and is negative (energy released) for Cu and Zn. The large PE values are a consequence of the separation technologies required for processing the ore to make the electrolyte for electrolysis. Therefore, there is a significant opportunity to reduce the overall energy requirements by developing new technologies not only for electrolysis, but also for pre-electrolysis steps.

A significant portion of the cell voltage of Zn, Cu, Ni, and Co electrowinning cells is the reversible voltage of the oxygen evolution reaction and the overpotential associated with it:

$$2H_2O \rightarrow O_2 + 4H^+ + 4e_o^- \qquad E_r = 1.23\ V \qquad (3.41)$$

Various alternate half-cell reactions, examined in an attempt to lower the overall energy consumption for the production of these metals, are:

$$CuCl_3^{2-} \rightarrow CuCl^+ + 2Cl^- + e_o^- \qquad E_r = 0.47\ V \qquad (3.42)$$

$$Fe^{2+} \rightarrow Fe^{3+} + e_o^- \qquad E_r = 0.77\ V \qquad (3.43)$$

$$H_2SO_3 + H_2O \rightarrow SO_4^{2-} + 4H^+ + 2e_o^- \quad E_r = 0.17\ V \qquad (3.44)$$

$$CH_3OH + H_2O \rightarrow CO_2 + 6H^+ + 6e_o^- \quad E_r = 0.03\ V \qquad (3.45)$$

$$H_2 \rightarrow 2H^+ + 2e_o^- \qquad E_r = 0.00\ V \qquad (3.46)$$

The cuprous/cupric chloride anode reaction is commercially practiced in the Duval electrowinning process. It is economical as it combines the low half-cell potential with one electron participation. This anode reaction, in addition to lowering the cell voltage, also regenerates $CuCl^+$ used in the process. The cell voltage of this electrowinning process is 1 kWh/kg of copper. Reactions (3.42) to (3.46) have been extensively studied to replace the oxygen evolution reaction. While this approach is theoretically attractive, none of them have become commercial as yet. Slow kinetics of reaction (cf. reaction 3.43) and poor cathode efficiency, because of the reduction of $Fe^{3+}$ during the reaction, were attributed as reasons for abandoning these two schemes. The energy savings with the other schemes appear to be relatively modest compared to the costs and operating problems associated with implementing these alternate anode reactions. It is important to note that studies with the hydrogen depolarized anode have shown a cell voltage lowering from 3.5 V to 1.55 V at 150 A/m$^2$, a reduction of 65% in energy savings for the zinc electrowinning process.[33]

## 3.6.    CORROSION INHIBITION/PASSIVATION

### 3.6.1.   Background

Corrosion of materials in buildings bridges, automobiles, ships, pipelines, etc. is very extensive and the expenses incurred in replacing these materials and in methods used to inhibit corrosion, including passivation, are enormous. In several countries these account for a few percent of the Gross National Product. In the USA, this expense is over $300 billion. An understanding of these phenomena and their technologies is most essential in fuel-cell research and development. Stability of the component materials for a fuel-cell power plant, from the single cell to the stack level, is vital for the attainment of the desired lifetimes. For the power generation application, the goal set for the operating lifetime of the electrochemical cell stack by the US Department of Energy is 40,000 h; for the transportation application it is about 3000 h, over a period of 5 years; and for the portable power it is in the range of 1000 to 10,000 h. The problem of corrosion is encountered in fuel cells because of the attack of electrodes and current collectors by the acidic, alkaline, or molten carbonate electrolytes, particularly in the oxidizing environment of the oxygen electrode. Even in the solid oxide fuel cells, corrosion problems are encountered.

There are excellent books, chapters in books, review articles, and original papers in journals about the topics of corrosion inhibition and passivation. To be more specific, the reader is referred to *Corrosion Engineering* by Green and Fontana; *Handbook of Corrosion* by Uhlig; *Handbook of Corrosion Engineering* by Roberge; Chapter 8 in the book *Surface Electrochemistry—A Molecular Level Approach* by Bockris and Khan; and a section of a chapter in *Electrochemistry* by Rieger.[34]

Within the limited space available for this Section, we shall deal with (i) the principles of the technologies dealing with corrosion inhibition and passivation and (ii) the benefits to and from fuel cell technologies. The economic aspects and other applications are dealt in the above-mentioned references in great detail.

## 3.6.2. Principles of Technologies

Corrosion is caused by the oxidation of metal. The cathodic reaction in the oxidation of a metal, like iron, is:

$$Fe \rightarrow Fe^{2+} + 2e_o^- \tag{3.47}$$

while the anodic reaction is either reduction of oxygen,

$$O_2 + 4H^+ + 4e_o^- \rightarrow 2H_2O \tag{3.48}$$

or the evolution of hydrogen:

$$2H^+ + 2e_o^- \rightarrow H_2 \tag{3.49}$$

The rates of these reactions are dependent on the environment that the metallic elements are exposed to. The pH of the medium and the oxygen concentration in the environment also have strong influences on the corrosion rate. The presence of halide (mostly chloride) ions accelerates the corrosion rate.

A most valuable method of assessing the stability of materials is by using Pourbaix diagrams. It presents a thermodynamic perspective of the stabilities of the metals and their oxides as a function of potential and of pH. A potential vs. pH diagram (Pourbaix diagram) for iron is illustrated in Figure 3.15. This figure shows the region of stability of the metal and its oxides as a function of the potential and

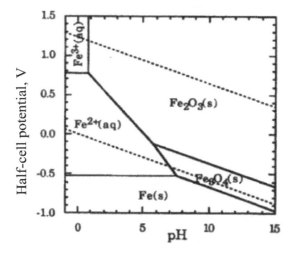

**Figure 3.15.** Potential vs pH diagram (Pourbaix diagram) for iron. Reproduced with permission from Reference 34.

the pH. To construct these diagrams, arbitrary values are assumed for the metal-ion concentration ($10^{-6}$ m/$l$) and the pressures of oxygen and hydrogen (1 atm). There will be a shift in these curves with the variations of these concentrations.

The corrosion rates, however, depend on the kinetics of the metal dissolution reaction and the rates of either the hydrogen evolution or oxygen reduction. If the reversible potential for the metal dissolution reaction is considerably more cathodic than that for the hydrogen evolution reaction, the anodic reaction becomes dominant. The reaction rate also depends on the pH of the liquid medium that the metal or alloy is exposed to. Most metals are stable at higher pH. But then, there is a take-over by the oxygen reduction as the cathodic reaction and corrosion of the metal or alloy will still occur.

As stated in Chapter 1 Section 1.3, a *mixed potential* is set up when corrosion occurs. At this potential, the rate of metal dissolution is equal to the rate of the oxygen reduction. The mixed potential ($V_{corr}$) is expressed by the equation:

$$V_{corr} = \frac{RT}{F} \ln \left\{ \frac{i_{O,H} \exp\left(\dfrac{V_{r,H}F}{2RT}\right) + i_{O,M} \exp\left(\dfrac{V_{r,M}F}{2RT}\right)}{i_{O,H} \exp\left(\dfrac{-V_{r,H}F}{2RT}\right) + i_{O,M} \exp\left(\dfrac{-V_{r,M}F}{2RT}\right)} \right\} \tag{3.50}$$

where $i_{O,H}$ and $i_{O,M}$ are the exchange current densities for the hydrogen evolution and metal dissolution reaction, respectively, $V_{r,H}$ and $V_{r,M}$ are the reversible potentials for these reactions, and the transfer coefficient ($\alpha$) of both the reactions is assumed to be ½. The corrosion current density is expressed by the equation:

$$i_{corr} = i_{O,M} \left( \left\{ \frac{i_{O,H} \exp\left(\dfrac{-V_{r,H}F}{2RT}\right) + i_{O,M} \exp\left(\dfrac{[V_{r,M}-V_{r,H}]F}{2RT}\right)}{i_{O,H} \exp\left(\dfrac{-V_{r,H}F}{2RT}\right) + i_{O,M} \exp\left(\dfrac{-V_{r,M}F}{2RT}\right)} \right\}^{-\frac{1}{2}} \right.$$

$$\left. \left\{ \frac{i_{O,H} \exp\left(\dfrac{-V_{r,H}F}{2RT}\right) + i_{O,M} \exp\left(\dfrac{[V_{r,M}-V_{r,H}]F}{2RT}\right)}{i_{O,H} \exp\left(\dfrac{-V_{r,H}F}{2RT}\right) + i_{O,M} \exp\left(\dfrac{-V_{r,M}F}{2RT}\right)} \right\}^{\frac{1}{2}} \right) \tag{3.51}$$

It is only when $V_{r,H} > V_{r,M}$ that $i_{corr}$ is positive.

From an experimental point of view, it is simple to measure the corrosion potentials and the corrosion current densities. One may use either the linear polarization method or the extrapolation of the Tafel lines for the anodic and cathodic reactions. A third method is to use electrochemical impedance spectroscopy (see Chapter 6).

For corrosion inhibition, several procedures have been employed. One method is by *cathodic protection*—i.e., making the potential of the method more negative by passing a small current between the metal and an auxiliary electrode. Another method is to use a metal with a more negative reversible potential (e.g., zinc) as the counter electrode (*sacrificial anode*) and connecting the two electrodes, as in a galvanic cell. This method is widely used for corrosion protection of bridges on roads and on railway lines and of ships. A second method of corrosion inhibition is by use of *protective coatings*—e.g., paints, plastics, ceramics, or coating of a metal like nickel or chromium, which form insulating oxides or gold films, which are quite stable. A third method is by using organic compounds that adsorb strongly on the metal. These *inhibitors* have the advantages of being able to greatly reduce the rates of the metal dissolution, oxygen reduction, and hydrogen evolution reaction. Typical inhibitors are aliphatic or aromatic amines. Only small quantities of the inhibitor are necessary.

*Passivation* of metals is another important electrochemical phenomena. It occurs on metals or alloys that form stable oxides on the surface. In most cases, these oxides are insulators, but in some cases, even if these passive layers are electronically conducting, they prevent the dissolution of the metallic substrate. The passivation of a metal can best be illustrated by the potential vs. current density plot on a metal, when it is polarized anodically (Figure 3.16). As the potential increases, the current increases due to metal dissolution. A further increase of potential then causes a sharp decrease in the current density due to the formation of a stable oxide film. The potential at which the current starts to decrease is referred to as the Flade potential. There is then an appreciable region of potential when the current has a low constant value—the passivation region. A still further increase of potential causes a fairly rapid increase of the current—i.e., in the transpassive region. This is effectively caused by the passive oxide layer transforming to a higher oxide that is ionically conducting (sometimes even electronically conducting). This type of behavior typically occurs with some transition metal oxides—e.g., cobalt, nickel, chromium.

### 3.6.3.  Benefits to or from Fuel Cell Technologies

The competing reactions, which occur during corrosion of a metal are either/both the hydrogen evolution and oxygen reduction reactions. An understanding of the mechanism and kinetics of these reactions in Fuel Cell R&D can be beneficial to inhibit corrosion. For example, carbon monoxide and sulfur species inhibit the hydrogen oxidation reaction. Organic impurities also greatly slow

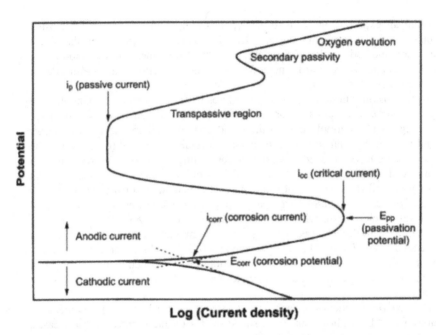

**Figure 3.16.** Typical potential vs current density plot for a metal or allow exhibiting the corrosion, passivation and transpassive regions – arbituary units. Reprinted from Reference 35, Copyright (1999) with permission from The McGraw-Hill Companies.

down the kinetics of the oxygen reduction reaction. The evaluation of these types of species in corrosion inhibition may be valuable.

One problem encountered in electrochemical-cell stacks is corrosion or passivation of metallic bipolar plates (e.g., steel in the former case or aluminum in the latter case). Gold-plating of these metallic elements has been evaluated. This procedure may be too expensive for the large-scale manufacture of fuel cells. Furthermore, if there are micropores in the gold film, local corrosion of the substrate could occur. An alternative is to use graphite with a binder for coating of these bipolar plates. One other problem found in fuel cells is the slow corrosion of the carbon support in the low and intermediate temperature fuel cells (e.g., PEMFC, AFC, PAFC). The corrosion rate is higher when the potentials of the oxygen electrode are closer to the open circuit potential. Several methods have been examined to reduce the corrosion rate—e.g., high temperature heat treatment to graphitize the carbon or treatment with nitric acid. It may be worthwhile using the former method for corrosion inhibition.

## 3.7.    ELECTROCHEMICAL-ENERGY STORAGE

### 3.7.1.   Background

Basically, there are two types of electrochemical-energy storage technologies: batteries and electrochemical capacitors. The common feature of both these technologies is that chemical energy is directly converted into electricity, just as in the case of fuel cells. The main difference between electrochemical-energy conversion (fuel cells) and electrochemical energy (batteries and electrochemical capacitors) is that in the former the oxidant and the reductant are fed into the electrochemical cell stack, while in the latter, both the reactants are in a sealed container, with a separator keeping the oxidant and the reductant apart. The distinctive features of both of these technologies are dealt with in Chapter 4, Section 4.2.

The battery technology is one of the oldest electrochemical technologies and dates back to the middle of the 19th century. Lead acid batteries had reached the commercialization stage in the latter part of this century. Batteries could be subdivided into four types:

- primary,
- secondary,
- reserve, and
- thermal batteries.

In this chapter we shall focus on the first two types. Short descriptions will be made on the other two types. As in all electrochemical cells, batteries (as well as capacitors) contain an anode, a cathode, and an electrolyte, with a separator containing the electrolyte between the anode and cathode to prevent the transport of the active materials (oxidant and reductant) from one electrode to the other. In essence, a battery is a galvanic cell, and it is desirable to have a high thermodynamic reversible potential. When power is needed from the battery, it is connected to an external load. The main difference between a primary and secondary battery is that in the former, it is the desired amount of chemical energy (and hence, the resultant electrical energy) that is stored in the active materials. In a secondary battery, once the oxidant and reductant of the active materials are depleted, their activities can be regenerated by charging the batteries with electrical energy.

Electrochemical capacitors are similar to secondary batteries (rechargeable). The main difference is that the amount of electric energy storage is considerably less than in secondary batteries. Electrochemical capacitors have the advantage of attaining very high rates of delivery of electrical power.

There are a multitude of types of primary and secondary batteries and probably several million publications on these technologies. One of the most valuable publications is the *Handbook of Batteries*,[36] which provides a comprehensive survey of all types of batteries. Within the space allocated for this Section, only a

**TABLE 3.19**
**Types of Batteries, Selected for Brief Analysis in Section 3.7**

| Primary | Secondary | Reserve |
|---|---|---|
| Zinc/Manganese | Lead Acid | Magnesium-water activated |
| Zinc/Air | Nickel/Hydrogen | |
| Lithium/Thionyl Chloride | Nickel/Cadmium Nickel/Metal Hydride Lithium/Ion | |

small fraction of these technologies is summarized. Table 3.19 presents the types of batteries chosen for this brief analysis. The main reason for their selection is that these types of batteries are relevant to fuel cells, in respect to technical, competing and synergistic (hybrid fuel cell/batteries) aspects. Section 3.7.6 is devoted to electrochemical capacitors.

According to a recent review,[7] the worldwide sales of batteries amounted to $39 billion in 2001; two thirds of this amount were from secondary batteries. Some interesting techno-economic aspects of the batteries chosen for the description in this chapter are found in Table 3.20.

**TABLE 3.20**
**Some Techno-Economic Aspects of Primary and Secondary Batteries.**

| Type of battery | Operating temperature (°C) | Cycle life | Cost ($/kW/h) | Worldwide sales (million $), (manufacturer's cost) |
|---|---|---|---|---|
| Primary | | | | |
| $Zn/MnO_2$ | 0 – 50 | N/A | 1000 | |
| Secondary | | | | |
| Lead Acid | -40 – 50 | 200 – 1500 | 200 | 13,000 |
| Alkaline Ni (Negative electrode Cd, MH) | -20 – 50 | 300 – 600 | 400 | 3,500 |
| Lithium-Ion | -20 – 40 | >1000 | 800 | 2,500 |

## 3.7.2.  Primary Batteries

### 3.7.2.1. Zinc/Manganese Dioxide.

(a)  *Background.* Historically, the Leclanche cell, based on zinc-manganese dioxide, was first invented. It had the active materials, separated by a gel electrolyte, containing ammonium chloride as the electrolyte. Since the 1950 s, the most commercialized zinc-manganese dioxide batteries have an alkaline electrolyte. These batteries contain electrolytically produced manganese dioxide (positive electrode), the needed polymer gelling agents for the potassium hydroxide electrolyte (8 M KOH) and powdered zinc for the negative electrode.

(b)  *Principles of technology.* The overall reaction, the standard thermodynamic reversible potential, the specific energy and energy density of alkaline Zn-MnO$_2$ batteries, along with those for the other batteries considered in this Chapter, are tabulated in Table 3.21. The half cell reactions are as follows:

Anode:       $Zn + 2\ OH^- \rightarrow ZnO + H_2O + 2\ e^-$                    (3.52)

Cathode:     $2\ MnO_2 + H_2O + 2\ e^- \rightarrow Mn_2O_3 + 2\ OH^-$           (3.53)

The cathode material generally contains manganese dioxide powder (70 to 80%), graphite powder (7 to 10%) and acetylene black (1 to 3%). The carbon materials enhanced the electronic conductivity of the electrodes. Binder materials (6% Portland cement and fibers) are included for mechanical strength. The separator material (in gel form, e.g., carboxymethylcellulose) has two functions—one to have good absorption of the KOH electrolyte for high ionic conductivity and the other to prevent zinc penetration. The zinc powder is contained in a polymer gel.

In respect to the design of the battery, the most common is the cylindrical one (types A, AA, AAA, B, C, D). Figure 3.17 illustrated in a review by Kordesch, is reproduced in this Chapter. These batteries have an open circuit potential of about 1.5 V, which is close to the reversible potential. The capacity is generally about 10 Ah for the D size battery. The other type of design is the prismatic one (Figure 3.18). A large percentage of batteries with this design have an open circuit potential of 9 V, which means that six cells are connected in series externally (a unipolar design). Typical *discharge plots* (cell potential–time plots) for the AA-cells under low and high rates of discharge are illustrated in Figure 3.19.

(c)  *Economics and applications.* The Zn-MnO$_2$ battery is a relatively inexpensive one and is the most widely used primary battery. However, the cost of these batteries as a function of their energy is high ($1/Wh),

**TABLE 3.21.**

**Preformance Characteristics of the Selected Batteries for Brief Analysis in this Chapter**

| Type of battery | Cell reaction | Standard thermodynamic reversible potential, V | Specific energy (attained), Wh/kg | Energy density (attained), Wh/l |
|---|---|---|---|---|
| **Primary** | | | | |
| Zn-MnO$_2$ | Zn + 2 MnO$_2$ $\rightarrow$ ZnO + Mn$_2$O$_3$ | 1.5 | 145 | 400 |
| Zn-Air | Zn + ½ O$_2$ $\rightarrow$ ZnO | 1.65 | 370 | 1300 |
| Li-SOCl$_2$ | 4 Li + 2 SOCl$_2$ $\rightarrow$ 4 LiCl + S + SO$_2$ | 3.65 | 590 | 1300 |
| **Secondary** | | | | |
| Lead Acid | Pb + PbO$_2$ + 2 H$_2$SO$_4$ $\rightarrow$ 2 PbSO$_4$+ 2 H$_2$O | 2.1 | 35 | 70 |
| Ni-H$_2$ | H$_2$ + 2 NiOOH $\rightarrow$ 2 Ni(OH)$_2$ | 1.5 | 55 | 60 |
| Ni-Cd | Cd + 2 NiOOH $\rightarrow$ 2 Ni(OH)$_2$ + Cd(OH)$_2$ | 1.35 | 35 | 100 |
| Ni-MH | MH + NiOOH $\rightarrow$ M + Ni(OH)$_2$ | 1.35 | 75 | 240 |
| Lithium Ion | Li$_x$C$_6$ + Li$_{1-x}$CoO$_2$ $\rightarrow$ LiCoO$_2$ + C$_6$ | 4.1 | 150 | 400 |
| **Reserve** | | | | |
| Mg-water activated | Mg + 2AgCl $\rightarrow$ MgCl$_2$ + 2Ag | 1.4 – 1.6 | 130 | 200 |

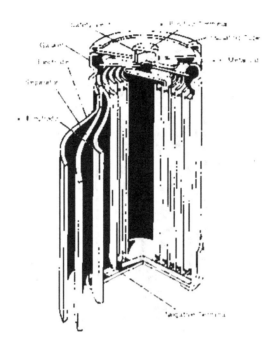

**Figure 3.17.** Schematic of a cylindrical design for a battery; this design is similar for primary and secondary batteries. Reprinted from Reference 36, Copyright (2002) with permission from The McGraw-Hill Companies.[36]

and they generally have rather low capacities. These batteries have a wide variety of applications (motor toys, intermittent lighting, radios, power-packs for soldiers, etc.). But due to their short lifetimes, there is interest in making these rechargeable and such types have reached the commercialization stage. There is also competition from primary lithium and zinc/air batteries for these applications.

(d) *Benefits to or from fuel-cell technologies.* At the present time, there is no benefit to or from fuel cell technologies from $Zn/MnO_2$ batteries. But it is worthwhile mentioning that for military applications (power-packs for soldiers), there is some interest in replacing these batteries with PEMFCs and DMFCs because of the considerably higher energy densities for the two latter type of batteries than that for the former.

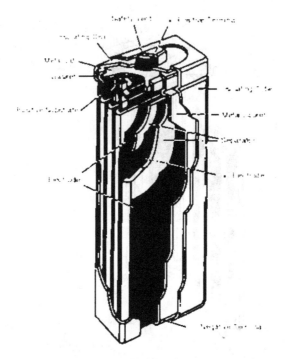

**Figure 3.18.** Schematics of a prismatic design for a battery. This design is similar for primary and secondary batteries. Reproduced from Reference 36, Copyright (2002), with permission of The McGraw-Hill Companies.

### 3.7.2.2. Zinc/Air

(a) *Background.* The most advanced of the metal-air batteries is the one with air for the positive electrode (as in the case of a fuel cell) and with zinc for the negative electrode. The second most advanced one is the aluminum-air battery. These batteries are attractive because they have relatively high energy densities.

(b) *Principles of technology.* A zinc/air battery contains a positive air electrode and a negative zinc electrode and an alkaline electrolyte (8 M KOH). The electrooxidation of zinc occurs according to the electrochemical reaction as represented by Eq. (3.52). The oxygen reduction reaction is the same as that in an alkaline fuel cell, i.e.,

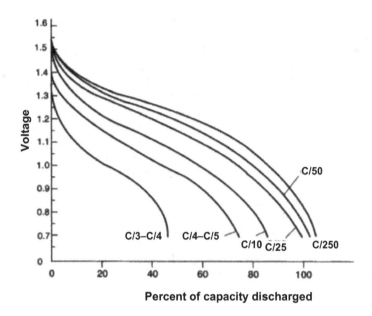

**Figure 3.19.** Typical discharge plots for an AA Zn/MnO$_2$ battery under different rates of discharge. Discharge rates are expressed as a fraction of C, defined as the discharge current in amps divided by the rated capacity of the battery in amp-hour. Reproduced from Reference 36, Copyright (2002), with permission of The McGraw-Hill Companies.

$$O_2 + 2H_2O + 4e^- \rightarrow 4\ OH^- \tag{3.54}$$

High surface area carbon has most widely been used as the electrocatalyst for oxygen reduction. But at this electrocatalyst, the predominant reaction leads to hydrogen peroxide, as represented by,

$$O_2 + H_2O + 2e^- \rightarrow\ OH_2^- + OH^- \tag{3.55}$$

The coulombic efficiency for this reaction is half that for oxygen reduction. Thus, in recent times, there has been a trend towards using high surface area platinum supported on high surface area carbon as the electrocatalyst. The electrocatalyst loading can be as low as 0.2 mg/cm$^2$. Alternative electrocatalysts that give a high performance are heat-treated organic macrocyclics (e.g., cobalt tetraphenyl porphyrin) and high surface area nickel-cobalt spinels.

In the original zinc-air cells, a large plate of amalgamated zinc was used for the negative electrode. Presently, the anode consists of zinc powder, potassium hydroxide contained in a polymeric gel, and a current collector. Initially, carbon plates were used for the cathode but currently with the advances in fuel cell technology, porous gas diffusion electrodes, as in alkaline fuel cells, are used. For the current collector, a metallic grid is used. A multi-layer polymeric absorbent is used for the separator containing the electrolyte. On the outer surface of the air electrode, a thin polymeric layer is formed to inhibit the transport of carbon dioxide and water into the cells. Original zinc/air cells contained a complex manifold and internal structure for airflow. An essential criterion in the design of the battery is to optimize the flow of air into the porous gas-diffusion cathode. In the miniature cells, the bottom of the container has air inlet holes.

An innovative design was researched for a zinc/air battery by Appleby, Jacqueline, and Pompon.[37] Powdered zinc, contained in the circulating electrolyte, was fed to a cell with a tubular design. The inner wall of the metallic tube served as the current collector for the zinc electrode. The outer wall contained a separator material and the air electrode. Thus, this cell functioned very much like a fuel cell with the electric power being generated only when the reactants were delivered to the cell. The cell performed reasonably well. The tubes had a constant cross-section. The main overpotential losses were due to activation overpotential at the air electrode and ohmic resistance. One reason for the high ohmic resistance was the high contact resistance (electronic) of the slurry particles with the current collector. A bundle of tubes were connected in series and in parallel to attain the desired potentials of the cell stacks and total currents. It was proposed that the zinc oxide produced by the anodic reaction could be reduced externally (electrochemically or using thermal energy). Specific energy of 110 Wh/kg was obtained at the 3h discharge rate. The energy efficiency of the stack was relatively low (about 40%) because of high activation overpotential and ohmic overpotential losses.

(c)   *Applications and economics.* Historically, zinc/air batteries were first used as power sources for railway signal and warning devices. These were soon replaced by the better performing, more reliable, and long lifetime nickel-iron batteries. Since the mid-fifties, the main application of the zinc/air batteries has been for hearing aids. Button-type cells are used for this purpose. The tubular design zinc/air batteries were proposed as attractive batteries for electric vehicles. As stated in Principles and Technology of Zn-Air Batteries, the energy density and efficiency are far too low for this application. Aluminum/air batteries, with a higher energy density, have also been proposed for this application. On a $/kWh basis, the costs for the portable electronic application is high ($\approx$ $1000/kWh), but it has been projected that for the automobile power application, the cost will be about $400/kWh.

(d)  *Benefits to or from the fuel-cell program.* From the preceding description of this type of battery, what is most visible is that the development of highly efficient air electrodes for alkaline fuel cells has greatly benefited the state-of-the-art zinc/air batteries.

### 3.7.2.3. Lithium

(a)  *Background.* Lithium is the most positive element and it is the lightest of all metals. If it is coupled with fluorine, the most electronegative and the lightest of all the group VII elements in the chemical periodic table, this battery will exhibit the highest thermodynamic reversible potential, as well as the highest energy densities. It is practically impossible to develop a lithium battery in aqueous media because of its rapid reaction with water to produce hydrogen. Hence, these batteries have been developed using organic and solid electrolytes. These electrolytes have to be compatible with lithium for chemical stability. The commonly used electrolytes are the lithium salts of perchloric acid, hexafluorophosphoric acid, hexafluoroarsenic acid, tetrafluoroboric acid, and trifluoromethane sulfonic acid. These are dissolved in aprotic solvents such as ethylene carbonate, propylene carbonate (or sometimes the mixture of the two), dimethyl formamide, tetrahydrofuran, and several others. The cathode materials could be in the solid or liquid state. Examples of the former are carbon monofluoride, manganese dioxide, copper oxide, iron sulfide, and silver chromate. Figure 3.20 illustrates the significantly higher specific energies of some of these lithium batteries in comparison with the well-developed alkaline $Zn/MnO_2$ battery.

Primary lithium batteries have also been developed and commercialized with liquid cathodes. The most common liquid cathode cells use sulfur dioxide with thionyl chloride. The liquid cathodes also serve as the solvent. A third type of lithium battery has a solid electrolyte (e.g., lithium iodide) with an organic additive (e.g., polyvinyl pyridine). The electrolyte is mixed with the cathodic reactant (iodine) to form a charge transfer complex. For the purpose of this discussion of a primary lithium battery, we have chosen the lithium-thionyl chloride battery.

(b)  *Principles of technology.* The lithium-thionyl chloride battery has the highest energy density. The reactions occurring in the cell are:

Anode:      $Li \rightarrow Li^+ + e^-$                                              (3.56)

Cathode:    $2\ SOCl_2 + 4\ e^- \rightarrow S + SO_2 + 4\ Cl^-$                       (3.57)

Overall:    $2\ SOCl_2 + 4\ Li \rightarrow S + SO_2 + 4\ LiCl$                        (3.58)

In the negative electrode, a lithium sheet is pressed onto an expanded nickel screen, which serves as the current collector. The anode and

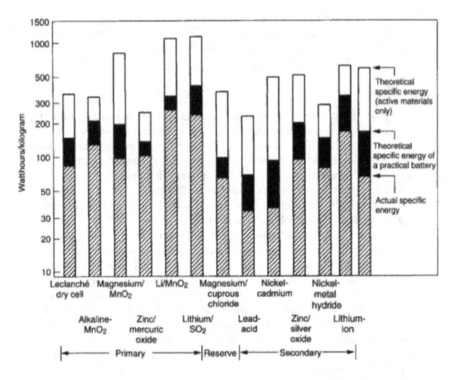

**Figure 3.20.** Specific energies for some primary, secondary, and reserve batteries. Reproduced from Reference 36, Copyright (2002), with permission of The McGraw-Hill Companies.

cathode are separated by glass filter papers; fluorocarbon plastics are used as insulators. In respect to the designs of the cells, the most common are the cylindrical cells (types AA, C, D as in the alkaline $Zn/MnO_2$ cells). For low capacity cells, bobbin and button type cells are used. It is interesting to compare the cell potential vs. capacity plots for typical $Li/SOCl_2$, $Zn/MnO_2$, and Zn/Air batteries (Figure 3.21). The significant advantage of the lithium battery is that its plateau potential is about 3.5 V, as compared with about 1.4 V for the $Zn/MnO_2$ battery.

There are safety measures that have to be taken into consideration in the manufacture of lithium batteries. Water impurities have to be completely avoided. The combination of $SOCl_2$ and water can produce hydrochloric acid and sulfuric acid. Direct contact of the anodic and cathodic reactants can lead to explosions.

(c) *Applications and economics.* Because of the high energy densities and cell voltages, the US Department of Defense uses this type of battery for the same applications as with $Zn/MnO_2$ batteries. Such batteries are also used in watches. The lithium batteries are at least five times as expensive as $Zn/MnO_2$ batteries on a $/kWh basis.

(d) *Benefits to or from fuel-cell technologies.* Since these batteries are of the non-aqueous type, there is no common feature in its development with fuel cell technologies. One noteworthy mention is that for some portable applications, the lithium batteries have an advantage over fuel cells because each cell has an operating potential of over five times as that of a hydrogen/air fuel cell.

### 3.7.3. Secondary Batteries

#### 3.7.3.1. Lead Acid

(a) *Background.* The lead acid battery is the oldest and most commercialized of all types of batteries. Plante invented it in 1859. Its most widespread application is as a starter battery for transportation vehicles.

(b) *Principles of technology.* The overall reaction in a lead acid battery and the standard thermodynamic reversible potential are as stated in Table 3.21. The half cell reactions are expressed by:

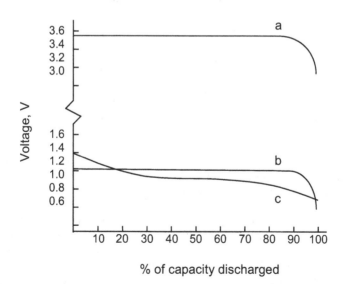

**Figure 3.21.** Typical discharge plots for AA primary batteries (a) $Li/SOCl_2$ (b) Zn-Air (c) $Zn-MnO_2$. Reproduced from Reference 36, Copyright (2002), with permission of The McGraw-Hill Companies.

Positive Electrode:

$$PbO_2 + H_2SO_4 + 2\ H^+ + 2\ e^- \Leftrightarrow PbSO_4 + 2\ H_2O \qquad (3.59)$$

Negative Electrode:

$$Pb + H_2SO_4 \Leftrightarrow PbSO_4 + 2\ H^+ + 2\ e^- \qquad (3.60)$$

From the above equations, it can be seen that during discharge, lead sulfate is the main product for both the cathodic and anodic reactants. Sulfuric acid serves as a reactant as well as an electrolyte in this battery. Water is also a product. Thus, during discharge, there is a dilution of the sulfuric acid electrolyte. The active materials formed during charging are lead, lead dioxide, and sulfuric acid. During charging, there is an increase in the electrolyte concentration. Hence, we can use the change in concentration of the electrolyte as a measure of the state of charge. Because of the highly positive reversible electrode potential of the positive electrode ($V_r$ = 1.65V) and of the high negative reversible electrode potential of the negative electrode ($V_r$ = -0.35V), and those being above and below the reversible potentials for oxygen evolution and hydrogen evolution, respectively, these gases are evolved during charging of the lead acid battery. However, these amounts are relatively small because $PbO_2$ and Pb have very low electrocatalytic activities for these electrochemical reactions. The hydrogen evolution reaction is the main contributor to the self-discharge of the lead acid battery.

The most common cell design has a plate construction. Plate constructions are of two types: grid plate or pasted plate. A grid network contains the active material in pellet form. Lead oxide (PbO) is the starting material for the preparation of the plates. The active materials (Pb and $PbO_2$) in the plates are produced electrochemically. The plates are then assembled in the cell. Individual cells are connected in series to attain the desired battery voltages. Several types of materials have been used for the separators between the anode and cathode—e.g., cellulose, microglass, sintered polyvinyl chloride, synthetic wood pulp. Microporous separators are preferred for starter batteries. The basic functions of the separator material are to retain the electrolyte in their microporous structure and to prevent electronic contact between the positive and negative electrodes.

Auxiliaries include vent caps to exhaust acid fumes and valve plugs to release gas pressure. Small amounts of palladium are used to recombine hydrogen and oxygen produced during charging. Automatic refilling systems are used in traction batteries.

(c) *Applications and economics.* The most common application worldwide of lead acid batteries is as a starter battery for transportation vehicles. These are commonly referred to as SLI batteries (starter-lighter-ignition). There has been great interest in using lead acid batteries for traction. In fact, the electric automobiles in the beginning of the 20th century used lead acid batteries as the power source. But due to the rapid progress made in internal combustion engine and diesel powered engine vehicles, the interest was greatly diminished. There was a revival in the development of lead-acid battery technology for electric vehicles since the late 1980's to the mid 1990's but due to the low range of those vehicles (less than 150 km) and long time (5 to 7 h) needed for charging, the interest for developing such vehicles greatly decreased. However, as seen from the Ragone plot (Figure 3.22), even though lead acid batteries have a low energy density, they could attain high power densities. Thus, these batteries are still being considered as the second power source with the primary internal combustion engine (ICE) or diesel engine power plants for hybrid electric vehicles. But here too, because of the comparably higher energy densities attained with Ni/MH$_x$ and lithium ion batteries, the latter types of batteries are the desired batteries for hybrid electric vehicles (HEVs) at the present time. At a considerably higher energy storage level, lead acid batteries are used for standby/emergency power by telephone companies, hospitals and electric power generation companies.

As indicated in Table 3.20, lead acid batteries have the predominant share of the battery market. The cost of the battery is variable, depending on its design and capacity. The SLI battery costs about $100/kWh, while

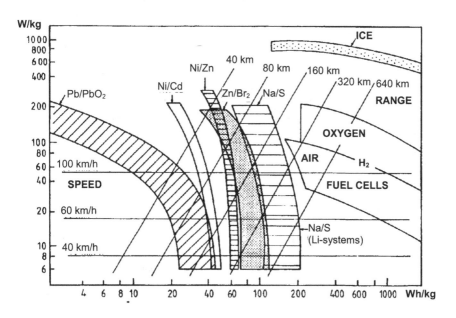

**Figure 3.22.** Ragonne plots for some batteries and fuel cells, an electrochemical capacitor and a PEMFC. Modified from Cited Reference 42.

batteries with relatively low capacity may have a cost five to ten times higher. One attractive feature of the lead acid battery is that practically 100% of its lead content can be recycled. This is a significant feature, considering the fact that lead is a toxic element and thus an environmental pollutant.

(d)  *Benefits to or from fuel-cell technology.* The Ragone plot (Figure 3.22) clearly show that lead acid batteries have a low specific energy density but can attain a high power density while it is the reverse in the case of fuel cells. Thus, in order for the fuel-cell powered vehicles to be competitive with IC or diesel engine powered vehicles, it is necessary for the former to have a hybrid power source. The lead acid battery could serve this function quite well. It still faces stiff competition from $Ni/MH_x$ and lithium ion batteries for this function. But from an economic point of view, the lead acid battery is the more favored choice (see Table 3.20). Another relevance to fuel cells is that fuel cell electrocatalysts could be used for the recombination of hydrogen and oxygen evolved during overcharging of lead acid batteries.

### 3.7.3.2. Nickel/Hydrogen

(a)  *Background and applications.* The $Ni/H_2$ battery contains the best of both worlds in respect to the positive and negative electrodes—one of the best battery electrodes, i.e., the nickel oxide electrode, and the ideal hydrogen electrode, as in a hydrogen oxygen fuel cell. Both these electrodes have very low activation overpotentials, in the charging and discharging modes. Furthermore, the cycle life of the battery can be as high as tens of thousands of cycles. It is for this reason that it is mainly used in satellites in the low earth (LEO) and geosynchronous earth orbits (GEO).

(b)  *Principles of technology.* The overall reaction in a $Ni/H_2$ battery, its standard thermodynamic reversible potential and its specific energy and energy density are as stated in Table 3.21. The half cell reactions are as follows:

Positive Electrode:
$$2\ NiOOH^- + 2\ e^- \rightarrow 2\ NiO + 2\ OH^-$$ (3.61)

Negative Electrode:
$$H_2 + OH^- \rightarrow 2\ H_2O + 2\ e^-$$ (3.62)

During overcharge, oxygen is evolved at the nickel oxide electrode;

$$4\ OH^- \rightarrow O_2 + 2\ H_2O + 4\ e^-$$ (3.63)

The oxygen, however, is reduced chemically and electrochemically at the positive electrode as:

$$2\ H_2O + O_2 + 4\ e^- \rightarrow 4\ OH^- \qquad (3.64)$$

and

$$2\ H_2 + O_2 \rightarrow 2\ H_2O \qquad (3.65)$$

Even though some of the hydrogen evolved during charging has free access to the nickel oxide electrode, it does not cause any problem because of the slow kinetics of electrooxidation of hydrogen on the nickel oxide electrode. For this reaction, it is necessary to have metallic nickel for the initial step of dissociative adsorption of hydrogen. The electrolyte in the cell is 8 M KOH. The cells operate in the temperature range of about 10 to 50 °C. The hydrogen evolved during charge is pressurized

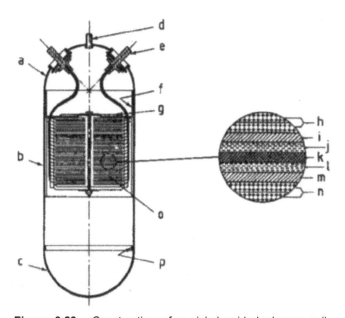

**Figure 3.23.** Construction of a nickel oxide-hydrogen cell, showing details of the cell construction. a) Container top hemisphere; b) Container cylindrical section; c) Container bottom hemisphere; d) Filling tube; e) Terminal; f) Welding ring; g) Pressure plate; h) Positive electrode; i) Separator; j) Negative electrode; k) Gas diffusion screen; l) Negative electrode; m) Separator; n) Position electrode; o) Electrode stack; p) Welding ring. Reproduced from Reference 38, Copyright (1985), with permission from Wiley-VCH.

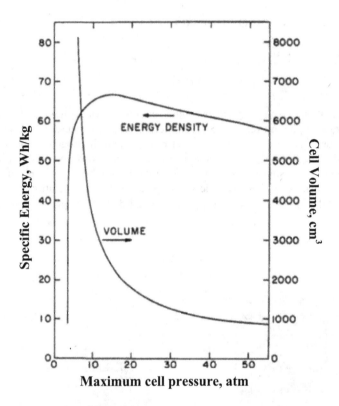

**Figure 3.24.** Dependence of specific energy and cell volume on maximum pressure for a Ni/H₂ battery. Reprinted from Reference 39.

(~ 50 atm). The cells operate in the pressure range 3 to 50 atm. The battery may be in a common pressure vessel or it may consist of cells in individual pressure vessels. A schematic of a battery in a common pressure vessel is illustrated in Figure 3.23. The dependence of specific energy and cell volume on maximum pressure is shown in Figure 3.24. It is assumed in this figure that the minimum pressure during operation of the battery is 3–4 atm. For the satellite application, the Ni/H₂ battery is charged with photovoltaic systems during daylight and the battery generates electric power during the dark period. Typical discharge plots for the Ni/H₂ battery are as illustrated in Figure 3.25. These plots clearly demonstrate that it is an ideal battery with minimal overpotential losses.

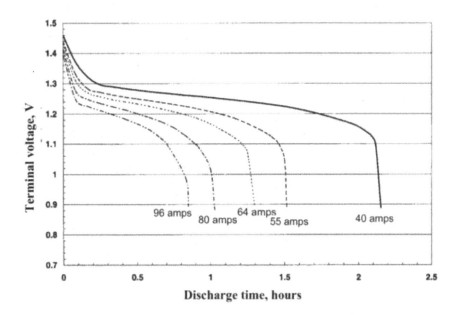

**Figure 3.25.** Discharge plots for a 90 Ah Hubble Space Telescope Ni/H$_2$ battery at different rates. Reproduced from Reference 36, Copyright (2002), with permission of The McGraw-Hill Companies.

To reduce the pressure in a Ni/H$_2$ battery, there have been several efforts to store the hydrogen as a metal hydride (e.g., LaNi$_5$H$_x$) or use hydride electrodes. In the latter case, the battery will be of the nickel/metal hydride type (see Section 3.7.3.4).

(c)  *Applications and economics.* The main application of this battery is for satellite power. The capacity of the battery is in the range 90 Ah. The battery voltage is about 1.2–1.3 V during discharge. For the LEO application, charge/discharge cycles are of equal duration per day, while for the GEO cycles, the charging time is about 20 h and discharge time is 1–2 h. The Ni/H$_2$ battery uses conventional nickel oxide electrodes as in Ni/Cd and Ni/MH$_x$ batteries (see next two Sections) for the positive electrode and highly Pt loaded fuel cell electrodes (more than 4 mg/cm$^2$) for the negative electrode. The latter makes the battery cost quite high ($1000/kWh). If the cost of the battery can be significantly reduced, it can be considered for terrestrial application, e.g., vehicle propulsion, standby/energy power. One drawback of the battery is that the self-discharge rate is about 10%/d at 20 °C.

(d) *Expected benefits to or from fuel-cell technology.* The Ni/H$_2$ battery technology development has obtained maximum benefits from fuel cell technology in respect to the production of high performance hydrogen electrodes that perform most efficiently in the charging and discharging modes.

### 3.7.3.3. Nickel/Cadmium

(a) *Background.* Ni/Cd batteries were invented by Jagner long after the lead acid batteries were developed. It was originally proposed that the Ni/Cd battery could be a power source for electric vehicles. This battery uses a nickel oxide positive electrode, as in all other alkaline nickel batteries, and a negative cadmium electrode. The Ni/Cd battery is better than the Ni/Zn or Ni/Fe batteries, which were discovered in the same time period. The reason for this is because the Cd/Cd(OH)$_2$ system is quite stable (Cd(OH)$_2$ is formed during discharge), while the products of zinc or iron electrooxidation can dissolve to a significant extent in the potassium hydroxide electrolyte. The latter problem causes morphological changes during charging of the battery. Other advantages of the Ni/Cd battery are its high rate capability, long cycle life, and good low temperature behavior.

(b) *Principles of technology.* The overall reaction in a Ni/Cd battery, its reversible potential, and its attained energy density are as stated in Table 3.21. The half-cell reactions, during charging and discharging of the nickel electrode, are the same as in the Ni/H$_2$ battery (Eq. 3.61). The half-cell reaction at the negative Cd electrode is Cd + 2 OH- $\Leftrightarrow$ Cd(OH)$_2$ + 2 $e^-$.

The thermodynamic reversible potential for this electrode is –0.809 V. Taking into account that the standard reversible potential for the positive nickel oxide electrode is 0.490 V, the standard reversible potential for the cell is 1.229 V. The theoretical specific energy of the cell is about 200 Wh/kg, based on the mass of the active materials, but as seen from Table 3.21, the maximum attained value is less by a factor of four to five, due to some irreversibilities and ohmic losses. The charge/discharge cycles are around 70% of the theoretical capacities.

In respect to electrode structures and materials, a porous sintered nickel plaque is used as the substrate for the deposition of the active materials (positive and negative electrodes). It has a porosity of 80% and contains a nickel screen as a current collector for the positive electrode; the active material is then deposited by using a vacuum impregnation method. The impregnation is carried out using 4 M Ni(NO$_3$)$_2$. An alternative approach is via the electrochemical impregnation of Ni(OH)$_2$ by cathodizing the plaque in a nickel nitrate solution. The cathode current reduces the nitrate ion to ammonia. The simultaneous increase in the pH precipitates Ni(OH)$_2$ in the pores of the plaque. The cadmium electrodes

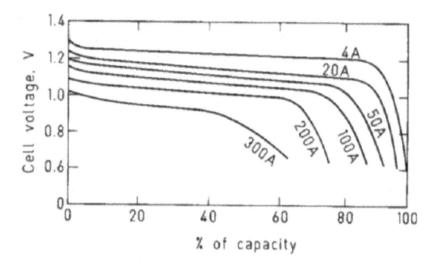

**Figure 3.26.** Typical discharge behavior at different rates for Ni/Cd batteries. Reproduced from Reference 38, Copyright (1985), with permission from Wiley-VCH.

are produced in a similar manner. Another approach for the manufacture of the cadmium electrodes is by binding cadmium oxide with polytetrafluoroethylene. This layer is then rolled onto a nickel current collector.

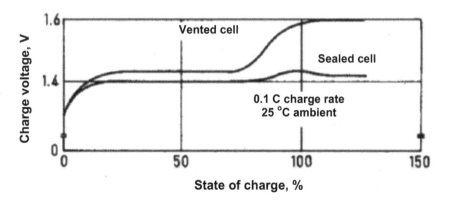

**Figure 3.27.** Charging behavior of vented and sealed Ni/Cd batteries. Reproduced from Reference 38, Copyright (1985), with permission from Wiley-VCH.

There are two types of cell design: vented cells and sealed cells. The former are flat-plate prismatic cells and the latter have a cylindrical configuration. In the vented cells, microporous propylene or radiation-grafted polyethylene membranes are used as separators. The electrolyte is 8 M KOH with small amounts of LiOH. The latter enhances the capacity of the nickel electrode. The sealed cells have a Swiss-Roll type construction. The separator material is woven nylon in non-woven polyamide felt. The electrolyte is embedded in the separator. Gas-venting is provided to inhibit pressure build up during charging.

The batteries of choice are the Ni/Cd batteries for high rates of charge and discharge. Typical discharge behavior at different rates is illustrated in Figure 3.26. There is some significant drop in capacity only above the 5 °C rate. The charging behaviors for a vented and a sealed cell are shown in Figure 3.27. Up to about a 75 % state of charge, the charge voltage is constant, while there is a rise in it above this voltage. This is because of some hydrogen and oxygen evolution reactions that have higher activation overpotentials.

The charge-retention characteristic of the vented Ni/Cd battery is excellent, over 70% retention after six months of charging. The self-discharge rate in sealed Ni/Cd batteries is somewhat higher. Likewise, the cycle life of this battery is very high. It is more than a few thousand cycles for a charge exchange of 70%. For aerospace applications at lower rates of charge and discharge, the cycle life range is in the tens of thousands.

(c) *Applications and economics.* For a considerable length of time, Ni/Cd batteries have had the second highest share of the secondary battery market. The main applications have been for standby or emergency power, aircraft auxiliary power, power sources for portable equipment (calculators, tools, laptop computers, video cameras, toys, etc.). Until recently, it also was used extensively in satellites, but due to the superior behavior of Ni/H$_2$ batteries, Ni/Cd batteries are being gradually displaced. Ni/Cd batteries were also researched on to power electric vehicles. For most of the above-mentioned terrestrial applications, the Ni/Cd batteries have been replaced by Ni/MH$_x$ batteries, which have somewhat similar performance characteristics because of the toxicity problems of cadmium. There are environmental legislations in several countries, particularly Japan, to terminate the production of Ni/Cd batteries. Thus, the costs of Ni/Cd batteries have significantly decreased to about $ 400/kWh.

(d) *Benefits to or from fuel-cell technology.* There are no clear benefits to or from fuel cell technology in the case of Ni/Cd batteries. However, one application which has been proposed for a long time, was to use the Ni/Cd battery as the hybrid power source with fuel cells for electric vehicles. The main reasons are because of its higher specific energy, specific power, and longer cycle life than the lead acid battery.

### 3.7.3.4. Nickel/Metal Hydride

(a) *Background and rationale.* Since the early 1960s, storing hydrogen in metals and alloys has attracted much attention. The pioneering work was carried out by researchers at Brookhaven National Laboratory in Upton, NY and the Philips Research Laboratory in Eindhoven, Netherlands. In principle, there are several metals and alloys that can adsorb and desorb hydrogen to the extent of one atom of hydrogen for each atom of the metal. The extent of hydrogen storage is about 1% for transition metals like iron and about 6% for lighter metals such as magnesium. Alloys have been found to be superior to the pure metals for fast adsorption/desorption kinetics. Conventionally, two types of alloys $AB_5$ (e.g., $LaNi_5$) and $AB_2$ (e.g., $VTi_2$) have been used for the metal-hydride electrodes. However, for hydrogen-storage applications, the alloy formulations (atomic %) are much more complex, as shown below:

- $AB_5 - La_{5.7}Cl_{8.0}Pr_{0.8}Nd_{2.3}Ni_{59.2}Co_{12.2}Al_{5.2}$
- $AB_2 - V_{18}Ti_{15}Zr_{18}Ni_{29}Cr_5Co_7Mn_8$

The benefits for incorporating more elements than only two in each alloy are:

- better kinetics for adsorption/desorption of hydrogen,
- inhibition of passivation of active components on the surface, and
- facilitation of dissociative adsorption of hydrogen

to name a few.

From the time period that metal hydrides were researched on for hydrogen storage, there has been great interest in these alloys for nickel/metal hydride batteries. The reason is that nickel/metal hydride batteries exhibit performance characteristics practically identical to nickel/cadmium batteries. The advantages of the former over the latter are:

- nearly the same reversible potential for the cell (about 1.35V),
- similar charge-discharge characteristics,
- higher energy density and specific energy, and
- most importantly, the toxic element cadmium is replaced by practically harmless metals.

Since the early 1990s, Japanese companies (Matsushita, SONY, Sanyo) have been developing and commercializing $Ni/MH_x$ batteries. But more recently, the technology and commercialization have spread to several other countries—e.g., Korea, Taiwan, and China. A bulk of the batteries is used for portable applications—laptop computers, cell phones, and other consumer electronics applications. There are ongoing efforts to use higher capacity batteries for hybrid electric vehicle applications.

(b)   *Principles of technology.* The active material in the positive electrode (NiOOH) is the same as in the Ni/H$_2$ and Ni/Cd batteries, as discussed in the two preceding Sections. The active material in the negative electrode is the alloy hydride. The half cell reaction is:

$$MH + OH^- \rightarrow M + H_2O + e^- \qquad\qquad E^0 = 0.83 \text{ V} \qquad\qquad (3.66)$$

where $M$ represents the alloy. In practically all the batteries, AB$_5$ alloys are being used, even though the AB$_2$ alloys have higher specific energies and energy densities. The advantages of the former are:

- higher charge/discharge rates and
- better performances at higher and lower temperatures than 25 °C.

A desired formulation for the multicomponent AB$_5$ alloy has been quoted above. However, in the large-scale manufacture of the batteries, a naturally occurring ore, misch metal, which has most of the desired constituent alloys, is used with some modification to enhance kinetics and improve stability and the cycle life.

As to the designs of the batteries, they are of the cylindrical (Figure 3.17), button, and prismatic cell types (Figure 3.18). The positive electrodes have the same structure and composition as in the Ni/H$_2$ and Ni/Cd batteries. The negative electrode has a highly porous structure, with a perforated nickel foil or grid, containing the plastic bounded hydrogen storage alloy. A typical multi-cell battery (9 V) is illustrated in Figure 3.28.

The discharge behavior of a Ni/MH battery is shown in Figure 3.29. There is some decrease in performance at temperatures below about 10 °C and above 45 °C. The region in between is the ideal one for battery operation. The cycle life of the battery has improved over the years. The charge retention rate of the battery is quite satisfactory; it is better at close to room temperatures.

(c)   *Applications and economics.* For the portable applications, the types are practically the same as in the primary Zn/MnO$_2$ and secondary Ni/Cd batteries (e.g., A, AA, AAA, C, D, 9 V). One significant advantage of this battery is that its sealed cell uses an oxygen recombination mechanism to prevent pressure build up during cell charging and overcharging. For this purpose, the negative electrode has a higher effective capacity than the rated one. Thus, the oxygen, evolved during overcharge at the positive electrode, diffuses through the separator to the positive electrode and combines with the MH to form water, while the hydride electrode is never overcharged because of the higher capacity. The larger batteries are similar in design and construction to the Ni/Cd batteries. These are the

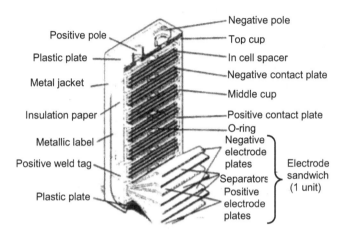

**Figure 3.28.** Typical design of a multicell Ni/MH battery. Reproduced from Reference 36, Copyright (2002), with permission of The McGraw-Hill Companies.

hybrid power sources, with the IC engines in the Toyota Prius and Honda Civic vehicles. For the portable electronics application, the cost of the batteries is comparable to that for the Ni/Cd batteries; for the hybrid electric vehicle operation, the projected cost is $300–400/kWh.

(d) *Benefits to or from fuel-cell technology.* The optimization of the composition of the MH electrode has benefited from the development of fuel cell electrocatalysts for the hydrogen electrode, particularly in alkaline medium, as for instance, incorporating nickel or cobalt. These elements enhance the activities for dissociative adsorption of hydrogen and/or inhibit formation of passive films or corrosion of the alloys. It must be noted that the standard reversible potential for the $Ni/MH_x$ battery is the same as for the hydrogen electrode in fuel cells. It is for this reason that there has been some developmental work of metal hydride/air batteries. The hydrogen storage and the hydrogen electrode charging/discharging characteristics are carried out in the sealed secondary battery.

### 3.7.3.5. Lithium–Ion

(a) *Background and rationale.* At the present time and for the foreseeable future, the rechargeable lithium-ion batteries will be in the forefront for consumer electronic (cell phones, laptop computers) and military electronic (radios, mine detectors, thermal weapon sites) applications. Potential applications are for aircraft, spacecraft, and for hybrid electric vehicles (IC or diesel/battery). The attractive features of this battery are:

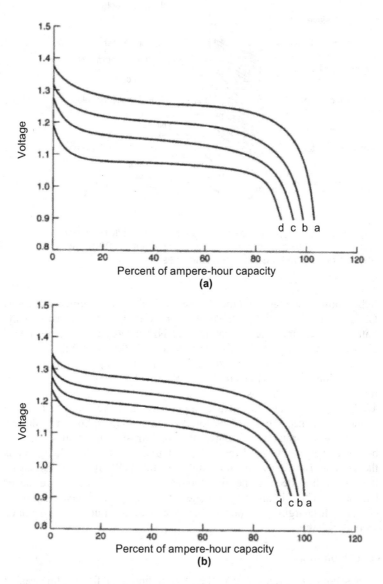

**Figure 3.29.** Charge/discharge behavior of a nickel metal hydride battery at (a) 25 $^\circ$C and (b) 45 $^\circ$C. The symbols a, b, c, and d represent discharge rates of C/5, C, 2C, and 3C, respectively, with C Being defined as the discharge current in amps divided by the rated capacity of the battery in ampere-hour. Reproduced from Reference 36, Copyright (2002), with permission of The McGraw-Hill Companies.

- high energy density,
- high power density, and
- low self-discharge rate.

A significant advantage of this battery is that its range of operating potential is from about 2.5 to 4.2 V; this is about three times higher than that for the competing secondary batteries, such as nickel/hydrogen, nickel/cadmium, and nickel/metal hydride. Thus, to operate in this potential range for any application, only one-third the number of Li-ion cells is required in a stack, as compared with that for the other above-mentioned batteries. It is noteworthy to mention that the fundamental research studies on the electrodes and electrolytes have been carried out, over the last 20 years, at universities and national and industrial laboratories in the USA, Canada, and France; soon after, this was followed by intensive R&D efforts in Japanese laboratories. However, the mass production of these batteries (over 90%) for the consumer and military electronic applications is by Japanese companies (e.g., SONY, Matsuchita, Sanyo, to name a few).

(b) *Principles of technology.* The lithium-ion battery has a unique characteristic: it operates by transport mechanism of a lithium ion, from the positive electrode to the negative electrode during charging and vice versa during discharging. For both electrodes, intercalation materials are used. The present batteries mostly use coke for the negative electrode material and lithium cobalt oxide ($LiCoO_2$) for the positive electrode materials. Alternatives are graphite for the former and lithium manganate ($LiMn_2O_4$), lithium-nickel-cobalt oxide ($LiNi_{1-x}Co_xO_4$) for the latter. Since lithium is a highly electropositive material, an aqueous electrolyte medium cannot be used (due to the rapid rate of hydrogen evolution). Thus, one has to employ organic solvents or ceramic electrolytes. The former are considerably better and are used in practically all the commercial batteries. The organic solvents include ethylene carbonate (EC), propylene carbonate (PC), dimethyl carbonate (DMC), acetonitrile (AN), tetrahydrofuran (THF), and several others.

In order to have ionic conduction of lithium ions in the electrolyte, inorganic or organic lithium salts are dissolved in the organic solvent. The most widely used material is lithium hexafluorophosphate ($LiPF_6$). Several other materials have been evaluated and some are already in use in lithium-ion batteries, e.g., lithium tetrafluoroborate ($LiBF_4$), lithium perchlorate ($LiClO_4$), lithium hexafluoroarsenate ($LiAsF_6$), and lithium triflate ($LiSO_3F$). Safety considerations have to be taken into account in the selection of both the salt and the solvent. The electrolyte can be in liquid or gel form (the latter types of batteries are commonly referred to as lithium polymer batteries). Lithium-ion batteries with the liquid electrolyte have higher ionic conductivities ($10^{-3}$ to $10^{-2}$ S/cm); the conductivities of the lithium polymer batteries are at least an order of magnitude lower. For the lithium-ion cells with a liquid electrolyte, a separator material is required to prevent contact between the positive and negative electrodes; microporous films (10 to 30 μm) of polyolefin materials (polyethylene, polypropylene, or laminates of the two) are used for this purpose. A commercial material used in some batteries is Celgard.

The most common designs of the cell are of the spiral-wound cylindrical and flat-plate prismatic types. This is for capacities ranging from 0.1 to 160 Ah. The batteries are prepared in the discharged state—for this purpose the positive electrode material is the lithium metal oxide and the negative electrode material is coke. But more recently, the manufacturers are increasingly using graphitic materials for the latter because of their higher specific capacity, cycle life, and rate capability. Both of these materials in powder form are bonded on to a metallic foil or a screen (aluminum for the positive electrode and copper for the negative electrode). The cell is then assembled with the electrolyte and the cell is charged and lithium ions are transferred from the positive to the negative electrode. Care must be taken during the first charging not to exceed about 4.2 V. Above this voltage, side reactions such as oxidation of the solvent may occur. During charging/discharging, the potential range has to be regulated between 2.5 and 4.2 V to avoid reaching the maximum and minimum voltage.

The reactions, which take place at the electrode/electrolyte interfaces and in the single cell, can be represented by the equation:

Positive electrode:
$$LiCoO_2 \rightarrow Li_{1-x}Co_xO_2 + x\ Li^+ + x\ e^- \tag{3.67}$$

Negative electrode:
$$C + x\ Li^+ + x\ e^- \rightarrow Li_xC \tag{3.68}$$

Overall:     $$LiCoO_2 + C \rightarrow Li_{1-x}Co_xO_2 + Li_xC \tag{3.69}$$

During charging all the lithium from the $LiCoO_2$ is not released. Customarily, x reaches a maximum value of about 0.8 (a capacity equivalent to about 80% of the theoretical capacity is attained). It has been demonstrated that the ratio of lithium to carbon atoms has a maximum value of 1:6. Thus, the theoretical capacity of the lithium electrode is 372 mA h/g, which is about 10% of that if pure Li were used for the negative electrode. Pure lithium cannot be used as the positive electrode material because the interaction of the metal with the solvents can lead to explosions.

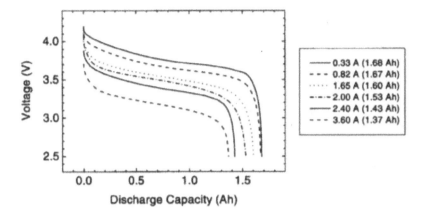

**Figure 3.30.** Typical discharge behavior of a Li-ion battery. Reproduced from Reference 36, Copyright (2002), with permission of The McGraw-Hill Companies.

The highly advantageous performance characteristic of a lithium-ion battery is reflected in its charge/discharge behavior (Figure 3.30); it has a far superior behavior, in comparison with that of nickel/hydrogen, nickel/cadmium, and nickel metal hydride batteries (see the three preceding Sections). The discharge potentials are relatively flat as a function of discharge capacity and are about 3 times higher for the lithium-ion batteries, as compared with those for the nickel-batteries. The lithium-ion batteries can also be operated at high charging/discharging rates (up to about 3–5 °C). The cycle life of the lithium battery has greatly improved during the last 5 years, over 1000 cycles. This is still a factor of 10 less than that attained for the aforementioned alkaline nickel batteries. However, for the portable consumer and military applications, the attained cycle life is adequate. For the transportation, space, and satellite applications, a longer cycle life is essential.

(c) *Applications and economics.* The demand for the lithium-ion batteries for the portable consumer/military electronic products has been growing exponentially since the mid 1990's. The number of units of lithium-ion batteries will be about 1.3 billion in the year 2007.[7] This amount is about twice that sold in the year 2000. Other interesting data are the enhancements made in the specific energies and energy densities of the battery—e.g., see Figure 3.31 for a much marketed lithium-ion cylindrical battery, since the early 1990s until the present time, in comparison with rechargeable Ni/Cd and Ni/MH batteries.

(d) *Benefits to or from fuel-cell program.* The fuel cell technology is markedly different from the lithium-ion technology and thus, the reader

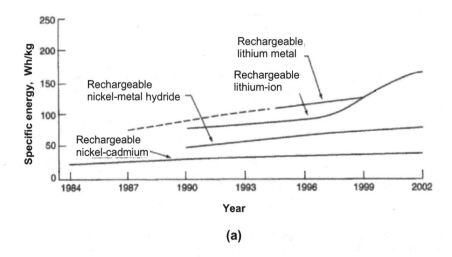

**(a)**

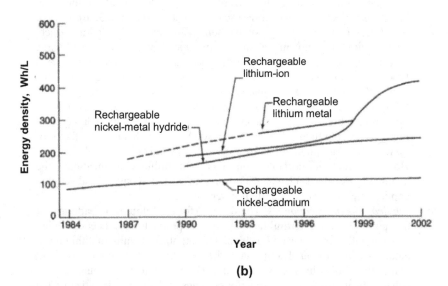

**(b)**

**Figure 3.31.** Enhancement in performance of portable rechargeable batteries (a) specific energy (b) energy. Reproduced from Reference 36, Copyright (2002), with permission from The McGraw-Hill Companies.

may not be able to visualize any technical benefits. However, the lithium-ion battery will be a strong contender for powering hybrid electric vehicles. In the long term, it may be more advantageous than the nickel-metal hydride batteries because of the higher-performance characteristics of the lithium-ion battery, achieved and projected. In the long term, the fuel cell will be the primary hybrid power source and will replace the IC or diesel engine.

## 3.7.4. Reserve Batteries

3.7.4.1. *Background and Rationale.* Reserve batteries have been developed and utilized for special applications. These applications require one or more of the following specifications:

- long shelf life,
- low self-discharge rate,
- high power density and specific power for short times, and
- high energy density and specific energy.

The applications include their uses in:

- torpedoes, missiles, and other weapons,
- lifeboat emergency equipment in commercial airlines,
- radios,
- sono buoys, and
- balloon transport equipment for high altitude and low ambient temperature operation.

There are four types of reserve batteries:

- water activated (by fresh or sea water),
- electrolyte activated by its injection,
- gas activated (e.g., air), and
- heat-activated (e.g., a solid salt electrolyte is connected to the molten state to become ionically conducting).

Water or seawater activated batteries are the most common types of reserve batteries. The active components of the reserve batteries are:

- corrodible metals like lithium, magnesium, aluminum, or zinc for the negative electrode, and
- silver chloride, cuprous chloride, lead chloride, manganese dioxide, or cuprous iodide for cathode materials. In some cases, air is the cathodic reactant.

Because of space limitations in this book, we shall present a short summary of magnesium-seawater activated batteries using silver chloride as the cathodic reactant. The reader is referred to *the Handbook of Batteries* edited by Linden and Reddy for more detailed descriptions.

### 3.7.4.2. Mg-Seawater Activated Batteries

(a)  *Principles of technology.* These batteries use magnesium or a magnesium alloy for the negative electrode and silver chloride, cuprous chloride, lead chloride, cuprous iodide, or magnesium dioxide for the positive electrode. Batteries have also been developed with air as the cathodic reactant. In the case of the Mg/AgCl battery, the half cell reactions are:

Negative Electrode
$$Mg \rightarrow Mg^{2+} + 2\ e^{-} \qquad\qquad E_{ox} = 2.37\ V \qquad (3.70)$$

Positive Electrode
$$2\ AgCl + 2\ e^{-} \rightarrow 2\ Ag + 2\ Cl^{-} \quad E_{red} = 0.80\ V \qquad (3.71)$$

The overall reaction is represented by the addition of the two above-mentioned half-cell reactions, as seen in Table 3.20. Batteries are produced with the anode, cathode, and separator in the dry state. The battery is activated by introduction of the electrolyte. There are several types of these water activated batteries: two of these are (i) immersion batteries which are activated by immersion in water or seawater, and (ii) forced-flow batteries, used for torpedoes, in which sea water is forced through the battery as the torpedo is driven through the water. A multi-cell stack in the battery can produce about 500 kW of power in about 15 min. Attained values of specific energy and energy density are 130 Wh/kg and 200 Wh/l.

(b)  *Applications and economics.* The main application of the Mg/AgCl batteries is in torpedoes. Other types of applications for magnesium-sea water batteries are as described in Section 3.7.4.1. The Mg/AgCl battery is quite expensive because of the high cost of silver and it being non-reversible after use. Thus, instead of AgCl as the cathodic material, $Cu_2Cl_2$, $Cu_2I_2$, $PbCl_2$, and $Cu_2(CNO)_2$ have been used but the energy densities and specific energies are considerably less in these cases.

(c)  *Benefits to or from fuel-cell technology.* At the present time there are no benefits of the reserve batteries to or from the fuel cell technologies. One exception may be for using better electrodes for oxygen reduction, as developed for alkaline fuel cells, in the batteries which use air as the cathodic reactant.

## 3.7.5. Thermal Batteries

3.7.5.1. *Background and Rationale.* This short Section on thermal batteries is included in this Chapter only for the sake of providing the reader with a perspective of all types of batteries. For more details, the reader is referred to an excellent review on thermal batteries by Klasons and Lamb in Chapter 21 of the *Handbook of Batteries.* Thermal batteries are akin to reserve batteries, discussed in the previous Section. The main difference is that during the design and construction of the batteries, the electrolytes are non-conducting, inorganic solids (at room temperature) and interfaced with the anodes and the cathodes. A pyrotechnic material is also incorporated in the material, which when activated, melts the electrolyte and makes it conducting. Immediately after activation and when the battery is connected to a load, the battery discharges for short periods of time. The discharge time depends on the capacity of the anode and cathode and more so falls off with time due to cooling of the battery and solidification of the electrolyte.

Thermal batteries were first developed in Germany, mainly for weapons applications. In order to attain the power levels required, the geometric areas are increased and cell stacks developed. Some examples of military applications are for use in missiles, bombs, and torpedoes. These batteries have also been used for space exploration. Some advantages of these batteries are:

- long shelf life (~ 25 years) without degradation in performance and no self-discharge,
- start-up time in milliseconds,
- peak power of about 10 $W/cm^2$,
- high reliability, and
- wide range of operating temperature.

The disadvantages of these batteries are:

- short periods of power generation (~ 10 min),
- low to moderate specific energies and energy densities,
- surface temperatures reach over 200 °C,
- power output decreases with time of discharge, and
- use for only one time, like a primary battery.

3.7.5.2. *Principles of Technology and Performance Characteristics.* The anode materials are the highly electropositive alkaline or alkaline metals or their alloys (e.g., Li, Si, Mg, Cu) and the cathode materials are strong oxidants (e.g., $K_2Cr_2O_7$, $CaCrO_4$, $V_2O_5$, $FeS_2$). The electrolytes are alkali halides (e.g., LiCl-KCl, LiBr-KBr, LiBr, KBr-LiF). The conducting species in the electrolyte is the lithium ion. There are two types of pyrotechnic heat sources—heat paper and heat pellets. The heat paper typically contains zirconium and barium chromate powders, impregnated in an inorganic fiber matrix. Heat pellets typically contain a mixture of iron powder and potassium perchlorate. When the power from the battery is required, bridge wires and a heat-sensitive pyrotechnic material are used to ignite the pyrotechnic material, which in turn rapidly melts the electrolyte. The discharge voltage range for the batteries using the anode and cathode materials mentioned above is 1.6 to 3.3.

The most widely used thermal battery is the $Li/FeS_2$ battery, with LiCl-KCl or LiCl-LiBr-LiF as the electrolyte. The overall reaction is:

$$3\ Li + 2\ FeS_2 \rightarrow Li_3Fe_2S_4 \qquad E_r = 2.3\ V \qquad (3.72)$$

Cell designs are of three types: cup cells, open cells, and palletized cells. A cup cell contains an anode having the active anode material on both sides of a current collector. On the external sides of the anode material is positioned a glass tape pad impregnated with the eutectic electrolyte. On the external sides of the electrolyte pad are placed the cathode material in an inorganic fiber matrix. Current collectors are placed on the external sides of these fiber matrices. Open cells are similar in design to the cup cells. These types of batteries are used for short time applications and pulse power. Open cells have rapid heat transfer and short activation times. Cell stacks have been designed to meet the power requirements. The cells are connected externally in series. For the above applications, the number of cells is in the range 14 to 80 cells and the discharge voltage is in the range 28 to 140 V. There has been an application for which a stack voltage was 400 V and the number of cells in the stack was 180. For a battery with a volume of 1300 $cm^2$, the performance characteristics were as follows:

- power density: 170 $mW/cm^2$,
- energy density: 85 Wh/$l$, and
- activated life: 1800 seconds.

In order to advance the Thermal Battery technology, efforts are being made to:

- enhance the specific energy and energy density by using light-weight materials, like aluminum and composites, instead of stainless steel for container materials; and
- use of plasma-spraying techniques for deposition of thin films of $FeS_2$. Molten nitrates have also been researched as electrolytes with lithium anodes, to enhance cell voltages and reduce operating temperatures by more than 200 °C. Efforts to increase discharge times have been made by using more efficient thermal insulation and lower melting point electrolyte compositions.

## 3.7.6. Electrochemical Capacitors

*3.7.6.1. Introduction.* One of the innovations in the field of electrochemical science is the development of electrochemical capacitors, EC, which exhibit a 20–200 times greater capacitance than conventional solid state capacitors. The first electrochemical capacitor was described in 1957 by Becker and Ferry[40] and it was based on porous carbon. Twelve years later, Sohio developed a carbon-based capacitor in a non-aqueous medium that could be charged to about 3 V. The recognition of the pseudocapacitative behavior during the charging of $RuO_2$ electrodes, by Trasatti and Buzzanca,[41] triggered a flurry of activity towards developing electrochemical capacitors based on double layer and adsorption pseudocapacitance. Since then there has been a major focus to commercialize these concepts as energy storage and delivery devices, complementary to that by batteries. A thorough discussion of the subject of electrochemical capacitors is available in the authoritative monograph by Conway[42] and in the proceeding volumes of the annual symposia held by Wolsky and Marincic from 1991 onwards, covering the ongoing progress in this area.[43]

*3.7.6.2. Principles of Technology.* The capacitance of a condenser, $C$, is expressed by the relationship:

$$C = \frac{A\varepsilon\varepsilon_0}{d}$$ 

(3.73)

where $A$ refers to the area of the plates, $d$ to the distance between them, $\varepsilon$ to the dielectric constant of the medium, and $\varepsilon_0 = 8.84 \times 10^{12}$ F/m the dielectric permittivity of the free space. The capacitance of the condenser can be increased by increasing the values of the parameters $A$ and $\varepsilon$ and by decreasing the distance between the plates. The dielectric constant can be increased by forming thin oxide films on the plates made of metals such as Al, Ta, Ti, and Nb. These oxide films, having high dielectric constants, enhance the overall capacitance; these component-devices are called the *electrolytic capacitors*. The energy stored in these capacitors is the free energy, $G$, which is given by the expression:

$$G = \frac{CV^2}{2}$$ 

(3.74)

where $V$ refers to the voltage applied to the capacitor.

When a metal is immersed in an ionic solution, an electrostatic potential is established at the metal/solution interface (see Chapter 2). This leads to the formation of a double layer as a result of the charge separation occurring on either side of the interface, which is manifested as a displacement current. Note that there is no charge transfer across the interface during this process. The capacitance of the electrical double layer varies in the range of 16–40 $\mu F/cm^2$ depending on the nature of the metal surface and the nature of the solvent and ions present in the system. The reason for high specific capacitance of the electrical double layer at the metal/solution interface is the small distance of separation of charges of 0.3 to 0.5 nm, compared to 10–100 nm in electrolytic capacitors and $> 10^{-3}$ nm with thin mica and polystyrene dielectric film capacitors.

In contrast, when adsorbed intermediates are involved during the course of a Faradaic reaction of the type $O_{ads} + ne^- \rightarrow R_{ads}$, where O and R refer to the oxidant and reductant respectively, the charge exchanged during the course of the reaction is stored in the form of an adsorbed species. This leads to a pseudocapacitive behavior (see Chapter 2). The capacitance associated with this process is large and is Faradaic and not electrostatic or non-Faradaic in origin. It can be shown from first principles that the generalized impedance, Z, for this reaction is given as: [44]

$$Z(p) = \left( pC_d + \frac{pC_\phi}{pC_\phi R_t} \right)^{-1}$$

(3.75)

where $p$ is the Laplace Transform variable, $R_t$ is the charge transfer resistance, $C_d$ is the double layer capacitance, and $C_\phi$ is the adsorption pseudocapacitance. Hence, the equivalent circuit can be represented as shown in Figure 3.32. The capacitance of this system is expressed by the relationship:

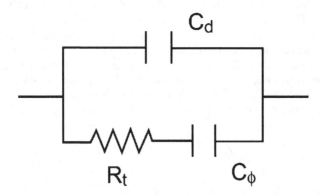

**Figure 3.32.** Equivalent circuit for a Faradaic reaction exhibiting pseudocapacitance behavior (see text for definition of symbols). Reprinted from Reference 44, Copyright (1996), by permission of The Electrochemical Society, Inc.

$$C = \frac{(C_d + C\phi)^2 + \omega^2 C_d^2 C_\phi^2 R_t^2}{C_d + C_\phi + \omega^2 C_d^2 C_\phi^2 R_t^2} \qquad (3.76)$$

Thus, $C = C_d$ as $\omega \to \infty$ and $C = C_d + C_\phi$ as $\omega \to 0$. However, when $R_t = 0$, the overall capacitance will be the sum of the double layer and adsorption pseudocapacitance and is frequency independent. However, if the same reaction is conducted on a porous electrode, the overall capacitance will be frequency dependent and only a fraction of the double layer capacitance is available at short times. As $t \to \infty$, all the charge associated with the double layer and adsorption pseudocapacity becomes accessible. This is a result of the distributed $RC$ effects in the porous matrix, exhibiting multiple time constants. However, if the electrolyte resistance and the pore length are significantly lowered, then it possible to realize high capacitances at very short times. The maximum pseudocapacitance available with the above reaction is given by the relationship:

$$C_\phi = \frac{n^2 F^2}{RT} \left( \frac{1}{\Gamma_O^0} + \frac{1}{\Gamma_R^0} \right)^{-1} \qquad (3.77)$$

where $\Gamma$ refers to the surface excess of the species in the subscript. $C_\phi$ can be as large as 8 F/cm$^2$ following the above equation with $\Gamma_O^0 = 10^{-8}$ mol/cm$^2$ and n = 2 with a roughness factor of 100.

Electrochemical capacitors behave like rechargeable batteries in terms of energy storage and delivery, although the mechanisms of charge storage are different from those operating in batteries. The amount of energy stored in the ECs is equal to $CV^2/2$ or $0.5qV$, where $q$ is the amount of charge stored. However, for a battery, the corresponding stored energy is $qV$, which is twice as much as that for a capacitor charged to the same voltage, since the battery has a thermodynamically constant voltage during charge and discharge except when the state-of-charge approaches 0 or 100%. It may be noted that the energy stored in a capacitor increases with the square of the cell voltage as charge is accumulated.

A wide variety of materials were examined for their capacitive characteristics and a listing of some of them are available in references.[42-44] These can be broadly split into two categories: double-layer capacitors and pseuocapacitance-based capacitors.

(a) *Double-layer capacitors.* The first EC employed activated carbon as the electrode and it exhibited capacitance values of 2 F/cm$^2$ in sulfuric acid solutions. However, carbon was found to suffer from slow oxidation in addition to having a high equivalent series resistance due to poor particle-to-particle contact of the agglomerates and high ionic resistance in the micropores of carbon. Various techniques have been examined to achieve high surface area with low matrix resistivity. These include, use of carbon

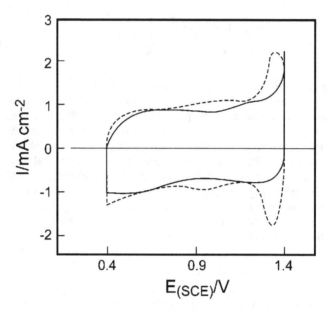

**Figure 3.33.** Cyclic voltammogram of a ruthenium oxide in acid and alkaline media, exhibiting pseudocapacitance behavoir. Reprinted from Reference 45, Copyright (1990) with permission from Elsevier.

foam and paste electrodes, pyrolysis of carbon-based polymers, heat treatment in the presence of additives, addition of Ru, Rh, Pd, Os, Ir, Co, Ni, Mn, Pt, Fe and combinations there of to carbon.

(b) *Pseudocapacitance-based capacitors.* Cyclic voltammetric studies showed a broad range of compositions exhibiting adsorption pseudocapacitance. Of all these, $RuO_2$-based EC systems showed high specific capacitance due to the $C_\phi$ arising from the surface reaction:

$$2\,RuO_2 + 2\,H^+ + 2\,e^- \leftrightarrow Ru_2O_3 + H_2O \qquad (3.78)$$

Capacitances as high as 2.8 F/cm$^2$ were reported with compositions based on $RuO_2$, mixtures of $RuO_2 + Ta_2O_5$, and mixed oxides of Mo,W,Co, and Ni. $RuO_2$ bonded solid ionomer membranes exhibited 6 to 10 F/g of active material with excellent charge-density delivery in pulse applications. The cyclic voltammogram of a $RuO_2$ electrode in acidic and alkaline solutions[45] presented in Figure 3.33 shows significant $C_\phi$, which is almost constant over a voltage range of 1.4 V. Some other transition metal oxides behave similarly but only over a small operating voltage

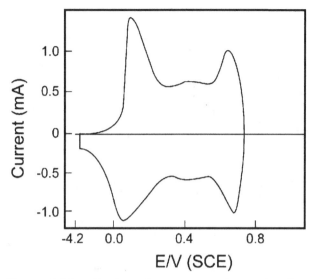

**Figure 3.34.** Cyclic voltammogram of a polyaniline electrode exhibiting pseudocapacitance behavior.[42]

range of about 0.6–0.8 V. The reversibility of this system is excellent as it has a cycle life of over several hundred thousand cycles. Electrochemical capacitors based on the $RuO_2$ electrodes are mainly used in military applications.

Another type of material exhibiting quasi-redox behavior that is highly reversible is the family of conducting polymers such as polyaniline[46] or derivatives of polythiophene. While they are less expensive than $RuO_2$, they are less stable giving only thousands of cycles, between 0.8 V and 3.0 V depending on the material. Figure 3.34 illustrates the cyclic voltammetry of polyaniline, and Figure 3.35 depicts the reversible behavior of $RuO_2$ and the irreversible nature of $Pb/PbCl_2$ electrodes.[42]

It is important for the ECs to exhibit high $C_\phi$ over a large voltage region. However, the voltage window available for the EC device is dictated by the operating pH and the thermodynamic stability of the pertinent species in the solution. Thus, in aqueous solutions, Ni and Co oxides are not suitable in acidic solutions and $RuO_2$ is unstable in alkaline media at high anode potentials. Pt-H is stable in acid and alkaline solutions and $MnO_2$ exhibits high charge storage capability only in unbuffered solutions, as the pH swings allow enhanced proton diffusion in and out of the $MnO_2$ lattice. The $C-H_2SO_4$ system undergoes corrosion beyond 0.8 V but up to 1 V; the corrosion rate is minimal so that a practical device can be built. To realize a large voltage window, a wide variety of solvents have

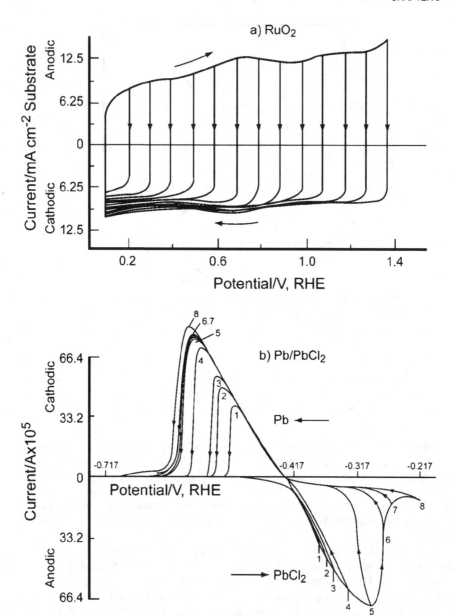

**Figure 3.35.** Cyclic voltammograms illustrating the reversible behavior of a ruthenium oxide electrode and the irreversible behavior of a lead/lead dichloride electrode. Reproduced by permission of the author.[50]

**TABLE 3.22.**
**Some Manufacturers of Electrochemical Capacitors**

| Aqueous medium | Non-aqueous medium |
|---|---|
| Symmetric design | |
| ECOND | Montena |
| ELITE | Maxwell |
| NEC | NESS |
| (PRI) | EPCOS |
| | Panasonic |
| | Okamura |
| Asymmetric design | |
| ESMA | Telecordia |
| Mega-C | |

been examined. These include solid electrolyte such as $RbAg_4I_5$, proton conducting ionomer membranes, solvents such as acetonitrile, propylene carbonate with alkali metal, or tetraalkyl ammonium cations with $CF_3SO_3^-$, $N(CF_3SO_2)^{-2}$, $BF_4^-$, $PF_6^-$, $AsF_6^-$ and $ClO_4^-$ anionic groups, and quaternary phosphonium salts in organic solvents. Based on lithium battery work, gel electrolytes (polyacrylonitrile or polyethylene oxide with an electrolyte salt cast into a film) were also studied. Of these, solid polymer electrolytes appear promising in terms of assembly and manufacturing ECs.

3.7.6.3. *Applications and Economics.* Electrochemical capacitors find applications in three major areas: consumer electronics, the automotive industry, and the industrial sector.[47] Consumer electronics can employ low-voltage ECs for digital wireless devices, memory back-up (e.g., clocks, VCRs), rechargeable tools, and appliances. Automotive applications include combined starter-generators, hybrid power systems, electric power steering, electric fork lifting, and use mid-voltage

**TABLE 3.23.**
**Commercially Available Electrochemical Capacitors**

| Supplier | Voltage (V) | Capacity (F) | Resistance (mΩ) | Time constant (s) | Wh/kg |
|---|---|---|---|---|---|
| Skelton Technology | 3 | 47 | 5.2 | 0.24 | 10 |
| Maxwell | 3 | 2700 | 0.5 | 1.35 | 4.8 |
| Ness | 3 | 2650 | 0.25 | 0.65 | 5.1 |
| Panasonic | 3 | 1200 | 1 | 1.2 | 4.2 |
| Montena | 3 | 1800 | 1 | 1.8 | 5.6 |

**TABLE 3.24.**
**Cost Comparison of Electrochemical Capacitors***

|                              | $/F           |
| ---------------------------- | ------------- |
| Large ECs(>20 kJ)            | 0.3 – 2       |
| Small ECs                    | 1 – 20        |
| Al electrolytic capacitors   | 100 – 300     |
| ECs-military                 | 300 – 700     |
| Tantalum capacitors          | 8000 – 13000  |

* commercial 5–10 V capacitors

ECs. High voltage ECs are for industrial uses such as magnetic actuators, battery load-leveling, and power-line conditioning. There are several companies throughout the world that are involved in the development and manufacture of ECs having a wide range of voltage and capacitance characteristics. Table 3.22 provides a list of some of these manufacturers,[48] and Table 3.23 lists some prototype and commercially available carbon-based ECs.[49] Table 3.24 provides a comparison of the costs of the ECs.[47-49] The present costs of aqueous ECs are about $ 100/kJ or $ 359/kWh, and it is the goal of the EC manufacturers to cut the costs down to less than $ 10/kWh.

*3.7.6.4. Benefits to or from fuel-cell technology.* The high power-density of ECs has made them attractive for use in hybrid configurations with batteries and fuel cells in electric vehicles and other load-leveling applications.[50] The EC, when coupled with a fuel cell or battery, provides the required power density for up-hill or accelerated driving while regenerative breaking provides recharging. Two-types of hybrid systems are available:

- $RuO_2$-based EC/electrolytic capacitor, and
- Carbon-based EC/$PbO_2$/$PbSO_4$ or $NiOOH$/$Ni(OH)_2$. These are called asymmetric capacitors.

Figure 3.22 illustrates the Ragone plot, comparing the energy density and power density of fuel cells and batteries. Electrochemical capacitors are not viewed as replacements for either fuel cells or batteries but as complementary energy storage and delivery systems.

## 3.8. ELECTROCHEMICAL SENSORS

### 3.8.1. Background and Rationale

Chemical and electrochemical sensors were invented more than 50 to 60 years ago. These sensors may be defined as small portable devices that are used for measurement of concentrations of species, over a wide range ($10^{-6}$ $\mu$M to a few molar); these species are, in general, in the presence of many others. There are a large number of applications of sensors in university, government, and industrial research laboratories. A few examples are in analytical chemistry, environmental monitoring and remediation, the automotive industry, clinical diagnostics, and biomedical research. This Section is devoted only to electrochemical sensors; more details of sensor technology can be found in the references in Suggested Reading.

### 3.8.2. Principles of Technology

There are principally, four types of electrochemical sensors:

- ion-selective electrodes,
- sensors for measurements of gas analysis in liquids,
- electroanalytical sensors for measurements of concentrations of metallic ions in electrolytes, and
- sensors using enzyme electrodes, used in biochemistry and clinical studies.

The first ion selective electrode was for the measurement of the pH of an electrolyte. It utilizes a $H^+$ selective glass electrode, in the form of a membrane. This was followed by development of other glass electrodes, selective to ions such as $Li^+$, $Na^+$, $K^+$, $Cs^+$, $Ag^+$, $Tl^+$, $NH_4^+$ and also anions such as the halides. These electrodes are based on equilibrium potential, obeying the Nernst equation:

$$E = E_r + \frac{RT}{nF} \ln a$$

$$(3.79)$$

where $E_r$ is the standard reversible potential and a is the activity of the ionic species. Since there is a logarithmic relationship between $E$ and $a$, the sensitivity is not as high as with amperometric sensors.

Some of the electrochemical sensors, used for analysis of trace gases in electrolyte use fuel cell principles. To quote some examples, these are used for analysis of gases such as $H_2$, $O_2$, $CO$, $H_2S$, and $Cl_2$. These sensors are used for measurements of limiting currents at electrodes for the oxidation or reduction of the above mentioned gases. This method is quite sensitive because the concentration of the gas, dissolved in the electrolyte is linearly related to the limiting current density.

A similar kind of electrochemical device is the voltammetric sensor. It is used widely for measurements of concentration of ionic species (metallic and negative ions) in electrolytes. This device measures the current at set potentials. Here again, the concentration of the ions is linearly related to the current. An advantage of this method is that it can be used for detection of several ionic species in the electrolyte. This is done by varying the set potentials to different values that correspond to the regions of potentials where oxidation or reduction of the ionic species can occur. One difficulty of the method is that electrodes can be poisoned by impurities in the electrolyte. But there are means of cleaning up of the electrolyte to get rid of mainly organic impurities or to reactivate the surface of the electrode.

The fourth type of electrochemical sensors, enzyme-modified electrodes, has been making the most advances in the recent years. These are extensively used in biomedical applications, both in vitro and in vivo. One of the simplest ones will be briefly described. This type is used for measurement of glucose concentration in electrolytes, including blood. In one design, glucose oxidase is covalently bonded to a $Ru^{2+}$ pentamine complex to electrochemically link the internal flavin group with the external redox mediators. A biocompatible gel (an epoxy resin) is used to bind the proteins and other electroactive materials. Other enzymes are also introduced to prevent interference from organic species such as ascorbate and urate.

Rapid advances have been made in chemical/electrochemical sensor technology using micro-fabrication and miniaturization methods. These include photo-lithographic techniques to design a surface with thin film deposition and oxide growth and silicon doping for fabrication of microelectronic circuits. Since the sensing element and signal conditioning electronics are on a single structure, the signal to noise ratio can be enhanced. Sensors have also been designed and developed for multi-elemental/ionic species analysis.

### 3.8.3.   Applications and Economics

Several of these have been mentioned in the first Section. In respect to economics, sensors are small/ultra small devices. The manufacturing scale is very high.

### 3.8.4.   Benefits to or from Fuel-Cell Technology

One of the most relevant benefits to fuel cells is the solid state sensor used for measurements of oxygen concentration in the exhaust gases from an automobile. It uses a solid oxide electrolyte, practically the same as in a SOFC, and it operates at a temperature of about 800 °C. The thin film technology, being developed for SOFCs, would be beneficial for further improvements in these sensors. Another similarity between the two technologies is in respect to measurements of dissolved gas as $H_2$, $O_2$, CO, and $H_2S$ in electrolytes. Advances made in the composition and structure of fuel cell electrodes could benefit the development of new sensor electrodes.

## Suggested Reading

1. J. S. Sconce, *Chlorine, Its Manufacture, Properties, and Uses* (Reinhold Publishing Corp., New York, 1962).
2. M. O. Coulter (Ed.), *Modern Chlor-Alkali Technology* (Ellis Horwood, London, 1980).
3. C. Jackson (Ed.), *Modern Chlor-Alkali Technology* (Ellis Horwood, Chichester, 1983), Vol. 2.
4. K. Wall (Ed.), *Modern Chlor-Alkali Technology* (Ellis Horwood, Chichester, 1986), Vol. 3.
5. N. M. Prout and J. S. Moorhouse (Eds.), *Modern Chlor-Alkali Technology* (Elsevier Applied Science, 1990), Vol. 4.
6. T. C. Wellington (Ed.), *Modern Chlor-Alkali Technology* (Elsevier Applied Science, 1992) Vol. 5.
7. R. W. Curry (Ed.), *Modern Chlor-Alkali Technology* (Royal Society of Chemistry, Cambridge, 1995), Vol. 6.
8. S. Sealey (Ed.), *Modern Chlor-Alkali Technology* (Royal Society of Chemistry, Cambridge, 1998), Vol. 7.
9. J. Moorhouse (Ed.), *Modern Chlor-Alkali Technology* (Blackwell Science, Oxford, 2001), Vol. 8.
10. K. Köster and H. Wendt, in *Comprehensive Treatise of Electrochemistry*, edited by J. O'M. Bockris, B. E. Conway, E. Yeager, and R. E. White (Plenum Press, New York, 1981), Vol. 2, p. 251.
11. H. Lund and Ole Hammerich (Eds.), *Organic Electrochemistry* (4th Edition, Marcel Dekker, New York, NY, 2001).
12. J. D. Genders and D. Pletcher (Eds.), *Electrosynthesis–From Laboratory to Pilot to Production* (Electrosynthesis Company, Lancaster, N.Y., 1990).
13. D. E. Danly, *Emerging Opportunities for Electroorganic Processes* (Marcel Dekker, New York, NY, 1984).
14. *Proceedings of the Annual Forum on Applied Electrochemistry*, Electrosynthesis Company, Lancaster N.Y., 1987–, all 15 Vols.
15. Electrolytic Processes–Present and Future Prospects, Electrosynthesis Company and Dextra Associates, Report Number TR-107022, EPRI, Palo Alto, California, December 1997.
16. T. R. Beck, R. Alkire, and N. L. Weinberg, A Survey of Organic Electrolytic Processes, DOE Contract Number 31-109-4209, ANL/OEPM-79-5, November 1979.
17. F. Hine, *Electrochemical Processes and Electrochemical Engineering* (Plenum Press, New York, NY, 1985).
18. D. Pletcher and F. C. Walsh, *Industrial Electrochemistry* (2nd edition, Chapman and Hall, London, 1990).
19. M. G. Fontana, N. D. Green, *Corrosion Engineering* (McGraw Hill, New York, 1989).
20. R. W. Revie (Ed.), *Uhlig's Corrosion Handbook* (2nd edition, Wiley, New York, 2000).
21. J. O'M. Bockris, S. U. M. Khan, *Surface Electrochemistry: A Molecular Level Approach* (Plenum, New York, 1993).

## Cited References

1. http://www.cia.gov/cia/publications/factbook/fields/2042.html
2. R. A. Papar and P. E. Scheihing, *Chemical Processing*, Feb (2001).

3. R. L. LeRoy, C. T. Boen, and D. J. LeRoy, *J. Electrochem.Soc.* **127**, 1954 (1980).
4. R. L. LeRoy, *J. Electrochem.Soc.* **130**, 2159 (1983).
5. D. Pletcher and F. C. Walsh, *Industrial Electrochemistry* (Chapman and Hall, London, 1982).
6. R. E. Sanders, Jr., Aluminum and Aluminum Alloys, in *Kirk-Othmer Encyclopedia of Chemical Technology* (John Wiley & Sons, New York, NY, 2002).
7. P. Arora and V. Srinivasan, *J. Electrochem Soc.* **149**, k1 (2002).
8. V. DeNora, *Interface*, 11(4), 20, The Electrochemical Society, 2002.
9. T. R. Beck, Final Report on Improvements in Energy Efficiency of Industrial Processes, ANL/OEPM-77-2, Argonne National Laboratories, Argonne, IL, 1977.
10. V. A. Ettel, Electrometallurgy and Energy Crunch, in *Energy Considerations in Electrolytic Processes 1* (Society of Chemical Industry, London, 1980).
11. F. R. Tuler and R. Scott-Taggert, *Encyclopedia of Materials: Science and Technology,* **3**(11), 2001.
12. T. V. Bommaraju, B. Lüke, G. Dammann, T. F. O'Brien, and M. C. Blackburn, in *Kirk-Othmer Encyclopedia of Chemical Technology*, electronic version (John Wiley & Sons, New York, NY, 2002).
13. L. C. Curlin, T. V. Bommaraju and C. B. Hansson, in *Kirk-Othmer Encyclopedia of Chemical Technology* (6th Edition, John Wiley & Sons, New York, NY, 1991), Vol. 1, p. 938,
14. P. Schmittinger (Ed.), *Chlorine–Principles and Industrial Practice* (Wiley-VCH, Weinheim, 2000).
15. H. S. Burney, N. Furuya, F. Hine, and K.-I.Ota (Eds.), Chlor-Alkali and Chlorate Technology, in *R. B. Macmullin Memorial Symposium*, (The Electrochemical Society Inc., Pennington, NJ, 1999), PV 99-21.
16. D. Bergner, M. Hartmann, and R. Staab, Entwicklungsstand der Alkaili-Chlorid Elektrolyse, Teil1: Zellen, Membranen, Elektrolyte, Produkte, *Chem.-Ing.-Tech.* **66**(6), 783-791 (1994).
17. K. Schneiders, A. Zimmermann, and G. Henßen, Membranelektrolyse – Innovation für die Chlor-Alkali-Industrie, *Forum Thyssen Krupp*, Vol. 2, 2001.
18. M. Faraday, *Poggendorfs Ann.Phys.Chem*, **33**, 438 (1834).
19. K. Köster and H. Wendt, in Comprehensive Treatise of Electrochemistry, edited by J. O'M. Bockris, B. E. Conway, E. Yeager, and R. E. White (Plenum Publishing Corporation, New York, NY, 1981), Vol. 2.
20. N. L. Weinberg, *Electroorganic Synthesis*, ACS Audio Course, 1979 [It is also found in N. L. Weinberg and B. V. Tilak (Eds.), *Techniques of Electroorganic Chemistry* (Wiley Interscience, New York, NY, 1982), Part III.]
21. Workshop on the Status of Industrial Organic Electrochemistry, SRI International, EPRI EM-2173, Project 1086-9, Proceedings, Research Reports Center, Box 50490, Palo Alto, CA, 1981.
22. Y. C. Yen and S- Y. Wu, Process Economics Program Report, #54B, SRI International, Menlo Park, CA, September 1987.

23. R. W. Spillman, R. M. Spotnitz, and J. T. Lundquist Jr., *Chem Tech*, (March) 176 (1984).

24. J. C. Trocciola, U.S. Patent 4,566,957 (1986).

25. N. L. Weinberg, J. D. Genders, E. A. George, P. M. Kendall, D. J. Mazur, and G. D. Zappi, Proceedings of the Symposium on Fundamentals and Potential Applications of Electrochemical Synthesis, edited by R. D. Waevert, F. Fisher, F. R. Kalhammer, and D. Mazur (The Electrochemical Society, Inc., Pennington, NJ, 1997), Proc. Vol. 97-6, p.69.

26. F. E. Goodwin, Zinc and Zinc Alloys, in *Kirk-Othmer Encyclopedia of Chemical Technology*, electronic version, (John Wiley & Sons, NewYork, 1998).

27. K. J. A. Kundig and W. H. Dresher, in *Kirk-Othmer Encyclopedia of Chemical Technology*, electronic version (John Wiley & Sons, New York, 2001).

28. J. H. Tundermann, J. H. Tien, and T. E. Howson, in *Kirk-Othmer Encyclopedia of Chemical Technology*, electronic version (John Wiley & Sons, New York, 2000).

29. V. A. Ettel and B. V. Tilak, in *Comprehensive Treatise of Electrochemistry*, edited by J. O'M. Bockris, B. E. Conway, E. Yeager, and R. E. White (Plenum Publishing Corporation, New York, 1981), Vol 2, p. 327.

30. A. K. Biswas and W. G. Davenport, *Extractive Metallurgy of Copper*, 3$^{rd}$ edition (Pergamon Press, New York, NY, 1994), p. 458.

31. J. A. Gonzalez (Cominco Metals) (private communication, 2003).

32. V. A. Ettel, *Energy Considerations in Electrolytic Processes* (Society of Chemical Industry, London, 1980), p. 1.

33. G. M. Cook, Energy Reduct. Tech. Met. Electrochem. Processes, Proc. Symp, edited by R. G. Bautista and R. J. Wesley, (Metall. Soc, Warrendale, Pa, 1985), p. 285.

34. P. H. Rieger, *Electrochemistry*, 2$^{nd}$ edition (Chapman and Hall, London, 1994).

35. P. R. Roberge, *Handbook of Corrosion Engineering* (McGraw Hill, New York, NY, 1999).

36. D. Linden and T. Reddy (Eds.), *Handbook of Batteries*, 3$^{rd}$ edition (McGraw-Hill, New York, NY, 2002).

37. A. J. Appleby, J. Jacquelin, J. P. Pompon, *Society of Automotive Engineers* (technical paper), 9 (1977).

38. Wolfgang Gerhartz (Ed.), *Ullman's Encyclopedia of Industrial Chemistry*, (5th edition, VCH Publishers, U.S.A., 1985), Vol. 6.

39. J. McBreen, in *Comprehensive Treatise of Electrochemistry*, edited by J. O'M. Bockris, B. E. Conway, E. Yeager, and R. E. White (Plenum Publishing Corporation, New York, NY, 1981), Vol. 3, p. 324.

40. H. L. Becker and V. Ferry, U.S. Patent 2,800,616 (1957).

41. S. Trasatti and G. Buzzanca, *J. Electroanal. Chem.*, **29**, APR1 (1971).

42. B. E. Conway, *Electrochemical Supercapacitors: Scientific Fundamentals and Technological Applications* (Kluwer Academic/Plenum Publishers, New York, NY, 1999).

43. *International Seminar on Double Layer Capacitors and Similar Energy Storage Devices*, symposium organized by S. P. Wolsky and N. Marincic, 1991 to 2002, Florida Educational Seminars, Boca Raton, FL.
44. S. Sarangapani, B. V. Tilak, and C. -P. Chen, *J. Electrochem Soc.* **143**, 3791 (1996).
45. S. Ardizzone, G. Fregonara, and S. Trasatti, *Electrochim. Acta*, **35**, 263 (1990).
46. D. McDonald and S. Narang, SRI final report, Contract # DAA L01-88-C-0840, 1991.
47. http://www.powerpulse.net/powerpulse/archive/aa_101898c8.stm.
48. J. D. Miller, Electrochemical Society short course on Electrochemical Capacitors, Paris, 2003.
49. A. F. Burke, *J. Power Sources*, **91**, 37 (2000).
50. B. E. Conway, Electrochemical capacitors: Their Nature, Function and Application, http://electrochem.cwru.edu/ed/encycl/art-c03-elchem-cap.htm

## PROBLEMS

1. Calculate the thermodecomposition and thermoneutral voltage of diaphragm, mercury, and membrane cells at various temperatures. Discuss whether or not these cells can be operated at the thermoneutral voltage, and if so, why so and if not, why not?

2. Describe the anode, cathode and the overall reactions involved in electrolytic production of aluminum by the Hall-Hèrault process?

3. Calculate the operating and capital costs for producing 1000M tons of copper per year. What is the % of electricity and material costs used for producing one ton of copper?

4. What are the various anodic and cathodic reactions that can supplant the hydrogen evolution and oxygen evolution reactions to achieve energy savings in electro-organic processes?

5. Calculate the maximum capacitance that can be realized with pseudocapacitance based capacitors?

# PART II:

# FUNDAMENTAL ASPECTS FOR RESEARCH AND DEVELOPMENT OF FUEL CELLS

# CHAPTER 4
# FUEL CELL PRINCIPLES

## 4.1.  SCOPE OF CHAPTER

Knowledge of basic and applied electrochemistry, as well as of electrochemical engineering is important for an understanding of the *modus operandi* of fuel cell power plant/power sources. It is for this reason that the first three chapters focused on evolution of electrochemistry, charge transfer reactions at electrode/electrolyte interfaces, and leading electrochemical technologies and applications. Chapter 3 was included because the electrochemical industries consume about 5% of the electrical energy consumed in the USA, and several of these technologies are in some way relevant to fuel cells. As stated in Chapter 3, there are three types of electrochemical technologies—electrochemical synthesis, electrochemical energy conversion, and electrochemical energy storage. Even though this chapter is entitled "Fuel Cell Principles," Section 4.2 is included to familiarize the reader with the basic similarities and dissimilarities among the three types of electrochemical technologies. Section 4.3 discusses the evolutionary aspects of fuel cells and their classification and highlights the inventions and demonstrations in the 20th and 21st centuries. Detailed technical descriptions on the six leading fuel cell technologies and abbreviated descriptions on other types of fuel cells, which are still in a development stage, are presented in Chapter 9. Sections 4.4 and 4.5 concentrate on the thermodynamic and electrode kinetic aspects of fuel cells. These two sections, though relatively short, are essential for an analysis of the performance characteristics of fuel cells. Thermodynamic and electrode kinetic parameters govern the performance of a fuel cell, as demonstrated in Section 4.6. This analysis logically leads to Section 4.7, which deals with the influence of the electrode

This chapter was written by S. Srinivasan, L. Krishnan and C. Marozzi.

189

# TABLE 4.1
## Basic Similarities and Dissimilarities among the Three Types of Electrochemical Technologies

| Electrochemical energy conversion (fuel cells) | Electrochemical energy storage | | | Electrochemical synthesis |
|---|---|---|---|---|
| | Batteries | | Electrochemical capacitors | |
| | Primary batteries | Secondary batteries | | |
| • Spontaneous cell<br>• Consumes fuel and oxidant<br>• Generates electricity | • Spontaneous cell<br>• Consumes chemicals<br>• Generates electricity | • Driven/spontaneous cell<br>• Consumes electricity<br>• Generates/consumes chemicals<br>• Generates electricity | • Same as secondary batteries | • Driven cell<br>• Consumes electricity<br>• Generates chemicals |
| • Thermodynamic reversible potentials and overpotential losses determine efficiency for conversion of chemical to electrical energy<br>• Activation overpotentials predominate at low current densities<br>• Mass transport overpotentials at higher current densities | • Thermodynamic reversible potentials and ohmic overpotentials determine efficiency for energy conversion<br>• Activation overpotentials significant in metal-air batteries | | • Redox potentials for positive and negative electrode reactions determine efficiency of cells<br>• Ohmic overpotentials could be significant | • Thermodynamic reversible potentials and overpotential losses determine efficiency of cell<br>• Ohmic overpotential predominates |
| • Cathode $\rightarrow$ positive electrode<br>• Anode $\rightarrow$ negative electrode | • Cathode $\rightarrow$ positive<br>• Anode $\rightarrow$ negative | Discharge:<br>• Cathode $\rightarrow$ positive<br>• Anode $\rightarrow$ negative<br>Charge:<br>• Cathode $\rightarrow$ negative<br>• Anode $\rightarrow$ positive | • Same as secondary batteries | • Cathode $\rightarrow$ negative<br>• Anode $\rightarrow$ positive |
| • High true surface area of electrocatalysts (i.e., high roughness factors) is essential | • Similar to fuel cells, more so for gas-consumption/gas-evolution electrode reactions | | • Same as fuel cells | • Same as fuel cells, more so for gas-evolution reactions |

kinetic parameters on the performance of proton exchange membrane fuel cells (PEMFCs) and direct methanol fuel cells (DMFCs).

## 4.2 BASIC SIMILARITIES AND DISSIMILARITIES AMONG THE THREE TYPES OF ELECTROCHEMICAL TECHNOLOGIES

Before we address the main topics in this chapter, we will summarize the distinguishing features of the three types of electrochemical technologies—i.e., electrochemical energy conversion (fuel cells), electrochemical energy storage (batteries and electrochemical capacitors), and electrochemical synthesis. The three types of electrochemical technologies may be broadly defined as follows:

(a)  *Electrochemical energy conversion.* It is defined as a spontaneous reaction in an electrochemical reactor that consumes a fuel and an oxidant, and their reactions at the anode and the cathode generate electricity, heat, and the products of the reaction.

(b)  *Electrochemical energy storage.* There are basically two types: (i) batteries and (ii) electrochemical capacitors. The batteries may be further subdivided into two types, i.e., primary and secondary batteries (rechargeable). In the case of primary batteries, the chemicals in a sealed container react at the individual electrodes (anode and cathode), set up a potential difference, and when connected to an external load, generate electricity spontaneously. In the case of a secondary battery, electrical energy is stored as chemicals in an oxidized form and a reduced form when the battery is charged (a driven cell); when the electrical energy is required, it is connected to an external load (a spontaneous cell). Electrochemical capacitors are akin to secondary batteries, which are used to store electrical energy, as in solid state capacitors. The main difference from the latter is that by using high surface area electrodes (such as carbon, noble metal oxides, conducting polymers), the amount of energy storage can be considerably higher than in solid state capacitors. In addition to the double layer capacitances, the pseudocapacitances at the individual electrodes further enhance the energy storage capacity. (For more details, refer to Chapter 3).

(c)  *Electrosynthesis.* This electrochemical reaction occurs in a driven cell. Electric energy delivered to the electrochemical cell generates chemicals at the individual electrodes by the decomposition of the reactants. The electrodes may play an active or a passive role depending on the types of reactions. (Further details are presented in Chapter 3).

An attempt is made in Tables 4.1 and 4.2 to summarize (i) the basic similarities and dissimilarities and (ii) performance characteristics, economics, and applications of the three types of electrochemical technologies. As exemplified in chapter 3, electrochemical synthesis and battery technologies have reached the most advanced state since the latter half of the 19th century. Even though fuel cell technology was

**TABLE 4.2**

**Typical Performance Characteristics, Applications, and Economics of the Three Electrochemical Technologies**

| Performance characteristics and economic applications | Electrochemical energy conversion (fuel cells) | Electrochemical energy storage | | Electrochemical synthesis |
|---|---|---|---|---|
| | | Batteries | Electrochemical capacitors | |
| Thermodynamic reversible potential ($E_r$), V | 1.2–1.35 | 1–6 | 1–1.5 | 1–6 |
| Stack design | Mostly bipolar | Mostly monopolar, some bipolar | Mostly monopolar | Monopolar/bipolar |
| Operating temperature range ($T$), °C | 0–1,000 | (−30)–550 | 0–100 | 1–1,000 |
| Overpotential losses: Activation | High for low-intermediate temperature | Low for most batteries | Low | High for gas-evolution/gas-consumption reaction |
| Ohmic | High in intermediate and high c.d. range | High for metal-air batteries | Low to high | Could be high |
| Mass transport | High in high c.d. range | Not significant | Not significant | Not significant |
| Voltage efficiency, % | 30–60 | 60–70 | 60–70 | 40–80 |
| Power density (output or input) based on geometric area of electrodes, mW/cm² | 0.1–600 | < 0.1 | 2–10 | 100–500 |

**TABLE 4.2. (Continuation)**

| Performance characteristics and economic applications | Electrochemical energy conversion (fuel cells) | Electrochemical energy storage | | Electrochemical synthesis |
|---|---|---|---|---|
| | | Batteries | Electrochemical capacitors | |
| Specific power for system, W/kg | 100–500 | 1–1,000 | 100–1,000 | 100–500 |
| Power density for system, W/l | 10–600 | 200–300 | Could be as high as 10 kW/l | N/A |
| Specific energy for system, Wh/kg | 10–600 | 10–200 | 10–50 | N/A |
| Lifetime of electrochemical stack, years | 0.5–5 | 0.1–10 | 0.1–1 | 1–10 |
| Capital cost, $ | 50–10,000/kW | 10–1,000/kWh | 100–1,000/kWh | 2,000–5,000/kW |
| Operating and maintenance cost, $/kWh | 0.1–1 | ~0 | ~0 | 0.1–1 |
| Commercialized/demonstrated applications | Auxiliary power for space vehicles since 1961 Power generation, co-generation, portable power, power range from 10W–10MW | Electric-utility energy storage, standby/emergency Starter batteries for transportation vehicles Power source for tools, computers, cell phones, pacemakers, defibrillators | Peak power | Aluminum and chlor-alkali production Electrowinning of metals Electroorganic synthesis Water electrolysis Corrosion protection/ passivation Bioelectrochemistry |

invented in the 19$^{th}$ century and developed in the 20$^{th}$ century, it has yet to find large-scale applications, except for fuel cells serving as auxiliary power sources in space flights. Electrochemical capacitors are in the infant stage of development, demonstration, and commercialization. To make it easier for the reader, summarizing remarks of Tables 4.1 and 4.2 are presented as follows.

## 4.2.1.  Table 4.1

As stated above, fuel cells, primary and rechargeable batteries, and electrochemical capacitors (both in the charging mode) are driven cells. This is also the case in electrochemical synthesis. The thermodynamic reversible potentials $(E_r)$ for all these electrochemical cells govern the potential at which these reactions occur. The range of $E_r$ for fuel cells is small (1.0 to 1.35 V); the $E_r$ values for the different types of batteries are quite high, i.e., 1.0 to 6.0 V, the latter value belonging to a lithium-fluorine battery. The $E_r$ range for pseudocapacitors is limited, as in the case of fuel cells; and in electrosynthesis, this range is high, e.g., for metal deposition-dissolution reactions it could be small, whereas for the production of lithium and aluminum, $E_r$ values are high. Overpotential losses at the electrode-electrolyte interfaces in the active layer of the electrode and in the electrolyte are the causes of efficiency losses in electrochemical cells. In the case of low and intermediate temperature fuel cells—proton-exchange-membrane fuel cell (PEMFC), direct-methanol fuel cell (DMFC), alkaline fuel cell (AFC) and phosphoric-acid fuel cell (PAFC)— activation overpotentials are most significant; ohmic and mass transport overpotentials could be as high in the intermediate and high current density region. Since the operating temperatures of molten-carbonate fuel cells (MCFCs) and solid-oxide fuel cells (SOFCs) are quite high, activation overpotentials are markedly reduced. But for these types of fuel cells, $E_r$ values are lower by about 0.1 to 0.2 V because of the decrease of the free energy change of the fuel cell reaction (see Section 4.4) with an increase in temperature. Ohmic and mass transport overpotentials are relatively high at intermediate current densities for MCFCs and SOFCs. There is also a Nernstian loss of $E_r$, which depends on the percent utilization of the reactants in these types of fuel cells.

Activation overpotential losses are relatively small in batteries, electrochemical capacitors, and electrochemical synthesis, the exceptions being for some gas-evolution/gas-consumption reactions (e.g., oxygen evolution/reduction) and in electroorganic synthesis. Ohmic overpotentials could be relatively high for these electrochemical technologies. Mass-transport overpotential losses are relatively low but reach appreciable values in gas-evolution/gas-consumption reactions. The cathode is the positive electrode and the anode is the negative one when an electrochemical cell is operating in the spontaneous mode. The reverse is the case for the driven cell. High surface areas of the electrodes (i.e., high roughness factors, which is defined as the ratio of the true surface area to the geometric surface area of the electrode) are vital for fuel cells, some batteries, and in gas dissolution/evolution reactions. Desirable values of roughness factors are in the range of 50 to 100. The main purpose of these high roughness factors is to minimize the activation overpotentials. High values of roughness factors are achieved by using supported electrocatalysts and porous electrodes. The latter is essential to increase the three-dimensional reaction zone for the electrochemical reaction. Reasonably high values of surface areas are required for all the electrochemical technologies. This is made possible by using rough or sintered electrodes.

## 4.2.2.  Table 4.2

An attempt is made in this table to analyze the performance characteristics, economics, and applications of the three types of electrochemical technologies. The values of the parameters in the table are reasonably correct and comparable for the commercialized technologies but these are projected for the technologies that are still under development or being demonstrated. With respect to stack design, fuel cells have, in general, a bipolar construction (see Chapter 9). One exception is in the case of the most advanced Siemens-Westinghouse SOFC, which has a unipolar design. The main advantage of the bipolar design over the unipolar one is that the electrons from the whole area of the anode in one cell are transported to the whole surface area of the next cell via the bipolar plate. The thickness of the bipolar plate can also be minimized to a few millimeters. Thus, the ohmic overpotential is greatly reduced. In a cell stack with the unipolar design, there is edge collection of electrons from the anode and there is an external cable connection to the cathode of the next cell. Thus, in this type of cell stack the electron flow is longitudinal over the length of the electrode via a relatively short cross-sectional area. The ohmic overpotential could therefore be relatively high. One advantage of the unipolar design is that if one cell in an electrochemical cell stack is malfunctional, it could be replaced relatively easily. This is not the case for an electrochemical cell stack with bipolar construction—the cell stack will have to either be disassembled and the damaged cell replaced or the entire stack may have to be replaced. Most electrochemical technologies generally have a monopolar construction. Notable exceptions are some kinds of batteries and some electrosynthesis technologies.

All the electrochemical technologies, except electrochemical capacitors, operate over a wide range of temperatures, as is necessary. The operating-current density range is also wide for all these technologies. In the case of electrochemical capacitors, very high current densities are achieved because the discharge occurs over a short period of time.

The topic of performance characteristics is presented in the summarizing statement for Table 4.1. Because of the overpotential losses, there could be significant departure of the efficiency from the theoretical value. As seen from Table 4.2, this is more so for fuel cells than for the other electrochemical technologies because of the high values of activation overpotentials for the low and intermediate temperature fuel cells. Values of power densities ($mW/cm^2$) are approximately in the same range for all the electrochemical technologies; for batteries and electrochemical capacitors higher values could be reached for peak power demands. The specific power (W/kg) for the system is higher for some batteries and electrochemical capacitors, when there are peak power demands. The low specific power of fuel cells is because of the considerable weights of fuel processors or of hydrogen storage. The power density of an energy conversion/storage system (W/l) also follows the same trend. Optimizing the values of these parameters is vital for electrochemical energy conversion and storage devices, particularly for the transportation and portable power applications. This is also the case with specific energy and energy density. A problem with fuel cells is that even though the energy density and specific energy could be higher than for energy storage devices by a factor of about 10, the weight of the fuel processor or of the hydrogen storage system are so high that the specific energy and energy density are somewhat lower than the batteries.

Lifetimes and cycle life are fairly well established for batteries. They are remarkably high for some nickel-based batteries, lead acid batteries, and silver-zinc batteries. The cycle life is also high for an electrochemical capacitor. The challenge is to enhance the lifetimes for fuel cells. The targeted value for fuel cell systems is about 40,000 h for the power generation/cogeneration application—this is for the electrochemical cell stack. This value has been demonstrated for the 200 kW PAFC system and developed and commercialized by UTC Fuel Cells. The main causes for performance degradation are connected with corrosion problems and to some extent poisoning of electrodes.

The values quoted for capital and operating costs in Table 4.2 are reasonably reliable for batteries and for electrosynthesis. This is not so for fuel cells and electrochemical capacitors because they are in the developmental stage except for the PAFC system. In the latter case, the capital cost is $4000/kW, which is higher at least by a factor of four, than the competing energy conversion technologies. These topics are dealt with in more detail in Chapters 10 and 11. The commercialized/demonstrated applications are specified in Table 4.2. Only the leading technologies are taken into account.

## 4.3.  TYPES OF FUEL CELLS

### 4.3.1.  Evolutionary Aspects of Fuel Cells and their Classifications

The fuel cell is a 19$^{th}$ century invention and a 20$^{th}$ century high technology development. The R&D efforts in these two periods will lead to reaching the era of clean and efficient power generation in the 21$^{st}$ century. The evolutionary aspects of fuel cells have been excellently dealt with in books and review articles (see Suggested Reading). Thus, in this chapter, an effort is made to present the historical sequence of fuel cell inventions in the 19$^{th}$ century (see Table 4.3). The notable features are:

- Sir William Grove had great insight in demonstrating a $H_2/O_2$ fuel cell in acid electrolytes;

## TABLE 4.3
## Historical Sequence of Fuel Cell Inventions in the 19th Century

| Year | Inventor(s) | Invention | Comments |
|------|-------------|-----------|----------|
| 1839 | W. R. Grove | $H_2/O_2$ fuel cell with spongy platinum electrodes and sulfuric acid electrolyte<br>Four cells connected in series and DC power used to electrolyze water | Grove aware of need for three phase contact |
| 1889 | L. Moud and C. Langer | Scaled-up cell area to 700 $cm^2$, used Pt foil and Pt black electrode<br>Sulfuric acid electrolyte in diaphragm (e.g., plaster of Paris, asbestos) | Current density: 2.5–3.5 $mA/cm^2$, 50% efficiency |
| 1894 | W. Ostwald | Advantage of chemical to electrical energy to attain high efficiency, "no smoke, no soot and no fire" | First analysis showing advantages of fuel cells over thermal engines |
| 1877 | A. C. Becquerel and A. E. Becquerel | Fuel-carbon rod<br>Molten-nitrate electrolyte<br>Platinum or ion crucible served as counter electrode | Direct utilization of carbon as fuel |
| 1896 | J. J. Jacques | Coal used as fuel<br>Molten-potassium or -sodium hydroxide electrolyte<br>Iron vessel served as cathode<br>Largest fuel cell stack built,1.5 kW | Current density 100 $mA/cm^2$ at 1 V |
| 1897 | W. Nernst | Nernst glower or lamp using high-temperature anion conducting solid $ZrO_2$ with 15% $Y_2O_3$<br>Also built fuel cell in 1900 and redox cell | Same electrolyte still used in leading SOFC |

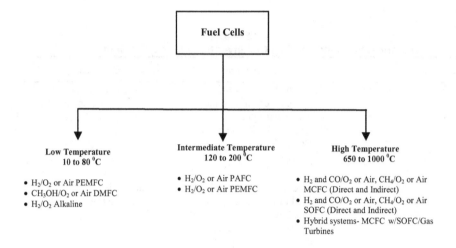

**Figure 4.1.** Classification of fuel-cells into leading technologies.

- Mond and Langer paved the way for developing acid electrolyte fuel cells, phosphoric acid, and proton exchange membrane fuel cells;

- Jacques was the first investigator to design, construct, and demonstrate an AFC power plant (1.5 kW);

- The pioneering work of Bacon from 1932-1952 in developing a 5 kW AFC led to its further development and application as an auxiliary power source for NASA's Apollo Flights. This system was developed by Pratt and Whitney, followed by International Fuel Cells (IFC), both divisions of United Technologies Corporation;

- General Electric Company was the first to develop, demonstrate, and find an application for PEMFCs (originally named Solid Polymer Electrolyte Fuel Cells). This type of fuel cell was the first to find an application—an auxiliary power source (1 kW) for NASA's Gemini Space Flights;

- Investigations by Bauer et. al, followed by Davtyan and coworkers and Broers and Ketelar provided great insight for the development of the MCFC technology;

- Nernst and Schottky invented the electrolyte in the first SOFC. This was followed by the R&D work of researchers of Westinghouse Corporation in developing the "bell and spigot" type SOFCs;

- Researchers at Allis Chalmers, Exxon, Shell, and Hitachi carried out the first investigations to demonstrate the DMFC. Sulfuric acid was used as the electrolyte in these systems.

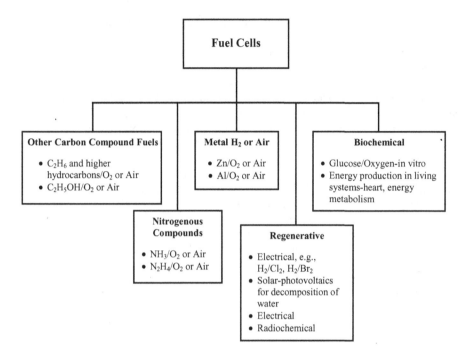

**Figure 4.2.** Classification of other types of fuel cells investigated.

In previous publications, there are probably as many classifications of fuel cells as types of fuel cells. The most common classifications have been according to (i) the operating temperature, (ii) the fuel used for direct utilization, and (iii) the electrolyte in the fuel cells. In this chapter and in Chapter 9 we deal with the leading types and other types of fuel cells. Figure 4.1 represents the leading types of fuel cell technologies, which have been explored since the 1960s (the first era in fuel cell R&D). Most of these investigations had been at a fundamental level until the early 1970s; but many of them have reached the stage of developmental technology at present. Several other types of fuel cells have been investigated (Figure 4.2), but practically all these types of fuel cells have mainly been investigated only at a research level. Detailed description of the leading fuel-cell technologies, which have progressed from half cell→single cell→cell stack→system are presented in Chapter 9, which also summarizes the other types of fuel cells.

## 4.3.2. Inventions and Demonstrations in 20<sup>th</sup> and 21<sup>st</sup> Century

The following Sections deal with the significant inventions and demonstrations of the leading types of fuel cells in the 20<sup>th</sup> and 21<sup>st</sup> century.

**4.3.2.1.** *1904-1907.* The first evidence for an indirect fuel cell was found when the anodic reaction in a fuel cell with C (C→H$_2$→H$_2$O) and Pd foil as the diffusion hydrogen electrode was invented. The temperature and pressure effects of the fuel cell reaction led to focus on high temperature fuel cells.

**4.3.2.2.** *1910-1939.* E. Baur and coworkers used molten sodium and potassium carbonates in a porous ceramic (MgO) as an electrolyte with molten Hg as the oxygen cathode and a carbon rod, Pt-Co, or Pt-H$_2$ anode. The open circuit potential recorded (1 V) at a temperature of 1000 $^0$C with H$_2$ + CO fuel had the same value as the thermodynamic reversible potential. This led to the idea of using matrix/ceramics to retain electrolyte in present day MCFCs.

**4.3.2.3.** *1932-1952.* F. T. Bacon first demonstrated the alkaline fuel cell system (5 kW). He was the pioneer of modern fuel cell research, which gave rise to the birth of Pratt and Whitney's AFCs for Apollo Space Shuttle Flights. Double-porosity sintered-nickel electrodes were used in place of noble metal electrocatalysts. The electrodes were made corrosion-resistant by the impregnation of the nickel oxide cathode with LiOH to form lithiated-nickel oxide. The cell was operated at a temperature of 200 $^0$C and a pressure of 5 bar.

**4.3.2.4.** *1938-1971.* O. K. Davtyan and coworkers produced fuel by a coal gasification procedure by melting a mixture of monazite sand, tungsten oxide, and soda glass. A clay matrix was used for the Na$_2$CO$_3$ electrolyte. At 70 $^0$C, the cell potential at 20 mA/cm$^2$ was 0.79 V, and the open circuit voltage was 0.85 V.

**4.3.2.5.** *1935-1937.* Several solid electrolytes like clay and kaolin with CeO$_2$, lithium silicate, and the Nernst electrolyte (ZrO$_2$ with 15% Y$_2$O$_3$) were investigated for SOFC technology. The best cell stack constructed consisted of a magnetite anode and a coke iron power cathode. The invention of this stack led to the modern SOFC technology development.

**4.3.2.6.** *1948-1975.* E. W. Justi invented the double-skeleton fuel cell electrodes in 1948. An AFC was built using the Raney nickel electrocatalyst with 30-50% KOH as the electrolyte. At a cell temperature of 67 $^0$C, the cell potential at 250 mA/cm$^2$ was 0.65 V. A lifetime of 1.5 years was achieved with this fuel cell at a temperature of 30–35 $^0$C, and Varta AG and Siemens subsequently continued this work.

4.3.2.7. *1950-1965*. The foundation for modern MCFC technology was laid during this period. Porous sintered MgO discs, impregnated with $Li_2CO_3$-$K_2CO_3$ eutectic electrolyte and a ternary electrolyte with addition of $Na_2CO_3$ were investigated. Thin layers of metal (Ni for anode and Ag for cathode) covered by metal gauze were used as electrodes. The cells operated with $H_2$, CO, and natural gas plus steam fuels. The operating cell temperature was 700 $^0$C and a voltage of 0.7–0.8 V was obtained at 50 mA/cm$^2$.

4.3.2.8. *1938-1965*. The Bell and Spigot tubular design SOFC was first designed by J. Weisbart et al. at Siemens-Westinghouse Electric Corporation. The electrolyte used was calcia-stabilized zirconia with sintered Pt as electrodes. Ni-Zirconia cermet anodes and electronically conducting oxide cathodes were used later. The typical operating temperature of the SOFC was 1000 $^0$C, and this SOFC was the origin of modern Siemens-Westinghouse technology.

4.3.2.9. *1959-1982*. Investigators from General Electric and Dupont paved the way for modern PEMFC technology development. The first PEMFCs were demonstrated using unsupported Pt electrocatalyst and polystyrene sulfonic acid (PSS) membranes. The PSS membranes were subsequently replaced with Dupont's Nafion (a perflurosulfonic acid). The first PEMFC was used in the NASA's Gemini Space Flights. The 1 kW PEMFC system served as an auxiliary power source and drinking water source for the astronauts. The operating temperature of the PEMFCs was around 40-80 $^0$C.

4.3.2.10. *1965-1995*. PAFC power plants (10 kW-10 MW) were first demonstrated by UTC Fuel Cells, General Electric Company, Exxon, Toshiba, and several other companies and organizations. Teflon bonded electrodes (carbon supported electrodes with Pt electrocatalysts) are the characteristic features of the PAFC systems. The commercially available 200-kW PAFC units are used for terrestrial applications and cogeneration (electricity and heat) purposes. The electrolyte is 100% $H_3PO_4$ at operating temperatures of 80–200 $^0$C. The commercial 200-kW unit consists of the fuel processor/ electrochemical cell stack and power conditioner, and has demonstrated a lifetime of 40,000 h. Intermediate temperature PAFC units were also demonstrated and used reformed natural gas/methanol as a fuel with Pt or Pt alloy catalysts supported on high-surface area carbon. The CO tolerance level was 1–2%.

4.3.2.11. *1958-present*. Alkaline-fuel cells (AFC) with power levels of 0.5–10 kW operating with $H_2/O_2$ or air were developed in the late 1950s. Non-noble metal and oxide electrocatalysts, low loading of supported Pt catalysts, and conductive plastic bipolar plates are the significant features of the multi kW AFC units. The AFCs operated best with $H_2/O_2$. When air was used as the cathodic reactant, $CO_2$ level had to be reduced from 350 ppm to practically zero values.

4.3.2.12. *1965-Present.* Tape-casting methods were used to prepare layers of electrodes and electrolyte for MCFC power plants. Modules of up to about 300 kW and 1.3 kW MCFC/gas-turbine power plants were constructed, demonstrated, and commercialized for cogeneration applications. Methanol was used as a direct or indirect fuel. Due to high temperature operation of the MCFCs, CO served as a fuel and not a poison, and relatively inexpensive component materials were used. The MCFC power plants were most beneficial for dispersed electric and heat power generation including chemical industries.

Siemens-Westinghouse developed the most advanced SOFC technology. Both tubular design and bipolar design SOFCs were used to design power plants ranging from 1–100 kW. This power plant is an attractive two-phase system with no need for management of liquid electrolytes. Corrosion problems are less severe than the other types of fuel cells. Lower operating temperature is preferred and solid oxide ion conductors are now under investigation for this purpose.

## 4.4.    THERMODYNAMIC ASPECTS

### 4.4.1.  Standard Free Energy and Enthalpy Change of a Fuel Cell Reaction

The fuel-cell reaction is a chemical process separated into two electrochemical half-cell reactions. The simplest and common reaction encountered in fuel cells is:

$$H_2 + \frac{1}{2}O_2 \rightarrow H_2O \tag{4.1}$$

From a thermodynamic point of view, the maximum-electric work obtained from the above reaction corresponds to the free-energy change (available energy in an isothermal process) of the reaction. Gibbs-free energy is more useful than the change in Helmholtz-free energy, since it is more practical to carry out chemical reactions at a constant temperature and pressure rather than constant temperature and volume.

The above reaction is spontaneous and is also thermodynamically favored because the free energy of the products is less than that of the reactants. The standard free energy change of the fuel cell reaction is represented by the equation:

$$\Delta G = -nFE_r \tag{4.2}$$

where $\Delta G$ is the free energy change, $n$ is the number of moles of electrons involved, $E_r$ is the reversible potential, and $F$ is Faraday's constant.  If the reactants and the products are in their standard states i.e., at a temperature of 25 $^0$C and 1 atm pressure, the equation can be rewritten as:

$$\Delta G^0 = -nFE_r^0 \tag{4.3}$$

Thus, for the reaction (4.1), $\Delta G$ is –229 kJ/mol, n = 2, F = 96,500 C/eq and, hence, the calculated value of $E_r$ is 1.29 V.

Water is the product of the fuel cell reaction and can be produced either as liquid water or steam. The higher-heating value (HHV) corresponds to the released heat when water is produced as liquid water and the lower-heating value (LHV) when water is produced as steam. The difference in the HHV and LHV is the heat required to vaporize the product water.

The enthalpy change ($\Delta H$) of a fuel cell reaction represents the entire heat released by the reaction at constant pressure. The cell potential based on $\Delta H$ is defined as the *thermoneutral potential*, $E_t$,

$$\Delta H = -nFE_t \tag{4.4}$$

where $E_t$ has a value of 1.48 V for the reaction represented by the reaction in (4.1).

Table 4.4 presents the thermodynamic parameters for a hydrogen-oxygen fuel cell operating under standard conditions. Also shown in this table is the theoretical efficiency of fuel cells based on the higher and lower heating value. These values are considerably higher than those for thermal power plants.

## TABLE 4.4
### Thermodynamic Parameters for an $H_2/O_2$ Fuel Cell Based on the Higher and Lower Heating Values (HHV and LHV). Reactants and Products are in Standard States

| Parameter | HHV | LHV |
| --- | --- | --- |
| Gibbs free energy, kJ/mol | – 286 | – 242 |
| Enthalpy, kJ/mol | – 237 | – 229 |
| Reversible potential, V | 1.23 | 1.18 |
| Theoretical efficiency, % | 83 | 94 |

## 4.4.2. Effect of Temperature and Pressure on the Thermodynamic-Reversible Potential, $E_r$

According to Eq. (4.2), the reversible potential, $E_r$, of the cell varies directly with the Gibbs's free energy. The effect of temperature and pressure on the Gibb's free energy leads to changes in the reversible potential ($E_r$). Since

$$\Delta G = \Delta H - T\Delta S \qquad (4.5)$$

the variation of free energy change with temperature at a constant pressure is given by:

$$\frac{\partial \Delta G}{\partial T} = -\Delta S \qquad (4.6)$$

because the variation of $\Delta H$ with temperature is negligible.

Substituting Eq. (4.6) in Eq. (4.2) leads to the equation

$$\Delta G = \Delta H + T\frac{\partial \Delta G}{\partial T} \qquad (4.7)$$

Using Eqs. (4.2) and (4.5) we have

$$\frac{\partial E_r}{\partial T} = \frac{\Delta S}{nF} \qquad (4.8)$$

From the previous equation, the reversible potential at any temperature can be calculated using the entropy change of the fuel cell reaction.

The thermodynamic-reversible cell potential can also be expressed as a function of pressure as:

$$E_P = E_{P_0} - \frac{1}{nF}\int_{P_0}^{P} \Delta V dP \qquad (4.9)$$

The variables $E_P$ and $E_{P_0}$ are the cell potential at pressure $P$ and $P_0$, respectively, and $\Delta V$ is the volume change of the reaction. When gaseous reactants and products are involved in a chemical reaction (e.g., $H_2/O_2$ fuel cell), pressure effects are significant. Temperature and pressure coefficients of standard reversible potentials can be calculated from the above equations.

### 4.4.3.  Thermodynamic Data for Some Fuel Cell Reactions

Hydrogen-oxygen fuel cells are considered for analysis in the above subsections. However, there are several other fuels investigated for direct energy conversion purposes. Thermodynamic data for a selected number of fuels are included in Table 4.5. The reactions represented in Table 4.5 produce both electrical energy and heat. The maximum useful work obtainable from the above fuels is the free energy change of the reactions and can be represented by Eq. (4.2). Spontaneous reactions are favored only when the free energy change between the products and the reactants is negative. The enthalpy change is the total thermal energy available while the free energy is available for obtaining useful work from the system. The entropy term represents the state of the disorder of the system. At equilibrium (under reversible operating conditions), the thermal energy available is $T\Delta S$.

As seen from Table 4.5, the entropy change of a chemical reaction could be zero, positive, or negative. There are reactions that can serve as examples for these three cases as explained below. The entropy change of a chemical reaction is zero if the number of moles of the products and reactants in the gas phase are equal. An example of this type of reaction is $C_{(s)} + O_{2(g)} \rightarrow CO_{2(g)}$. The entropy change of the reaction $C_{(s)} + \tfrac{1}{2} O_{2(g)} \rightarrow CO_{(g)}$ is positive, while that for the reaction $H_{2(g)} + \tfrac{1}{2} O_{2(g)} \rightarrow H_2O_{(g)}$ is negative.

**TABLE 4.5**
**Thermodynamic Data for Some Fuel Cell Reactions**

| Reaction | $-\Delta G^0$ kJ/mol | $-\Delta H^0$ kJ/mol | $n$ | $E_r^0$ V | $\partial E_r/\partial T$ mV/°C | $\partial E_r/\partial \log P$ mV |
|---|---|---|---|---|---|---|
| $H_2 + \tfrac{1}{2} O_2 \rightarrow H_2O$ | 237 | 285.58 | 2 | 1.23 | -0.84 | 45 |
| $CH_4 + 2 O_2 \rightarrow CO_2 + H_2O$ | 817.19 | 889.50 | 8 | 1.06 | -0.31 | 15 |
| $CH_3OH_{(l)} + 3/2 O_2 \rightarrow CO_2 + H_2O$ | 701.86 | 725.94 | 6 | 1.21 | -0.13 | 5 |
| $NH_3 + \tfrac{3}{4} O_2 \rightarrow \tfrac{1}{2} N_2 + 3/2 H_2O$ | 355.46 | 382.22 | 3 | 1.23 | -0.31 | 25 |
| $N_2H_4 + O_2 \rightarrow N_2 + 2 H_2O$ | 601.2 | 621.52 | 4 | 1.56 | -0.18 | 15 |
| $C + \tfrac{1}{2} O_2 \rightarrow CO$ | 137.12 | 110.43 | 2 | 0.71 | 0.46 | -15 |
| $C + O_2 \rightarrow CO_2$ | 394 | 393.13 | 4 | 1.02 | 0 | 0 |
| $CO + \tfrac{1}{2} O_2 \rightarrow CO_2$ | 256.86 | 282.69 | 2 | 1.33 | -0.44 | 15 |

## 4.5.    ELECTRODE KINETIC ASPECTS

### 4.5.1.    Single Cell: the Heart of the Fuel Cell

The single cell is the heart of a fuel cell power plant/power source. The cell potential vs. current density behavior of the single cell predominantly determines the performance of fuel cell systems. There are minor variations in the performances of the single cells in a cell stack, as compared with that in a single cell, because of (i) scale-up, (ii) variations in temperature from cell to cell (iii) variation in flow patterns of reactant gases from cell to cell, (iv) problems of product water rejection with accompanying problems of flooding, i.e. water droplets in the substrate diffusion layers, and (v) uneven flow of gases in the channels of bipolar plates, mainly due to water blockage problems. The last two characteristics are particularly observed in PEMFCs, DMFCs, and AFCs. However, many of these problems have been overcome by advanced engineering technology development, and the cell potential vs. current density behavior in a stack is almost identical with that in a single cell.

### 4.5.2.    Role of Electrode Kinetics and Electrocatalysis on the Performance of Fuel Cells

Electrode kinetics plays a vital role in determining the performance of fuel cells. The series of steps involved in electrode reactions in the fuel cells are:

(a)    dissolution of the reactant gases in the electrolyte
(b)    diffusion of the dissolved reactant gases to the active sites in the electrode
(c)    adsorption of reactants and/or intermediate species formed by dissociative adsorption on the electrode from the electrolyte
(d)    charge transfer from reactant in electrolyte or from the above mentioned adsorbed species to the electrode
(e)    diffusion of species away from the electrode
(f)    transfer of conducting ions from one electrode to the other through the electrolyte
(g)    transfer of electrons from one electrode to the other through the external load.

The single cell potential ($E$), during operation of a fuel cell, is diminished by the losses in overpotential at the anode, cathode, and in the electrolyte. Thus, one may express $E$ by the equation:

$$E = E_r - \eta_{act,a} - \eta_{act,c} - \eta_{ohm} - \eta_{mt,a} - \eta_{mt,c} \qquad (4.10)$$

The variable $E_r$ is the thermodynamic reversible potential, $\eta_{act,a}$ and $\eta_{act,c}$ are the activation overpotentials at the anode and cathode, respectively, $\eta_{mt,a}$ and $\eta_{mt,c}$

are the mass-transport and concentration overpotentials at the anode and cathode, respectively, and $\eta_{ohm}$ is the ohmic overpotential in the cell. Bockris and Srinivasan[1] used simplified expressions to determine the effects of electrode kinetic parameters on activation, mass transport, and ohmic overpotentials, assuming that the electrodes are planar. However, porous-gas diffusion electrodes are always used in fuel cells and thus the reactions occur in 3 dimensional reaction zones on the electrodes. The theoretical treatment of the electrode kinetics of reactions at porous electrodes is quite complex and involves setting up of second order differential equations to determine the half cell potential (as well as the cell potential) vs. current density behavior and the current distribution within the electrodes (see Chapter 7).

### 4.5.3.  Cell Potential-Current Density Behavior of Fuel Cells

The efficiencies attained in fuel cells are 30-40% lower than the theoretical values due to overpotential losses A typical cell potential vs. current density plot of a fuel cell is shown in Figure 4.3. The regions for the losses in cell potential are noted in Table 4.6.

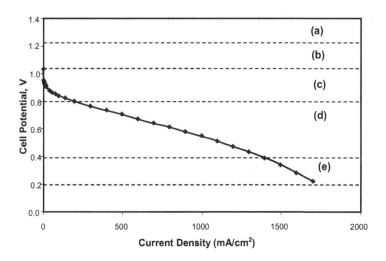

**Figure 4.3.** Typical cell potential vs. current density plot for a $H_2/O_2$ PEMFC operating at $80^0$ C and 3 atm. Please refer to Table 4.6 for an explanation of the losses in cell potential marked with the letters in parenthesis.

The open-circuit potential is lower than the theoretical value of 1.23 V (calculated from $\Delta G$) due to (i) the mixed potential at the oxygen electrode due to platinum oxidation, (ii) fuel crossover, and (iii) sluggish electrode kinetics of the oxygen reduction reaction, accounting for a 200 mV overpotential loss in low and intermediate temperature fuel cells. Three types of overpotential losses are encountered in fuel cells (regions c, d, and e in Figure 4.3) and these can be explained as follows: (i) activation overpotential mainly occurs due to the slower kinetics of electro reduction of oxygen and competing electrochemical reactions, (ii) ohmic overpotential at intermediate current densities due to resistance to proton transfer in the electrolyte, and (iii) mass transport losses at higher current densities due to lower concentrations of reactant gases at the electrode/electrolyte interfaces. In region (c), there is only activation overpotential at the oxygen electrode. Region (d) is predominantly ohmic but there are contributions from activation overpotential at the hydrogen and oxygen electrodes, and region (e) is controlled by mass transport overpotentials, with contributions from activation and ohmic overpotentials.

## 4.6.    ANALYSIS OF FUEL CELL PERFORMANCE CHARACTERISTICS

### 4.6.1.   Cell Potential

The analytic expression that best describes the behavior of PEMFCs in the entire current density range is given by the equation:[2]

$$E = E_0 - b\log(i) - Ri - m\exp(ni) \qquad (4.11)$$

**TABLE 4.6**
**Overpotential Losses Encountered in Fuel Cells as shown in Figure 4.3**

| Region | Losses encountered in cell potential |
|--------|--------------------------------------|
| (a) | Entropy change of reaction |
| (b) | Mixed potential of oxygen electrode and low rate of $H_2$ crossover to the cathode |
| (c) | Activation overpotential, predominantly at the oxygen electrode |
| (d) | Ohmic overpotential, predominantly in electrolyte |
| (e) | Mass transport overpotential |

where

$$E_0 = E_r + b \log i_0 \qquad (4.12)$$

where $E_r$ is the reversible potential, $b$ is the Tafel slope, and $i_0$ is the exchange-current density for the oxygen reduction reaction. The parameter $R$ is predominantly the ohmic resistance in the cell with a small contribution from the charge transfer resistance of the anodic reaction and the electronic resistance in the electrodes and cell fixtures, and $m$ and $n$ are parameters governing the semi-exponential decrease of cell potential with high current density due to mass transport overpotential.

## 4.6.2. Efficiency

Fuel-cell power plants are more efficient (by about a factor of 2) than the conventional thermal power plants because they are not limited by the Carnot cycle. The predominant reaction in most types of fuel cells is the production of water represented by the following two half-cell reactions.

Anode: $$H_2 \rightarrow 2H^+ + 2e_0^- \qquad (4.13)$$

Cathode: $$\frac{1}{2}O_2 + 2H^+ + 2e_0^- \rightarrow H_2O \qquad (4.14)$$

The operating efficiencies of the fuel cells are lower than the theoretical values due to activation, ohmic, and mass transport overpotentials. All the available energy from the above reaction can be converted to electrical energy in an ideal process. Considering that $\Delta H$ is the total energy available in the process and $\Delta G$ is the theoretical energy that can be converted to electrical energy, the theoretical or reversible efficiency of the fuel cell reaction is then expressed as:

$$\varepsilon_r = \frac{\Delta G}{\Delta H} \qquad (4.15)$$

Here, $\Delta G$ is less than the enthalpy change because of the entropy of the reaction, which generates heat rather than electricity. The efficiencies based on the lower heating value and higher heating values of the reaction are 94% and 83%, respectively. High temperature fuel cells such as MCFCs and SOFCs exhibit minimal activation overpotentials because of enhanced electrode kinetics at operating temperatures of 650-1000 $^0$C. The main loss encountered in MCFCs and SOFCs are due to ionic resistance of the electrolyte and mass transport limitations.

The voltage efficiency of a fuel cell is given by the expression:

$$\varepsilon_V = \frac{E(i)}{E_r} \qquad (4.16)$$

where $E(i)$ is the cell potential at a given current density ($i$) and $E_r$ is the reversible potential.

Another type of efficiency encountered is the current efficiency ($\varepsilon_f$) and is defined as:

$$\varepsilon_f = \frac{i_f}{i_t} \qquad (4.17)$$

This efficiency can be less than 100% due to incomplete oxidation of the reactant fuels and due to crossover of the fuel from the anode directly to the cathode resulting in a chemical oxidation at the cathode. In Eq. (4.17), $i_f$ represents the measured current obtained in the fuel cell, and $i_t$ is the theoretical current if all the fuel is completely oxidized at the anode. The problem of crossover is encountered mostly in DMFCs where methanol is used as a feed. Methanol is highly soluble in the fuel-cell electrolyte, causing crossover to the cathode. The fuel, which crosses over from anode to cathode, is thus chemically oxidized without generating any electricity. The overall efficiency ($\varepsilon$) is expressed as the product of the above-mentioned efficiencies.

$$\varepsilon = \varepsilon_r \varepsilon_V \varepsilon_f \qquad (4.18)$$

### 4.6.3. Differential Resistance

The differential resistance of a the fuel cell is given by the equation:

$$\frac{dE}{di} = -\frac{b}{i} - R - mn\exp(ni) \qquad (4.19)$$

As seen from this equation, the initial steep decrease in differential resistance is due to the charge transfer resistance in the low current density region. It is followed by a constant value in the intermediate region and a semi-exponential increase in the mass transport region.

### 4.6.4. Power Density

The power density is defined by the equation:

$$P = Ei \qquad (4.20)$$

Thus, substituting $E$ from Eq. (4.1) leads to:

$$P = [E_0 - b\log(i) - Ri - m\exp(ni)]i \tag{4.21}$$

From the cell potential vs. current density behavior (Figure 4.3), it is clear that the $P$ vs. $i$ plot will be pseudo-parabolic (Figure 4.4). It must be stressed that if mass transport overpotential is significant at higher current densities, there will be a rapid decrease of power density with current density after the maximum power is attained.

### 4.6.5.  Rate of Heat Generation

There are two sources of heat generation in fuel cells—one is due to the thermodynamic-entropic losses and the other is due to overpotential losses in the fuel cell.

The heat generation rate ($\overset{\bullet}{Q}$) can thus be expressed by:

$$\overset{\bullet}{Q} = -\frac{4.18T\Delta S}{nF}i + i\sum_{j}\eta_{j} \tag{4.22}$$

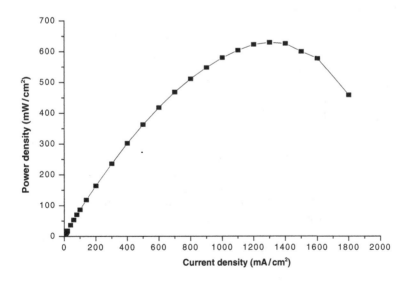

**Figure 4.4.**  Plot of power density vs. current density for a PEMFC under similar operating conditions as in Figure 4.3

A simplified manner for calculating the heat generation rate is to use the expression:

$$\overset{\bullet}{Q} = (E_t - E)i \qquad\qquad (4.23)$$

where $\overset{\bullet}{Q}$ can then be expressed as a function of current density by using the Eqs. (4.11) and (4.23):

$$\overset{\bullet}{Q} = [E_t - E + b\log(i) + Ri + m\exp(ni)]i \qquad\qquad (4.24)$$

At low values of $i$, $\overset{\bullet}{Q}$ varies pseudo-linearly with current density, but in the mass transport region the variation is semi-exponential.

## 4.7.   EFFECTS OF ELECTRODE KINETIC PARAMETERS ON PERFORMANCE CHARACTERISTICS OF PEMFCS AND DMFCS

### 4.7.1.   Rationale for Selection of PEMFCs and DMFCs

Typical cell potential vs. current density plots for a PEMFC and a DMFC are shown in Figure 4.5. Also shown in the figure for comparison are the plots for AFCs, PAFCs, MCFCs, and SOFCs. The effect of electrode-kinetic parameters are investigated for PEMFCs and DMFCs. These types of fuel cells were chosen for this study because they exhibit all forms of overpotential losses (activation, mass transport, and ohmic). The analysis is also valid for AFCs and PAFCs over a limited range of current densities (1 mA/cm$^2$ to a few hundred mA/cm$^2$), where only the activation and ohmic overpotentials are predominant and the mass transport region is not reached. In the case of MCFCs and SOFCs that operate at high temperatures (about 650 $^0$C for the former and 1000 $^0$C for the latter), the activation overpotentials are relatively small. This is because of two reasons: (i) the high values of exchange current densities, and (ii) the mass transport of the reactants to the active sites of the electrodes are greatly accelerated.

As a result, the $E$ vs. $i$ plot is linear, as in the case for ohmic overpotentials. The slope of the $E$ vs. $i$ plot reflects the sum of the charge transfer, mass transfer, and ohmic resistances for the half cell reactions, as well as the ohmic resistance of the electrolyte. Due to the consumption of the fuel and the oxidant, a Nernstian effect (i.e., change in reversible potential) is present, and there is a slow decrease of $E_r$ with increasing current density. This is also included in the slope of the $E$ vs. $i$ plot.

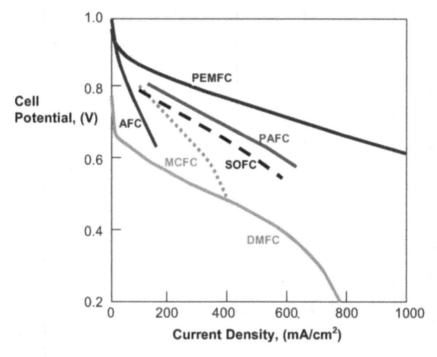

**Figure 4.5.** Plot of cell potential vs. current density for the leading types of fuel cells. Reprinted from S. Srinivasan, R. Mosdale, P. Stevens, and C. Yang, *Fuel Cells: Reaching the Era of Clean and Efficient Power Generation in the Twenty-First Century*, Annual Review of Energy and Environment, Volume 24, Copyright (1999) with permission from Annual Reviews, www.annualreviews.org.

### 4.7.2. Methodology for Analysis

Table 4.7 lists the electrode kinetic parameters used for the theoretical analysis to determine their effects on the performance characteristics of PEMFCs and DMFCs. The influence of a specific parameter was found by varying the values for that parameter, while the other parameters were kept constant. In all the figures, the arrows indicate the increasing value of the parameter, for which the effect is analyzed. The $H_2/O_2$-fuel cell was selected to illustrate the effect of cathodic exchange current density, cathodic Tafel slope, overall ohmic resistance, and the parameters related to mass transport phenomena. For DMFCs, due to activation overpotentials for electrooxidation of methanol being as high as that for the

**TABLE 4.7**

**Electrode Kinetic Parameter Values for Theoretical Analysis for the Fuel Cells Described in Figures 4.6 and 4.10-4.12.**

| Curve | $T$ °C | $P$ atm | $i_{o,c}$ A/cm² | $i_{o,a}$ A/cm² | $b_c$ mV/dec | $b_a$ V/dec | $R$ Ω cm² | $m$ V | $n$ cm²/A |
|---|---|---|---|---|---|---|---|---|---|
| | | | | | Figure 4.6A | | | | |
| a | 80 | 3 | $10^{-6}$ | 1 | 60 | 0 | 0.2 | $3 \times 10^{-4}$ | 3 |
| b | 80 | 3 | $10^{-5}$ | 1 | 60 | 0 | 0.2 | $3 \times 10^{-4}$ | 3 |
| c | 80 | 3 | $10^{-4}$ | 1 | 60 | 0 | 0.2 | $3 \times 10^{-4}$ | 3 |
| | | | | | Figure 4.6B | | | | |
| a | 80 | 3 | $10^{-6}$ | $10^{-6}$ | 60 | 120 | 0.2 | $3 \times 10^{-4}$ | 3 |
| b | 80 | 3 | $10^{-6}$ | $10^{-5}$ | 60 | 120 | 0.2 | $3 \times 10^{-4}$ | 3 |
| c | 80 | 3 | $10^{-6}$ | $10^{-4}$ | 60 | 120 | 0.2 | $3 \times 10^{-4}$ | 3 |
| | | | | | Figure 4.10 | | | | |
| b | 80 | 3 | $10^{-6}$ | 1 | 30 | 0 | 0.2 | $3 \times 10^{-4}$ | 3 |
| c | 80 | 3 | $10^{-6}$ | 1 | 120 | 0 | 0.2 | $3 \times 10^{-4}$ | 3 |
| | | | | | Figure 4.11 | | | | |
| b | 80 | 3 | $10^{-6}$ | 1 | 60 | 0 | 0.1 | $3 \times 10^{-4}$ | 3 |
| c | 80 | 3 | $10^{-6}$ | 1 | 60 | 0 | 0.3 | $3 \times 10^{-4}$ | 3 |
| | | | | | Figure 4.12A | | | | |
| b | 80 | 3 | $10^{-6}$ | 1 | 60 | 0 | 0.2 | $6 \times 10^{-4}$ | 3 |
| c | 80 | 3 | $10^{-6}$ | 1 | 60 | 0 | 0.2 | $9 \times 10^{-4}$ | 3 |
| | | | | | Figure 4.12B | | | | |
| b | 80 | 3 | $10^{-6}$ | 1 | 60 | 0 | 0.2 | $3 \times 10^{-4}$ | 6 |
| c | 80 | 3 | $10^{-6}$ | 1 | 60 | 0 | 0.2 | $3 \times 10^{-4}$ | 9 |

oxygen reduction reaction, the effect of anodic exchange current density was also included in the analysis. The Tafel slope for methanol electrooxidation reaction on the best-known electrocatalyst is 120 mV/decade, and hence the effect of the anodic Tafel slope was not considered. The $E$ vs. $i$ plots for a DMFC do not show a well-defined activation controlled region because of the cross-over of methanol from anode to the cathode (Figure 4.5); the linear region is significant and due to the high solubility of methanol in water, the mass transport region is not accessible.

### 4.7.3. Effect of Exchange-Current Density

Activation overpotential mainly depends on the exchange-current density for the oxygen reduction at the cathode. For the case of the $H_2/O_2$ fuel cell, it can be seen from Figure 4.6A that as a result of the increase of the exchange current density for the oxygen reduction reaction from $10^{-6}$ to $10^{-4}$ A/cm$^2$, there are parallel displacements of the cell potential vs. current density plots, but the displacements are not so large because of the semi-logarithmic dependence of the cell potential on the exchange current density. In all these cases, it was assumed that $i_{o,a}$ for the electrooxidation of hydrogen is 1 A/cm$^2$. Thus, the variation of the half-cell potential of the hydrogen electrode with current density will be linear over the entire current density range. The semi-exponential behavior of the cell potential vs. current density plot at low current densities is due to the slow kinetics of the oxygen electrode reaction.

On the contrary, for the case of the methanol/$O_2$ fuel cell (Figure 4.6B) because of the slow electrochemical kinetics of the anodic reaction, the anodic activation overpotential has to be taken into account along with that of the cathode. In this case it was assumed that the exchange current density for the oxygen reduction reaction is constant ($10^{-6}$ A/cm$^2$), while for the methanol oxidation $i_{o,a}$ was varied from $10^{-6}$ to $10^{-4}$ A/cm$^2$. Taking into the account the Tafel type behavior and the Tafel slope for methanol oxidation as 120 mV/decade and for oxygen reduction as 60 mV/decade, the decrease in $E$ with increasing $i$ is more significant. There is a parallel shift in the linear regions in both the curves shown because $R$ was assumed to be 0.2 ohm cm2. The higher value of the Tafel slope for methanol oxidation with respect to the oxygen reduction, is reflected in a bigger separation between the two $E$ vs. $i$ plots. Furthermore, because of the low values of exchange current density for both the half-cell reactions, the cell potentials for the DMFCs reach zero values in the linear region without reaching the mass transport region. This is not so for the PEMFC, because only the oxygen electrode reaction is highly irreversible and the hydrogen electrode reaction is quite reversible.

The dependence of the fuel cell efficiency on current density (Figures 4.6A and 4.6B) is similar to that of cell potential on $i$. Thus, variations of $E$ and $\varepsilon_v$ with $i$ are identical, except for a shift in the scale of the ordinate-axis. The efficiency of the $H_2/O_2$-fuel cell is about four times higher than that of the methanol/$O_2$ fuel cell over the entire range of current density. The other aspects are the same as those mentioned above.

The differential resistance does not change significantly with increase of the anodic or cathodic exchange current density in these two cases (Figures.4.7A and 4.7B). The initial steep drop has an inverse dependence on current density and is due to the influence of the semi-logarithmic activation overpotential behavior. After this initial drop there is an intermediate constant region governed by the ohmic overpotential and finally a semi-exponential increase of the differential resistance because of the mass transport phenomena. In spite of the fact that the cell potential values are higher for the PEMFC than those for the DMFC, the differential resistance values are quite similar in the two cases. The only difference is that in Figure 4.7B, the mass transport region of the plot is not present as the cell potential reaches zero in the ohmic part of the $E$ vs. $i$ plot.

According to Eq. (4.21), the power density is zero at the reversible potential, passes through a maximum and drops to zero again at the limiting current. Figures 4.8A and 4.8B show that the power increases with the increase of $i_{o,c}$ and $i_{o,a}$ respectively. However, for the DMFC, even though the absolute values are smaller, the relative power increase is higher than that for the PEMFC due to the twice as high Tafel slope for the anodic reaction in the former case. In both these cases, the current density corresponding to the peak of maximum power is shifted to higher values with increase of the anodic or cathodic exchange current density. It can be seen from Figure 4.8A, that because of mass transport limitations, the power density decreases rapidly after reaching the maximum, while in Figure 4.8B, the curves are parabolic due to the absence of the mass transport region.

With respect to the rate of heat generation, Figure 4.9A shows an initial linear part, but with increasing current densities, this is followed by a parabolic behavior. At high current densities, the mass transport limitations give rise to a semi-exponential increase. On the contrary, it can be inferred from Figure 4.9B, that the dependence on the current density for the DMFC is almost linear. This fact correlates with the absence of the mass transport region in the $E$ vs. $i$ plot. In concordance with the lower power values for the DMFC, and consequently the lower values for its efficiency, the heat generation rate is up to about six times higher for the DMFC than the PEMFC. Also, the relative difference between the maximum values reached in both cases is higher for the DMFC due to the $b_a$ value of 120 mV/decade.

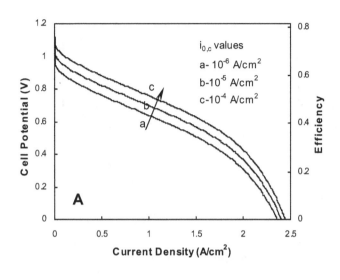

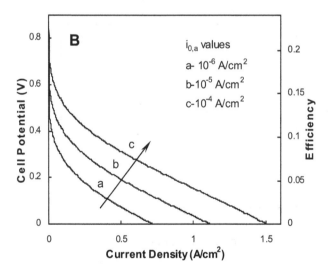

**Figure 4.6.** Dependence of the cell potential on the current density at different values of exchange-current density. (A): PEMFC, $i_{0,c}$ = cathodic-exchange current density, $T = 80\ {}^0C$, $P = 3$ atm, $b = 60$ mV/decade, $R = 0.2\ \Omega cm^2$, $m = 3 \times 10^{-4}$ V and $n = 3$ cm$^2$/A. (B) DMFC, $i_{0,a}$ = anodic-exchange current density, $T = 90\ {}^0C$, $P = 1$ atm, $b = 120$ mV/decade, $R = 0.2$ $\Omega cm^2$, $m = 3 \times 10^{-4}$ V and $n = 3$ cm$^2$/A.

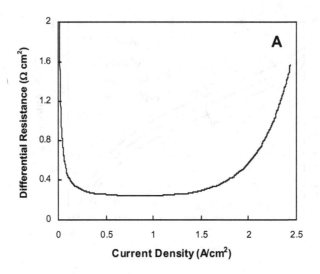

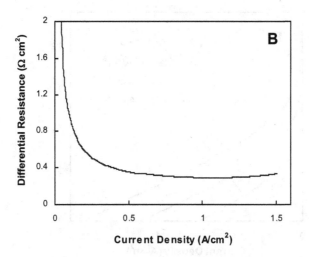

**Figure 4.7.** Dependence of the differential resistance on the current density at different values of (A) cathodic-exchange current density, PEMFC, $T = 80\ ^0$C, P = 3 atm, $b = 60$ mV/decade, $R = 0.2\ \Omega$cm$^2$, $m = 3 \times 10^{-4}$ V, and $n = 3$ cm$^2$/A, and (B) anodic-exchange current density, DMFC, $T = 90\ ^0$C, $P = 1$ atm, $b = 120$ mV/decade, $R = 0.2\ \Omega$cm$^2$, $m = 3\times10^{-4}$ V, and $n = 3$ cm$^2$/A.

218 CHAPTER 4

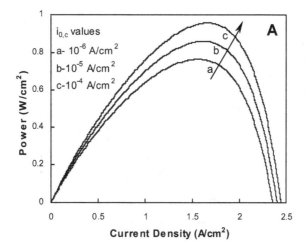

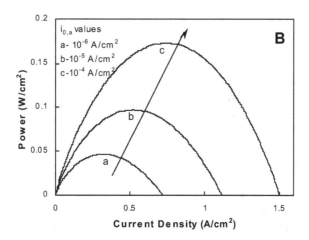

**Figure 4.8.** Dependence of the power on the current density. (A) PEMFC, $i_{0,c}$ = cathodic-exchange current density, $T$ = 80 $^0$C, P = 3 atm, $b$ = 60 mV/decade, $R$ = 0.2 $\Omega cm^2$, $m$ = 3 x $10^{-4}$ V, and $n$ = 3 $cm^2$/A. (B) DMFC $i_{0,a}$ = anodic-exchange current density, $T$ = 90 $^0$C, $P$ =1 atm, $b$ = 120 mV/decade, $R$ = 0.2 $\Omega cm^2$, $m$ = 3x$10^{-4}$ V, and $n$ = 3 $cm^2$/A.

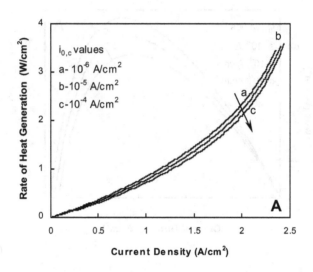

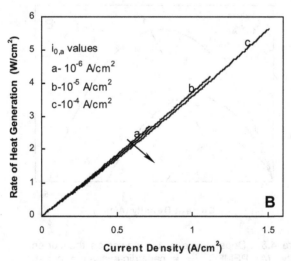

**Figure 4.9.** Dependence of the heat-generation rate on the current density. (A) PEMFC, $i_{0,c}$ = cathodic-exchange current density, $T$ = 80 $^0$C, P = 3 atm, $b$ = 60 mV/decade, $R$ = 0.2 $\Omega$cm$^2$, $m$ = 3 x $10^{-4}$ V, and $n$ = 3 cm$^2$/A. (B) DMFC $i_{0,a}$ = anodic-exchange current density, $T$ = 90 $^0$C, $P$ =1 atm, $b$ = 120 mV/decade, $R$ = 0.2 $\Omega$cm$^2$, $m$ = 3x$10^{-4}$ V, and $n$ = 3 cm$^2$/A.

### 4.7.4.  Tafel Slope

At a constant exchange current density, the different Tafel slope values for the anodic or the cathodic reaction affects the activation overpotential on the corresponding electrode. However, in this case the effect of $b$ on $E$ is more pronounced because of the linear relation between $E$ and $b$, as opposed to the semi-logarithmic dependence between $E$ and $i_o$ (Eq. 4.10). This is demonstrated in Figure 4.10A, which shows that the performance of the PEMFC is significantly reduced with the increase of the Tafel slope for oxygen reduction from 30 to 120 mV/decade. In this figure, the cathodic exchange current density was kept constant at $10^{-6}$ A/cm$^2$. In the intermediate ohmic region, the plots are almost parallel because the overall resistance is constant with a value of 0.2 ohm cm$^2$. In the case of curve d with air as the oxidant, the current density at which the cell potential reaches zero is shifted from the mass transport region to the intermediate region, where the ohmic behavior is predominant. The decrease of the values of $E$ with increase of Tafel slope leads to a corresponding decrease in the efficiency of the fuel cell. This illustrates the importance of improving the activity of the electrocatalyst for enhancing the performance of PEMFCs and DMFCs.

The influence of the Tafel slope for the oxygen reduction reaction is important only in the initial part of the differential resistance plot, i.e., the region where the activation overpotential plays the major role (Figure 4.10B). Higher $b_c$ values increase the differential resistance values only to a smaller extent in this region, and in the intermediate ohmic region, its influence is slowly diminished. In the cases where the mass transport region is reached, there is a significant increase of the resistance at high current density values.

The power density delivered by a PEMFC is significantly reduced from a maximum of 1 W/cm$^2$ to less than 0.3 W/cm$^2$ when the cathodic Tafel slope is increased from 30 to 120 mV/decade (Figure 4.10C). Besides, the current density corresponding to the peak power is shifted to lower values and the curve approaches a parabolic behavior, as the mass transport region is smaller. In agreement with the above conclusion, Figure 4.10D shows the increase in the heat generation rate with higher $b_c$ values.

### 4.7.5.  Ohmic Resistance

The linear region in the $E$ vs. $i$ plot includes the overall fuel cell resistance, the predominant contributor being the ohmic resistance of the proton conducting membrane. Electronic resistance of the fuel cell fixtures and the charge transfer resistance of the hydrogen electrode also contribute toward the fuel cell resistance in PEMFCs. It can be observed from Figure 4.11A that there is a significant reduction in cell potential and efficiency when the $R$ value increases from 0.1 to 0.3 ohm cm$^2$. The most important effect is the change in the slope of the intermediate linear region. The increase in $R$ leads to smaller current density values and the cell potential can reach zero in the linear region when $R$ is quite high, i.e., before the mass transport region is reached (e.g., Figure 4.11A, curve c).

The change of the overall-ohmic resistance causes a parallel displacement of the differential-resistance vs. current-density plots in the intermediate linear region (Figure 4.11B). The higher the $R$ value, the higher the differential resistance is, according to Eq. (4.19). In the initial activation controlled region, there is a small difference between the curves shown, and in the mass transport region the differential resistance increases in a semi-exponential manner.

Figure 4.11C shows the effect of increasing $R$-values on the fuel-cell power density vs. current-density plots. The current density corresponding to the maximum power shifts to smaller values and the shape of the curve changes from one with a steep decrease of the plot (due to the presence of mass transport limitations), to another with a nearly parabolic shape (due to the absence of mass transport limitations).

On the other hand, the rate of heat generation is significantly increased as the overall ohmic resistance increases. In Figure 4.11D, curve b is nearly linear at low current densities, parabolic at intermediate values of $i$, but it becomes semi-exponential at higher current density values due to mass transport overpotential problems.

## 4.7.6.   Mass-Transport Parameters

In the semi-empirical Eq. (4.10), the two parameters $m$ and $n$ account for the mass-transport phenomena in PEMFCs and DMFCs. This term describes the third region of the $E$ vs. $i$ plot, where the cell potential departs from the linear region and falls exponentially because of mass transport limitations. A theoretical interpretation of these parameters is still needed.

The dependence of cell potential on $m$ is linear, while that of $E$ on $n$ is exponential. Hence, the effect of $n$ is more pronounced in the $E$ vs. $i$ plot than $m$. This can be verified by examining Figures 4.12A and b. In Figure 4.12A, the increase of $m$ from $3 \times 10^{-4}$ to $9 \times 10^{-4}$ causes a decrease in the cell potential. The intermediate ohmic region is not affected by the change in $m$. On the contrary, by changing the value of $n$ from 3 to 6, the fuel cell performance is dramatically affected in the high current density region, with a higher loss in cell potential and efficiency, both in the mass transport and ohmic region. Figure 4.12B shows that, with the increase of the parameter $n$, the shape of the $E$ vs. $i$ curve approaches that of a planar electrode, where the existence of a limiting current density gives rise to a steep drop in the cell potential after the linear region. Moreover, the slope of the linear region changes to higher values, in spite of the $R$ value remaining constant in these cases, because of the apparent inclusion of mass transport resistance at higher

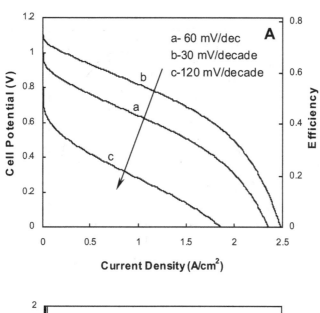

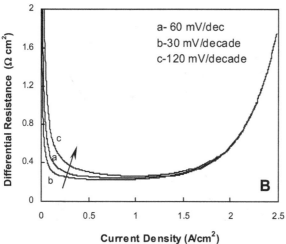

**Figure 4.10** Influence of the Tafel slope on (A) cell potential and efficiency, (B) differential resistance, (C) power, and (D) rate of heat generation for the oxygen-reduction reaction in a PEMFC; a, b and c represent cathodic Tafel slopes at 80 $^0$C, and 3 atm; $i_{0,c} = 10^{-6}$ A/cm$^2$, $R = 0.2$ $\Omega$cm$^2$, $m = 3 \times 10^{-4}$ V, and $n = 3$ cm$^2$/A.

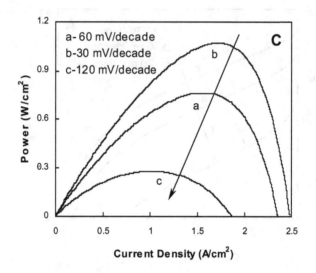

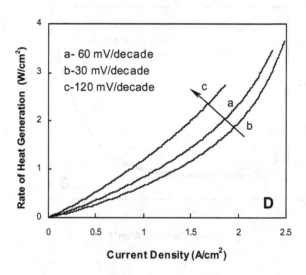

**Figure 4.10.** Continuation

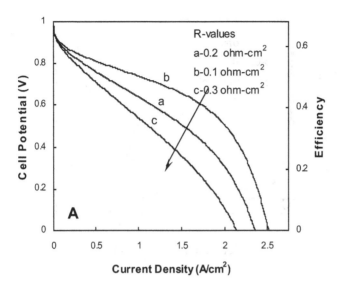

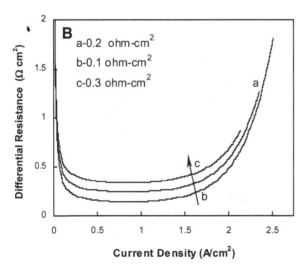

**Figure 4.11** Influence of ohmic resistance on (A) cell potential and efficiency, (B) differential resistance, (C) power, and (D) rate of heat generation in a PEMFC; a, b and c represent different ohmic resistance values at 80 $^0$C, and 3 atm; $i_{0,c} = 10^{-6}$ A/cm$^2$, $b = 60$ mV/decade, $m = 3 \times 10^{-4}$ V, and $n = 3$ cm$^2$/A.

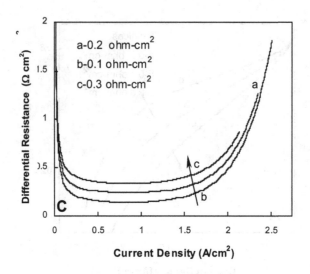

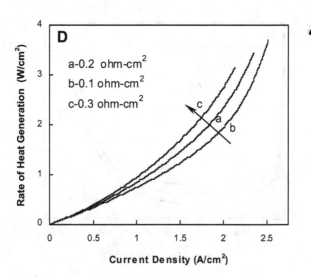

**Figure 4.11.** Continuation

current densities. This type of behavior is often encountered when air is used instead of pure oxygen as the cathodic reactant. The range of useful current densities for practical fuel cell applications is enormously reduced because of the greater effect of the parameter $n$ on the fuel cell performance.

In agreement with the $E$ vs. $i$ plots, the corresponding differential-resistance curves show an influence of $m$ in the mass transport region. The initial decrease and the constant intermediate zones are about the same, but the final exponential increase becomes more pronounced for increasing $m$ values (Figure 4.13A). In a similar manner, with an increase in $n$, the initial steep fall does not alter in the low-current density region, but the current density in the constant ohmic region is strongly diminished. Further increase in current density causes significant increase in the differential resistance (Figure 4.13B).

The influence of both mass-transport parameters on power density is illustrated in Figures 4.14A and 4.14B. The major changes are the smaller power density and limiting current density values as the $m$ and $n$ parameter values increase, while the initial part remains unchanged. This effect is more pronounced in the case of parameter $n$, because of its stronger effect on the diffusion phenomena. The maximum power density decreases with an increase of $m$ and $n$ and the peak power shifts to smaller current density values. The plots deviate slightly from the parabolic behavior due to a great decrease in power as a result of mass transport overpotentials after the maximum power is attained.

The heat-generation rate (Figures 4.15A and B) also increases as the values of the m and $n$ parameters become higher, but the relative effect is higher again in the case of $n$ than of $m$ values. Also, all the plots maintain qualitatively the same shape, having an initial part with a nearly linear tendency and a final region with a semi-exponential dependence, due to mass transport limitations.

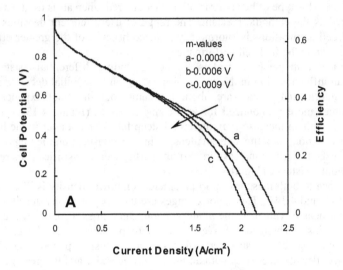

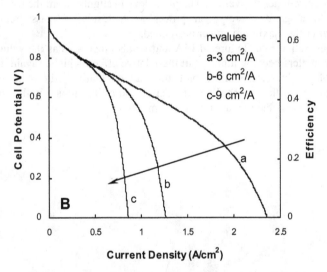

**Figure 4.12** Influence of (A) the mass-transport parameter $m$ with $n = 3$ cm$^2$/A, and (B) the mass transport parameter $n$ with $m = 3 \times 10^{-4}$ V on cell potential and efficiency at different current-density values for a PEMFC; a, b and c represent different $m$ or $n$ values at 80 $^0$C and 3 atm; $i_{0,c} = 10^{-6}$ A/cm$^2$, $b = 60$ mV/decade, and $R = 0.2$ $\Omega$ cm$^2$.

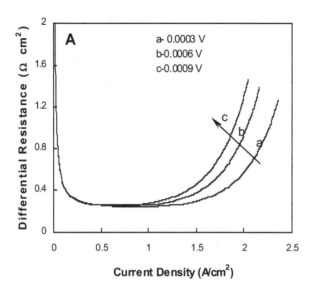

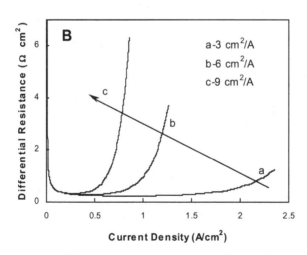

**Figure 4.13** Influence of (A) the mass-transport parameter
$m$ with $n = 3$ cm$^2$/A, and (B) the mass transport parameter $n$
with $m = 3 \times 10^{-4}$ V on the differential resistance at different
current-density values for a PEMFC; a, b, and c represent
different $m$ or $n$ values at 80 $^0$C and 3 atm; $i_{0,c} = 10^{-6}$ A/cm$^2$,
$b = 60$ mV/decade, and $R = 0.2\ \Omega$ cm$^2$.

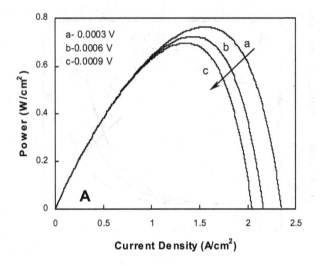

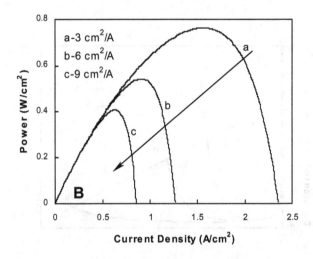

**Figure 4.14** Influence of (A) the mass-transport parameter
$m$ with $n = 3$ cm$^2$/A, and (B) the mass transport parameter $n$
with $m = 3 \times 10^{-4}$ V on the power at different current-density
values for a PEMFC; a, b and c represent different $m$ or $n$
values at 80 $^0$C and 3 atm; $i_{0,c} = 10^{-6}$ A/cm$^2$, $b = 60$
mV/decade, and $R = 0.2$ $\Omega$ cm$^2$.

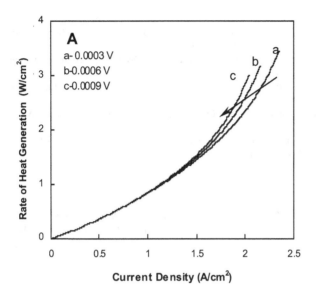

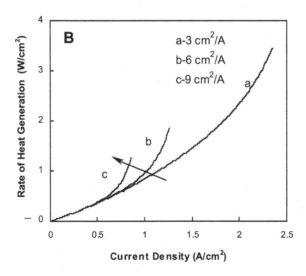

**Figure 4.15** Influence of (A) the mass-transport parameter *m* with *n* = 3 cm$^2$/A, and (B) the mass transport parameter *n* with *m* = 3 x 10$^{-4}$ V on the rate of heat generation at different current-density values for a PEMFC; a, b and c represent different *m* or *n* values at 80 $^0$C and 3 atm; $i_{0,c}$ = 10$^{-6}$ A/cm$^2$, *b* = 60 mV/decade, and *R* = 0.2 Ω cm$^2$.

## Suggested Reading

1.   J. O'M. Bockris and S. Srinivasan, *Fuel Cells: Their Electrochemistry*, McGraw Hill Publishing Company, New York, NY, 1969.
2.   L.J.M.J. Blomen and M.N.Mugerwa, Eds. Fuel Cell Systems, Plenum Publishing Corporation, New York, NY, 1994.
3.   K.Kordesch and G.Simander, Fuel Cells and their Applications, John Wiley & Sons Ltd. New York, NY, 2000.
4.   J. Larminie and A.Dicks, Eds. Fuel Cell Systems Explained, John Wiley & Sons Ltd. UK, 1999.
5.   W. Vielstich, A. Lamm and H. Gasteiger, Eds Handbook of Fuel Cells-Fundamentals, Technology and Applications, John Wiley & Sons, New York, NY, 2003.

## Cited References

1.   J. O'M. Bockris and S. Srinivasan, *J. Electroanal Chem.* **11**, 350 (1966).
2.   J. B. Kim, S. Srinivasan and C. Chamberlin, *J. Electrochem.Soc.* **142,** 2670 (1995).

## PROBLEMS

1.   What are the three types of electrochemical technologies? Which of these involve spontaneous electrochemical reactions generating electrical energy and which of these need electrical energy for chemical synthesis and storing chemical energy? What are the specific characteristics of rechargeable batteries? What are the similarities and dissimilarities among secondary batteries, electrochemical capacitors, and fuel cells?

2.   It was stated in Section 4.4, that the research and development activities in the first half of the 20$^{th}$ century led to technology development and are still continuing. Taking into consideration the summarizing statements made in this section and reviewing the literature for more details, prepare tables of the chronological inventions and discoveries that led to the present status of the technologies of the six leading types of fuel cell technologies.

3.   What predominantly restricts the temperature range of operation of each type of fuel cells listed on Problem 2? Prepare a table of the specific conductivities of the electrolytes in each type of fuel cell at the desired temperature. Take into consideration the thickness of the electrolyte layer and that its ohmic resistance is 50% higher because of use of porous matrices. Calculate the ohmic resistance per cm$^2$ for each type of fuel cell. The thickness of the electrolytes necessary to make these calculations can be obtained from Chapter 9.

4. Using the data in the following table, calculate the reversible potentials for the fuel cells under the following operating conditions:

| Type of fuel cell | Oxidant | Temperature, $^0C$ | Pressure, atm |
|---|---|---|---|
| PEMFC | $O_2$ | 25 | 1 |
| PEMFC | $O_2$ | 50 | 1 |
| PEMFC | $O_2$ | 80 | 3 |
| PEMFC | $O_2$ | 120 | 3 |
| PEMFC | air | 80 | 3 |
| PEMFC | air | 120 | 3 |
| PAFC | $O_2$ | 200 | 1 |
| PAFC | $O_2$ | 200 | 3 |
| PAFC | air | 200 | 1 |
| PAFC | air | 200 | 3 |
| MCFC | air | 650 | 1 |
| SOFC | air | 1000 | 1 |

5. For what types of fuel cells will the entropy change of the reaction be (i) negative, (ii) zero, and (iii) positive? Clearly state reasons.

6. What are the reasons for PEMFCs being able to attain the highest current densities and power densities at reasonable high efficiencies (~ 50%) than the other types of fuel cells?

7. Assume a Tafel equation for oxygen reduction in a PEMFC, $E = E_r - RT/F \ln i/i_0$, $E_r = 1.2$ V, and $i_0 = 10^{-6}$ A/cm$^2$. If the rate of hydrogen crossover is 1 mA/cm$^2$, calculate the open circuit potential for the fuel cell. Assume the anodic reaction (electrooxidation of hydrogen) has a high value and its open circuit potential (half cell) has the reversible value.

8. Consider the five-parameter equation $E = E_0 - b \log i - Ri - m \exp (ni)$. Assuming the values $E_0 = 1$ V, $b = 0.06$ V/decade, $R = 0.2 \, \Omega \, cm^2$, $m = 3 \times 10^{-4}$ V and $n = 3$ cm$^2$/A, plot $E$ vs. $i$ for current densities in the range 1 mA/cm$^2$ to 2 A/cm$^2$. Note that $E$ varies logarithmically with the current density initially and that you have to use at least 10 values of c.d. in this range. Rationalize the three regions of current densities in terms of predominantly activation, ohmic, and mass transport overpotential losses. After drawing the plot, make plots of the following as a function of current density: (i) differential resistance ($dE/di$), (ii) efficiency ($\varepsilon$), (iii) power density ($P$), and (iv) rate of heat generation ($\dot{Q}$). State reasons for $E$ vs. $i$ plots and $\varepsilon$ vs. $i$ plots being similar. Explain the behavior of the above parameters in the entire current density range.

## CHAPTER 5

# ELECTROCATALYSIS OF FUEL CELL REACTIONS

## 5.1.   INTRODUCTION

An understanding of the electrocatalysis of fuel-cell reactions is central to the development of fuel cells as a commercial product. The primary emphasis here will be on low temperature fuel cells PEMFC, AFC, and PAFC but a brief description of the catalysts/materials used in mid- to high- temperature fuel cells, MCFC and SOFC will also be presented along with a bibliography that will allow the reader to investigate those topics more fully on their own. The electrocatalysis of fuel-cell reactions is a prominent area of research for both academics in the university and applied research engineers working in industry.

The development of fuel cells for electric-power generation has certainly fostered renewed interest in electrocatalysis and the patent literature is filled with descriptions of materials that can enhance the reaction rate for a given electrochemical reaction. Many times these patents have been developed by industrial companies and have sparked the interest of academics, who have gone on to develop an understanding of how the new material may work to enhance the reaction rate. This cross-fertilization has led to new materials with improved properties. The observations by applications engineers have also brought about some conflicts between the two groups since the controlled research on single crystal materials of academics has not always agreed with the claims of the industrial engineers for the observed activity of the "technical catalyst." In the long

This chapter was written by F. J. Luczak (Glastonbury, CT 06033) and S. Sarangapani (ICET, Inc., Norwood MA 02062).

run, the results of the different experiments and procedures have broadened the overall understanding in the field of electrocatalysis.

The emphasis in this Chapter is to acquaint the experimentalist with the research developed thus far in the field of electrocatalysis, specifically directed towards fuel cell reactions. There is no "complete" theory of electrocatalysis. One cannot choose a reaction and then predict the optimum catalyst based purely on theoretical grounds. Advances in electrocatalysis have emerged because of experimental evidence and complimentary theoretical studies. This will probably continue to be the case for the near future. The analysis of electrochemical reactions, which typically are composed of several individual steps where the state of the catalyst and the environment are changing, is a very complex undertaking. We hope to provide the interested researchers with sufficient information and reference sources so that they will be able to contribute to the advancement of fuel cell electrocatalysis.

### 5.1.1.  Kinetics

Increasing the rates of fuel-cell reactions is central to developing highly efficient commercial fuel cells. Advances in experimental techniques have led to improved kinetic studies to determine the reaction rate, any intermediate reactions, and the rate determining step (rds) for a given electrochemical reaction. Once the rate-determining step is established, one can probe the chemical species involved and determine how they are changed by a modification of the electrocatalyst. Correlations between the rds characteristics and various electrocatalysts can then be utilized to identify the most active catalyst for a given reaction. Ideally, this is the way to develop an understanding of the electrocatalyst and a reaction. Unfortunately, it is not always easy to identify the mechanism and the rds because of experimental difficulties. The experimentalist resorts to hypothesizing reaction steps and measuring key kinetic parameters.

The key parameters that influence the rate of an electrochemical reaction at a given potential are the *exchange current density*, $i_0$, and the *Tafel slope*, b. Electrochemists strive to identify materials that will have a very large value for the exchange current so that the reaction will proceed at a very high rate at high voltage efficiency. The second experimentally measured parameter of interest is the Tafel slope. The rate of reaction is directly related to the exchange current and exponentially related to the Tafel slope. A detailed description of the rate equation and the exchange current density is provided in Chapter 2.

It is desirable to minimize the Tafel slope so as to achieve high voltages at high operating current densities. To do this, one must alter the rds controlling the rate of the reaction. This is not an easy change to accomplish. When one studies the literature, the Tafel slope values obtained for a given catalyst do not always appear to be consistent. For instance, the quoted results for technical platinum catalysts do not always agree with the results from experiments with a single-crystal platinum surface plane. The experimental conditions can also strongly influence the observed values.[1]

## 5.1.2. Electrocatalysis

Chapter 2.8 of this book gives a general introduction to electrocatalysis with a discussion of distinctive features of this type of catalysis. As part of the research into electrocatalysis, quite often the activity or performance of a series of catalysts is plotted versus some associated physical or chemical property. This is a useful tool to discover if there is any correlation of the activity with the proposed experimental parameter. If the plot shows a maximum in the activity or performance for a given value of the property with roughly linear decreases on either side of the maximum, this type of experimental curve is known as a volcano plot. The trends discovered in these types of correlations have been useful in identifying new catalysts.

The rate of an electrochemical reaction, for a given electrocatalyst, is dependent on two basic properties of the electrocatalyst. These properties are associated with the electronic and geometric character of the electrocatalyst. The geometric effect correlates to the surface structure of the catalyst and its relationship to the reactant species. The separation of the catalyst atoms on the surface may relate to the bond length of the reactant molecule. For instance, for platinum, which has a face-centered-cubic lattice structure (fcc), the lattice parameter is 3.9231Å and the nearest neighbor distance is 2.774 Å. The geometry of the platinum atoms on the primary surfaces of platinum {<100>, <110>, and <111>} are different. If for example, the reactant molecule is oxygen, where the bond length between oxygen atoms is 1.208 Å, the *bonding energy* between the oxygen atoms in a molecule and the platinum surface atoms will be different depending on the exposed platinum surface geometry. This can, in the extreme, result in either no adsorption or chemisorption of the oxygen, either of which would be expected to diminish catalytic activity. As we will show later, if the platinum catalyst is alloyed with another metal atom, the geometry will change and the catalytic activity will also be affected.

The other primary way to influence the activity of a catalyst is through a change in the electronic factor. The electronic structure of the catalyst atoms is described by the electron orbitals of the atoms. In particular, the outer or valence electrons, have the most impact on the interaction with neighboring atoms. When one considers a technical catalyst (high surface area/small particle sizes), the electronic structure of the surface atoms changes. The surface atoms for these high surface area materials do not have a full complement of nearest neighbors, and so the valence electrons are not fully satisfied in terms of their electronic bonding as they would be in the bulk. These *dangling* electrons are available to interact with a reactant molecule. In addition, since some of the small particles are not perfect crystals, there may be steps or kinks in the particle structure that will also alter the electronic bonding of the atoms in the particle. Experiments on bulk platinum and other metal catalysts suggest that the work function of the material changes again depending on the exposed crystallite face. It is clear that these factors, geometric and electronic, are related and it may not be possible to totally separate them. Significant research into the effect of these parameters on catalytic activity is underway utilizing new techniques such as X-ray Absorption Spectroscopy (XAS), where the two factors can be studied in the same experimental environment. It is extremely advantageous that they can be studied simultaneously especially during the actual operation in a fuel cell. It is essential to consider how all the experimental parameters (electrolyte, electrode potential, and concentration of reactants and products) affect the electrocatalyst in order to arrive at an understanding of electrocatalysis.

### 5.1.3.  Electrocatalysts

Most of the electrocatalysts utilized for low temperature fuel cells are metals or more specifically the noble metals. They include platinum, iridium, ruthenium, palladium, gold and silver. Some of the non-noble elements have also demonstrated catalytic activity, often in conjunction with the noble elements, and they include nickel, iron, cobalt, chromium, vanadium, molybdenum, tin, tungsten and others. Some organic materials have also demonstrated catalytic activity and they include the metal pthalocyanines, metal porphyrins and others. The exact nature of the catalytic activity of these organics is still not established but there are reasons to believe that a 4-nitrogen (N-4) bonding structure with the metal atom is the key requirement for their catalytic activity. This conclusion is inferred from heat treatment experiments on the metal porphyrins where the original organic structure is altered leaving only the N-4 bonding structure. The key feature here is the apparent final electronic state of the metal that interacts with the reactant.

The noble metals, as blacks, i.e., unsupported high area materials, were initially utilized as catalysts for fuel cells. They were evaluated in the form of single elements or as combinations of two or more elements, noble or non-noble (e.g. W). The combinations include alloying of the multi-metal combinations as well as simple mixtures of the elements. They proved to be active as catalysts but very expensive since high metal loadings were required to achieve high performance and efficiency in a fuel cell. The electrochemical reactions in a fuel cell take place on the surface of the catalyst. This makes the development of high surface area catalysts imperative for commercial systems since they are dependent on costs, in terms of kW/$.

This led directly to the development of supported catalysts, in particular those where the support was carbon; the relative cost of carbon is an important consideration. From a technical viewpoint, carbon is an attractive support in that it possesses satisfactory electronic conductivity and is obtainable in many different high surface area forms, $> 1000$ m$^2$/gm. The high surface-area carbon support provides sites for the platinum catalyst particles to reside. Since the possible sites are so numerous, the platinum will be crystallized on the many different sites and thereby not grow to very large individual crystallites. It is possible, depending on platinum loading, i.e., weight percent (Pt/C), to achieve platinum surface areas greater than 100 $m^2$/gm by applying different methods of catalyzation. This high surface area implies particle sizes less than 20 Å; the available catalyst area for a given mass loading, e.g. 1mg $Pt/cm^2_{real}$, can be quadrupled or more, compared to the platinum black catalyst. This is very attractive to those companies attempting to commercialize fuel cells and minimize capital costs. When one achieves such a high surface area for the catalyst particles, it is clear that the platinum particles would not always make physical contact with one another. This makes good electronic conductivity of the support a key parameter. It must be high to limit the ohmic losses within the catalyst layer. Another important property of carbon is that it has reasonable corrosion resistance in the electrolyte. In some cases, a partial graphitization of the carbon support for the cathode catalyst has been necessary to provide adequate resistance to corrosion for long life, e.g. PAFC.[2]

While platinum supported on carbon catalysts are interesting, higher activities for the fuel cell electrochemical reactions are often achieved by using supported multi-component metal catalysts. These can be fabricated on carbon rather easily and are commonly used in acid electrolyte fuel cells (PEM, PAFC) to provide higher cell performances.

## 5.2.  FUEL-CELL REACTIONS IN ACID AND ALKALINE FUEL CELLS

The electrochemical reactions of oxygen reduction, hydrogen oxidation, and direct methanol oxidation on different electrocatalysts will be described in the following Sections. The primary fuel cells considered will be PEMFC, AFC and PAFC. It is very interesting to note the effect of pH and how it affects the rate of oxygen reduction on different catalysts. The fact that gold is such an excellent catalyst in base electrolytes but is very poor in acid electrolytes is intriguing. Platinum is very good in both electrolytes but is not as stable in base electrolytes. This area has been investigated in detail and the reasons for differences are addressed (but not fully understood) in an overall theory of oxygen reduction.

### 5.2.1.  Oxygen Reduction in Acid Electrolytes

There are two primary reactions for the oxygen-reduction reaction (ORR) in acid electrolytes, a two-electron and a four-electron process, which can be written as:

$$O_2 + 4H^+ + 4e^- \rightarrow 2H_2O \tag{5.1}$$

$$O_2 + 2H^+ + 2e^- \rightarrow H_2O_2 \qquad \text{(Peroxide Path) (5.2)}$$

A condensed scheme for the reaction pathway is:[3]

$$O_2 \longrightarrow (O_2)_{ads} \underset{k_4}{\overset{k_2}{\rightleftharpoons}} (H_2O_2)_{ads} \qquad (5.3)$$

This reaction scheme shows that oxygen can be electrochemically reduced to water directly ($4e^-$ process) with a rate constant, $k_1$, without the intermediate $(H_2O_2)_{ads}$. Alternatively it can be reduced to $(H_2O_2)_{ads}$ ($2e^-$ series process) with a rate constant, $k_2$. The adsorbed peroxide can be electrochemically reduced to water with the rate constant, $k_3$, (series $4e-$ process), catalytically decomposed on the electrode surface with a rate constant, $k_4$, or desorbed into the bulk electrolyte with a rate constant, $k_5$.[4] For platinum and platinum alloy catalysts it appears that the series pathway via the adsorbed peroxide intermediate is the operative reaction path.[5]

One of the main problems associated with the ORR in fuel cells is the low $i_0$ for the ORR, $\sim 10^{-10}$ A/cm$^2$. High area catalysts employed in fuel cells have very high real surface area (per geometric unit area) and thus the apparent exchange current density becomes adequate to make use of this reaction in practical applications. In order to achieve even better performance in fuel cells, one must find a way to increase the intrinsic exchange current significantly. For fuel cell and similar practical applications, one needs to consider both the exchange current and the Tafel slope. Higher Tafel slopes result in higher overpotentials at fuel cell operating current densities. However, Tafel slopes are dependent on the mechanism and cannot be easily changed or reduced. The Tafel slope for oxygen reduction on high area platinum catalyst is of the order of 65–90 mV/decade, but it is 120 mV/decade on bulk platinum. A full understanding of these differences is still not available. The activation overpotential for the cathode reaction typically starts out at 200–300 mV, and thus contributes to the highest individual performance loss in the cell. Rarely is the observed open-circuit potential much above 1.0 V although the theoretical thermodynamic potential is much higher at about 1.226 V. These characteristics of the ORR made the development of high activity electrocatalysts a necessary condition for the commercialization of fuel cells. This realization resulted in the development of practical high surface area electrocatalysts, primarily platinum, which paved the way for the commercialization of fuel cell technology.

The electrolyte dependence on the ORR has been studied in several papers by the Ross group.[6-8] The structural sensitivity and anion adsorption effect almost always have negative effects on the reaction rate. The ORR on certain platinum crystallite planes, cf. <111> face, has been found to be adversely affected by the bisulfate ions in sulfuric acid while there are minimal changes and adsorption with perchloric acid. The final result appears to be a blocking of the initial adsorption of the $O_2$ molecule rather than any change in pathway for the reaction. In addition, no change in the Tafel slope was observed. Since the sulfonate ion is known to be non-adsorptive in PEM fuel cells, an anion adsorption effect is not expected to be a factor. The phosphate anion adsorption in PAFC is known to be a large negative factor for the higher temperature acid fuel cells.

For acid electrolytes (PEMFC and PAFC) platinum and several platinum alloys have been reported to give the highest measured activity. The parameter, catalyst surface area, is a significant factor since it is the high area of the Pt catalyst that made possible the practical application of the very low exchange current density ORR (Figure 5.1). While the original fuel cell catalysts were platinum blacks with areas about 30 $m^2$/gm, cost conscious engineers developed carbon supported platinum catalysts with areas approaching 150 $m^2$/gm. Researchers worked very hard to develop efficient catalyst preparation methods that could raise the surface area and thereby lower the average particle size of the platinum that translates into lower initial capital costs for catalysts.

Some early research by Bregoli,[9] on carbon supported platinum electrocatalysts in phosphoric acid electrolyte, suggested that the catalyst activity did not continuously increase as the particle size was reduced. There was an apparent limit to the achievable activity.

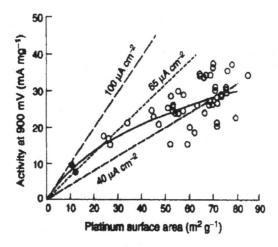

**Figure 5.1.** Activity of platinum for the electrochemical reduction of oxygen as a function of surface area at 177 °C in 99 wt% phosphoric acid. o, platinum supported on Vulcan XC-72; •, platinum black blended with Vulcan XC-72. The solid line represents a fit of the data; the dashed lines are for constant specific activity. Reprinted from Reference 10, Copyright (1978) with permission from Elsevier Science.

This experimental result stimulated numerous investigations and remains a high visibility research topic in electrocatalysis to this day. Various interpretations of the results of the observation by Bregoli have been instrumental in leading to a greater understanding of the effect of particle size on activity. One important observation was that the shape of the catalyst particle changes as the particle size was made smaller. As the particle size is enlarged to the 20–50 Å range, the particle assumes a cubo-octahedral shape, based on surface free energy considerations and this results in a surface that has a large percentage of the total area associated with the crystallite planes (<100>, <110>).[11,12] Single crystal and supported catalyst studies suggest that the <100> surface is one of the most active surfaces for oxygen reduction.

The structural sensitivity of the ORR on the nature of the electrolyte is well established.[13] The thin-film rotating ring-disk electrode method has been utilized with good results. This work suggested that for PEM fuel cells, the presence of any chlorides, either from residues from membrane-electrode-assemblies (MEA) production or contamination from reactant streams, could lead to poor cell performance. The Cl⁻ species has very strong negative effect on the platinum catalysts. It appears to not only block catalyst sites but also enhances peroxide formation that can then eventually cause degradation of the proton exchange membrane. The halide adsorption onto the <100> and <110> platinum surfaces is very strong.

Catalysts prepared by using carbon as the support results in higher surface area materials for platinum. The effect of this procedure appears to be strictly an increase in the number of available reactive sites but no change in rate determining step for the ORR. However, this appears to be not the case with supported alloy catalysts. In this regard, International Fuel Cells (IFC) developed a series of carbon supported platinum alloy catalysts that demonstrated high activity and performance in phosphoric acid fuel cells.[14] Various non-noble metals were used to alloy with the platinum and a correlation of specific activity with nearest platinum neighbor distance was observed.[15] (Figure 5.2) It was interpreted that the bonding of the oxygen molecule was dependent on the platinum spacing in the catalyst. Earlier hypotheses had already suggested that the adsorption model for oxygen on platinum metals was a bridge mode with two bonds and on two adjacent catalyst sites.[16] The breaking of the oxygen bond as the rate determining step for the ORR had previously been proposed.[17] Since the bond energy could be expected to depend on the relationship between the oxygen molecule and the distance between individual catalyst atomic sites, the slow breaking of oxygen bond may be a plausible explanation for the increase of activity associated with platinum alloys. The platinum nearest neighbor spacing can be changed by the incorporation of an alloying atom. The size of the alloying atom and its concentration will determine the extent of platinum lattice contraction or expansion.

Interest in this observation led to further research on single crystals[19] of platinum-chromium systems. The initial experimental results reported suggested that the increase in activity was simply due to a roughening of the catalyst surface, i.e., essentially an increase in the effective platinum surface

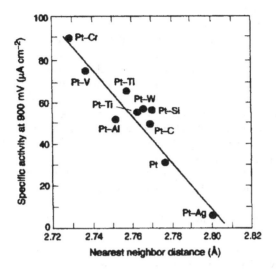

**Figure 5.2.** Reduction of oxygen in phosphoric acid on various platinum alloy catalysts: specific activity for oxygen reduction vs. electrocatalyst nearest-neighbor distance. Test conditions: 100% $H_3PO_4$ at 200 °C using the floating half-cell technique with a dynamic hydrogen reference electrode. Reproduced from Reference 18, Copyright (1983) with permission of The Electrochemical Society.

area activated by the chromium. The activity increase was not associated with any alloying of platinum and chromium.

The applied alloy catalyst work at IFC continued with the goal of achieving a catalyst with optimum platinum nearest neighbor distance to maximize the activity. Three different alloy systems ($Pt_xCr_y$, $Pt_xNi_y$, $Pt_xCo_y$) were investigated with various non-noble metal concentrations, in order to vary the platinum nearest neighbor distance parameter.[20] The peak in specific activity vs. platinum distance, was determined to be different for each alloy (volcano plots). If the peak in specific activity had occurred at the same nearest platinum neighbor distance for each of the alloys, one could ascribe the activity changes directly to a geometric effect. The result then suggests that another factor must be operable, an electronic factor introduced by the different alloying elements.

A good summary of the structural and anionic effects on catalyst activity is given in a review by Adzic.[21] The slow kinetics and complexity of the ORR continues to be a barrier to commercial development of the fuel cell. The lack of agreement on the reaction mechanism continues at this writing. This may be resolved at some point by including the effects of PtOH, anions and water at the catalyst surface.

Platinum-alloy catalysts for phosphoric acid fuel cells do not have unlimited stability at the typical operating temperatures, $\sim 200\ ^\circ$C; in fact, even platinum dissolves depending on the particular operating conditions. Investigations directed towards improving both the chemical and physical stabilities led to work on developing catalyst crystal structures that were ordered fcc crystals. The stability of such systems[22,23] is still open to question but the associated performance benefits for oxygen reduction for both ordered and disordered platinum alloy catalysts appear to be generally accepted.[24] The activity advantages of the platinum-alloy catalysts have also been observed in the lower temperature PEMFCs. The lower operating temperature may lead to better corrosion properties for the alloys but could ultimately lead to membrane problems if the non-noble elements dissolved into the membrane.

The significance of the platinum valence electrons, the platinum 5d-electronic orbitals, has already been identified as an important factor in electrocatalysis. This was investigated in a comprehensive study (XAS) of the alloy catalyst behavior.[25] The study suggested that the oxygen adsorption was related to the unpaired d-electrons in the catalyst.

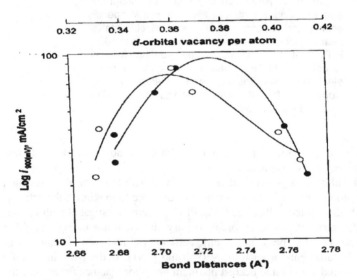

**Figure 5.3.** Correlation of oxygen electrode performance (Log $i$ @ 900 mV, mA/cm$^2$) of Pt and Pt alloy electrocatalysts in proton-exchange membrane fuel cell with Pt-Pt bond distance: •, and the d-orbital vacancy of Pt; o, obtained from in-situ XAS. Reprinted from Reference 27, Copyright (1995) with permission from The Electrochemical Society.

The analytical technique of XAS allowed an investigation of the alloy systems in-situ. The investigation probed the chemistry and physical structure of platinum catalyst surface atoms. In addition, the number of platinum 5d-electrons was directly measured by observing the platinum $L_3$ edge of the absorption spectrum. Several different platinum alloys (Pt-Cr, Pt-Mn, Pt-Fe, Pt-Co, Pt-Ni) were investigated using this technique in PEM cell conditions and the results are instructive.[26] The study showed that all of the alloys have higher platinum 5d-electron vacancies than unalloyed platinum and furthermore that each alloy had a reduction in platinum nearest neighbor spacing. A relationship of specific activity with the d-electron density was observed for supported platinum and the platinum alloy catalysts and is reproduced in Figure 5.3.

The interaction between the catalyst and the carbon support may also affect the way the catalyst behaves. In an attempt to study the platinum-carbon support interaction, for the ORR, the functionality of the carbon surface[28] was altered. When the support was functionalized with sulfur, the initial catalytic activity increased due to an increase in catalyst surface area. Life testing showed that the rate of decay in surface area was also increased. There was no net benefit realized except that associated with the higher initial catalyst surface area with the treated support. The only differences in catalyst activity could be attributed to the differences in particle size of the different catalysts. No change in Tafel behavior was identified. This particular negative result does not necessarily mean that a strong metal substrate interaction (SMSI) is not possible, since SMSI has been observed in heterogeneous catalysis for many years.[29]

For acid fuel cells (PEM and PAFC) supported platinum and supported platinum-alloy catalysts are commonly used for the oxygen reduction catalyst. The various alloy systems (PtV, PtCr, and PtCrCo) have shown from 1.5X to 2.5X the catalyst activity of unalloyed platinum catalysts. The increase in activity of the alloy systems apparently is not due to any change in mechanism between the alloy and non-alloy catalysts. Although there remain some questions in the interpretation of the data, it appears that the mechanism is a *series* pathway via an $(H_2O_2)_{ad}$ intermediate as shown in equation 5.3.[30] The conclusion reached by the Ross's group suggests that there is a structural effect of the catalyst that leads to preferred platinum surfaces that are more active and also inhibit anion adsorption. For the alloys, they suggest that PtOH formation is inhibited by the metal oxides surrounding the platinum. These conclusions represent current thinking on this important area of research but it is by no means the final word.

## 5.2.2  Oxygen Reduction in Alkaline Electrolytes

The overall ORR on the cathode in base electrolytes can be written as:

$$O_2 + 2\,H_2O + 4\,e^- \rightarrow 4\,OH^- \qquad (5.4)$$

The kinetics of the ORR is more rapid in the alkaline fuel cell, when compared to acid electrolytes. The exchange current in alkaline systems is about 10–100 higher than that in acid systems. Gold[31] and platinum[32] are the most active electrocatalysts for the reaction and both show a structural dependence. Gold is unique as a catalyst for the ORR in that it appears to support both a 2-electron and a 4-electron reduction. The 4-electron reduction is seen only on the Au(100) face[33] and is the most active surface for oxygen reduction in alkaline solutions. The reaction mechanism has been studied by Adzic et. al.[34,35] and they have determined that the first charge transfer is the rate limiting step for all the low index planes. The Tafel slope is −120 mV/decade and the reaction order is one.

One of the main problems associated with AFCs is that the electrolyte is intolerant to carbon dioxide. This has been *the* reason why they have not been a prime area for applied research, cf. PEMFC, PAFC. The AFC fuel cells suffer large performance losses when the electrolyte is exposed to fuels containing $CO_2$ or even air with its associated low $CO_2$ content. The reaction of $CO_2$ with the electrolyte can be written as:

$$CO_2 + 2\ OH^- \rightarrow CO_3^{2-} + H_2O \qquad (5.5)$$

The effects of the $CO_2$ in alkaline media are:

- a reduction in $OH^-$ concentration and lowering of the kinetics,
- an increase in electrolyte viscosity with lowered diffusion coefficients,
- eventual precipitation of carbonate salts in the electrode pores,
- lowered oxygen solubility, and
- lowered electrolyte conductivity.

While this is not primarily an electrocatalytic problem, it does impact on the suitability of the AFCs for commercial applications. Some well-known applications, such as the electric power plants used in the United States manned space program, surmount this problem by using pure reactant gases, hydrogen and oxygen. This technology, while interesting, employs materials and catalysts that would be prohibitively expensive for consumer applications.

## 5.2.3.  Hydrogen Oxidation in Acid Media

The hydrogen oxidation reaction has a very high exchange current density and platinum is an excellent catalyst for the reaction. If it were not for the practical problems associated with fuels for commercial applications, platinum catalysts would be more than satisfactory to meet the needs for the hydrogen oxidation reaction. This is because with neat hydrogen as a fuel, the overpotential associated with the hydrogen electrode is very low even at very high current densities. It is only when the hydrogen is diluted with other gases such as $CO_2$ and CO that the anode overpotential increases and leads to a reduction in cell performance. The thrust of research on hydrogen oxidation catalysts has been towards finding catalysts with higher exchange currents per unit cost and resistance to poisoning. Although the exchange current for platinum is high, the surface area of the anode catalyst plays an important role. This is because, in practical systems, the effects of operating conditions (impure fuels, shutdowns, etc.) can lead to degradation of the area. The practical challenge for PEM fuel cells is the identification of a stable catalyst that will oxidize hydrogen at a very high rate in the presence of carbon oxides (CO, $CO_2$).

The electrochemical reaction can be written as:

$$H_2 \rightarrow 2\,H^+ + 2\,e^- \tag{5.6}$$

or stepwise:

$$H_2 + 2\,C_s \rightarrow 2\,C_sH \tag{5.7}$$

$$C_sH \rightarrow C_s + H^+ + e^- \tag{5.8}$$

where $C_s$ is the catalyst site.

For single crystal platinum catalysts in 0.05 M $H_2SO_4$ electrolyte at 333 °K, the $i_0$ for the hydrogen-oxidation reaction is about 1 mA/cm$^2$. The activation energy for the reaction ranges from about 10 to 20 kJ/mol depending on the particular exposed crystal face.[36]

The reaction is highly reversible and proceeds at high current densities with low polarization losses. There is no problem in achieving high performance on the anode until one considers practical fuels such as those coming from a reformate where some C-containing species are always a part of the generated fuel. The fuel can be *cleaned* to remove these species but at a cost and an increase in system complexity. The development of an anode catalyst that will tolerate a limited amount of CO either under steady state operation or under conditions of high CO transient content is the challenge for the low temperature PEMFC. The effect on the anode catalyst of $CO_2$ is not insignificant; it is more than just a dilution effect. A reverse gas-shift reaction can take place that will generate some CO depending on the fuel composition. The $CO_2$ effect on PEM performance has been investigated and reviewed by a number of researchers.[37-39] Generally the performance of the cell is adversely affected by the presence of $CO_2$ even without the presence of CO. Platinum-ruthenium catalysts appear to tolerate $CO_2$ better than platinum catalysts and an increase in cell temperature improves performance. Since the result of $CO_2$ in the fuel is similar to the catalyst problem with CO, we will limit our discussions to the CO effect.

The preferred catalysts for the hydrogen-oxidation reaction in the presence of CO include platinum and various platinum alloys or mixtures of platinum with a noble or non-noble constituent. The alloying components most used include ruthenium, tin, and molybdenum. The problem with using platinum alone in a low temperature PEM fuel cell is that CO adsorption on the platinum is very strong, and even a low level of a few parts per million of CO in the hydrogen stream will cause substantial performance losses on the anode. The mechanism for the performance loss is a significant reduction in the number of active catalyst sites available for hydrogen oxidation that results in lower fuel cell performance when operating at high current densities.

Research on alternative catalysts has led to the development of bimetallic catalysts that are based on the hypothesis that a bi-functional catalyst can both adsorb and oxidize the hydrogen while simultaneously oxidizing the CO component. The additional metal, M, is generally chosen based on its oxophilic property, to provide an oxygen containing species to sites adjacent to adsorbed CO and thereby remove the CO by an oxidation process. This oxidation step can be described by the reaction shown below:[40]

$$Pt\text{--}CO_{ad} + M\text{--}OH_{ad} \rightarrow Pt + M + CO_2 + H^+ + e^- \qquad (5.9)$$

The above reaction has been found to be dependent on the electrolyte type since the adsorption of the OH species may be inhibited by the electrolyte anion. The relationship between CO oxidation and methanol oxidation reaction is noted here.

Since the performance of the anode catalyst for hydrogen oxidation is dependent on the total active surface area, carbon supported catalysts have been developed to maximize catalyst surface area. Some results of CO oxidation experiments on platinum suggested that there was a linear bonded and a bridge bonded CO adsorption that corresponded to the two observed voltage peaks. Recent research suggests that perhaps the CO oxidation peaks may be associated with the extent of CO coverage. Some of the conclusions reached from experiments are based on 100% coverage of the catalyst sites with CO; this is questionable. The time, CO concentration, and rest potential applied during CO adsorption may not lead to 100% coverage. This may lead to the twin peaks observed during the potential sweeps to oxidize the CO. The catalyst regions that are fully covered with CO could be associated with the high peak potential while the sparsely covered regions, could be oxidized at a lower potential.[41] Since the experimental data in most of the experiments is analyzed with the assumption that there is full coverage of the catalyst sites, the conclusions reached may not be valid. Simply using long CO residence times prior to oxidation experiments is insufficient to quantify the coverage levels (assumed full coverage state) and the state of the absorbent species.

An interesting observation for the Pt-Ru system is that the CO oxidation peak structure changes with the ruthenium content of the catalyst.[42] As one approaches pure ruthenium only one oxidation peak is observed. The adsorbing potential of the CO also has an effect on the subsequent oxidation rate.[43] The extent of OH concentration, or other oxygen containing species, is also dependent on the potential of CO adsorption and therefore the final observed rate of CO oxidation would be different depending on this adsorbing potential.

It is well known that the typical supported platinum catalyst has very low CO tolerance because of the adsorption onto the platinum, although certain unique physical structures may lead to a change in that perception. Nano-structured platinum was prepared by a chemical reduction of hexachloroplatinic acid dissolved in aqueous domains of the liquid crystalline phases of oligoethylene oxide surfactants. The material has high surface area and a periodic mesoporous nanostructure. When tested, it exhibited higher CO tolerance at potentials higher than the hydrogen adsorption-desorption range.[44] These types of structures are provocative and suggest future binary adaptations that may lead to further increases in CO tolerance. However, platinum catalysts in conjunction with other elements have been the preferred catalyst system for development of practical fuel cells.

The platinum-ruthenium system has received the most attention for hydrogen oxidation in the presence of carbon monoxide. This catalyst system is able to tolerate low ppm levels of CO without excessive polarization losses. Other systems include simple binary mixtures of the two elements, alloyed material, and decorated platinum or ruthenium with either ruthenium or platinum. Platinum-ruthenium alloys and mixtures with the same atomic compositions (50:50) were evaluated to determine the CO oxidation potential. They were also tested for hydrogen oxidation in the presence of 100 ppm of CO.[45] The results suggested that the alloy was a better catalyst for the reaction. The explanation for the difference in behavior was based on the intimate contact of the two metals and an apparent change in electronic state of the platinum in the alloy.[46] Supported platinum-ruthenium alloys are generally used for commercial fuel cell systems to oxidize low CO concentrations in reformed fuels. Since practical systems experience transient CO concentrations in the fuel that are too high for this catalyst to handle, and hence air bleed systems are commonly utilized. This system allows a controlled amount of oxygen (air) to be mixed with the fuel; the CO is oxidized by the oxygen and this minimizes cell performance losses.

Two newer developments in catalyzation techniques have led to novel *platinum decorated ruthenium* and *ruthenium decorated platinum* catalysts. These materials are of very high surface area and the coverage of the decorating element can be much less than a monolayer. An interesting ruthenium supported platinum catalyst has been investigated.[47] Catalysts were prepared by depositing platinum on nano-particles of ruthenium on a carbon substrate via a *spontaneous deposition* technique. This resulted in a high surface area system of small platinum islands on the ruthenium nano-particles. The Pt/Ru atomic ratio was 1:20. The activity for hydrogen oxidation and tolerance for 100 ppm CO was much better than for a commercial PtRu catalyst with 3X the platinum content. The improved performance results were interpreted as an electronic effect caused by a d-band shift that lowers the CO adsorption energy. In a similar manner, a ruthenium decorated platinum catalyst was prepared by the spontaneous deposition method.[48] The activity for CO oxidation stems apparently from the platinum-ruthenium interfaces with surface diffusion playing a part. The CO tolerance of this catalyst is better than most other Pt-Ru binary systems and fuel cell development studies are underway.

The bi-functional catalyst mechanism used to describe the observed CO tolerance phenomena has led to the investigation of other metals that might provide an oxygen species for CO oxidation. Bulk and supported platinum-molybdenum alloys were prepared and tested for hydrogen oxidation and CO tolerance in a rotating thin-layer electrode system (RTLE) and also in full cell configurations.[49] Various alloy compositions were evaluated and a $Pt_3Mo$ alloy was found to be most active. This catalyst performed better than commercial platinum-ruthenium systems evaluated under similar conditions. In contrast to the platinum-ruthenium catalyst, the molybdenum species does not specifically adsorb CO. This is advantageous, compared to the ruthenium systems, where the ruthenium does adsorb CO. This adsorption of the CO on the non-platinum component leads to a competition between CO and OH species adsorption and therefore lowers the number of available sites for oxygen species adsorption and hence, the oxidation rate for CO.

Another binary system that acts as a bi-functional catalyst is the platinum-tin system. The CO oxidation potential on $Pt_3Sn$ (hkl) surfaces has been found to be greatly reduced when compared to pure platinum surfaces.[50] This was attributed to the unique states of CO and OH at the platinum-tin interface. The experimental data suggests that the bi-functional catalyst mechanism would be consistent with the data. A continuous production of $CO_2$ from very small CO content fuels was observed even at the low operating potential of 100mV. Tin is similar to molybdenum in the respect that CO is not adsorbed as in the platinum-ruthenium catalyst. The challenge associated with platinum-tin catalysts is achieving a low enough potential required to continuously oxidize the CO species at a high rate. The turnover number is very low for potentials lower than 0.2 V. In addition, hydrogen does not adsorb on tin. So while the rate of oxidation of hydrogen in the presence of CO is very high for platinum-tin catalysts, it occurs at too high a potential for practical fuel cell use. One report suggests that the platinum-tin system is chemically unstable[51] at operating temperatures above 85 °C.

The literature suggests that at least two primary properties of platinum binary catalysts are involved with the hydrogen-oxidation reaction in the presence of CO. The action of the added metal, in terms of its properties for oxygen species absorption and the changes of electronic character of the platinum remains to be understood. Any development of a new catalyst would benefit from in-situ XAS examinations. The details of catalyst electronic and geometric structure can be studied as a function of operating potential to determine the effect on performance for hydrogen oxidation of the chemistry and composition of the particular catalyst. The nature of the CO bonding can also be probed by this technique since nearest neighbor species can be determined. It also would be useful to study the catalyst during operation with FTIR and EMIRS techniques. The details of the intermediate pathways could be studied to identify the adsorbent species and it's bonding to the catalyst. Changes in adsorbent species can also be investigated by these techniques in conjunction with potential step experiments.

These experiments can provide important information about both technical and model (single crystal) catalysts even though there may be differences in the reaction mechanism between the two. It is also important now with the evidence of past experiments, to conduct new investigations in a consistent manner so that *pretreatments* of the catalyst are understood and kept constant when comparing results. One of the key observations in this regard is that the potential at which the

CO is adsorbed may interfere with OH adsorption and this must be considered when analyzing the data.

### 5.2.4.  Hydrogen Oxidation in Alkaline Media

The reactions for hydrogen oxidation in alkaline systems can be written as:

$$H_2 \rightarrow 2\,H_{ad} \tag{5.10}$$

$$H_{ad} + OH^- \rightarrow H_2O + e^- \tag{5.11}$$

As we have noted earlier, although the AFC has the advantage of having the highest efficiency of all fuel cells, it works well only with pure reactant gases. Consequently this requirement has limited its commercial development. Depending on the operating conditions, non-noble elements such as nickel have been used for the oxidation of hydrogen. Platinum catalysts have demonstrated high performance and in some cases platinum binary systems, e.g. platinum–palladium, have also been utilized.

Because of the limitations on commercial development the high performance of the reaction, the area of hydrogen oxidation reaction has not been explored to the extent that has been done on acid systems. Interested readers are referred to the papers written by Kordesch his collaborators, and others who have championed this area for decades.[52-56]

The hydrogen electrochemistry on platinum single crystal surfaces was investigated by Ross and his associates.[57] They looked at the hydrogen oxidation and the hydrogen evolution reactions in alkaline media utilizing rotating disk techniques. Their results, for the first time, show that the kinetics on platinum surfaces is sensitive to surface structure. They interpret the kinetics of the reaction based on the differences in the sensitivity of $H_{upd}$ and $OH_{ads}$ on the platinum (hkl) surfaces.

### 5.2.5.  Methanol Oxidation in Acid Media

Direct methanol oxidation has been researched for the past several decades in PEMFCs, AFCs and PAFCs. This liquid fuel is very attractive in that 6 electrons can be oxidized per molecule of methanol and the energy density of the fuel is very high. Considerable controversy has existed in the literature concerning the rds for the reaction, the identity of the intermediates and poisons, the role of the promoter metals and the overall mechanism of the reaction on promoted platinum catalysts. Early mechanistic studies have focused mainly on platinum, primarily because oxidation occurs only on the platinum sites. Mechanistic investigation of platinum-ruthenium binary systems, especially on well characterized surfaces, has been conducted over the past decade and platinum-ruthenium is perhaps the only catalyst

### TABLE 5.1
### Parameters for the Methanol-Oxidation Reaction

| Kinetic Parameter | Value | Ref. |
|---|---|---|
| Tafel slope (Pt) | 55–60 mV/decade (0.42 V < E < 0.55 V)<br>110 mV/decade (0.55 V < E < 0.72 V) | 58,59 |
| Activation energy, apparent | 60 kJ/mole (Ru-OH formation?)<br>40–45 kJ/mole (CO diffusion?) | 60,61 |

system that has been well characterized. The basic understanding of the mechanism of the reaction is a very active area of study. The differences observed in catalyst behavior (for the same catalyst), for oxidation of $CO/H_2$ and for methanol (with intermediate CO-like poisons, is an area where fundamental understanding is lacking.

There is a consensus on the Tafel slope of the oxidation reaction but some differences exist in the reported values of activation energy; the exchange current density is rarely reported. Table 5.1 shows some of the basic parameters for the methanol-oxidation reaction on Pt-Ru.

The two distinct Tafel slopes, for the low and high potential regions, correspond to the changeover of the mechanism from water activation to C-H activation. Similarly, the higher slope of activation energy is associated with Ru-OH formation and the lower value represents CO diffusion.

The reaction scheme for methanol oxidation can follow either a serial pathway[62] or a parallel pathway as shown in Figure 5.4. The serial path involves the sequential deprotonation of methanol with an end product, $CO_{ads}$, that is finally oxidized by water (or an OH species) to $CO_2$. The serial mechanism assumes three electrons are involved in the chemisorption step; the adsorbed intermediate is coordinated to three platinum sites and it is proposed that oxidation of the Pt-C-OH is the rate determining step. If this mechanism were true, the only product identified during methanol oxidation would be $CO_2$. However, as early as 1963, a product

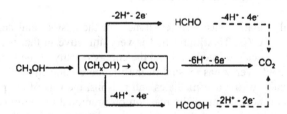

**Figure 5.4.** Schematic representation of the parallel pathways for methanol oxidation on platinum based catalysts. Reproduced from Reference 66, Copyright (1992) with permission from Elsevier Science.

analysis performed by Schlatter,[63] showed the presence of formaldehyde, formic acid and methyl formate, during the controlled potential oxidation of methanol. A scheme of parallel reactions leading to the generation of 6 electrons in the oxidation of methanol, was suggested by Breiter as early as 1967. Recently Smotkin and coworkers[64] showed the formation of methyl formate using a tandem FTIR technique. This and other studies[65] clearly indicate that the oxidation of methanol involves both the direct oxidation of methanol to carbon dioxide and through the formation of formaldehyde and formic acid intermediates.

Over the past several years, many new catalysts have been evaluated for methanol oxidation; a list of those catalysts is given in a paper by Sarangapani et al.[67] Recently, Pt-Ru-Os[68] and Pt-Ru-Os-Ir,[69] platinum-nickel[70] and platinum with macrocyclic promoters[71] have been proposed as catalysts. These redox promoters use various ligands coordinated to metal sites such as Ru, Os, Ir, Sn, Mo, Mn, Fe and Co. These complexes are:

- characterized by well defined oxidation states at a given potential that have affinity for CO-like species and/or oxygen,
- coordinated to water, *hydroxo* or *oxo* groups that can catalytically oxidize poisoning impurities, and
- electrochemically reversible.

In spite of extensive work into various catalysts, Pt-Ru, (1:1 atomic ratio) appears to be the best available catalyst.

The direct oxidation of methanol has been studied primarily on platinum and platinum-ruthenium systems and has been extensively discussed in the literature. The effect of the promoter on oxidation has been attributed to both an electronic effect as well as a geometric effect. The modification of electronic property may result in either the activation of water molecule or a decrease in binding energy. The geometric effect, on the other hand, targets the three adjacent sites that are needed for the platinum, making it difficult to adsorb methanol and provide adjacent ruthenium (or some other oxophilic metal) sites where the water can be activated and thereby provide the source of the oxygenated species for CO oxidation. Recent hypotheses tend to combine the two mechanisms.

The surface vs. bulk composition of the catalysts has been investigated for both unsupported and supported catalysts. Surface composition changes with the type of support, method of preparation and pretreatment.[72,73] For example, if the catalyst is heated in air, the surface is enriched with ruthenium. This has a profound impact on the kinetics and mechanism, since kinetic studies from different groups may lead to different interpretations for the same bulk composition. The oxidation state of ruthenium has also been a subject of several investigations. The key oxidation state Ru(IV) has been proposed based on XPS studies,[74] although, other oxidation states have also been reported.[75] Of course, the actual state of the catalyst in an operating fuel cell is dependent on a number of factors and may not be sufficiently controllable to achieve maximum performance. Although earlier studies have noted the absence of methanol adsorption on ruthenium sites, more recent studies have clearly demonstrated that there is adsorption of methanol on ruthenium at temperatures $\geq 60\ °C$.[76]

The platinum-tin catalyst is another system that has been investigated for methanol oxidation. Both increases as well as decreases in activity for this catalyst have been reported in the literature. Morimoto and Yeager's[77] investigation appears to clarify the controversy. They conclude that on high area platinum, tin is more stable and exhibits enhanced oxidation for an extended time period; on smooth platinum, however, tin dissolves from the surface, and results in a short-lived enhancing effect.

Based on the study of literature through the year 2000, there are several open questions pertaining to the kinetics and mechanism of methanol oxidation in acid medium, which include:

(a) What is the nature and distribution of products – on bulk platinum, and promoted platinum, and technical catalyst – 50/50 Pt/Ru?
(b) Is it necessary to form a single phase alloy for the best catalytic activity or will a multiphase system or even a mixture work equally well?
(c) What species are *intermediates* and what is the *poison*?
(d) Is the oxidation of the poisoning residue brought about by $H_2O^*$ or $OH_{ads}$?
(e) What is the rate determining step – on platinum, on promoted platinum?
(f) Does the rate determining step change from *poison oxidation* to C-H activation at higher potentials?
(g) What is the influence of temperature on the mechanism?

The literature, starting in 2000, appears to answer some of these questions providing directions for the development of new and more efficient electrocatalysts. The relevant studies and their conclusions are summarized in the following paragraphs.

Earlier studies focused on the methanol-oxidation products on a bulk platinum catalyst. Product identification with PEMFC *membrane-electrode assemblies* (MEA) under real fuel cell operating conditions is important:

• to establish the mechanism of methanol oxidation in a fuel cell, and
• to determine whether there is an environmental discharge of poisonous exhaust from a DMFC.

Sanicharane et. al.,[78] published results of such a product analysis using the tandem FTIR analysis technique. In this technique, two FTIR instruments are deployed simultaneously, one to analyze the intermediates on the surface of the electrode, and the second to identify the exhaust gas composition. Methyl formate was identified, for anode potentials $\geq 0.5V$ (vs. RHE), as a reaction product as well as $CO_2$. It has to be emphasized that, although this study employed fuel cell MEAs, the reactant feed was a mixture of methanol and water vapors, not aqueous methanol. By comparing the peak positions for $CO_{ads}$ for an arc-melted Pt-Ru alloy and the fuel cell electrode, the authors inferred that the high surface area platinum-ruthenium catalyst that they used in the fuel cell had a platinum rich surface.

Lei et al.[79] conducted a detailed investigation of the kinetic-isotopic effect on the methanol-oxidation reaction using high area platinum, platinum-ruthenium, and platinum-ruthenium-osmium-iridium catalysts packed in a recessed microelectrode assembly. Deuterated methanol (with various levels of deuteration), in 0.5 M $H_2SO_4$ or $D_2SO_4$, was used as the fuel in the study. Based on an analysis of the isotopic effects, the authors estimated the crossover potential at which the rate determining step for methanol oxidation shifted from water activation to C-H bond activation. Interestingly, the mean values of crossover potentials for the catalysts platinum, platinum-ruthenium, and platinum-ruthenium-osmium-iridium are 0.65 V, 0.34 V and 0.33 V respectively, suggesting that the observed limitation of methanol oxidation reaction may be due to C-H activation. This result is supported by some data from fuel cell experiments with a Pt-Ru catalyst, where the performance of an anode with 0.5 M methanol becomes transport limited at anode potentials around 400 mV.

Evidence for C-H bond activation as the rate-determining step is also found in the work of Batista et. al.[80] They investigated the oxidation of methanol on UHV cleaned Pt-Ru (85:15 and 50:50) alloys using FTIR and chronoamperometry. Chronoamperometric measurements were carried out as a function of temperature, over the range, 27–60 °C, by a dynamic variation of the electrode temperature at the rate of 6.75 °K/min. The FTIR data shows that below 0.5 V, the rate of $CO_2$ formation is independent of CO surface concentration. Above this potential, the production of $CO_2$ is higher for the platinum rich alloy and below 0.5 V, water activation is the rate determining step. Above 0.5 V potential, rate of methanol adsorption (C-H bond activation) is the rate-determining step. Furthermore, their results suggest that the reaction pathways may be different for the two catalysts; for the 85:15 alloy there are higher yields of HCHO and HCOOH than for the 50:50 alloy where the reaction appears to proceed via the sequential deprotonation mechanism leading to $CO_{ads}$.

Dubau et al.[81] investigated whether an alloy of platinum-ruthenium was required for the efficient oxidation of methanol. They prepared platinum and ruthenium colloids separately from the corresponding salts; the platinum-ruthenium alloy colloid was formed by starting with a mixture of the platinum and ruthenium salt solutions. A mixture of the individual platinum and ruthenium colloidal solutions resulted in an intimate dispersion of the two colloids that was adsorbed onto carbon. The other colloidal solutions were also adsorbed on carbon. A physical mixture was prepared by separately mixing the Pt/C and Ru/C catalysts. Each of the above catalysts was heat-treated at 300 °C. Methanol oxidation, CO stripping and FTIR spectroscopic measurements were carried out on glassy carbon electrodes which were deposited with each of the catalysts described above. CO coverage from methanol dissociation was higher on the unalloyed Pt + Ru/C than the other catalyst preparations. The results from the FTIR experiments clearly show that the highest quantities of $CO_2$ are formed at all potentials for the Pt + Ru/C catalyst. The experimental results also indicate that the activity relationship for the series is:

$$Pt/C < (Pt/C + Ru/C) < Pt\text{-}Ru/C < Pt + Ru/C \qquad (5.12)$$

The Pt + Ru/C refers to the preparation involving colloidal mixture of platinum and ruthenium adsorbed onto carbon. The authors claim that superior performance of the colloid mixture shows strong interaction between platinum and ruthenium arising from the decoration of platinum particles by ruthenium. In the physical mixture, on the other hand, there is no interaction between platinum and ruthenium and that is why the mixed catalyst has poor activity in terms of CO tolerance. The authors attribute the lower activity of alloyed Pt-Ru/C to the surface enrichment of the alloy with platinum; they also infer that the electronic interactions are different between platinum and ruthenium atoms in the decorated catalyst and in the alloy. They concluded that the alloy catalyst is not the most active and oxidation of CO by adsorbed OH is not the rate-determining step. The diffusion of CO from platinum to ruthenium sites as the rate determining step is also ruled out based on independent studies of diffusion rates. Therefore, they conclude that dehydrogenation of methanol appears to be the rate determining step, which is in agreement with the conclusions made by Lei, et. al.,[79] and Batista, et. al.[80] The case for C-H activation as the rate determining step for methanol oxidation on Pt-Ru for potentials > 0.35 V is compelling in view of the results from this investigation.

The influence of catalyst structure and the need for a better understanding of the geometric effect on the catalytic activity is highlighted in a report using a novel nano-structured platinum morphology that is capable of sustained activity towards methanol oxidation.[82] TEM and SEM studies show that this catalyst possesses a novel periodic mesoporous structure with high surface area. This novel structure of platinum and the reported data given in this publication argues against the conventional wisdom that platinum is incapable of oxidizing the CO intermediate, without the presence of an oxophilic metal. These investigators prepared a mesoporous platinum through a directed template technique. The resulting catalyst had a surface area, 34 $m^2$/g, an average pore size of 3.6–3.8 nm, and an internal to external surface area ratio in the range, 17–73. In Figure 5.5, the activity of this catalyst is compared to a Johnson-Matthey

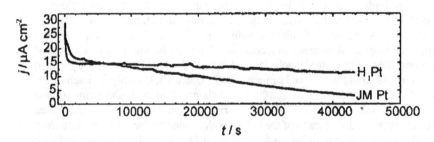

**Figure 5.5.** Durability of current density measured at 0.476 V on a mesoporous Pt catalyst ($H_1$-Pt) and a J-M platinum electrode. Reprinted from Reference 82, Copyright (1998) with permission from Elsevier Science.

platinum catalyst. After an initial drop in activity, the residual activity of the mesoporous Pt remains constant over a test period of twelve hours, while the commercial Johnson-Matthey platinum's activity declines to nearly zero.

Based on an analysis of the current transients as a function of potential, the authors hypothesize that for these mesoporous structures, methanol oxidation follows a parallel mechanism. The major path involves oxidation of methanol to a soluble species such as formic acid or formaldehyde with only a minor path yielding a CO-like species. Due to the high area mesoporous nature of the catalysts, the soluble intermediates are oxidized to $CO_2$ before they have a chance to diffuse away. This conclusion is substantiated with calculations showing a 75% current decay in short times if CO-species formation is the main pathway; such a large current decay is not seen with this catalyst. In fact, after an initial decay, the current is constant over the twelve hour test period. The enhanced poison tolerance of this catalyst is attributed to the unique nano-structure morphology of the material.

The question as to what are the active sites for poison removal was addressed in some work by Waszcuk et al.[83] The hypothesis that a monophasic alloy is not necessary and may even be undesirable in a methanol-oxidation catalyst is supported from the studies of platinum nanoparticles decorated with ruthenium, a method originated by Andrej Weikowski, who prepared such catalysts and studied hydrogen and methanol oxidation as well as CO stripping. The decorated catalysts demonstrated significant hydrogen adsorption-desorption activity compared to a commercial catalyst, platinum-ruthenium. Normalizing the activity on a per real surface area basis, the decorated platinum was more active than a commercial Johnson Matthey catalyst by approximately a factor of two. Through an analysis of the experimental data, the authors claim that the interior of the ruthenium islands are inactive to methanol decomposition, while the edges are the most active for the oxidation of CO to $CO_2$. They also point out that the ruthenium edge atoms are under-coordinated (by ruthenium and/or platinum surface atoms) and may be more active for water activation than the platinum-ruthenium sites in the alloy. The authors conclude that the enhanced activity of the decorated catalyst is attributable to ruthenium edges. This observation  may pave the way to designing improved methanol oxidation catalysts.

In the introduction to this methanol-oxidation section, several questions related to methanol oxidation that were unanswered, as of year 2000 were identified. More recent work in this area suggests that answers are emerging for the following topics:

- the nature and distribution of products on bulk platinum, promoted platinum, and technical catalyst,
- the role of a single-phase alloy vs. multiphase system for the best catalytic activity,
- intermediates vs. poison effects,
- mechanistic pathway and rate determining step on platinum-ruthenium.

The question of whether water or $OH_{ads}$ is the mediator in the oxidation of the poisoning residue, although addressed in the literature, is not yet conclusively answered. Similarly, work on the influence of temperature on the mechanism is emerging but needs to be considered in more detail, especially due to the increased recent interest in the micro-fuel cells as possible battery replacement power packs. The advances in understanding of the methanol oxidation reaction have taken place because of new experimental techniques that have allowed more detailed information to be gleaned from the experiments. Studies of new catalyst systems coupled with the new results suggest that direct methanol oxidation fuel cells may indeed fulfill the great promise for commercial applications in the not-so-distant future. Beyond the technology of methanol fuel cells however, it is expected that the development and application of new techniques and novel catalysts will bring further success to all of the fuel-cell technologies.

### 5.2.6.   Methanol Oxidation in Alkaline Media

This again is an area that has received little attention because of the problem of carbonation of the electrolyte. This area is covered in detail by Koscher and Kordesch.[84] The overall reaction can be written as:

$$2 \, CH_3OH + 3 \, O_2 + 4 \, OH^- \rightarrow 2 \, CO_3^{2-} + 6 \, H_2O \qquad (5.13)$$

It is evident that not only oxygen and methanol are consumed in the reaction but two $OH^-$ ions are used per methanol molecule. This requires an electrolyte replenishment scheme that could be easily handled by adding electrolyte with the methanol feed. Several different catalysts have been investigated as oxidation catalysts for the reaction. They include platinum,[85] platinum-ruthenium,[86] and platinum-palladium[87] but no detailed kinetic studies have been published at this point. This is an area that may receive more attention because of resurgence in interest in the AFC systems.

### 5.3.   HIGH-TEMPERATURE FUEL CELLS

There are two main types of high temperature fuel cells, the molten-carbonate fuel cell (MCFC) and the solid-oxide fuel cell (SOFC). These fuel cells operate at temperatures above 650 °C and are quite different than the low temperature fuel cells that we have already discussed. The advantage of high temperature with increased reaction rates is counterbalanced by the associated materials challenges. The electrocatalysis of the reactions at the electrodes is not fully understood because of the complexity of the reactions and the difficulty of performing in situ experiments that allow one to monitor the reactions under controlled experimental conditions. Beyond this, however, because of the temperature constraints on the possible materials that can be utilized, the challenge centers on the identification of materials that will be compatible in the electrolyte of choice. The anode polarization losses are minimal when compared the cathodic polarizations and therefore much of the research in improving the cell performance has been on the oxygen reduction electrode.

## 5.3.1.  Molten-Carbonate Fuel Cell (MCFC)

The electrochemical reactions occurring in MCFCs are:

$$H_2 + CO_3^{2-} \rightarrow H_2O + CO_2 + 2e^- \tag{5.14}$$

$$\tfrac{1}{2} O_2 + CO_2 + 2e^- \rightarrow CO_3^{2-} \tag{5.15}$$

with the overall reaction as:

$$H_2 + \tfrac{1}{2} O_2 + CO_2 \text{ (cathode)} \rightarrow H_2O + CO_2 \text{ (anode)} \tag{5.16}$$

The anode is typically a low surface area porous nickel metal doped with chromium. For an internal reforming system, the nickel can be supported on a stable metal oxide such as MgO or $LiAlO_2$. This catalyst would not only oxidize the hydrogen but also reform a fuel such as methane to provide hydrogen for the anodic reaction.

The electrolyte is a $LiCO_3/K_2CO_3$ melt held in a $LiAlO_2$ matrix. The electrolyte ionic conductivity is a large factor in cell performance losses. The thickness of the matrix also plays a part in the overall ohmic loss associated with the electrolyte. Changes to the electrolyte have been investigated that could lead to a reduction in ohmic losses and also to reduce the dissolution of the cathode catalyst.

The major obstacle to the commercial development of MCFCs is the dissolution of the cathode. A lithiated nickel oxide cathode (NiO) is the current state-of-the-art material. The catalytic activity is reasonable but the stability in the very corrosive molten carbonate environment for a proposed 40,000-hour life is a major limitation.[88] The material is porous and must provide sufficient sites for the reactions. The porosity of the material makes studying the reaction difficult. Generally the cathode operates in an acidic region of the solubility curve, so that the dissolution can be written as:

$$NiO + CO_2 \leftrightarrow Ni^{2+} + CO_3^{2-} \tag{5.17}$$

The dissolution of the cathode material has been minimized by changes in operating conditions and also by altering the composition of the melt. In spite of the extensive search for alternate cathode materials, NiO remains the material of choice. Modeling and operational changes have suggested that this material can meet the lifetime goals of a MCFC power plant. This is for an atmospheric pressure power plant; if a high-pressure operating system is considered then another material will have to be developed because of the dependence on $CO_2$ pressure of the nickel oxide dissolution.

The molten-carbonate electrolyte in an oxidizing environment is extremely corrosive. This corrosive environment makes it difficult to know the exact active surface area of the electrocatalyst and current methods of determining the various reactant species in the environment are not available. Half-cell testing with noble metals has been utilized to study the reaction in a variety of melt and gas environments.[89] However, it is not clear what relevance they have to the actual porous nickel-oxide catalyst of the fuel cell where the oxide surface clearly generates and absorbs many unique intermediate reactants or reaction products.

The overall electrochemical reaction includes various chemical reactions that can impact the reaction rate. The chemical reactions take place in any number of sites not at the three-phase boundary of electrolyte, catalyst and reactant. These reactions tend to create a number of different oxygen species that will affect the reaction rate. The oxides, peroxides and superoxides can play an important part in the reaction. In fact even the oxygen dissolution into the melt has been not only thought of as the first step of the reduction process but also as the rate determining step. Tomczyk et. al.[90] measured the catalytic activity of various cathode materials using linear scan voltammetry. They found that the differences in exchange current for the materials studied, including NiO, was not very large—not orders of magnitude—however, even a factor of two to three would make a large impact on the performance of a MCFC system.

The main steps for the oxygen reduction in MCFC include gas dissolution, dissolved as transport, electron transfer, ionic transport and chemical reactions. Much of the work studying these reactions have centered on those reactions involving $CO_2$. There are two main steps involving $CO_2$: the initial dissolution in the melt and the chemical reaction of oxide ion with the $CO_2$.

There are two approaches to solving the cathode dissolution problem: changing the molten carbonate composition to reduce dissolution or using alternative cathode materials.[91] Attempts to change the molten carbonate by reducing the basicity have not led to significant improvements.[92] The second choice has led to the work on the materials $LiFeO_2$ and $LiCoO_2$ as candidates to replace NiO.[93,94] Ternaries containing all three materials in suitable concentrations have also been investigated.[95] Materials stability and performance results are still being determined for these catalysts and will continue as long as MCFC power plants remain as an efficient way to generate large amounts of electrical power.

## 5.3.2.  Solid-Oxide Fuel Cell (SOFC)

Solid-oxide fuel cells are another very attractive means of generating power. Again because of the high operating temperatures their development has been a little slow in terms of commercialization. The operating temperature for the SOFCs is ~ 1,000 °C.

There is an intermediate temperature version of the SOFC that is also very attractive because it allows the use of lower cost components, reduces degradation, and makes possible shorter warm-up and cool down times. The lower temperature means that thinner components are necessary to reduce the ohmic losses.

The SOFC electrolyte is a solid and therefore the extensive problems associated with electrolyte migration are not a concern here. In addition, there is no re-cycle of $CO_2$ required as in the MCFC. The main electrochemical reactions for the SOFC are:

$$H_2 + O^{2-} \rightarrow H_2O + 2\ e^- \tag{5.18}$$

$$\tfrac{1}{2}\ O_2 + 2\ e^- \rightarrow O^{2-} \tag{5.19}$$

or an overall cell reaction of

$$H_2 + \tfrac{1}{2}\ O_2 \rightarrow H_2O \tag{5.20}$$

The high temperature solid electrolyte is usually yttria stabilized $ZrO_2$. The electrode materials include a nickel $ZrO_2$ cermet for the anode and an electronically conducting perovskite or a mixed ionic-electronic conductor for the cathode. Again the main cell performance loses occur on the cathode side and they include ohmic losses, gas diffusion losses and cathodic polarization losses, i.e., electrochemical reduction of oxygen together with associated reaction and transport processes.

The requirements for a good cathode material are high chemical and physical stability, high electronic conductivity and chemical and thermal compatibility with the electrolyte. Of course, physical stability is a key requirement for any fuel cell that has periodic shutdowns, i.e., extreme temperature cycling from operating temperature to near room temperature. The expansion/contraction characteristics of the cell materials must match closely in order to keep the cell from physically disintegrating. The materials compatibility concerns are paramount for commercial development.

There are two different types of cathode electrocatalysts. One is an electronic conductor and the second is a mixed electronic-ionic conducting material. Fundamental studies on noble metals such as platinum have been undertaken but there is no consensus on the details of the reaction: charge transfer, oxygen intermediate species, or reaction pathways, involved in the reaction.[96,97] Perovskites are the most utilized materials for the cathode electrocatalyst. These can either be electronic conductors or mixed ionic–electronic conductors.

The electronic conductors are of the form $La_{1-x}\ Sr_x\ MnO_3$; the details of the transport of oxygen through these materials remain an open question. The mixed conductors are also perovskites of the form $ABO_3$, A being Ln (mostly La) doped with alkaline earth metals (Sr, Ca), B being transition metals (M: Cr, Mn, Fe, Co, Ni). These later materials are the most attractive for SOFC because of their chemical stability, chemical compatibility with the electrolyte and similar expansion characteristics with the electrolyte.

In SOFCs as well as the MCFCs, the cathode material makes a difference in the reaction scheme. The surface reactions of the oxygen molecule (adsorption, dissociation, ionization), surface and bulk transport, and incorporation into the electrolyte are a complex series of events that are not only dependent on the intrinsic properties of the electrocatalyst but also on the details of its structure. Therefore, some conclusions regarding the details of the oxygen reduction mechanism are not universally valid. This is a very daunting field of research but remains probably one of the most fascinating areas of study in fuel cells.

The electrolyte of choice for the intermediate temperature SOFCs is usually one of the following: $Zr(Sc)O_2$, $Ce(Sm)O_2$, and $(La,Sr)(Ga, Mg)O_3$.[98,99] The temperature of operation for these fuel cells is typically near 700 °C. This temperature and the ability to use metals and cheaper components has made this fuel cell an attractive alternative to the $ZrO_2$-based SOFCs. Again the cathode catalyst/electrode is the area where higher performance can be achieved with the proper material. The materials of choice for the cathode include $La(Sr)CoO_3$, $Sm(Sr)CoO_3$, $La(Sr)FeO_3$ and $LiNi(Fe)O_3$. The fabrication of these materials in a form that will have the proper characteristics for performance (conductivity, oxygen diffusion, catalytic activity and chemical compatibility) is a task in itself. Testing them in a fuel cell is another hurdle.

Materials research in both the high and intermediate temperature SOFC is likely to continue in the near and far term. The attractiveness of the SOFCs in terms of performance and efficiency as well as fuel choice will continue to make this a desirable method of electrical energy generation.

## Cited References

1. K. C. Neyerlin, W. Gu, H. A. Gasteiger, to be published in The Journal of Electrochemical Society, 2006.
2. D. Landsman and F. Luczak, in *Handbook of Fuel Cells*, edited by W. Vielstich, A. Lamm, and H. Gasteiger (John Wiley & Sons, Ltd., Chichester, NY, 2003), Vol. 4, p. 811.
3. H. S. Wroblowa, Y. C. Pan, and G. Razumney, *J. Electroanal. Chem* **69**, 195 (1976).
4. N. Markovic, T. Schmidt, V. Stamenkovic, and P. Ross, *Fuel Cells 2001* **1**, 105 (2001).
5. Same as reference 4
6. N. M. Markovic and P. N. Ross, Jr., in *Interfacial Electrochemistry – Theory, Experiments and Applications*, edited by A. Wieckowski (Marcel Dekkar, NY, 1999).
7. B. N. Grgur, N. M. Markovic, and P. N. Ross, Jr., *Can. J. Chem.* **75**, 1465 (1997).
8. N. M. Markovic, H. A. Gasteiger, and P. N. Ross, Jr., *J. Phys.Chem.* **99**, 3411 (1995).
9. L. Bregoli, *Electrochimica Acta* **23**, 489 (1978).
10. Same as reference 9
11. B. C. Beard and P. N. Ross, Jr., *J. Electrochem. Soc.* **130**, 3368 (1990).

12. K. Kinoshita, *J. Electrochem. Soc.* **137**, 845 (1990).
13. R. N. Ross, in *Handbook of Fuel Cells*, edited by W. Vielstich, A. Lamm, and H. Gasteiger (J. Wiley & Sons, Ltd. Chichester, NY, 2003), Vol. 2, p. 464.
14. V. Jalan and D. A. Landsman, U.S. Patent 4,186,110 (January 29, 1980).
15. V. Jalan and E. J. Taylor, *J. Electrochem. Soc.* **130**, 2299 (1983).
16. E. B. Yeager, *Electrochim. Acta* **29**, 1527 (1984).
17. W. Vogel and J. Baris, *Electrochim. Acta* **22**, 1259 (1977).
18. Same as reference 15
19. J. T. Glass, G. L. Cahen, G. L. Stoner, and E. J. Taylor, *J. Electrochem. Soc.* **134**, 58 (1987).
20. J. A. S. Bett, *Electrochemical Society Proceedings Volume,* **92-11**, 573 (1992).
21. R. Adzic, *Recent Advances in the Kinetics of Oxygen Reduction in Electrocatalysis*, edited by J. Lipkowski and P. Ross (Wiley VCH 1998), p. 197-242.
22. F. J. Luczak and D. A. Landsman, U.S. Patent 4,677,092 (June 30, 1991).
23. M. Watanabe, K. Tsurumi, T. Mizukami, T. Nakamura and P. Stonehart, *J. Electrochem. Soc.* **144**, 2659 (1994).
24. Same as reference 23
25. M. L. B. Rao, A. Damjanovic, and J. O'M. Bockris, *J. Phys. Chem.* **67**, 2508 (1963).
26. S. Mukerjee, S. Srinivasin, M. P. Soriaga, and J. McBreen, *J. Electrochem. Soc.* **142**, 1409 (1995).
27. Same as reference 26
28. S. C. Roy, P. A.Christensen, A. Hamnett, K. M. Thomas, and V. Trapp, *J. Electrochem. Soc.* **143**, 3073, (1996).
29. J. Schwank, A. G. Shastri, and J. Y. Lee, *ACS Symposium Series 298*, *(Amer. Chem. Soc.*, Washington DC, 1986), p. 182.
30. N. M. Markovic, T. J. Schmidt, V. Stamenkovic, and P. N. Ross, *Fuel Cells 2001* **1**, 105 (2001).
31. R. Adzic, *Recent Advances in the Kinetics of Oxygen Reduction in Electrocatalysis*, edited by J. Lipkowski and P. Ross (Wiley VCH, 1998), p. 197–242.
32. N. Markovic, H. Gasteiger, and P. Ross, *J. Phys. Chem.* **99**, 3411 (1995).
33. Same as reference 31
34. R. R. Adzic, J. Wang, and B. M. Ocko, *Electrochim. Acta.* **40**, 83 (1994).
35. R. R. Adzic and N. M. Markovic, *J. Electroanal. Chem.* **138**, 443 (1982).
36. N. Markovic, in *Handbook of Fuel Cells – Fundamentals, Technology and Applications,* edited by W. Vielstich, H. Gasteiger, and A. Lamm, (John Wiley & Sons, Ltd. 2003). Vol. 2 Ch. 26.
37. F. deBruijn, D. Papageorgopoulos, E. Sitters, and G. Janssen, *J. Power Sources* **110**, 117 (2002).
38. R. Bellows, E. Marucchi-Soos, and D. Buckley, *Ind. Eng. Chem. Res.* **35**, 1235 (1996).
39. M. Wilson, C. Derouin, J. Valerio, and S. Gottesfeld, *28th IECEC*, 11203, 1993.

40. M. T. M. Koper, A. P. Jamsen, R. A. van Santen, J. J. Lukkien, and P. A. J. Hibers, *J. Chem. Phys.* **109**, 6051 (1998).
41. J. Jiang and A. Kucernak, *J. Electroanal. Chem.* **533**, 153 (2002).
42. G. -Q. Lu, P.Waszczuk, and A. Wieckowski, *J. Electroanal. Chem.* **532**, 49 (2002).
43. N. P.Lebedeva, M. T. M. Koper, J. M. Feliu, and R. A. van Santen, *J. Electroanal. Chem.* **524-525**, 242 (2002).
44. J. Jiang and A. Kucernak, *J. Electroanal. Chem.* **520**, 64 (2002).
45. T. R. Ralph and M. P. Hogarth, *Platinum Metals Rev.* **46**, 117 (2002).
46. J. McBreen and S. Mukerjee, *J. Electrochem. Soc.* **142**, 3399 (1995).
47. S. R. Brankovic, J. X. Wang, Y. Zhu, R. Sabatini, J. McBreen, and R. R. Adzic, *J. Electroanl. Chem.* **524-525**, 231 (2002).
48. Same as reference 42
49. B. N. Grgur, N. M.Markovic, and P. N. Ross, *J. Electrochem. Soc.* **146**, 1613 (1999).
50. H. A. Gasteiger, N. M. Markovic, and P. N. Ross, *J. Phys. Chem.* **99**, 8945 (1995).
51. S. Lee, S. Mukerjee, E. Ticianelli, and J. McBreen, *Electrochimica Acta* **44**, 3283 (1999).
52. K. Kordesch, J. Gsellmann, M. Cifrain, S. Voss, V. Hacker, R. Aronson, C. Fabjan, T. Hejze, and J. Daniel-Ivad, *J. Power Sources* **80**, 190 (1999).
53. K. Kordesch, V. Hacker, J. Gsellmann, M. Cifrain, G. Faleschini, P. Enzinger, R. Fankhauser, M. Ortner, M. Muhr, and R. Aronson, *J. Power Sources* **86**, 162 (2000).
54. E. Gulzow and M. Schulze, *J. Power Sources* **127**, 243 (2004).
55. E. Gulzow, *J. Power Sources* **61**, 99 (1996).
56. M. Cifrain and K. Kordesch, *J. Power Sources* **127**, 234 (2004).
57. N. Markovic, S. Sarraf, H. Gasteiger, and P. Ross, *J. Chem. Soc., Faraday Trans.* **92**, 3719 (1996).
58. V. S. Bagotzky, and Yu. B. Vasiliev, *Electrochim. Acta* **12**, 1323 (1967).
59. M. W.Breiter, *Electrochim. Acta* **12**, 1213 (1967).
60. H.A.Gasteiger et al., *J. Electrochem. Soc.* **141**, 1795 (1994).
61. P. A. Christenson, A. Hamnett, and G. L. Troughtohn, *J. Electroanal.Chem.* **362**, 207(1993).
62. M. W. Breiter, *Electrochimica Acta* **12**, 1213 (1967).
63. M. J. Schlatter, in *Fuel Cells*, edited by G. J.Young (Reinhold, NY, 1963), p. 199.
64. E. A. Batista, H. Hoster, and T. Iwasita, *J. Electroanal. Chem.* **554-555**, 265 (2003).
65. S. Sanicharane, A. Bo, B., Sompalli, B. Gurau, and E.S., Smotkin, *J. Electrochem. Soc.* **149**, A554 (2002).
66. Same as reference 64
67. S. Sarangapani, P. Lessner, J. Kosek, and J. Giner, Methanol fuel cell, in *Proceedings of the Workshop on Direct Methanol-Air Fuel Cells*, Washington,

D. C., ECS Proceedings, Vol 92-14, p161 (The Electrochemical Society, Pennington, NJ, 1992).

68. K. Ley, R. Liu, C. Pu, Q. Fan, N. Leyarovska, C. Segre, and E.S. Smotkin, *J. Electrochem.Soc.*, **144**, 1543(1997)

69. E. Reddington, A. Sapienza, B. Gurau, R. Viswananthan, S. Sarangapani, E. S. Smotkin, and T. E. Mallouk, "Combinatorial Electrochemistry: A Highly Parallel, Optical Screening Method for the Discovery of Better Electrocatalysts", *Science,* **280,** 1735 - 1739 (1998).

70. M. Goetz, and H., J. Wendt. *Appl. Electrochem.* **31**, 811 (2001).

71. Sarangapani, Shantha., U.S.Patent 5,683,829 (1996).

72. H. Miura, T. Suzuki, Y. Ushikubo, K. Sugiyama, T. Matsuda, and R. D. Gonzalez, *J. Catal.* **85**, 331(1984).

73. B. D. McNicol and R.T. Short, *J. Electroanal. Chem.* **81**,249 (1977).

74. A. Hamnett, B.J. Kennedy, and F.E. Wagner, *J. Catal.* **124**, 30 (1990).

75. H. Kim, I.R. de Moraes, G. Tremiliosi, R. Haasch, and A. Wiekowski, *Surf. Sci.* **474**, L203 (2001).

76. H. A. Gasteiger, N. Markovic, P.N. Ross, and E. J. J. Cairns, *J.Electrochem. Soc.* **141**, 1795 (1994).

77. Yu Morimoto, and E. B. Yeager, *J. Electroanal. Chem.* **444**, 95 (1998).

78. S. Sanicharane, A. Bo, B. Sompali, B. Guru, and E. S. Smotkin, *J. Electrochem. Soc.* **149**, A554 (2002).

79. H. W. Lei, S. Suh, B. Gurau, B. Workie, R. Liu, and E.S. Smotkin, *Electrochim. Acta.* **47**, 2913 (2002).

80. Same as reference 64

81. L. Dubau, F. Hahn, Coutanceau, J.M. Leger, and C. Lamy, *J. Electroanal. Chem.* **554-555**, 407 (2003).

82. J. Jiang and A. Kucernak, *J. Electroanal. Chem.* **533**, 153 (2002).

83. P. Waszczuk, J. Solla-Gullon, H.S. Kim, Y.Y. Tong, V. Montiel, A. Aldaz, A. Wieckowski, *J. Catal.* **203**, 1(2001).

84. G. Koscher and K. Kordesch, in *Handook of Fuel Cells – Fundamentals, Technology and Applications*, edited by Vielstich, H.A.Gasteiger and A.Lamm (John Wiley & Sons, 2004), Vol. 4, Ch. 80.

85. W. Vielstich, in *Hydrocarbon Fuel Cell Technology*, edited by B. S. Baker (Academic Press, New York, 1965), p. 79.

86. S. Sarangapani, unpublished work, (2005).

87. S. Sarangapani, same as reference 86.

88. K. Tanimoto, M. Yanagida, T. Kojima, H. Matsumoto, and Y. Miyazaki, *J. Power Sources* **72**, 77 (1998).

89. K. Ramaswami and J. Selman, *J. Electrochem. Soc.* **141**, 619, 622, 2338, (1994).

90. P. Tomczyk, M. Mosialek and J. Oblakowski, *Electrochim. Acta B* **6**, 945 (2001).

91. A. Wijayasinghe, B. Bergman, and C. Lagergren, *J. Electrochem. Soc.* **150** A558 (2003).

92. T. Brenscheidt, F. Nitschke, O. Sollner, and H. Wendt, *Electrochim. Acta* **46**, 783 (2001).

93. L. Plomp, J. Veldhuis, E. Sitters, and S. van der Molen, *J. Power Sources* **39**, 369 ( 1992).

94. A. Lundblad, S. Swartz, and B. Bergman *J. Power Sources* 90, 224 (2000).

95. Same as reference 91

96. F. van Heuvlen, H. Bouwmeester, and F. van Berkel, *J. Electrochem. Soc.* **144**, 126 (1997).

97. A. Mitterdorfer and L. Gauckler, *Solid State Ionics* **117**, 203 (1999).

98. T. Ishihara, H. Matsuda, and Y. Takita, *J. Am. Chem. Soc.* **116**, 3801 (1994).

99. M. Feng and J. Goodenough, *J. Solid State Inorg. Chem.* **32**, 663 (1994).

## CHAPTER 6

# EXPERIMENTAL METHODS IN LOW TEMPERATURE FUEL CELLS

Experimental methods practiced in the conduct of fuel cell R&D vary from basic electrochemical methods—e.g., rotating-disk electrode (RDE) for evaluation of catalyst activity—to downright applied methods such as electrode preparation. Several review articles in the literature cover specific aspects of such experimental methods. In this Chapter, we attempt to cover many of the experimental methods commonly practiced in a low-temperature fuel cell laboratory, with emphasis on applied aspects. Some mention of techniques that are useful in the investigation of various aspects of fuel cells, but are specialized research topics, will be presented but not described in detail. Reference articles will be noted to allow the reader to study them if desired. This Chapter is organized beginning with the preparation and characterization of the components of a fuel cell and progresses through the fabrication and evaluation of a full cell. The outline of the Chapter is shown in Table 6.1 for ease in finding those topics of particular interest to the reader.

This Chapter is devoted mainly to those methods used in proton-exchange-membrane fuel cells (PEMFC); however, many of the methods described can be utilized for other low-temperature fuel cells such as the alkaline fuel cell (AFC) and the phosphoric-acid fuel cell (PAFC). Some of the specific differences for these other fuel cells will be pointed out in the text where they would be instructive.

---

This Chapter was written by S. Sarangapani (ICET, Inc., Norwood, MA 02062) and F. J. Luczak (Glastonbury, CT 06033).

**TABLE 6.1**

**Specific Sections of Interest**

| Component | Fabrication | Characterization | Diagnostics |
|---|---|---|---|
| Catalyst | 6.1.1 | 6.1.2 | |
| | | 6.1.3 | |
| Electrode | 6.2.1 | 6.2.2 | |
| | | 6.2.3 | |
| MEA and full cell | 6.3.1 | 6.3.4 | 6.3.7 |
| | 6.3.2 | 6.3.5 | 6.3.9 |
| | 6.3.3 | 6.3.6 | 6.4.1 |
| | | 6.3.8 | |

## 6.1. CATALYST

Catalysts employed in fuel cells range from pure metals (Pt, Ag etc.) to alloys (Pt-Ru, Pt-Cr, Pt-Co-Cr, Pt-Pd), mixtures (Pt black and carbon, Pt black and Pt-Ru black) and non-noble metal catalysts (spinels, porphyrins, pthalocyanins) The following discussion pertains mainly to the preparation of metals and alloys. U.S.Patents by Kampe et al., Kordesch, Kordesch et al.,[1] and a comprehensive report from Diamond Shamrock Corporation and Case Western Reserve University[2] discuss preparatory methods for spinels. General preparatory methods for organometallics can be found in a publication by Jahnke et al.[3]

### 6.1.1. Preparation Techniques

There are many techniques described in the literature for preparing precious-metal catalysts. There are numerous variations of these techniques that may be listed in the literature by different names, but they are essentially similar and in general can be classified as variations of the basic procedures (Table 6.2). These techniques can generally be used to make either supported catalysts or simple metal blacks, with some exceptions; for example, the Adams method would not be the method of choice to make a carbon supported platinum catalyst. Some of these methods can be combined to make an alloy or a doped catalyst. For instance, one can prepare a carbon-supported platinum–cobalt alloy catalyst by using a formaldehyde reduction of chloroplatinic acid followed by the incipient wetness technique to deposit a cobalt salt. The resulting Pt/C powder, impregnated with cobalt salt, can then be treated at high temperature in an inert atmosphere, where the cobalt would be reduced carbothermally, to create the supported bimetallic alloy.

It is important to use proper judgment, based on a thorough knowledge of the particular chemical processing techniques, to arrive at a fabrication procedure that will result in a catalyst with properties that maximize the rate of the electrochemical reaction.

The *basic* procedures for preparing catalysts are listed in Table 6.2. The initial metal compounds, solvent/dispersing fluid, process and reduction technique need to be identified before a catalyst can be prepared. For instance in preparing a Pt-Mo catalyst, the Watanabe method fails because the molybdenum species created dissolves during processing. Similarly in preparing an osmium-based catalyst, the Adams method would fail because the product osmium tetroxide has very high vapor pressure.

There are some relatively new techniques for catalyst preparation, namely the *sol-gel* and *nano-technology* methods. These methods are very attractive since some new materials with very interesting properties have been fabricated in these ways. The methods of catalyst preparation are always expanding and the interested researcher must watch the literature to keep aware of any new methods that can enhance the performance of the catalyst.

Bimetallic catalysts can be prepared using the methods listed in Table 6.2, by employing a mixture of the metal salts in proportion to the desired alloy composition. For example, Pt-Ru black can be prepared by Adam's, Watanabe's, or dithionate methods by employing a mixture of chloroplatinic acid and ruthenium trichloride; most of the methods listed in Table 6.2 are amenable for the preparation of Pt-Ru. Supported Pt-Ru can be prepared by the aqueous based methods, provided carbon is added to the solution prior to the start of the reduction process. The Tanaka method is especially suitable for the preparation of high loaded Pt-Ru. Several *binary catalysts* (also referred to as *alloys* or *intermetallics* in the literature) have been advocated for improved oxygen cathode performance for both phosphoric acid and PEM fuel cells, e.g., Pt-Cr, Pt-V, Pt-Co-Cr etc.

Landsman and Luczak[4] prepared Pt-Cr alloy by starting with Pt on graphitized carbon black dispersed in water using the following steps: adjusting the pH to 8 by the addition of ammonium hydroxide, adding desired quantity of ammonium chromate, bringing the pH down to 5.5 to enable the adsorption of chromium species onto the catalyst, filtering, drying and heat treating at 1700 °F. Preparation of several other bimetallic catalysts can also be found in an EPRI report authored by Ross.[5]

## 6.1.2.  Catalyst Characterization: Physical

Physical characterization methods for supported and unsupported catalysts include electron microscopy for morphology and surface area, BET for surface area and X-ray absorption spectroscopy (XAS) for catalyst structure. A detailed presentation of various TEM techniques for supported catalysts is given by Yacaman.[17]

Average-particle size can be estimated using TEM, by estimating the particle size of about 200 particles in the TEM image. For more precise particle size distribution, techniques such as Coulter counter may be employed.

**Table 6.2**
**Basic Catalyst Preparation Techniques**

| Method | Precursor | Solvent | Conditions | Other agents | Form | Surface Area $m^2/gm$ | Reference |
|---|---|---|---|---|---|---|---|
| Borohydride | Chlorides | Water | Room temperature | Caustic | Metal | Low: ~ 20 | Carter et al.[6] |
| Watanabe | Chlorides | Water | Room temperature | Sulfite, peroxide | Oxide | Black: 30–40; /C: 60–90 | Watanabe et al.[7] |
| Formaldehyde | Chlorides | Water | Boiling | Caustic | Metal/oxide | Low: ~ 30 | Giner et al.,[8] Luczak patent[9] |
| Dithionate | Chlorides | Water | Room temperature | $Na_2S_2O_4$ and peroxide | Metal | 30–50 | Jalan[10] |
| Adams | Chlorides | Nitrate | 500 °C | None | Oxide | High: 80–100 | McKee[11] |
| Nano-technology (template) | Chlorides | Alcohol | Room temperature | Surfactant | Oxide | High: > 100 | Attard et al.[12] Kresge et al.[13] |
| Sol-gel | Chlorides | $CH_3OH$ | Various | Carrier ($TiO_2$) | Oxide | Medium: ~ 40 | Schneider et al.[14] |
| Incipient wetness | Chlorides | Water/organic | High temperature | Support, reducing agent | Metal | Medium: ~ 40 | Adams at al.[15] |
| Tanaka-CO processing aid | Chlorides | Water | Low temperature < 60 °C | Various reducing agents | Metal | High: 100+ | Tsurumi et al.[16] |

The surface area of catalysts can also be determined using the BET method. However, with supported catalysts, the surface area of high area supports will prevent the determination of the area of the catalyst itself. In such cases, the determination of particle size using TEM (as outlined above) is helpful in determining the catalyst surface area. A rough estimate of average catalyst surface area can be made from the particle size data, assuming a spherical geometry for the particles. If the catalyst is non-spherical (e.g., raft like), then, appropriate consideration to the geometry should be given while calculating the surface area.

The method XAS is becoming a very useful research tool for studying the details of catalyst behavior for specific reactions. The electronic and geometric behaviors of the catalyst can be determined simultaneously by this in-situ technique. There are several technical papers describing the details of the technique and analysis.[18,19] For platinum or platinum alloy catalyst, one can determine the extent of d-band character of the platinum catalyst by observing the peaks for $L_3$ and $L_2$ edges in the X-ray absorption near-edge spectroscopy (XANES). An analysis of the extended X-ray absorption fine-structure data (EXAFS) will give information regarding the geometry of the catalyst structure and also the structure of the adsorbed reactant species. These techniques are very specialized and are outside the realm of a fuel cell laboratory, but are important to those scientists who want to understand the details of electrocatalytic behavior.

### 6.1.3. Catalyst Characterization: Electrochemical

6.1.3.1. *Cyclic Voltammetric (CV) Measurements*: This technique is valuable to study the catalyst behavior in different potential regions—the hydrogen adsorption/oxidation region and the OH⁻ adsorption/oxide formation region. The surface area of some platinum-based catalysts can be determined, redox behavior investigated and the presence of any poisons in the catalyst can be determined. Experimentally the CV of catalysts can be run in nitrogen saturated 0.5M $H_2SO_4$ or 1 M caustic at the desired temperature. In the case of supported catalysts, it is desirable to characterize the carbon support separately. Electrodes for the CV studies are normally made using only 5% Teflon T-30 binder (Dupont); this helps to ensure that the entire catalyst will be wetted by the electrolyte and therefore available for the electrochemical reactions. These electrodes can be prepared on a carbon paper (e.g., Toray H060), carbon cloth or gold plated nickel screen (for alkaline medium) current collector. An electrode holder that can be used for CV experiments is shown in Figure 6.1. In use, the electrode is mounted into the holder and immersed into a cell (Figure 6.2) containing the desired electrolyte. The cell can be heated by circulating water from a thermostat. A graphite rod can be used as a counter electrode and DHE or SCE reference can be placed in the separate reference compartment. The electrode needs to be wetted well to access all the available surface area. CV scanning is normally carried out at a sweep rate of 10 mV/s in the potential range of 0.0 to 1.2 V vs. a RHE reference electrode. Hydrogen adsorption/oxidation, redox behavior, and oxide formation activity can be

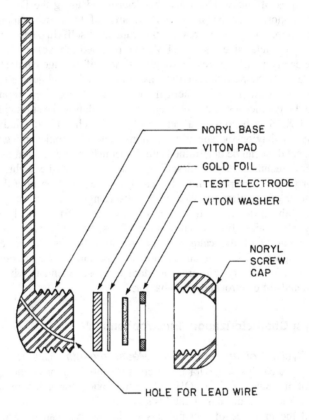

**Figure 6.1**. Electrode holder for conducting CV and half-cell experiments involving dissolved fuel in aqueous electrolyte.

monitored using cyclic voltammograms by choosing appropriate potential limits for the sweeps. A good description of CV measurements and their interpretation for surface reactions, in particular, are given by Conway.[20]

6.1.3.2.    *Rotating Disk Electrode (RDE) Method*: The activity of a catalyst for the oxygen reduction reaction and the methanol oxidation can be assessed using this method. High surface area catalysts have been applied in the form of a thin film to the rotating disk electrode in the past. However, Schmidt et al.[21] immobilized high area Pt/C dispersed on a glassy carbon (GC)  rotating disk electrode (RDE) with a thin layer of Nafion™. Mo et al.[22] showed that, even without the over layer of Nafion™, high area catalysts possessed good adherence to the GC surface.

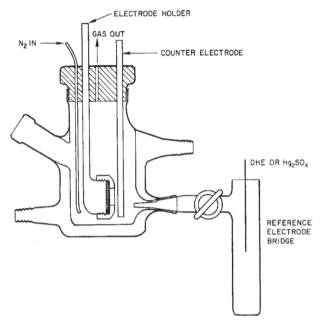

**Figure 6.2.** A water-jacketed glass cell for half-cell experiment.

Catalyst dispersions can be prepared by mixing 10 mg of catalyst and 1.0 ml of water under ultrasonic agitation for ~1 min. About 10 μl of the resulting suspension is then placed on the GC disk surface of the RDE and allowed to dry at room temperature, preferably under a slow, steady stream of nitrogen or argon. After rinsing with pure water, this loading procedure can be repeated two more times to attain a more uniform catalyst layer. The actual area of the platinum particles adhering to the glassy carbon disc surface can be determined by cyclic voltammetry, using the area under the hydrogen adsorption or desorption features. If one assumes that each platinum site is covered by one-hydrogen atom then the charge measured will correspond to the platinum area. The platinum surface is composed of three dominant crystal faces (<100> : $1.3 \times 10^{15}$ atoms/cm$^2$, <110> : $0.92 \times 10^{15}$ atoms/cm$^2$ and <111> : $1.56 \times 10^{15}$ atoms/cm$^2$). Based on an average surface density, the charge measured can be related to the surface area by the factor, 0.210 mC/cm$^2$.

A typical RDE plot for oxygen reduction is presented in Figure 6.3. The relationship between the current and rotation rate is given by[23]

$$\frac{1}{i} = \frac{1}{i_k} + \frac{1}{B\sqrt{\omega}} \qquad (6.1)$$

where $i_k$ is the kinetic current, $\omega = 2\pi f$ and $B$ is a constant defined by:

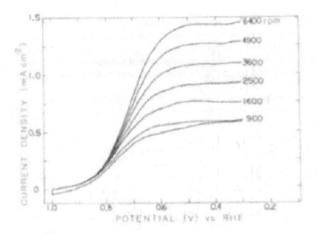

**Figure 6.3.** Typical RDE data for oxygen reduction. Sweep rate = 10 mV/s. Reprinted from Reference 23 (Figure 2), Copyright (1979), with permission from The Electrochemical Society.

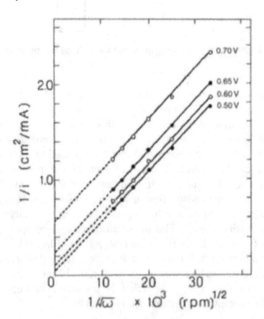

**Figure 6.4.** Further processing of the RDE data shown in Figure 6.3 to obtain the $n$ value. Reprinted from Reference 23 (Figure 3), Copyright (1979), with permission from The Electrochemical Society.

$$B = \sqrt{\frac{2\pi}{60}} nF v^{1/2} C_{O_2} \left[ 0.621 S^{-2/3} \left( 1 + 0.298 S^{-1/3} + 0.145 S^{-2/3} \right) \right] \qquad (6.2)$$

where $v$ is the kinematic viscosity, $F$ is the Faraday, $n$ is the number of electrons per mole of $O_2$, $C_{O2}$ is the concentration of $O_2$ in moles/cm$^3$, and $S$ is the Schmidt number given by $S = v/D$, with $D$ the diffusion coefficient. A plot of $i^{-1}$ vs. $\omega^{-1/2}$ for various potentials should yield straight lines with the intercepts corresponding to the kinetic current (Figure 6.4). The slope of the plot will give the $B$ values, from which the number of electrons transferred can be calculated using known values of solubility and the diffusion coefficient of oxygen in the medium under consideration.

## 6.2.    ELECTRODE

Electrode fabrication is a key element for creating a fuel cell that will perform at a high level. There are many factors involved with this process and they are strictly dependent on the catalyst, reaction and type of fuel cell. This is an area of active investigation and the latest information regarding new electrodes can be found in both the scientific and patent literature.

In this chapter, we will describe some electrode fabrication methods for low-temperature fuel cells,which include phosphoric acid, alkaline and proton-exchange-membrane fuel cells. The goal of electrode fabrication is to provide a structure that will maximize catalyst utilization and minimize performance degradation due to water/electrolyte *flooding* of the structure. Flooding can be defined as the process that leads to increased path lengths for the reactant gas to reach the catalyst sites. This usually occurs over extended periods of operation and ultimately leads to large mass transport losses in the electrode. Correspondingly, cell performance will also be reduced.

Gas-diffusion electrodes have been used in both acid and alkaline fuel cells. Gas-diffusion electrodes typically consist of:

(a) a *carrier* (or a support or substrate) that also acts as a current collector (e.g., carbon cloth/carbon paper in acid medium or a sintered nickel plaque/nickel expanded metal in alkaline media);

(b) an optional *base layer* that provides efficient gas distribution and acts as an anchor for the catalyst layer; in alkaline medium, the base layer is always present, whereas in acid medium (PEMFC), the base layer is optional. In the absence of base layer, the carrier also acts as a gas distribution layer;

(c) a *catalyst layer*.

With polymer-electrolyte membrane (PEM), earlier work used gas diffusion electrodes with the catalyst layer extended by a coating of Nafion™.[24] Recently, the technology has evolved into coating the catalyst with some Nafion™ (but without any PTFE) directly onto the membrane to form a catalyst coated membrane (CCM).

Gas distribution is accomplished in this case by the use of a separate gas diffusion medium, which is non-catalyzed and hydrophobic; this is basically a coating of uncatalyzed carbon black (with PTFE binder) on a PTFE-coated substrate (carbon paper or carbon cloth).

## 6.2.1.  Fabrication of Gas Diffusion Electrodes

There are various publications in the literature that describe the preparation of various types of electrodes. The PTFE-bonded gas-diffusion electrode is described in a patent by Neidrach and Alford.[25]

### 6.2.1.1.  Electrode Preparation

(a) *Dry-ice spread.*[26] The process involves mixing the catalyst with dry ice and spreading the resulting mix on a substrate and letting it dry. Dry ice is first ground to fine granular form using an industrial grinder or mortar and pestle. The desired quantities of catalyst and PTFE (in the powder form) are then mixed with the granular dry ice thoroughly in a mortar. The ratio of the quantity of dry ice to the catalyst has to be determined by experiment by each user, although a ratio of 3–5 may be a starting point. The mixture is then spread on a hydrophobized carbon paper or carbon cloth and the dry ice allowed to evaporate. The loose structure is then carefully pressed in a hydraulic press to compact the catalyst onto the substrate.

(b) *Spray.* An ink of catalyst, PTFE emulsion (e.g., T-30 of Dupont) is first prepared by stirring a mixture of the catalyst in an alcohol-water solution for several hours; addition of ~10% (of catalyst) Nafion solution helps in forming a smooth ink. PTFE emulsion is normally added immediately before the spray process. The ink is then sprayed using an artist's air brush (e.g., Pasche). For larger electrodes, air-less spray equipment or electrostatic spray equipment can be used.

(c) *Screen printing.* This method is the same as used commercially for screen printing logos on clothing (e.g., T-shirts). The ink for this method has to be fairly viscous. Various addition agents available for standard ink preparations have to be carefully assessed for their harmful effects on catalyst activity. This method is described in detail in several recent patents.[27] For example, Kawahara et al.[27] describe the preparation of a paste-like ink by mixing 1 g of the catalyst-carrying carbon (mean particle diameter: approximately 20 nm) that carries 20 percent by weight of platinum or an alloy of platinum and another metal as a catalyst with 3 ml of 5 percent by weight of Nafion solution, 5 ml of an alcoholic solvent (such as ethanol or isopropyl alcohol), and 2 g of camphor as the pore-forming agent and applying sonication so as to homogeneously disperse the contents. Addition of a gelling agent to increase the viscosity is described in the patent by Gevais et al.[27]

(d) *Dry-roll process.*[28] A mixture of PTFE, catalyst powder, and pore former (e.g., sodium carbonate) is first prepared by blending them in a blender using isopropyl alcohol; the blended mix is filtered and dried to remove the solvents and water. The dry mix may require some gentle (low energy) chopping prior to use. The mix is then fed to a pair of vertical rollers (heated to 100–150°C). The shear force applied between the rollers converts the powder mix into a sheet of uniform thickness. This sheet is then pressed onto the hydrophobic carbon paper or carbon cloth in a hydraulic press at 80–120°C. The resulting electrode is then soaked in warm water to remove the pore former.

(e) *Gel-roll process.*[29] This method consists of preparing a gel by mixing appropriate proportions of the catalyst powder (e.g., Pt/C), PTFE emulsion, toluene or xylene, and thoroughly mixing the contents. When the mix is worked well—preferably with the application of shear force—a gel results. This gel can then be rolled in a pinch roller to obtain electrode sheets of desired thickness.

(f) *Filter casting.*

There are several steps involved in the preparation of gas-diffusion electrodes:

(a) preparation of gas diffusion medium;
(b) preparation of catalyst-PTFE suspension;
(c) preparation of *green* (i.e., not heat-treated) electrode; and
(d) sintering of *green* electrode.

A gas-diffusion medium, which consists of a thin layer of uncatalyzed carbon and PTFE coating on a hydrophobized carbon paper or carbon cloth is available from commercial sources (e.g., E-Tek). Gas diffusion electrodes with standard Pt/C or Pt black as catalysts are also available commercially.

However, for non-standard catalysts, one has to start with the gas-diffusion medium and prepare the required electrodes. Step (b) involves the preparation of catalyst-PTFE suspension. The catalyst is first dispersed in sufficient quantity of distilled water; dispersion can be accomplished by a combination of stirring, sonication and/or homogenization. The PTFE emulsion is added per the desired composition, and the PTFE flocculated onto the catalyst powder. Improper flocculation results in a colloidal suspension of the catalyst-PTFE emulsion, which can pass through the filter or substrate; loss of catalyst occurs and the quantity of PTFE on the remaining catalyst becomes uncertain, but results in a much lower than the intended weight percent of PTFE. This process has been studied in detail[30] and the boundaries of flocculation are known for some carbon support materials. The pH of the catalyst has to be between 2.5 and 3.0 for proper flocculation to occur. However, many catalysts (carbon supported) as well as carbon supports exhibit pH values ranging from 2–11. The pH of a dispersed carbon support is determined primarily by the type and concentration of its surface groups. The catalyst/carbon, more often than not, has a low pH and this can be associated with insufficient washing during the catalyst processing procedures.

Step (c) is the filtration process to make the *green* electrode. The catalyst-PTFE suspension as prepared above can be directly filtered onto the carbon paper or filtered onto a thin porous paper such as medical examination paper and subsequently transferred onto the carbon paper (or cloth) by rolling. This step is usually followed by cold and hot pressing in a hydraulic press to ensure good adhesion of the catalyst layer to the substrate.

Variations in processing parameters are necessary for different catalysts, particularly on different carbons. For example, black-pearls carbon, particularly the heat treated BP-2000, can be a difficult carbon to make electrodes with; extensive cracking occurs during the green electrode preparation, and if not controlled, could result in the cracks remaining even after sintering. Such adjustments to the electrode making process are carried out on a trial and error basis, for a given new catalyst. Filter casting technology is still substantially an art. Even when many or all of the known process variables are controlled, the properties of the end product are not always reproducible. This is due to a lack of a thorough scientific understanding of many underlying variables and their effect on the final properties of the electrode.

Some parameters can be controlled on a scientific basis. A typical example is the flocculation of the catalyst-PTFE suspension. Crack propagation during step (c) above is a phenomenon for which complete explanation is not yet available. The *green state* can be defined as the structure of the catalyst layer directly after application to the substrate. Here, the rate of water evaporation is an important variable that can result in *mud-flat* cracking, much the same way observed in ceramics. To some degree, this can be controlled by the extent of filtering, rolling, cold and hot-pressing operations. The water absorption isotherms for different carbons are markedly different and are dependent on the micro- and meso-pore volumes of the carbon. In our experience, this phenomenon can be controlled only by a trial and error determination of the cycle times involved in the electrode preparation processes outlined above.

The sintering process (step d) consists of two steps, one a baking process for 10–15 minutes at 270°C (to remove most of the surfactants) and a sintering step for 10–20 minutes at 340–350°C. Not all PTFEs have surfactants; for example some of the powdered materials are free of surfactant so the intermediate temperature treatment can be eliminated for electrodes of this type. The sintering process allows for cohesive interaction between the carbon/catalyst particles and the PTFE, and imparts good mechanical strength to the electrodes. The baking and sintering temperatures depend on the type of PTFE used and should be decided in conjunction with the manufacturer's recommendations.

## 6.2.2. Physical Characterization

Clyde Orr gives a general description of physical characterization of porous electrodes.[31] It is important to know as much about the structure of the electrode as possible to determine how to improve its efficiency for carrying out the electrochemical reaction.

6.2.2.1. *Electronic Resistivity.* The overall electronic resistance of the electrode is an important quantity. Since most fuel-cell applications involve several individual electrodes, any large resistance losses will reduce the overall performance of a stack of cells. This measurement, while not a mainstream characterization method, is useful as a quality control or developmental tool. A four-probe resistivity method has to be employed to avoid contact resistance problems. ASTM C611 outlines the general procedure. The resistance can be measured using a variety of commercial LCR (inductance/capacitance/resistance) meters and the resistivity values can be calculated with the knowledge of the geometry of the probe employed. Many electronic supply houses sell probes for four point resistance measurements.

6.2.2.2. *Porosimetry.* This technique is valuable in determining the pore structure of the catalyst layer, especially for gas-diffusion electrodes, where a knowledge of the distribution of gas and liquid phase pores is essential for the optimization of performance. Mercury porosimetry has been employed and described in the literature to obtain this information. Qualitative indication of the porous nature can also be obtained by the bubble-pressure method, which involves recording of the pressure at which bubble(s) formation begins. Mercury porosimetry can distort the pore size due to the elastic nature of the carbon-PTFE composite; also for thin electrodes and for electrodes consisting of two or three layers of different porosity, this method is of limited application. Recently Jena and Gupta[32] have published a method that allows the determination of pore size distribution and porosity information by monitoring the through-plane and in-plane flow of a gas through dry and wet two-layer gas-diffusion electrodes.

Another method to determine pore size distribution of porous materials has been developed recently by Volfkovich et al.[33] In this method, discs of the test sample (0.1–3 mm thick) and two (porous) standards are first filled with a low contact angle liquid (e.g., octane) and weighed. The sample is then sandwiched between the standards and held in compression to attain capillary equilibrium. By sending a small volume of dry inert gas through the sandwich, some of the wetting liquid is evaporated and the liquid in the sample is allowed to reach capillary equilibrium with the standard. The discs are then taken apart and weighed. This process is repeated until all the liquid from the sample is evaporated. A *porogram* is obtained by comparing the liquid volume in the sample to the liquid volume in the standard. From this relationship, the pore size distribution of the sample is calculated using radial-distribution function.[*]

---

[*] An automated test apparatus using this principle is manufactured by Porotech, Inc., Ontario, Canada; www.porotech.net.

## 6.2.3. Electrochemical Characterization

6.2.3.1. *Cyclic-Voltammetric Evaluation.* Electrodes can be evaluated using the technique of cyclic voltammetry (CV). Catalyst surface area and poisoning effects can be assessed from the CV curves.

Before proceeding with the CV, it is important to wet the electrode thoroughly to access all the available catalyst. This can be done by soaking the sample electrode in 90% isopropanol/ 10% water for 2 hours. The alcohol solution is then decanted and the electrode rinsed thoroughly, to remove the alcohol. It is then immersed in distilled water and warmed to 50 °C, and set aside for 2 hours. The electrode is then equilibrated with 0.5 molar $H_2SO_4$ for several hours or preferably overnight. This procedure wets the catalyst and finally exchanges the wetting liquid with electrolyte.

The electrode is then mounted in the holder (Figure 6.1), placed in the cell shown in Figure 6.2, and swept from –0.02V to 1.2V in 0.5M $H_2SO_4$. After about ten sweeps the trace should stabilize; the area under the curve from the double layer to the second minimum in reduction current before the onset of hydrogen evolution is determined. Alternatively, the hydrogen desorption area up to ~0.3V (vs. RHE) can also be used for integration. The hydrogen adsorption charge calculated from the area under the curve is converted to surface area using the value of 210 μC (one H per Pt atom) per square centimeter of platinum. It is preferable to chemically analyze the sample for platinum content to ensure accurate surface area determination. An example of CV obtained for surface area determination is shown in Figure 6.5.

6.2.3.2. *Polarization Studies.* The performance of individual electrodes can be determined by tests in a half-cell. Here the individual electrode performances (anode and cathode) can be determined separately as opposed to full-cell experiments where it can be difficult to separate the performances of the individual components. The types of polarizations (activation, concentration, and resistance) can be evaluated and improved electrodes formulated and fabricated for future full cell testing.

6.2.3.3. *Half-Cell Testing with Dissolved Fuel.* For methanol, formic acid and other reactants soluble in water, the electrodes can be mounted in an electrode holder shown in Figure 6.1, and immersed in the glass cell shown in Figure 6.2. The cell is filled with the dissolved reactant and electrolyte and heated to the desired temperature.

6.2.3.4. *Gas-Diffusion Electrode Evaluation.* One type of half-cell test apparatus consists of a machined-Noryl-electrode holder (Figure 6.6) and a glass cell. This holder has the facility to input any desired gas (e.g., air, oxygen or hydrogen) and it is possible to maintain a slight pressure by pinching the gas outlet. This holder can be used for both gas-diffusion electrode and membrane–electrode assembly (MEA) characterization. A separate counter electrode has to be employed in the case of a gas-diffusion electrode. In the case of MEA, the electrode of interest is made the working electrode (i.e., exposed to the gas stream) and the opposite

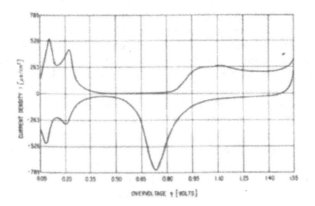

**Figure 6.5.** Cyclic voltammogram of Pt for the determination of the surface area. Reprinted from F. G. Will, *J. Electrochem. Soc.* **112**, 453 Copyright (1965) with permission from The Electrochemical Society.

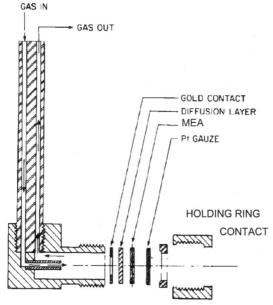

**Figure 6.6.** Electrode holder used in gas-diffusion electrode and MEA half-cell experiments.

electrode is made the counter electrode. Carbon fiber paper or cloth backing serves as both gas-diffusion layer and current collector for the working electrode. A gold ring on the working electrode and a platinum gauze on the counter electrode provide the electrical contacts. The whole package is held in place with a Noryl holding ring and screw cap. In use, the holder is immersed in the glass cell shown in Figure 6.2. An appropriate reference electrode is chosen depending on the choice of the electrolyte. The use of electrolyte is *only* to provide an ionic conducting path for the reference electrode; the Nafion™ membrane itself serves as the electrolyte for the reaction being investigated.

In half-cell experiments, the electrode is pre-treated in the galvanostatic mode using five one minute 'on'/ one minute 'off' cycles, over a current density range of $10$–$500$ mA/cm$^2$, in 14 steps. After break-in, the polarizations are recorded. Break-in regimes can vary, and can be simply repeated polarizations over the range of desired current density, or holding at a given current density for long duration (hours) or complex patterns such as trapezoidal waveforms. Each researcher has to conduct such experiments and choose a preferred break-in procedure.

6.2.3.5. *Floating-Electrode Method*. Another commonly used half-cell apparatus is the so-called *floating-electrode half-cell*. Here the electrode *floats* on the electrolyte, i.e., the catalyst layer contacts the electrolyte and the opposite side (gas side) is exposed to the reactant gas and is also contacted by an electronic conductor (typically a gold screen). This type of half-cell was first described by Giner and Smith[34] for use with alkaline electrolytes; Kunz and Gruver[35] adapted the original design for use with hot phosphoric acid electrolyte. By floating the electrode, the complications due to hydrostatic head (and the necessity to compensate with appropriate back pressure, particularly in alkaline electrolyte) are completely avoided.

Figure 6.7a shows a drawing of the construction of the floating-electrode apparatus and Figure 6.7b shows a photograph of the apparatus.[*]

## 6.2.4. Choice of Reference Electrode for Half-Cell Studies

Table 6.3 gives the various reference electrode choices and their standard electrode potentials for use in half-cell studies. The user has to choose a proper reference electrode based on the type of electrolyte, availability, and ease of use. All these reference electrodes can be placed in a separate compartment (as shown in Figure 6.2), with a Lugin-tip (shown in Figure 6.2) positioned as close to the working electrode as possible with the least amount of blocking of the working electrode. The reversible hydrogen electrode mentioned in Table 6.3 can also be placed in this separate compartment, with unit activity hydrochloric acid; under these conditions the RHE becomes the standard-hydrogen electrode (SHE).

---

[*] The apparatus can be ordered from Finkenbeiner, Inc. (www.finkenbeiner.com)

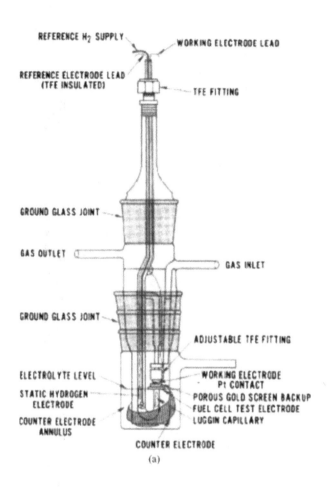

(a)

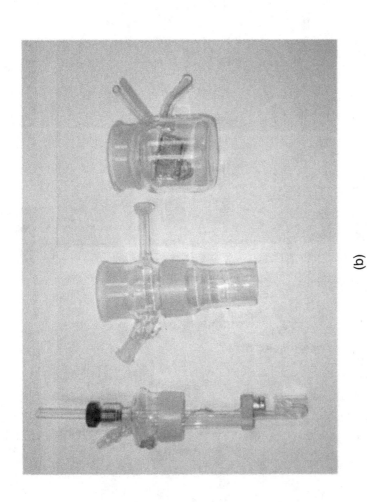

(b)

**Figure 6.7.** **(a)** Floating electrode half-cell apparatus. Reprinted from Reference 35, Copyright (1975) with permission from The Electrochemical Society. **(b)** A photograph of the floating electrode half-cell apparatus.

**TABLE 6.3**
**Reference Electrode Choices**

| Name | Preferred electrolyte contains | Standard electrode potential, $E^{o\prime}$ (V) vs. RHE @ 25 °C | Preparation, comments and references |
|---|---|---|---|
| $Hg/Hg_2SO_4$ | Sulfate | 0.615 | Can be purchased from many commercial sources. Preparation described in the textbook by Potter.[36] |
| $Hg/Hg_2HPO_4$ | Phosphate | 0.636 | $Hg_2HPO_4$ can be prepared by mixing aqueous mercurous nitrate with aqueous phosphoric acid. Further preparation follows the same procedure as for calomel electrode. |
| $Hg/Hg_2Cl_2$ | Chloride | 0.242 | Common name: saturated-calomel electrode (SCE). It can be purchased from many commercial sources. Preparation described in the textbook by Potter (op. cit). This electrode has to be used with a reference electrode bridge to avoid chloride contamination. |
| $Hg/HgO$ | Hydroxide | 0.926 | Commercially available. For technical measurements, a small quantity of HgO can be mixed with 10% graphite powder and ground into a paste with ~5 w% PTFE emulsion, and applied to a nickel screen or expanded metal grid with the aid of a nickel spatula. After drying at room temperature, it can be used as a reference electrode |
| DHE | Acid or alkaline solutions | Calibration required | Two small Pt black electrodes (or even platinized Pt) (2 mm x 2 mm or greater) held 1 cm apart with leads insulated. The electrode assembly is inserted into a PTFE or glass tube. A 45 V (five 9 V alkaline in series) battery with series resistors to control the current at 1 mA is connected such that the lower most electrode is negative. (this is to avoid oxygen gas from the anode contaminating the hydrogen electrode). Calibration against a standard calomel (from a commercial source) is required to find the correct potential. The cathode lead is also connected to the potentiostat or the galvanostat. For more details, see Giner.[37] Care should be exercised when using with near neutral solutions, due to pH changes towards alkaline region near the hydrogen electrode. |
| RHE | Acid or alkaline solutions | 0.0 | Hydrogen gas is passed over a platinized Pt or preferably a Pt black electrode immersed in the solution of interest. The potential is −0.059 pH. |
| Zn Wire | Hydroxide | 1.4 | Simple. Useful for experiments not requiring great accuracy. Frequent scrapping necessary to maintain reasonable accuracy. |

*Caution!*: Handling of mercury and its salts should be done with utmost care to avoid inhalation or skin contact; residues should be disposed of using a licensed service provider.

## 6.3.  FULL CELL

### 6.3.1.  MEA Fabrication

Table 6.4 identifies various methods available for the fabrication of MEAs and their advantages/disadvantages. The following description provides an overview of two common MEA fabrication methods:

- hot pressing gas-diffusion type electrodes to a membrane, and
- the decal-transfer method.

6.3.1.1.  *Hot-Pressing Method.* The steps involved in the preparation of MEAs are: pretreatment of membranes, painting of electrodes with Nafion solution, and bonding of electrodes to the membrane.

A full description of MEA fabrication together with the incorporation of Nafion within the electrode is given in the patent by Niedrach.[38]

(a) *Pre-treatment step.* The commercial membranes as received (e.g., Nafion 115 or Nafion 117) are in the $H^+$ form. Treatment with 3% hydrogen peroxide at 50 °C should remove any organic impurities and subsequent equilibration with 0.5M $H_2SO_4$ would ensure near 100% acid form.

(b) *Nafion™ Coating.* The painting of electrodes with Nafion™ solution accomplishes two functions: the bonding strength between the electrode and the polymer electrolyte membrane is strengthened due to the presence of an interpenetrating network of the catalyst and the polymer electrolyte; the reaction zone is also extended into the catalyst layer, thus providing better catalyst utilization.

(c) *Bonding.* The bonding of the electrodes to the polymer electrolyte can be accomplished by hot-pressing the electrodes to the sandwiched membrane at 400 psi. The hot-pressing step softens the polymer electrolyte and the pressure allows the penetration of the electrode into the membrane. Cooling results in the permanent bonding of the electrodes to the polymer electrolyte membrane.

6.3.1.2.  *Decal-Transfer Method.* This particular method was developed at the Los Alamos National Laboratory (LANL) by Mahlon Wilson.[39] The procedure is also described in detail by Wilson and Gottesfeld.[40] The general procedure is briefly described below and illustrated in Figure 6.8:

(a) An ink is formed by mixing (through sonication and stirring) the desired catalyst with sufficient quantity of Nafion™. A weight ratio of 1:3 is recommended.

(b) After mixing the catalyst and Nafion™ solution, add glycerol and water such that the ratio of carbon/water/glycerol is 1:5:20.

(c)    The ink is painted onto a blank PTFE film. Multiple coats are applied to build up the desired loading. The coating is allowed to dry at 130°C.

(d)    A Nafion™ film (117/115/112) is sandwiched between two such coated PTFE films and hot pressed at 145°C for 90 seconds at 200–400 psi.

(e)    The PTFE blank is peeled off after cooling the pressed MEA.

The above MEA can be used with an uncatalyzed-PTFE bonded gas-diffusion medium.

6.3.1.3    *Reference Electrode Incorporation in Full Cells.* A hydrogen electrode can be easily incorporated into an MEA, by applying a small patch of platinum ink used to make the MEA onto the PEM. A similar patch placed on the other side would act as a counter electrode, if a dynamic hydrogen mode of operation is desired. In both of these cases, the hardware for single cell tests has to be appropriately modified. For the reversible-hydrogen electrode, the hardware should allow a lead to be taken out and also a separate gas supply channel to supply hydrogen. Care should be exercised in ensuring the absence of leakage of the gas from the main electrode into the reference electrode compartment and vice versa. The use of a reversible-hydrogen electrode set-up for the performance evaluation of a $H_2/O_2$ fuel cell was demonstrated by Srinivasan et al.[41] Kuver et al.[42] describe the use of a dynamic hydrogen electrode approach for monitoring the performance of a direct-methanol fuel cell.

### 6.3.2.  Liquid-Electrolyte Cells

Typically, in the fabrication of liquid electrolyte cells such as phosphoric acid or alkaline fuel cells, the components (electrodes and matrix/electrolyte) are separate entities and form an integral part of the cell build, unlike in the case of PEM cells where the MEA, the heart of the system, is built in a separate step.

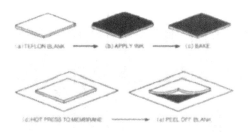

**Figure 6.8.**  Primary steps in the fabrication of thin film MEA. Reprinted from Reference 40, Copyright (1992).

**TABLE 6.4**
**MEA Fabrication Methods**

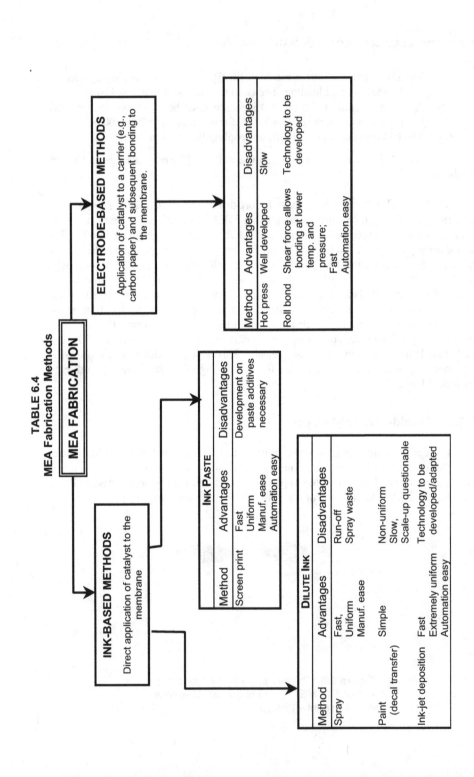

| | | MEA FABRICATION | |

**INK-BASED METHODS**
Direct application of catalyst to the membrane

**ELECTRODE-BASED METHODS**
Application of catalyst to a carrier (e.g., carbon paper) and subsequent bonding to the membrane.

**INK PASTE**

| Method | Advantages | Disadvantages |
|---|---|---|
| Screen print | Fast Uniform Manuf. ease Automation easy | Development on paste additives necessary |

**DILUTE INK**

| Method | Advantages | Disadvantages |
|---|---|---|
| Spray | Fast, Uniform Manuf. ease | Run-off Spray waste |
| Paint (decal transfer) | Simple | Non-uniform Slow, Scale-up questionable |
| Ink-jet deposition | Fast Extremely uniform Automation easy | Technology to be developed/adapted |

| Method | Advantages | Disadvantages |
|---|---|---|
| Hot press | Well developed | Slow |
| Roll bond | Shear force allows bonding at lower temp. and pressure; Fast Automation easy | Technology to be developed |

## 6.3.3.  Cell Hardware and Cell Build

The cell hardware, depicted schematically in Figure 6.9, consists of high-density graphite anode and cathode blocks held between metal end plates with stainless steel tie rods. Insulating bushings at either end ensure electrical isolation. An alternative design would be to use end plates of the same dimensions as the cell plates and use bolts through the entire stack, instead of tie-rods. In this design, the bolts will have to be insulated with FEP heat shrink. The blocks are machined with fluid plenums and flow channels for the distribution of reactants and the collection of the product exhausts.

To build stacks, the hardware must be capable of accommodating more than one MEA. The basic hardware is the same as shown for a single cell, except that bipolar plates are used to separate the individual MEAs. The bipolar plates are composed of the same material as the individual cell blocks, but are machined in anode flow configuration on one side, cathode on the other. When properly assembled, the plates form a set of four fluid plenums extending the length of the stack that provide parallel reactant supply and exhaust collection while the cells are connected electrically in series. Each bipolar plate has a voltage jack and thermocouple well so that individual cell voltages and temperatures can be monitored and recorded. For experiments with higher than ambient pressure, the cathode and anode blocks have to be made totally impervious by high pressure and/or vacuum-impregnation with phenolic resin.

Cell current is collected from solid copper plates situated between the cell blocks and end plates, while cell voltage sensing is accomplished with leads attached directly to the cell blocks (to eliminate the effects of current line drop on potential measurements). Temperature control is achieved with silicone-insulated resistive heating pads bonded to the metal end plates in conjunction with a PID-type microprocessor-based controller and a thermocouple placed in a well immediately behind the center of one electrode; cartridge heaters directly inserted into the cathode and anode plates can also be used. These cells employ flat-gasket seals; gasket materials with some elastic characteristic are preferred, e.g., silicone rubber, Viton® or EPDM. PTFE or FEP gaskets can be used for ambient pressure operations.

One of the important parameters in a cell build is the thickness of the gaskets. The gasket thickness determines how much the flow fields are allowed to pinch into the electrode. For a good contact (i.e., low contact resistance), it is essential that the ribs of the flow fields bite into the electrode providing good contact. Typical pinch values are 0.002–0.003" for a carbon paper backing and 0.010–0.015" for a carbon cloth backing.[*] Since electrode thickness for anode and cathode may vary, gasket thickness for each side is determined separately using the formula:

---

[*] These numbers are provided for guidance only. It is better to determine the pinch for individual situation with due consideration for the type of backing used and the nature of the gasket material (elastomeric materials, e.g., EPDM, tend to compress more than non-elastomeric films, e.g., PTFE.)

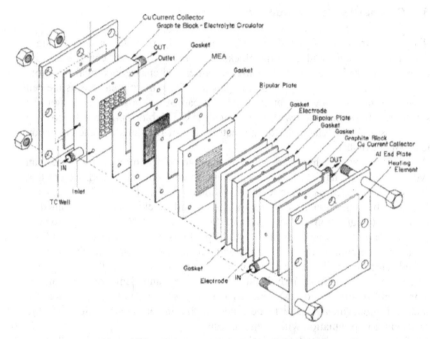

**Figure 6.9.** Exploded view of a fuel-cell stack.

Gasket thickness = (individual electrode thickness) – (desired pinch)     (6.3)

A single cell can be easily built by stacking the cathode block, cathode gasket, MEA, the anode gasket, and the anode block sequentially on an end plate (fitted with the necessary bolts) and finally completing the assembly with the other end plate. Current-collector plates, if different than the anode and cathode blocks, have to be inserted in the appropriate sequence. Once all the components are stacked, compression is applied by tightening the nuts sequentially in much the same way automobile tires are assembled onto the wheel. Spring washers provide more uniform force across the assembly. The end plate thickness is chosen with appropriate consideration for the mechanical properties of the material used. Compression is applied using a torque wrench; measurement of elongation of the bolt is a more sophisticated approach and should be followed if appropriate equipment is available. Assembly of a multiple-cell stack follows the same sequence except that one bipolar plate is inserted after each MEA. The effect of the compression pressure on the performance of a PEM fuel cell is reported for various types of gas-diffusion layers by Lee et al.[43] As would be expected the torque required for optimum performance varies with the type of the gas-diffusion layer; e.g., less compressible carbon paper substrate required much less force than more compressible carbon cloth. The optimum force is a balance between establishing a good electrical contact while not compromising the porous structure of the carbon paper.

The cell hardware for alkaline- and phosphoric acid- fuel cells are similar to the PEM cells but have some significant differences because of operating conditions. The liquid electrolytes pose an electrolyte migration problem, which has to be controlled to prevent electrolyte loss. These liquid electrolytes can migrate, degrade via corrosion, and evaporate. This is not the case with proton exchange membranes where the electrolyte is contained. Power plants with liquid electrolytes must reduce the factors leading to electrolyte loss in order to maintain high performance for extended time periods in order to meet commercial requirements. The methods of addressing these challenges are a significant effort but their details will not be addressed here.

Liquid electrolytes at operating temperatures above $100^{\circ}C$ will also cause corrosion of simple carbon plates. For stacks of cells in power plants, the usual bi-polar plates must be impervious to electrolyte passage, i.e., they must be dense and corrosion resistant. Any leaks in the plates that can allow reactants to mix will result in a serious failure of the power plant.

There is another factor involved with the liquid-electrolyte fuel-cell electrolyte loss via evaporation. Special operational schemes are required if the operating temperature is such that loss of electrolyte will take place because of evaporation.

## 6.3.4. Instrumentation

In order to test fuel cells, one must be able to provide reactants, control currents, and monitor and store data efficiently. While there are some differences depending on the type of fuel cell, they all require similar instrumentation. We will describe a set-up for a PEM cell test in the following paragraphs.

The fuel-cell test stand, depicted schematically in Figure 6.10, includes gas selectors and mixers, metering valves, and flow meters for the supply of reactants to the fuel cell anode and cathode. Reactants may be humidified at any temperature and pressure through use of the independently heated stainless steel humidifier bottles. Reactants can be supplied in liquid form by the special metering pumps that are capable of delivering pressurized liquid from a non-pressurized reservoir. Downstream of the cell, a pair of stainless steel backpressure regulators with stainless steel diaphragms and Vespel® (polyimide) seats allow for independent control of the anode and cathode pressures.

A computerized data acquisition and control system[44,45] has become the norm for fuel cell R&D. These systems automatically and continuously sample and store fuel cell current, voltage, cell temperature, humidifier temperatures, reactant flow rates and pressures on a 24-hour basis. Most systems also control cell discharge in either constant voltage or constant current mode and can control gas selection, reactant flow rates, and cell and humidifier temperatures. In the event of a power

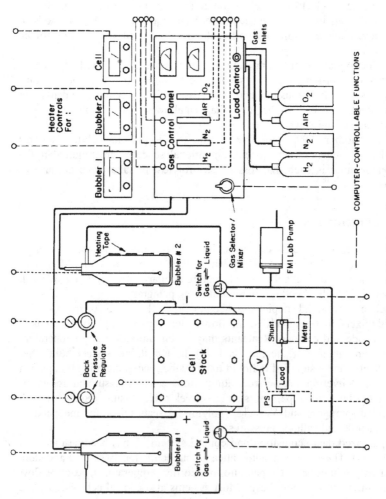

**Figure 6.10.** Schematic of a fuel-cell test stand.

failure or other dangerous condition—such as leakage, thermal runaway, running out of reactant, membrane perforation, or short circuit—such systems can also be programmed to perform a safe unattended shutdown by turning off the reactants, flooding the cell with nitrogen, disconnecting the load, and turning off the heaters. There are many commercially available fuel cell data acquisition and control systems of various complexities depending on the requirements of the end-user.

### 6.3.5. Cell Start-Up Procedures

6.3.5.1. *Wet-Up Process*. The first task in the start-up sequence is to wet-up the MEA in order the hydrate the membrane to its maximum water absorption capacity. Humidified reactants are applied to the cell for several hours, with the cell temperature maintained around 30 °C. The reactant saturators are kept at 60 °C. After a minimum of two hours, the air saturator is set to the same temperature as the cell and the hydrogen saturator is set 5 °C above the cell temperature. At this point, the cell can be heated to the desired temperature. After the cell reaches the desired temperature, the reactant flows are adjusted to the desired stoichiometric levels (usually 4X for air and 1.5–2X for hydrogen). The open circuit voltage of the cell is recorded approximately one hour after the flow rates are adjusted.

6.3.5.2. *Cross-Over Test*. The next task is to determine the absence of gas cross-over. The air supply is shut off, while maintaining a steady stream of hydrogen to the anode. The open circuit voltage is recorded for approximately 2 minutes. Any leak of hydrogen into the cathode chamber will react with the available oxygen thereby lowering the cell voltage, thus indicating a cross-over. If the cell design has open air channels, then, all open air channels must be closed (with tapes or gaskets) to prevent the entry of external air. The cell voltage will drop perhaps as much as 100 mV and level off. Figure 6.11 shows the open-circuit voltage of a 4-cell stack under the cross-over test. If the voltage continues to drop or quickly goes to zero, there is gas crossover through the membrane. No further fuel cell testing should be done until or unless gas crossover is eliminated.

6.3.5.3. *Break-In Procedure*. There are various break-in procedures practiced by the fuel-cell researchers. The most common ones are:

- constant-voltage break-in
- constant-current break-in.

In the constant-voltage break-in test, the cell voltage is set at a desired voltage (e.g., 0.5 V) and the cell is run until the current recorded reaches a plateau value that is acceptable for the specific experimental goal. The cell voltage can be set at a higher value initially and gradually lowered to 0.5 V. The constant current break-in is accomplished by applying a small initial current (e.g., 50 mA/cm$^2$) and then incrementally increasing the current density until the desired performance level is reached.

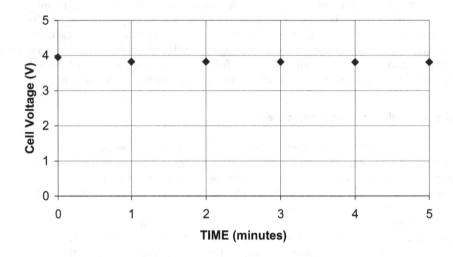

**Figure 6.11**. Cross-over test data for a 4-cell $H_2$/Air stack.
The drop in open-circuit voltage is only 35 mV/cell after 5
minutes of shut off of air flow.

### 6.3.6.   Cell Testing: $H_2/O_2$ Cells

In the following description of fuel-cell performance tests, the discussion
covers PEM, acid and alkaline fuel cells; however, the description sometimes may
be slanted heavily towards PEM fuel cells, due to the many specific challenges of
the membrane, e.g., water balance.

6.3.6.1.   *Open-Circuit Test.*   The open-circuit voltage of most low temperature
fuel cells operating with air and hydrogen should be in the range of 0.95–1.0 V. Any
lower value suggests either a cross-over or electronic short circuit through the
membrane or poisoning of the catalyst or the electrolyte. In the case of PEM fuel
cells, it can indicate total dehydration of the membrane in the case of PEM fuel cells.

6.3.6.2.   *Cell Voltage as a Function of Current Density.*   This test is the
most important performance test for all fuel cells. The data is obtained by
incrementally increasing the current density starting from a low value, e.g.,
5 mA/cm$^2$ to the maximum desired value, e.g., 1 A/cm$^2$; 5 to 7 data points are
collected per decade of current density. It is important to stay at each current value
for several minutes (except in the case of quick screening tests), to allow the voltage
to reach a stable value. Small incremental increases in current and adequate dwell

time at each point ensure good water equilibration within the cell and provide stable performance data.

6.3.6.3.    *Reactant Flow Rates.* Influence of the reactant flow rate on the performance provides some important diagnostic information on the status of the MEAs. For example, when performance improves with higher flow rates of air (> 2–3X stoichiometry), it implies a cathode-flooding problem; water generated at the cathode is not being removed expeditiously, resulting in excess water at the cathode. An increase in airflow can remove this water as water vapor and provide for better access of oxygen to the cathode.

Better performance with lower flow rates would imply that the MEA was operating on the dry side, and increased moisture retention arising out of lower gas flows improves the cell performance.

6.3.6.4.    *Humidification Level.* The diagnostics discussed above can also be obtained by the alteration of the humidification levels of the reactant gases. The reactant flow rate and the humidifier temperature together give us a handle on the water balance problem. Water balance is a complex parameter that is controlled by an interplay of the above two variables together with the cell temperature for any given operating current density. In all this discussion, it is assumed that the gas leaving the humidifier is saturated at the humidifier temperature and that gas entering the cell is at nearly the same temperature as the humidifier. This assumption is based on a good design of the humidifier and the maintenance of good temperature control in the pipes carrying the humidified reactants.

The water-balance problem mentioned above exists for all aqueous fuel cells. However, the problem is critical with proton-exchange-membrane fuel cells (as opposed to phosphoric-acid fuel cells). With the alkaline system, the electroosmotic drag is in the opposite direction (to that in acid fuel cells) and thus the design complication is somewhat lessened.

6.3.6.5.    *Effect of Reactant Pressure.* Increase of hydrogen and oxygen (or air) pressure leads to improvement in performance. It also allows higher temperature operation. However, the water balance problem still exists and has to be managed. With air, the pressure effect is more pronounced due to the lower concentration of oxygen in air.

### 6.3.7.  Diagnostic Methods: $H_2/O_2$ Cells

6.3.7.1.    *Resistance Measurement.* Resistance of a fuel cell consists of contributions from:

- the *electrical* resistance of the hardware, electrode substrate and the electrode,
- the resistance of the *electrolyte*, and

- various *contact* resistances—e.g., bipolar plate to the electrode, electrode to the electrolyte, etc.

In the case of liquid-electrolyte fuel cells, the contact resistance between the electrode and electrolyte is almost zero. What is normally measured is a combination of the ionic and electronic resistances. Resistance measurements are particularly valuable in PEMFC research, due to two factors:

- the steep dependence of the membrane resistance on the state of hydration of the membrane, and
- the characteristic of the contact at the membrane/electrode interface, which is being dependent upon the MEA fabrication process.

There are three different methods of measuring the resistance of the fuel cell: *polarization curves* at low current densities with hydrogen on both the electrodes (referred hereafter as $H_2/H_2$ polarization), the *current-interruption method*, and *AC methods*. Smith et al. have recently published a comparison of the later two methods.[46]

(a) *$H_2/H_2$ polarization method.* In this method, hydrogen oxidation and hydrogen evolution reactions are carried out at the two electrodes and the polarization behavior is followed at low overpotentials. Since both reactions are fast, the polarization resistance is negligible compared to the electrolyte (and contact) resistance for low currents.

To conduct this experiment, the cathode of the fuel cell is first switched to humidified nitrogen, and purged for at least 15 minutes at high-flow rate. The water in the humidifier is saturated with oxygen from air, and has to be de-aerated before putting hydrogen gas through. It is essential to remove all residual oxygen before switching over to hydrogen gas, to avoid a fire and/or damage to the catalysts. After purging with nitrogen for the required time, humidified hydrogen flow is maintained at both the cathode and anode chambers of the fuel cell. When a voltmeter connected to the cell shows nearly zero volts, the cell is ready for the polarization study. The fuel cell is then connected to a precision power supply and run as a hydrogen pump, while recording the current-potential data up to 200 mA/cm$^2$. The slope of the curve yields the resistance of the cell—which includes the electrolyte and contact resistances. The curve should be linear for some extended region—at least to 200mA/cm$^2$—and also pass through zero. Figure 6.12 illustrates this method with data from the authors' laboratories. Any deviation from linearity at high current densities is usually caused by a water concentration gradient in the membrane. Deviations at lower current densities may be caused by anode polarizations.

(b) *Current-interrupt method.* The principle behind this method is the behavior of the voltage response of the cell for a given step change of current flow. Such an interruption to current can be accomplished either through a fast switch or a superimposed square wave. There are three components that give rise to a voltage recovery (or decay) curve as a function of time: (i) IR drop, (ii) activation overpotential, and (iii) concentration overpotential.

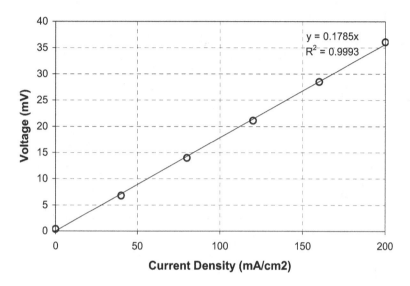

**Figure 6.12.** Resistance determination using $H_2/H_2$ polarization method. Temperature: 60 °C; 5 $cm^2$ area; Pt-Ru black anode; Pt black cathode; and Nafion 117 membrane.

On current interruption, the first component manifests as a jump or abrupt rise, since in the term IR, when I is set equal to zero (as during the current interrupt), IR becomes equal to zero. At that instant, the cell recovers a voltage value equivalent to the IR drop instantaneously, i.e., at $t = 0$. The second and third components of the voltage recovery take place in an exponential manner, and the exact relationship is governed by the respective current-potential relationships. By measuring the jump (or drop) at the zero time (or realistically within 10 microseconds), one can obtain the value of the IR of the cell. A schematic of the potential-time response for a current interrupt is shown in Figure 6.13.

(c) *AC methods.* The basic principle of AC methods is the same as used in the conductivity meters. A 1-kHz (or higher) sine wave is applied to the cell and the AC impedance measured. The resistance part is normally taken as the internal resistance of the cell. To be accurate, it is preferable to conduct this measurement at a fairly high frequency. However, measurements at lower frequencies can be used as a simple tool to diagnose problems with the cell. Many commercial AC-resistance meters would not function well in presence of a source voltage as is found in fuel cells and batteries. In such instances, special blocking circuitry has to be put in place to avoid errors. Alternatively, the $H_2/H_2$ technique described above can be used

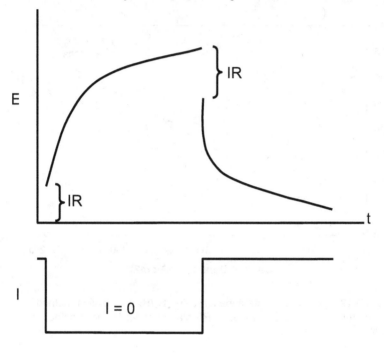

**Figure 6.13.** Typical current interrupt profile and the response waveform.

with these instruments, since, under $H_2/H_2$ conditions, the cell voltage is very close to zero at open circuit.

A better way to use the AC method is to employ a frequency response analyzer. With this instrument, one gets an impedance spectrum of the cell (i.e., data over a wide range of frequencies). This information can be used for various diagnostic and kinetic analysis of the experimental cell.

6.3.7.2.   *Polarization Data Analysis.* The polarization data obtained above can be processed using the following scheme:

(a)  plot of the hydrogen-air data uncorrected for resistance on a linear plot (V vs. I);

(b)  plot of the hydrogen-air data corrected for resistance loss on a semi-log plot (V vs. log I);

(c) plot of the hydrogen-oxygen data uncorrected for resistance on a linear plot, and

(d) plot the hydrogen-oxygen data corrected for resistance loss on the above semi-log plot with corrected hydrogen air performance.

The reader is referred to Chapter 4 of this book for further information on data treatment and analysis.

6.3.7.3.    *Catalyst Activity.*    In an electrochemical reaction, the exchange current density and Tafel slope together define the activity of a catalyst. Due to the difficulties and uncertainties in the determination of these parameters for technical electrodes, alternative ways of defining catalyst activity in fuel cells have emerged. These include: mass activity, specific activity, and catalyst utilization. A good discussion of the effects of catalyst surface area is given by Bregoli.[47]

(a) *Mass activity.* This term is given by the equation

$$A_m = \frac{i_{0.9}}{W} \qquad (6.4)$$

where $A_m$ is the mass activity of the catalyst, $i_{0.9}$ is the current density in mA/cm$^2$ at 0.9 V, and $W$ is the loading of platinum in mg/cm$^2$. The value of 0.9 V is chosen to avoid inclusion of any concentration polarization. This measure was used extensively in the PAFC; it is not widely used in PEMFC. The mass activity is quoted for operation under 100% $O_2$ and 100% $H_2$ as the reactants.

To calculate the mass activity, the polarization curve for a given electrode is established as outlined in Section 6.3.1. The current density at 0.9 V is obtained from the current-potential data. The weight loading of platinum in the electrode is determined by subtracting the weight of the PTFE binder and the substrate and gas diffusion layer weights from the weight of the electrode rationalized for the geometric area of the electrode. A more accurate method is to determine the Pt loading by a suitable analytical method (e.g., ICP).

(b) *Specific activity.* The specific activity is defined as:

$$A_s = \frac{i_{0.9}}{S_r} \qquad (6.5)$$

where $A_s$ is the specific activity of the catalyst and $S_r$ is the real (accessible) electrochemical surface area of the catalyst in the electrode in cm$^2$.

The accessible surface area of the catalyst is determined using the methods described in the next Section.

(c)  *Catalyst utilization.* Catalyst utilization can be defined as the ratio of the *electrolyte-connected* catalyst surface area in a fuel cell electrode to the raw catalyst surface area used in making the electrode. This number gives an idea about how good is the electrode making process and any possible contamination of the catalyst during the electrode preparation process. Raw catalyst surface area is available from the manufacturers as a BET surface area. However, the active catalyst area determination after the electrode is made uses the hydrogen stripping method. These two methods give somewhat different numbers for surface area. The difference may be attributed to a lack of electrolyte access to all the available catalyst particles, the re-crystallization phenomenon that occurs during electrode testing, and/or the blocking of some catalyst particles by PTFE or other catalyst particles. One tries to achieve a utilization of greater than 70% for an efficient electrode structure.

Catalyst utilization is particularly useful in PEM fuel cells, where the electrolyte contact with the bulk of the catalyst is not always optimum. Unlike in the liquid electrolyte fuel cells, the PEM cells depend on the coating of perfluorosulfonic acid on the electrode to make ionic contact throughout the catalyst network. Catalyst utilization data can be used as a diagnostic to identify any problems with the extension of the ionomer coating into the electrode structure.

6.3.7.4.  *In-Situ CV Experiment.* Cyclic voltammetry of the full cell can be performed one electrode at a time, by making the other electrode a pseudo-hydrogen reference. The electrochemical surface area can be obtained from the CV experiments, in addition to some qualitative information on poisoning.

- Electrochemical surface-area determination using hydrogen stripping: An example of a cyclic voltammetric data obtained with an MEA is shown in Figure 6.14. Poisoning effect due to adsorbed impurities would mask all the details in the hydrogen adsorption-desorption region and might even be manifested as bumps or peaks in the oxide formation region.
- Electrochemical surface-area determination by CO-stripping: For CO stripping, the electrolyte in the above experiment is saturated with CO[*]. After the electrolyte is saturated, the CO is adsorbed on the Pt catalyst by holding the electrode at a potential of +0.1 V vs. RHE for 30 minutes. The adsorption step is followed by a stripping cycle at 10 or 20 mV/s starting from +0.1 V and continuing to 1.2 V. The CO oxidation peak is then integrated to obtain the charge. From the charge under the CO oxidation peak, one can calculate the surface area of the

---

[*] Caution: CO is dangerous in ppm levels. This experiment has to be conducted in a fume hood, with adequate exhaust airflow

Pt. Interested readers are referred to a recent publication from the LBL group[48] for further insight on this technique.

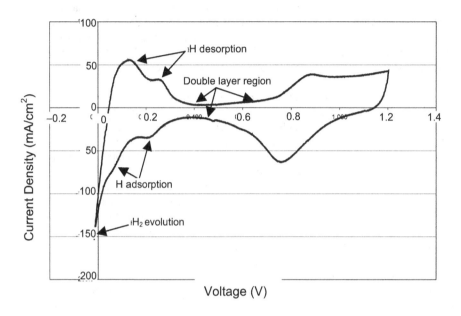

**Figure 6.14.** Cyclic voltammogram obtained on an MEA (5 cm$^2$) with the cathode as the working electrode and anode acting as a reference. Cathode was purged with nitrogen and anode with hydrogen. Sweep rate: 100 mV/s. Pt black electrodes, Nafion™ 112 membrane, 70 °C

- Catalyst poisons: CV experiments can also be used to identify possible poisoning of the catalyst, by observing the behavior of the hydrogen region of the voltammogram. With severe poisoning, small peaks may be observable in the potential region beyond 0.6 V (vs. RHE).

6.3.7.5. *Limiting Currents (N$_2$/O$_2$ vs. He/O$_2$).* For this discussion, it is assumed that the cathode is the limiting electrode. The anode performance is rarely limiting, except under circumstances of heavy poisoning. Limiting currents can be determined by running the polarization curves up to 0.2 V. While performing the limiting current determination, it is preferable to conduct the scan quickly to avoid permanent damage to the MEA. The value of the limiting current gives a clue to the presence of transport restriction in the air electrode. Lower limiting current values may be observed either due to a flooded cathode or due to inherent transport limitation in the gas diffusion layer. The nature of the problem can be identified by comparing the limiting current values obtained with air, pure oxygen, $N_2/4\%$ $O_2$ and He/4% $O_2$ mixtures. Under normal circumstances, the pure oxygen limiting current would be ~ 5X the limiting current observed with air. If the cathode were flooded, the limiting current observed with nitrogen/oxygen mixture would be the same as that observed with He/$O_2$ mixture. In the event of a restrictive gas diffusion layer, He/$O_2$ mixture will give a much higher limiting current than that for the nitrogen analog; this is due to the easier counter diffusion by He molecules. This particular diagnostic test is especially useful with phosphoric acid fuel cells.

   6.3.7.6.   *Impedance Spectroscopy.* The *AC* impedance technique was first applied to porous electrodes by deLevie.[49] Subsequent treatment of ac reponse to technical porous electrodes was published by Mund[50] and Raistrick.[51] Application of AC impedance to fuel cells has also reviewed more recently.[52] Several commercial instruments are available (e.g., Solartron Frequency Response Analyzer, Gamry Instruments Impedance Analyzer) for the experimental determination of impedance data on individual electrodes as well as full cells. Interpretation of AC response of porous electrodes is done by approximating the response to that of an equivalent circuit. By comparing the values of circuit elements obtained with an electrode before and after performance tests, one can obtain valuable information on catalyst utilization, increase in ionic resistance, mass transport limitation, etc.

   Mueller and Urban[53] have used AC impedance method to characterize DMFC. By comparing the cathode response in oxygen and air, they show the limitation in mass transport of oxygen in the cathode backing due to the trapped water in the cathode backing. Impedance spectroscopy is also one of the best methods to determine the resistance of the cell; the intersection point of the high frequency arc with the x-axis represents the cell resistance.

## 6.3.8.   Cell Testing: DMFC

   The mainstream work on direct-methanol fuel cells is confined to PEM type and alkaline fuel cells. The start-up procedure for PEM DMFC is the same as described for PEM $H_2$/air fuel cells. It is useful, if at all possible, to assess the $H_2$/air performance before proceeding with the assessment of methanol performance. Sometimes, if the anode is too wettable, it will not perform well even under 100% hydrogen and no diagnostic information is obtained. It is desirable to make the anode wettable in order to ensure that the methanol has access to the catalyst sites.

   6.3.8.1.   *Differences.* The main difference between $H_2$ and methanol fuel is the method of fuel feed. Methanol is fed into the fuel cell as an aqueous solution. Various concentrations have been used, although 0.5–1 M remains to be the most preferred concentration range. There has been limited work on the vapor feed, particularly with membrane systems operating at > 100 °C. The following are some additional differences between the $H_2$/air and methanol/air systems:

(a)  The fuel flow rate expressed in terms of number of times the stoichiometric requirement is not a concern in DFMC, since the fuel is typically re-circulated.

(b)  Fuel utilization is a concern due to the cross-over of methanol to the cathode side, which besides reducing the energy density of the system, causes system complications due to the additional water generation of the crossing over methanol.

(c)  At the present stage of development, high loadings of catalysts are required for DFMC unlike in hydrogen-air fuel cells. Also, the anode requires Pt-Ru (1:1) catalyst to prevent poisoning by methanol residues.

(d)  The preferred temperature range of operation is 50–70 °C; at temperatures less than 50 °C, the rate of methanol oxidation is very low, causing a significant reduction in energy density and at temperatures higher than 70 °C, system complications arise due to a combination of low boiling point of methanol, and higher methanol cross-over rates.

(e)  For the liquid feed PEM based DMFC, the anode is made very hydrophilic to allow methanol access to all the available catalyst sites.

(f)  In the alkaline DMFC, the fuel is mixed with the caustic and the mixture is re-circulated. With such systems, the catalyst layer is made very hydrophilic, while maintaining the hydrophobicity of the gas-diffusion layer.

(g)  For liquid feed DMFC, humidification of air is not necessary because of the typically high water content of the fuel.

6.3.8.2.  *Open Circuit Voltage.* Methanol fuel cells show low open-circuit voltage (0.7 V) due to the mixed potential established at the cathode; oxidation of the methanol crossing over takes place at the same time as the cathodic reduction of oxygen.[54] Adopting methanol tolerant cathode catalysts can increase the open circuit voltage, assuming they are completely immune to methanol.

6.3.8.3.  *Current-Voltage Performance.* The performance curves are obtained much the same way described in Section 6.5.1. An additional requirement with DMFC, is the assessment of performance as a function of methanol concentration. Optimization would also require varying the air flow rates, since, cross-over of methanol results in the flooding of the cathode, and additional air flow is necessary to remove the excess moisture in addition to supplying the necessary oxygen for the parasitic oxidation of methanol.

6.3.8.4.  *Cross-Over and Fuel Efficiency Determination.* The Faradaic fuel efficiency (fraction of consumed methanol converted to electric current) can be determined by $CO_2$ analysis at the cathode exhaust. This analysis can be made using a non-dispersive infrared $CO_2$ analyzer (e.g., Vaisala Oyj of Finland, www.vaisala.com) or an on-line gas chromatograph. Properly calibrated, this unit will continuously monitor the percentage of $CO_2$ in a gas stream. Accurate flow measurement with a bubble meter allows conversion of the measured value for the percentage of $CO_2$ to moles/sec and from Faraday's law to equivalent $mA/cm^2$ of methanol oxidized to $CO_2$ on the cathode side. $CO_2$ oxidized on the cathode side results in no useful electrochemical current. The fuel efficiency is then calculated by the simple relationship:

$$\xi_I = \frac{I_{EC}}{I_{EC} + I_{x/o}} \qquad (6.6)$$

where $\xi_I$ is the measured fuel efficiency, $I_{EC}$ is the electrochemical current density, and $I_{x/o}$ is the crossover rate expressed as equivalent current density.

This method of determining fuel efficiency assumes that methanol passing through the membrane is completely oxidized to $CO_2$ and that the only source of $CO_2$ is oxidation of methanol on the cathode side. Permeation of product $CO_2$ from the anode through the membrane can cause anomalously high estimates of equivalent methanol crossover. Any significant error of this type can be detected and corrected for by comparing the rate of $CO_2$ evolution from the anode to the theoretical rate calculated from Faraday's law. A mass balance on the operating fuel cell can be used to verify the assumptions used to assess fuel efficiency. Dohle et al.[55] found that $CO_2$ permeating through the membrane into the cathode becomes significant at higher current densities. They observed that at a current density of 100 $mA/cm^2$, the product $CO_2$ on the anode can result in an overstatement for the crossover of $\sim 25\ mA/cm^2$.

An electrochemical method of determining cross-over has been published by the LANL group.[56] In this method, the cathode side is flushed with humidified nitrogen instead of air. On the methanol anode side, protons migrating from the *nitrogen side* are reduced to hydrogen, thus providing a reference electrode. Methanol in the desired concentration is fed to the anode side and crosses over to the *nitrogen side* and is oxidized. A potential sweep or steady state potential step experiment can be performed to obtain a limiting current of methanol oxidation at the *nitrogen side*. The cross-over flux can be calculated from the observed limiting current. This method however does not take into account the reduced flux at the anode membrane interface especially at high current densities.

A review[57] of methanol cross-over in DMFC summarizes work in the literature on various methanol cross-over studies and the effect of methanol cross-over on the performance of PEM based DMFC.

## 6.3.9. Diagnostics: DMFC

When a mixture of hydrogen and methanol are fed to the cell, only hydrogen is oxidized at the anode; however, the crossing over methanol causes a decrease in cathode performance. This test can be used as a first order diagnostic to assess the cross-over effect. However, it has to be noted that the magnitude of cross-over is much less when methanol is oxidized at the anode due to the lower concentration at the anode-membrane interface.

An estimation of anode polarization for methanol (especially at low current densities) can be obtained by performing a current-voltage curve with pure oxygen at the cathode and comparing it with the polarization behavior of methanol + $H_2$ on anode and pure oxygen on cathode. At higher current densities, methanol cross-over is reduced as the methanol is oxidized at the anode; this may result in an under estimate of the anode polarization at higher current densities.

## 6.4.    POST CELL TEST ANALYSES

Very often, it becomes important to conduct various non-destructive and destructive tests on a cell build, to find out the causes of failure or poor performance or cause of degradation. The following Sections give a brief overview of some of these diagnostic tests.

### 6.4.1.   Catalyst Area

Change in catalyst area is one of the main reasons for performance degradation in fuel cells. Both electrochemical and physical methods can be employed to find the changes in surface area of the catalyst. It is preferable to preserve a small part of the original electrode used in the tests to facilitate the comparison.

6.4.1.1.   *Electrochemical Methods.*  Hydrogen stripping and CO stripping methods described above can be employed to obtain the electrochemical surface area of the catalyst.

6.4.1.2.   *Physical Methods.*[58-60]  TEM and SEM can be used to study the surface morphological changes to the electrode and also to determine surface area. For SEM, the electrode samples can be used as such, but for TEM, a thin slice of the electrode material is often required to get good images. Slicing can be done using skiving technique after first cooling the electrode in liquid nitrogen. Skiving results in a thin slice of the material, and the low temperature helps in preventing the fibrillation of PTFE binder. From the TEM images, one can estimate the catalyst surface area by counting the particles in a given grid and estimating the particle size. A detailed description of sample preparation for SEM and TEM analyses can be found in the article by Kampe.[34]

SEM of the used electrodes clearly show the surface morphological changes resulting in loss of porosity.

One of the most important aspects of PEM fuel cells is the property of water management for the membrane. The ionic conductivity of the membrane is strongly dependent on the water transport as well as the overall level of the water within the membrane. Although many articles on modeling the water transport and water management of a membrane have appeared in the literature, there has been little detailed experimental investigation because of the difficulty associated with the measurements. Some measurements of localized conductivity[61] have been made and the water content determined by inference. Recently, however, some new imaging techniques have been applied to determine the localized water content. Bellows et. al.[62] and Satija et al.[63] used neutron radiography to measure the water gradients in Nafion™ in an operating PEM fuel cell. The expected responses to changes in operating current and humidification levels were observed. More recently Tsuhima et. al.[64] and Feindel et al.[65] have used magnetic resonance imaging (MRI) to make a more detailed analysis of water transport in operating cells. Changes in operating current and time showed the water concentration gradients across a cell. The MRI has been able to provide detailed pictures of the spatial distribution of water in the membrane. The water dry out on the anode side was observed as the operating current density increased. The technique should prove valuable for developing high performance PEM cells.

## Cited References

1. U.S. Patent 4,438,216, March 20, 1984; 3,307,977, March 1967; 3,405,010, Oct 1968.
2. R. K. Sen and E. B. Yeager, *Annual Report on Contract* EC-77-C-02-4146, Submitted to DOE, Jan 15, 1980.
3. H. Jahnke, M. Schonborn, and G. Zimmermann, Organic Dyestuffs as catalysts for fuel cells, in *Topics in Current Chemistry* (Springer-Verlag, Heidelberg, 1975).
4. D. A. Landsman and F. J.Luczak, U.S.Patent 4,316,944 Feb 23, 1982.
5. P. N. Ross, Jr., EPRI Report EM-1553, September 1980.
6. J. L. Carter, J. A. Cusumano, and J. H. Sinfelt, *J. Catal.* **20,** 223 (1971).
7. M. Watanabe, M. Uchida, and S. Motoo, *J. Electroanal. Chem.* **229,** 395 (1987).
8. J. Giner, J. M. Parry, and S. Smith, *Adv. Chem. Ser.* No. 90, 151 (1969).
9. F. J. Luczak, U.S. Patent 5,013,618, May 1991.
10. V. M. Jalan, U.S.Patent 4,202,934, May 13, 1980.
11. D. McKee, *J. Catal.* **8,** 240 (1967).
12. G. S. Attard, P. N. Bartlett, N. R. B. Coleman, J. M. Elliott, J. R. Owen, and J. H. Wang, *Science,* **278,** 838 (1997).
13. C. Kresge, M. E. Leonowicz, W. Roth, J. Vartuli, and J. Beck, *Nature* **359,** 710 (1992).
14. M. Schneider and A. Baiker, *Catal. Rev. –Sci. Eng.* **37,** 515 (1995).
15. C. Adams, H. Benesi, R. Curtis, and R. Meisenheimer, *J. Catal.* **1,** 336 (1962).
16. K. Tsurumi H. Sugimoto, N. Yamamoto, T. Nakamura, and P. Stonehart, U. S. Patent 5,275,999, January 1994.
17. M. J. Yacaman, *Applied Catalysis,* **13,** 1 (1984).
18. A. N. Mansour, J. Cook Jr., and D. Sawyer, *J. Chem. Phys.,* **28,** 2330 (1984).

19. D. E. Sawyers, D. Bunker, in *X-Ray Absorption: Principles, Application, Techniques of EXAFS, SEXAFS, and XANES*, edited by D. C. Konigsberger and R. Prins (John Wiley & Sons , New York, 1998), p. 211.
20. B. E. Conway, *J. Electroanal. Chem.* **524**, 4 (2002).
21. T. J. Schmidt, H. A. Gasteiger, G. D. Stab, P. M. Urban, D. M. Kolb, and R. J. Behm, *J. Electrochem. Soc.* **145**, 2354 (1998).
22. Y. Mo, S. Sarangapani, A. Li, and D. A. Scherson, *J. Electroanal. Chem.* **538-539**, 35 (2002).
23. J. C.Huang, R. K. Sen, and E. Yeager, *J. Electrochem. Soc.* **126**, 786 (1979).
24. I. D. Raistrick, U.S.Patent, 4,876,115, Oct 24, 1989.
25. L. W. Niedrach et al., "Polytetrafluoroethylene Coated and Bonded Cell Structures", U. S. Patent 3,432,355, March 11, (1969).
26. A. J. Appleby and E. B. Yeager, in *Assessment of research needs for advanced fuel cells*, edited by S. S. Penner (Report Prepared for U.S.DOE, under contract No. DE-AC01-8ER30060, Nov. 1985), p. 141.
27. T. Kawahara et al., Method of manufacture of electrodes, U.S.Patent 6,653,252, (Nov 25, 2003); A. Datz et al., Screen printing paste and method of fabricating gas diffusion electrode, U.S.Patent 6,645,660 (Nov 11, 2003); W. Gervais et al., Aqueous ionomeric gels and products and methods, U.S.Patent, 6,679,979 (Jan 20, 2004); G. J. Goller et al., Screen printing method for making an electrochemical cell electrode, U.S.Patent 4,185,131 (Jan 22, 1980).
28. F. Solomon and C. Grun, U.S.Patent, 4,379,772 (April 12, 1983).
29. K. Kordesch and G. Simader, *Fuel Cells and their Applications* (John Wiley & Sons, NY, 1996).
30. S. Kratohvil and E. Matijevic, *J. Colloid Interf. Sci.* **57**, 104 (1976).
31. C. Orr, in *Porous Electrodes: Theory and Practice*, Proceedings Vol. 84-8, (The Electrochemical Society, Pennington, NJ, 1984), p. 278.
32. A. Jena and K. Gupta, *J. Power Sources*, **96**, 214 (2001).
33. Y. M. Volfkovich, V. S. Bagotzky, V. E. Sosenkin, and I. A. Blinov, *Colloids and Surfaces A* **187-188**, 349 (2001).
34. J. Giner and S. Smith, *Electrochem. Technology* **5,** 59 (1967).
35. H. R. Kunz and G. A. Gruver, *J. Electrochem. Soc.,* **122**, 1279 (1975).
36. E. C. Potter, *Electrochemistry: Principles and Applications* (Cleaver-Hume Press, London, 1956).
37. J. Giner, J. Electrochem. Soc., **111,** 376 (1964)
38. L.W. Niedrach, "Fuel Cell", U.S. Patent 3,134,697, May 26 (1964).
39. M. Wilson, U.S. Patent 5,211,984 (May 18, 1993); 5,234,777 (Aug 10, 1993).
40. M. S.Wilson and S. Gottesfeld, *J. Appl. Electrochem.*, **22**, 1 (1992).
41. S. Srinivasan, D. J. Manko, H. Koch, M. A. Enayetullah, and A. J. Appleby, *J. Power Sources*, **29**, 267 (1990).
42. A. Kuver, I. Vogel, and W.Vielstich, *J. Power Sources* **52**, 77 (1994).
43. W. K. Lee, C-H. Ho, J. W. Van Zee, and M. Murthy, *J. Power Sources* **84**, 45 (1999).

44. *Fuel Cell Magazine*, Februay/March 2004 issue, Webcom Communications Corp., Greenwood Village, CO.; http://www.fuelcell-magazine.com/eprints/free/advancedmeasurementsfeb04.pdf .

45. http://zone.ni.com/devzone/conceptd.nsf/webmain/8578FE9EAE5B7C4186256AA20054E325.

46. M. Smith, K. Cooper, D. Johnson, and L. Scribner, *Fuel Cell – A Webcom Publication* (April/May 2005) p. 26; also accessible at www.Fuelcell-Magazine.com.

47. L. J. Bregoli, *Electrochimica, Acta* **23**, 489 (1978).

48. B. M. Rush, J. A. Reimer, and E. J. Cairns, *J. Electrochem. Soc.* **148**, A137 (2001).

49. R. deLevie, in *Advances in Electrochemistry and Electrochemical Engineering*, edited by P. Delahay and C. W. Tobias (Wiley Interscience, N. Y., 1967) Vol. 6, p. 329.

50. K. Mund, *Siemens Forschungs-Entwicklungsber* **4** 68 (1975); K. Mund, M. Edeling, and G. Richter, in *Porous Electrodes, Theory and Practice*, PV84-8, (The Electrochemical Society, Pennington, NJ, 1984) p. 336.

51. I. D. Raistrick, *Electrochimica Acta* **35**, 1579 (1990).

52. J. R. Selman and Y. P. Lin, *Electrochimica Acta*, **3**, (1993); W. Jenseit et al., *Electrochimica Acta* **38**, 2115 (1993); M. Ciureanu, S. D. Mikhailenko, and S. Kaliaquine, *Catalysis Today* **82**, 195 (2003).

53. J. T. Mueller and P. M. Urban, *J. Power Sources* **75**, 139 (1998)

54. Z. Qi, and A. Kaufman, *J. Power Sources* **110**, 177 (2002).

55. H. Dohle, J. Divisek, J. Mergel, H. F. Oetjen, C. Zingler, and D. Stolten, *J. Power Sources* **105**, 274 (2002).

56. X. Ren, T. A. Zawodzinski, F. Uribe, H. Dai, and S. Gottesfeld, in *Proton Conducting Membrane Fuel Cells I, ECS Proceedings* (The Electrochemical Society, Pennington, NJ, 1995) Vol. 95-23, p. 284.

57. A. Heinzel and V. M. Barragan, *J. Power Sources* **84**, 70 (1999).

58. A. S. Arico, P. Creti, Z. Poltarzewski, R. Mantegna, H. Kim, N. Giordano, and V. Antonucci, *Materials Chemistry and Physics* **47**, 257 (1997).

59. V. Radmilovic, H. A. Gasteiger, and P. N. Ross, *J. Catal.* **154**, 98 (1995).

60. D. J. Kampe, Application of electron microscopy to electrochemical analysis, in *Comprehensive Treatise of Electrochemistry, Vol. 8: Experimental Methods in Electrochemistry*, edited by R. E. White, J. O'M. Bockris, B. E.Conway and E. Yeager (Plenum Press, New York, 1984) Ch. 10, p. 475.

61. M. Watanabe, H. Igarashi, H. Uchida, and F. Hirasawa, *J. Electroanal. Chem.* **399**, 239 (1995).

62. R. J. Bellows, M. Y. Lin, M. Arif, A. K. Thompson, and D. Jacobson, *J. Electrochem. Soc.* **146**, 1099 (1999).

63. R. Satija, D. L. Jacobson, M. Arif, and S. A. Werner, *J. Power Sources* **129** , 238-245 (2004).

64. S. Tsushima, K. Teranishi, and S. Hirai, *Electrochem. and Solid State Letters* **7**, A269 (2004).

65. K. Feindel, L. P.-A. LaRocque, D. Cao, R. Du, R. E. Wasylishen, and S. H. Bergens, *The 207ᵗʰ Meeting of ECS,* Quebec City, Canada, May 2005, Abstract 1570.

# PART III:

# ENGINEERING AND TECHNOLOGY DEVELOPMENT ASPECTS OF FUEL CELLS

## CHAPTER 7

# MODELING ANALYSES: FROM HALF-CELL TO SYSTEMS

## 7.1. GENERAL OVERVIEW OF MODELING ANALYSES

### 7.1.1. The Role of Simulation in Fuel Cell R&D

The importance of modeling analyses of fuel cells is twofold. First, it leads to a better understanding of the physicochemical phenomena occurring at the electrodes, in the fuel cell, and in the stack. Second, it provides a useful tool for optimization of fuel cell systems. The former determines physicochemical effects, and in particular, a selection of the phenomena that primarily influence its performance (that are then included in the model). After a selection of the physicochemical effects that have to be taken into account, the subsequent step is to describe these in terms of differential equations (1$^{st}$ and 2$^{nd}$ order). The set of equations can then be solved, rarely in an analytical fashion, but almost always through numerical integrations.

The next logical step is the comparison of the results of the modeling analysis with the experimental values. This provides model validation, that is, confirmation that the equations used and the values assigned to the model parameters are correct. The modeling analysis also provides, at a more fundamental level, reasonable support that the phenomena that have been taken into account in the model leads to an elucidation of the rate determining steps of the process. After validation, the model is useful to provide insights about the choice of materials, geometric parameters, and operating conditions to obtain the best performance in the fuel cell system, i.e. optimization studies.

Modeling of fuel cells addresses different levels of detail, such as:

This chapter was written by S. Srinivasan and P. Costamanga.

- the electrode and the electrolyte,
- the single cell,
- the stack (Figure 7.1 and Figure 7.2), and
- the system.

Obviously, at each step the simulation must take into account the results obtained in the previous step, often in a simplified or parameterized form. The same procedure applies to all types of fuel cells, even if the phenomena are different, due to the different nature of the materials, electrochemical reactions, and operating temperatures. In the following sections, the characteristic features of the simulation at the different levels of detail will be addressed, and in each section attempts will be made to consider all types of fuel cells, and to analyze the differences between them.

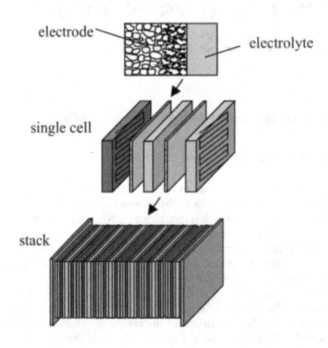

**Figure 7.1.** Modeling: from half-cell to stack level

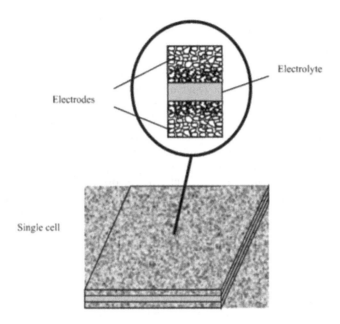

**Figure 7.2.** Illustration of a single cell in a fuel cell stack and of the morphology of a membrane/electrode assembly.

## 7.1.2.  Modeling of Electrode and Electrolyte Overpotentials

This type of simulation aims at evaluating the local relationship between current density and voltage, i.e., point by point on the cell surface.  This is, in practice, to ascertain the local kinetics of the fuel cell reactions at the half-cell level and it is, in principle, quite different from the cell potential vs. current density of the overall cell.  The starting point for the evaluation of the half-cell potential ($V_a$ or $V_c$) vs. local current density ($i$) is the equations given by (see Chapter 4):

$$V_a = V_{r,a} - \eta_{act,a} - \eta_{ohm,a} - \eta_{mt,a} \qquad (7.1)$$

$$V_c = V_{r,c} - \eta_{act,c} - \eta_{ohm,c} - \eta_{mt,c} \qquad (7.2)$$

where $\eta_{act}$, $\eta_{ohm}$, and $\eta_{mt}$ are the activation, ohmic, and mass transport overpotentials at the anode and cathode, respectively.

Thus, the local modeling is based on the expression for the thermodynamic voltage, $V_r$, and of $\eta_{act}$, $\eta_{ohm,}$ and $\eta_{mt}$ as a function of the operating variables (temperature, gas concentration, etc.) and of the geometric and material characteristics. The expression for the reversible potential is derived from thermodynamic considerations as described in Chapter 4. On the other hand, the evaluation of $\eta_{ohm}$, $\eta_{act}$, and $\eta_{mt}$ is a rather complex problem, and several considerations have to be made before analyzing them in more detail: (i) it is often difficult to distinguish the different types of contributions, since the transport of mass (related to $\eta_{conc}$) and charges $(\eta_{ohm})$ often occur together with the electrochemical reaction $(\eta_{act})$; (ii) ohmic overpotentials occur in both the electrode and in the electrolyte, while mass transport overpotentials often occur in the electrodes and only rarely in the electrolyte (as for example in PEMFCs); (iii) finally, activation overpotentials only occur at the electrode/electrolyte interfaces. While it is difficult to separate the different types of overpotentials by measuring the cell potential $E$ as a function of the current density $i$, it is relatively easy to make a net distinction between the overpotentials in the electrode and the electrolyte, and thus, Eqs. (7.1) and (7.2) can be modified to:

$$E = E_r - \eta_a^{tot} - \eta_c^{tot} - \eta_{ohm,el} \qquad (7.3)$$

where $E_r$ represents the reversible potential for the single cell, $\eta_a^{tot}$ and $\eta_c^{tot}$ include all forms of overpotentials (ohmic, concentration, and activation) in the anode and cathode, respectively, and $\eta_{ohm,el}$ represents the ohmic overpotential in the electrolyte. In the following subsections, a detailed discussion of each of these contributions, together with a discussion on the types of overpotential that have the most relevant effect, will be made. In addition, the specific characteristics of the different types of fuel cells (AFCs, PAFCs, PEMFCs, MCFCs and SOFCs) will be analyzed in detail.

## 7.2.  MODELING OF HALF-CELL REACTIONS: ELECTRODE POTENTIAL VERSUS CURRENT DENSITY BEHAVIOR AND CURRENT DISTRIBUTION IN ACTIVE LAYER

### 7.2.1.  Vital Need of Porous Gas-Diffusion Electrodes for Fuel Cells to Enhance 3-D Reaction Zone

In Chapter 2, the topic of the kinetics of charge transfer reactions at electrode/electrolyte interfaces was discussed in detail, assuming that the electrodes are smooth and planar, and accounting for activation, mass transport, and ohmic overpotentials. It was also stated that the anodic fuel cell reaction, i.e., electro-oxidation of hydrogen, is a fast and pseudo-reversible reaction, with an exchange current density $(i_o)$ of about $10^{-3}$ A/cm$^2$ in acid medium at low temperatures (< 100 °C). On the other hand, the cathodic-fuel-cell reaction, i.e., the electro-reduction of oxygen, is highly irreversible and has a lower $i_0$, about six orders of magnitude in the same medium. The latter is also the case for the electro-oxidation of the most electroactive fuel, methanol, in a low temperature fuel cell. With these low values of exchange current densities at low temperatures, in a $H_2/O_2$ fuel cell with smooth planar electrodes, the current density at a cell potential of 0.6 V will be only about $10^{-5}$ A/cm$^2$. However, these types of fuel cells exhibit current densities, which are higher by 4 to 5 orders of magnitude. If roughened electrodes–e.g., platinized platinum–instead of smooth platinum are used, the current densities in the fuel cells can be increased only by about an order of magnitude (i.e. $10^{-4}$ A/cm$^2$ at 0.6 V). However, with porous gas-diffusion electrodes it is possible to attain high current densities and power densities in fuel cells. One may question as to how the porous gas diffusion electrodes increase the current densities by three to four orders of magnitude. First one has to consider the high-surface area of the used electrocatalysts, which might range from nano- to micro-crystals in size. These electrocatalysts are located in a three-dimensional (3D) reaction zone within the porous electrode. Thus effectively, the roughness factor (the ratio of the electrochemically active surface area to the geometric surface area) of the electrodes is enhanced by more than 3 orders of magnitude. In addition, when small particles are used, the diffusion of the reactants to the electrocatalytic sites is greatly accelerated (i.e., spherical diffusion to the nano- to micro-crystals, instead of diffusion to a planar surface). Furthermore, since the porous gas diffusion electrodes in the low- and intermediate-temperature fuel cells contain nanocrystals of Pt supported on high-surface-area carbon particles, the platinum loading can be greatly reduced (i.e., to about 0.1 to 0.4 mg/cm$^2$). The structures of porous gas-diffusion electrodes are designed in such a way that they permit the entry of the fuel cell reactants and the electrolyte into the 3D reaction zone. It is in this manner that the surface area in the 3D-reaction zone is maximized, which permits the attainment of the highest level of performances based on the geometric surface area of the electrode. A schematic of such an electrode in contact with the electrolyte is illustrated in Figure 7.3.

The reactant gas enters the electrode from the substrate/diffusion layers (2 phase zone) and then reaches the active sites of the electrode through the active layer (3 phase zone); prior to this, the reactant gas is dissolved in the electrolyte. The same type of electrochemical reaction occurs at the electrocatalyst-particle/electrolyte interface as in the case of a smooth, planar electrode/electrolyte interface. The overpotential losses within the active layer are:

- activation overpotential for the charge transfer reaction;
- mass transport overpotential, more so for the diffusion of the reactants to the active site than for the products to diffuse away from the active site; and
- ohmic overpotential within the active layer, which is predominantly due to the transport to or from the active site (within the active layer) to the bulk electrolyte layer. The ohmic overpotential, due to electron transport within the electrode and to or from the current collector is generally negligible.

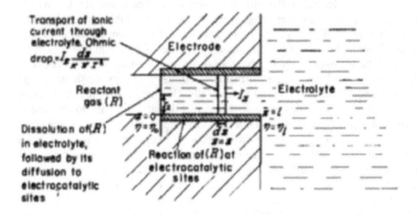

**Figure 7.3** Mode of operation in a single pore of a porous gas-diffusion electrode using the simple-pore model. Reprinted from Reference 2, Copyright (1967) with permission from the American Institute of Physics.

Considering all these rate-limiting processes, the next question is: How effective are all the sites within the active layer? To answer this question, one has to try to determine the current and potential distribution within the porous gas-diffusion electrode, as well as the total current density vs. potential relation for the fuel cell reaction at this electrode. It must be noted that, in general, the mass transport overpotential losses in the two-phase zones of the electrodes (i.e., substrate/gas diffusion layer) are negligible because the diffusion coefficients of the reactant gases in these media are about three orders of magnitude higher than in the liquid media. The structures of porous gas-diffusion electrodes are quite complex, with some unknown or immeasurable physicochemical parameters. Furthermore, theoretical treatments of current distribution, potential distribution, or total current density (based on geometric area of the electrode in contact with the electrolyte layer) vs. half cell potential do not in general yield analytic solutions when all forms of overpotential are present. Thus, except in some limiting cases, it is necessary to resort to numerical analyses. A brief synopsis of the evolutionary process of modeling analyses of fuel-cell electrodes is presented in Table 7.1 and is summarized in the following sub-sections.

**TABLE 7.1**
**Evolution of Modeling Analyses of Electrode Performance in Fuel Cells**

| Types of Fuel Cells | Model | Significant Findings | Refs. |
|---|---|---|---|
| Fuel cells with liquid electrolyte (PEMFC, AFC, PAFC, MCFC) | Simple Pore • parallel cylindrical pores • flat meniscus in pore (non wetting) | • Prediction of current and potential distribution in pore • Prediction of half-cell potential vs. current density | 1 to 9 |
| Ibid | Thin Film • parallel cylindrical pores • thin film electrolyte (wetting) | Ibid | 3, 4 |
| Ibid | Finite Contact Angle • wetting | Ibid | 5 |
| Ibid | Intersecting Pore • wetting | • Expression only for maximum current density | 6 |
| | Agglomerate with Micro and Macro pores | • Prediction of half-cell potential vs. current density | 7 |
| Fuel cells with liquid electrolyte (PEMFC, AFC, PAFC, MCFC) and with solid electrolyte | Macro-Homogenous Active Layer: a Homogenous Domain | • Prediction of current and potential distribution • Prediction of half-cell potential vs. current density | 8, 9 |

## 7.2.2. Evolution of Physicochemical Models for Porous Gas-Diffusion Electrodes and Performance Analyses

*7.2.2.1. Simple Pore Model with Parallel Cylindrical Pores.* This model was first proposed by Austin et al.[1] for an electrode of the non-wetting or hardly wetting type. As illustrated in Figure 7.3, the fuel-cell reactions occur by:

- dissolution of the reactant gas at the gas/liquid interface;
- transport of dissolved gas to the electrocatalytic sites assumed to be the walls of the single pore;
- charge transfer reaction at the electrocatalytic site/electrolyte interface;
- ion transport to/from these sites (in an acid-electrolyte fuel cell, it is proton transport) from/to the electrolyte layer; and
- product transport by diffusion to the backside of the electrodes in the fuel cell.

For a $H_2/O_2$ fuel cell in an acid electrolyte, the intermediate steps for the anodic reaction may be represented by:

(i) dissolution of $H_2$ in gaseous electrolyte:

$$H_{2,g} \rightarrow H_{2,e} \tag{7.4}$$

(ii) diffusion of dissolved gas to active site:

$$(H_{2,e})_{z=0} \rightarrow (H_{2,e})_{z=z} \tag{7.5}$$

(iii) charge transfer reaction at the active site/electrolyte interface:

$$(H_{2,e})_{z=z} \rightarrow (2H^+)_{z=z} + (2e_0^-)_{z=z} \tag{7.6}$$

(iv) proton transport in the electrolyte layer:

$$(2H^+)_{z=z} \rightarrow (2H^+)_{z=l} \tag{7.7}$$

(v) electron transport to the back of the electrode:

$$(2e_0^-)_{z=z} \rightarrow (2e_0^-)_{z=0} \tag{7.8}$$

A similar sequence of reactions could be expressed for the cathodic reaction. One additional step in this case will be the transport of product water to the back of the electrode. The position of the meniscus in the pore, according to this model, depends on a differential pressure, $\Delta P$, given by:

$$\Delta P = \frac{2\gamma \cos\theta}{r} \tag{7.9}$$

where $\gamma$ is the surface tension, $\theta$ is the contact angle between the gas/electrolyte and electrode/electrolyte interfaces, and $r$ is the pore radius.

In the theoretical analysis of this model carried out by Srinivasan, Hurwitz, and Bockris,[2] it was found that only numerical solutions to the second order differential equations for the current or potential distribution and total current density were possible, when all forms of polarization—activation, mass transport, and ohmic—are present. Analytical solutions were found for the limiting cases of activation/mass transport and activation/ohmic overpotentials. This analysis predicted a doubling of Tafel slope when passing over from the low to high current density region and was experimentally validated. The reader is referred to the original publication for the detailed mathematical treatment. The difficulty with this model was that it was too simplistic, with the assumptions of hardly wetting and parallel cylindrical pores. In addition, the predicted total current densities were considerably lower than the experimentally determined ones in fuel cells.

*7.2.2.2. Thin Film Model.* This model, first proposed by Will,[3] was of the wetting type. For a theoretical analysis of the current or potential distribution and of the total current density vs. potential relation, Srinivasan and Hurwitz[4] assumed uniform cylindrical pores as in the case of the simple pore model, and a uniform structure of a thin film of electrolyte within each pore, which was assumed to have a constant thickness (Figure 7.4). It was further assumed that the depth of penetration of the thin film had a constant length and that the charge transfer reaction occurred only in this region. As may be seen from Figure 7.4, the intermediate steps of (i) dissolution of reactant gas in the electrolyte according to Henry's law and (ii) diffusion of reactant gas to the electrocatalytic sites (i.e., the wall of the pore) were considerably faster for this model than for the simple pore model. However, the cross-sectional area for the ionic transport (i.e., proton transport from the electrocatalytic sites to the electrolyte layer) was only a fraction of the cross-sectional area of the pore; thus, the ohmic resistance was considerably higher in this case than for the simple pore model. For the theoretical analysis, it was assumed that the

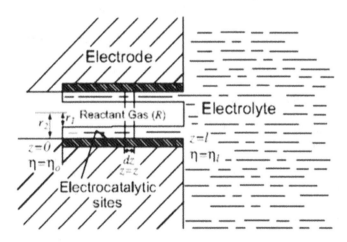

**Figure 7.4.** Schematic representation of a single pore of a porous gas-diffusion electrode using the thin-film model. Thickness of film is largely exaggerated. Reprinted from Reference 16, Copyright (1963) with permission from The McGraw-Hill Companies.

reactant gas had a flux from the gas/electrolyte interface to the electrocatalytic sites only in the radial direction. As in the treatment for the simple pore model, in the case where all forms of overpotential are present, the second order differential equation for the overpotential with respect to the length parameter of the films did not yield an analytical solution. A numerical analysis revealed: (i) high limiting current densities, independent of the exchange-current density, and (ii) the predominance of activation and ohmic overpotential at low to intermediate current densities. Here again, two-section Tafel slopes were predicted, the one at high current densities being double the one at lower current density, depending on the values chosen for the exchange current densities. The current distribution for this model was rather uniform, unlike in the case of the simple pore model up to the limiting current densities. The thin film model proved to be more realistic than the simple pore model, as demonstrated by the comparison of the predicted and the experimental potential vs. current density curves.

### 7.2.2.3. *Finite Contact Angle Meniscus Model.*

The pore and thin film models presented in the previous sections were made with rather simplistic assumptions, with respect to contact angles ($\theta$) for the meniscus of the liquid at its interface with the wall of the pore of the electrode (i.e., $\theta = 90$ for former and $\theta = 0$ for the latter models). A more realistic model was the one proposed by Bockris and Cahan,[5] the *finite contact-angle meniscus model* (Figure 7.5). Theoretical analysis carried out for this model, similar to the ones for the simple pore and thin film models, showed that the current was more localized near the top of the meniscus at any overpotential and the overpotential vs. current plots tended to be quite linear. With a well designed experimental set-up—i.e., an interface formed on a glass plate with platinum sputter-deposited on a tantalum-coated thin porous glass sheet in contact with the electrolyte, the experimental results validated the theoretical predictions. An interferometric technique was used to measure the contact angle of the electrolyte with the electrode. The fuel cell reactions (electro-oxidation of hydrogen and electro-reduction of oxygen) were carried out to determine the potential vs. current density behavior. Because of the rudimentary geometry of the electrode, the ohmic contribution predominated due to the high localization of the current generation at the top edge of the meniscus. It was also observed that there was a significant localized heating effect. On the positive side, this model appears quite close to reality in porous gas diffusion electrodes with intersecting macro- and micro-pores and meniscus formation in the complex structures.

### 7.2.2.4. *Intersecting Pore Model.*

The three models discussed in the preceding sub-sections, are based on designs of identical parallel cylindrical pores for porous electrodes. The modeling analyses were conducted for the configurations in single pores for the interfaces between (i) the reactant gas and electrolyte and (ii) the electrolyte and the electrocatalytic site. The transport of the dissolved

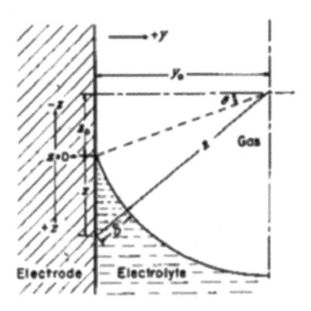

**Figure 7.5.** Finite contact angle meniscus model. A coordinate system for theoretical calculation is shown. Reprinted from Reference 5, Copyright (1969), with permission from the American Institute of Physics.

reactant to the electrocatalytic site and that of electrons from the latter to the back of the electrode were also taken into consideration. In all these cases, it was assumed that the walls served the function of electrocatalytic sites, at the interface, with the electrolyte. Burstein et al.[6], proposed a more realistic design for porous gas-diffusion electrodes. The basis of this model involves:

- multitude of intersections of micro and macropores (Figure 7.6),
- a surface of the macropore entirely covered with the electrolyte,
- micropores filled with electrolyte by capillary action and macropores filled with the reactant gases,
- dissolution of the reactant gas at the gas/electrolyte interface and diffusion to the electrocatalytic sites on the surfaces of the macropores where the reaction occurs, and
- ionic conduction through the electrolyte film in the macropores and then through the electrolyte in the micropores and electronic conduction via the solid phase.

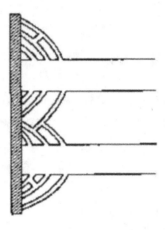

**Figure 7.6.** Intersecting Pore
Model. Two macropores are
shown.       Reprinted      from
Reference 6, Copyright (1964),
with permission from Elsevier.

For an optimum electrocatalytic activity, it is necessary to have a critical ratio
of the total cross-sectional area of micropores to that of the macropores. The
theoretical analysis of the performance of an electrode, by Burnstein et al., was
only limited to the determination of the optimum distribution of intersecting
micro and macropores in the electrode. Using a statistical method, the
expression for the maximum current density $(i_m)$ in a fuel cell was:

$$i_m = k\sqrt{\delta v^2_1 v^2_2} / \tau r_2 \qquad (7.10)$$

where $\delta$ is the length of the reaction zone, $v_1$ and $v_2$ are the porosities of the micro
and macropores, $\tau$ is the tortuosity factor in the micropores, $r_2$ is the radius of the
gas filled pore, and $k$ is a constant. This theoretical analysis did not take into
consideration the influence of activation overpotential on the current density.

   **7.2.2.5.  *Agglomerate Model.*** Of all the models proposed for porous gas-
diffusion electrodes, the most successful one was the agglomerate model of Giner
and Hunter.[7] Subsequently, there have been several publications on the theoretical
analyses of the performances of different types of fuel cells using this model.[8-13]

In the agglomerate model (Figure 7.7), the electrode pores were considered to be divided into two groups: micropores and macropores. Among them, the macropores contain no electrolyte and permit the fast flow rate of the gaseous reactant, while the micro-pores were flooded with electrolyte due to capillary forces; the microporous regions are referred to as *agglomerates*. The agglomerates were assumed to be pseudo-homogeneous and were usually modeled as cylinders or slabs. This model is applicable to fuel cells with liquid electrolyte operating at low or intermediate temperatures, like in PEMFC, AFC, and PAFC, and at high temperatures like in MCFC. In the former case, the electrodes use Teflon in both the substrate/diffusion and active layers primarily for introducing the desired amounts of hydrophobicity in the two layers. Teflon also serves as a binder in the gas diffusion and active layers. At the high operating temperatures of the MCFC (~ 650 $^0$C), one cannot use Teflon because of mechanical and chemical instabilities. Sintered electrodes are used in MCFCs with optimized structures-dual porosity electrodes with macropores for gas supply and micropores for the electrolyte. The electrode structure has to be carefully tailored to obtain and maintain three phase zones.

In both types of fuel cells, the active layer consists of agglomerates of the electrocatalyst. For the low and intermediate temperature fuel cells, these are almost always, at the present time, carbon-supported noble metal electrocatalysts, while for MCFCs, these are sintered nickel or nickel alloy particles (at the cathode of the MCFC), oxidized to nickel oxide, which, in turn by reaction with lithium from the electrolyte form lithiated nickel oxide particles.

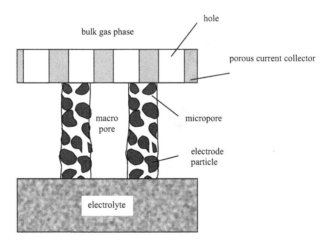

**Figure 7.7.** Schematic of the porous electrode geometry in the agglomerate model.

7.2.2.6. *Macro-Homogeneous model.* In this type of model, the active
layer is considered as a homogeneous domain, in which the various transport media
are superimposed to each other. While the agglomerate model has been
demonstrated to be most suitable for three-phase electrodes, the macro-
homogeneous model is the state-of-the-art model for SOFC electrode
simulation.[11-14]

## 7.2.3. General Treatment Based on Three Basic Equations

### 7.2.3.1. *Basic Aspects Relevant to All Types of Fuel Cells.* Figure 7.8
illustrates a typical structure of an electrode for a low- to intermediate-temperature
fuel cell. It shows a detailed schematic of a typical PEMFC electrode, which
displays three layers: (i) teflonized substrate (typically carbon cloth); (ii) a diffusion
layer generally formed by carbon particles with a size of about 0.1 μm along with
Teflon; and (iii) an active layer consisting of Pt electrocatalyst nanocrystals
(dimensions 20-40 Å) supported on carbon particles (Pt loading usually 0.4 mg/cm$^2$
or less) with Teflon or Nafion.

When using a liquid electrolyte like AFC or PAFC it penetrates the active
layer. For PEMFCs with only the supported electrocatalysts and Teflon in the
active layer, the solubilized perfluorosulfonic acid (e.g. DuPont's Nafion or Asahi
Chemical's Aciplex) is impregnated into it to enhance the three dimensional
reaction zone. Even if the void area is different in the various layers, still a high

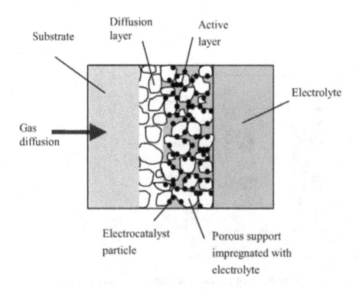

**Figure 7.8.** Schematic of the structure of a typical PEMFC
electrode.

porosity is characteristic of all the layers of the electrode. Conduction properties differ in the various layers too. The support and diffusion layers are electronic conductors, while the active layer, due to the overlapping of electrodes and membranes, features two parallel paths for proton and electron conduction (carbon particles and Pt grains form an interconnected electron-conducting network through the active layer). Generally, the thickness of the active layer is about 10 to 100 μm and the thickness of the overall electrode is 300-400 μm.

All types of fuel cells with liquid electrolytes have a rather similar electrode structure. On the other hand, cells with solid electrolytes (SOFCs) have a rather different electrode configuration (Figure 7.9).

In an SOFC, the electrodes contain only an active layer, without any support or gas diffusion layer; sometimes, a porous current collector is present at the back of the active layer. To increase the two-phase contact area,[9] it is necessary for the electrodes to be composed of cermet formed of a mixture with electronic conductor particles ($La_{0.85}Sr_{0.15}MnO_3$ for the cathode, Ni for the anode) and ionic conductor materials (usually yttria stabilized zirconia[12] or YSZ). It is very important that the composition and the granulometry of the electrodes are accurately chosen. The particles of the same type (electronic conductor or ionic conductor) have to touch each other so that a network is formed through the electrode. This is the condition under which high conductivity is attained. Moreover, adequate contact between the particles of different types ensures that a large active area is formed. All these properties depend on the composition and granulometry of the SOFC electrode, and more details about these aspects will dealt with in Section 7.2.3.3.

In spite of the differences in configurations among the different types of fuel-cell electrodes, it is possible to set up a physical model that can be applied with suitable simplifications and/or modifications. Indeed, the main process taking place in the fuel-cell electrode is the electrochemical reaction in the active layer of the electrode, the rate of which is governed by the Butler-Volmer equation as discussed in Chapter 2,

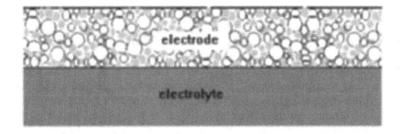

**Figure 7.9.** Schematic of structure of a typical SOFC electrode.

$$i = i_0 \left\{ \frac{c_r}{c_r^0} \exp\left( \alpha \frac{\eta F}{RT} \right) - \frac{c_p}{c_p^0} \exp\left[ -(1-\alpha)\frac{\eta F}{RT} \right] \right\} \qquad (7.11)$$

This equation represents the relationship between the current density ($i$) and the overpotential ($\eta$) for the charge transfer across the electrical double layer in the active layer/electrolyte interface. However, this equation is not sufficient to calculate the overall electrode overpotential, the reason being that the electrical double layer is widely extended through the electrode, and thus different values of reactant concentration, $c_r$, prevail at different points, causing different performances. For this reason, the Butler-Volmer equation must be coupled to the mass transport equation, in order to get a correct picture of the phenomenon. For this purpose, the Stefan-Maxwell equations are used to model the multicomponent diffusion in the porous electrode:

$$\nabla x_i = \sum_{j=1}^{n} \frac{RT}{pD_{ij}^{eff}} \left( x_i N_{j,g} - x_j N_{i,g} \right) \qquad (7.12)$$

where $n$ is the number of components of the mixture, $p$ is the operating pressure, $x$ is the molar fraction, $N_{i,g}$ is the superficial gas-phase flux of species $i$ averaged over a differential volume element, and the quantity $D_{ij}^{eff}$ is an effective binary diffusivity of the pair $i$-$j$ in the porous medium. These equations can be reduced to the Fick's first law of diffusion by simplifying (e.g., in case of binary mixtures of gases) as:

$$N_{i,g} = x_i \left( N_{i,g} + N_{j,g} \right) - cD_{ij}^{eff} \nabla x_i \qquad (7.13)$$

A further complication arises due to the fact that reactant diffusion does not often occur in the gaseous phase through the electrode; instead the reactant species dissolves in the electrolyte and then diffuses into the solution. Of course this is possible only with fuel cells with liquid electrolyte (AFC, PAFC, MCFC) or with polymeric electrolyte (PEMFC and DMFC), while it does not occur in fuel cells with a solid electrolyte (SOFC). The structure of the active layer of fuel cell electrodes is rather complicated with gaseous pores of different sizes crossing the electrode with different geometries. Whether diffusion occurs in the gaseous or in the dissolved phase has been a subject of several studies. The complicated structure of the electrode has been treated in different ways (flooded pore[1,2,14], thin-film[3,4], meniscus[5], and agglomerate models[8-15]). Another important consideration is that mass transport occurs in all the layers in the electrode; thus, Eqs. (7.12) and (7.13) must be applied not only to the active layer (where they are combined with the Butler-Volmer equation), but also to the substrate and diffusion layers, when present. Mass transport can become the rate-determining step of the reaction,

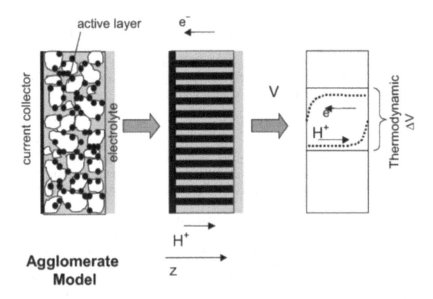

**Figure 7.10.** Schematic of charge transfer of electrons and protons in the active layer.

especially in the low temperature fuel cells operated at high current densities, and in this case the local $i$ - $V$ curve presents an abrupt and steep voltage drop occurring at a *critical-current density*. When this phenomenon occurs, the limiting current is reached where the transport of reactants through the electrode has reached a limiting value.

An additional phenomenon that must be taken into account when modeling the rate of reaction of an electrode is that of charge transport, which occurs through the entire active layer. Figure 7.10 shows how this phenomenon interferes with the evaluation of the overpotential at the electrode.

The charge transfer process occurs in the active layer (the mass transport effect discussed previously is not shown in this figure). The active layer in the electrode can be described as a homogeneous mixture of ionic and electronic conductors. In this model, not only do the charges flow along each conductor, but also there is an exchange of charges from one conductor to the other (electrochemical reaction). Figure 7.10 refers to a hydrogen electrode (it is analogous for all other types of fuel cell electrodes), where the electrochemical reaction leads to the formation of electrons (which migrate toward the current collector, on the left hand side of the figure) and protons (which then migrate toward the electrolyte on the right hand side). This charge migration gives rise to ohmic losses, $\Delta V_{ohm}$, given by

$$\Delta V_{ohm} = \rho^{eff} i \qquad (7.14)$$

where $\rho^{eff}$ is an effective conductivity, which takes into account the tortuosity of the conducting paths, and $i$ is the current density. Equation (7.14) holds for both the electronic and the ionic conducting phases.

Thus, at any point along the thickness of the electrode, the real potential of the charges (depicted by the dotted lines in Figure 7.10) is lower than the theoretical thermodynamic voltage, and the following relationship holds:

$$\eta = \Delta V_{th} - \Delta V_{real} \qquad (7.15)$$

where $\eta$ is the overpotential which is in the Butler-Volmer equation. Equation (7.11) will then have to be taken into consideration to determine (i) the dependence of the current density on overpotential and (ii) the current or potential distribution in the electrode.

In summary we can state that electrode modeling is based on three basic equations, i.e. Butler-Volmer, Stefan-Maxwell, and Ohm's law equations, which can then be modified as necessary for the particular type of electrode under study. Some further observations about the solutions to the overall model are that: (i) usually the solution is carried out only one-dimensionally, i.e., along the z-direction for the main flow of the electrical charges (see Figure 7.10); (ii) the complex electrode structure is taken into account via an appropriate geometrical schematization of the electrode (flooded pore, thin-film, meniscus, agglomerate and filmed agglomerate models); and (iii) the use of appropriate boundary conditions and effective transport parameters ($D^{eff}$, $\rho^{eff}$) take the electrode structure (tortuosity, porosity, etc.) into account.

In particular, depending on the particular model studied, the equations can be simplified if one of the three phenomena is non-limiting compared to the other ones. In such a case, analytical expressions could be derived for the current density as a function of the overpotential and for the current and potential distributions in the active layer from the second order differential equations.[16] In general, however, all three phenomena (activation, mass transport, and ohmic overpotential) play roles in all types of fuel cells, and the solution of the resulting system of equations is not analytical, but requires numerical integration. In the following Section, typical results obtained for the different types of fuel cells are presented.

*7.2.3.2. Application to 3-Phase Electrodes (PEMFC, AFC, PAFC, DMFC, and MCFC): the Agglomerate Model.* The equations presented in Section 7.2.3.1 have been applied to simulations of the PEMFC electrode in several theoretical studies. Among them, the model of Bernardi et al.[17,18] is one-dimensional in the direction perpendicular to the cell plane and considers the membrane to be always fully hydrated. The main results of their modeling analysis are presented in Figures 7.11 and 7.12. Figure 7.11 shows the concentration of oxygen dissolved in the membrane phase, the electrocatalytic layer, and the gas-diffusion layer. The gas diffusion region shows a profile that is almost flat, demonstrating that no gas-diffusion limitations take place in this area. The active-layer region exhibits a

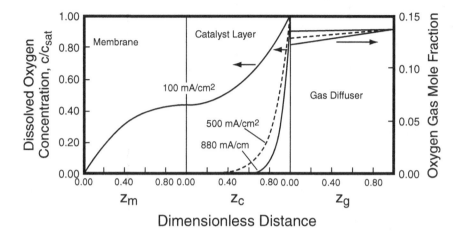

**Figure 7.11.** Modeling analysis results of concentration profiles for dissolved oxygen concentration in active layer of the electrode for three current densities based on the geometric area of the electrode. Reprinted from Reference 17, Copyright (1991) AIChE, with permission from The American Institute of Chemical Engineers. All rights reserved.

different picture. In the active layer, the dissolved-oxygen concentration is almost uniform at low current densities. At high current densities, the portion of the active layer located near to the boundary with the electrolyte is depleted of dissolved oxygen. Oxygen cannot diffuse fast enough to replenish what is consumed by the electrochemical reaction and a limiting current density is not reached until the oxygen concentration is zero in all regions of the electrocatalyst layer. Figure 7.12 illustrates the reaction-rate distribution throughout the electrocatalyst layer for the same current densities reported in Figure 7.11. As a consequence of the dissolved-oxygen concentration distribution, the reaction rate is almost uniform throughout the electrocatalyst layer only at low current densities (i.e.100 mA/cm$^2$); at higher current densities, the current is generated only in the part of the active layer closer to the interface with the gas diffuser.

Equations similar to the ones discussed above have been applied in many studies aimed at gaining insight on optimization of structure and composition of the electrode to obtain the maximum performance in the PEMFC.[19,20] As mentioned in Section 7.2.2, the agglomerate model is the most appropriate. In this model, the electrocatalyst particles are contained within the structure of the agglomerates (see Figure 7.8). The reactant gas diffuses through a porous backing layer, then into the

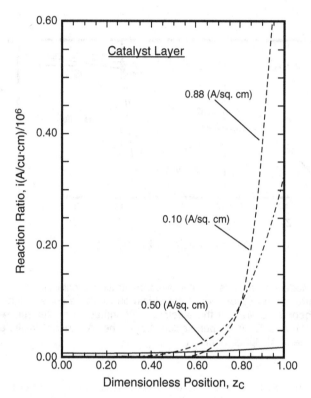

**Figure 7.12.** Modeling analysis results for variation of local current density in active layer at three current densities based on geometric area of electrode. Reprinted from Reference 17, Copyright (1991) AIChE, with permission from The American Institute of Chemical Engineers. All rights reserved.

active layer, and finally through the thin film of electrolyte to contact the Pt catalyst layer. Although the region of greatest interest in the electrodes is the active layer, it is important to take the gas diffusion layer into consideration when modeling the performance of the fuel cell.

The following analysis is based on a one-dimensional, steady state modeling study performed by Ridge and White[21] and Iczkowski and Cutlip.[22] In the case of humidified reactant gases, water vapor as well as the reactant gas are present in the electrodes and it is necessary to use the Stefan-Maxwell equation to describe the multi-component diffusion. Using Eq. (7.12) from the previous section and the ideal gas law, the one-dimensional diffusion equations for oxygen and water vapor transport through the gas diffusion layer can be expressed as:

$$\frac{dP_{O_2}}{dz} = \frac{RT}{P_{total}D_{O_2,H_2O}^{eff}}\left(P_{O_2}N_{H_2O} - P_{H_2O}P_{NO_2}\right) \qquad (7.16)$$

$$\frac{dP_{H_2O}}{dz} = \frac{RT}{P_{total}D_{O_2,H_2O}^{eff}}\left(P_{H_2O}N_{O_2} - P_{O_2}N_{H_2O}\right) \qquad (7.17)$$

where $P_{O_2}$, $P_{H_2O}$, and $P_{total}$ are the oxygen, water, and total pressures, $N_i$ is the flux of species i, and $D_{ij}^{eff}$ is the effective binary diffusivity of the pair i-j.

The effective diffusion coefficient that accounts for the electrode porosity and tortuosity is expressed at $D_{O_2,H_2O}^{eff}$. In order to simplify the PEMFC analysis, we can solve the equations for short times at low temperatures in which the net flux of water in the vapor phase can be considered to be negligible. In a more thorough analysis, evaporation of product water and condensation of water from the gas streams in the pores should be considered. In the case of a PAFC, the high temperatures do not allow this simplifying assumption because the product water readily evaporates and must be considered in the vapor phase. A similar one-dimensional analysis of the PAFC was performed by Iczkowski and Cutlip.[22]

The one-dimensional diffusion of oxygen through the macropores in the active layer can also be described using the Stefan-Maxwell equation as shown below:

$$\frac{dP_{O_2}}{dz} = \frac{RTN_{O_2}}{D_{O_2,H_2O}^{eff}}\left(P_{O_2} - P_{total}\right) \qquad (7.18)$$

After the oxygen has diffused through the gas diffusion and active layer pores, it must dissolve in the electrolyte film in the agglomerates where it contacts the electrocatalyst surface and undergoes an electrochemical reaction with the hydrogen ions. If the thickness of the film ($\delta$) is much less than the radius of the agglomerate ($r_a$), the curvature in the film can be neglected. The flux of oxygen through the thin film of electrolyte can then be described using Fick's law for a flat film. As shown in Eq. (7.19), $c_{O_2}^{diss}$ is the concentration of dissolved oxygen in the electrolyte when in equilibrium with oxygen with a partial pressure of one atm, $P_{O2}(z)$ is the partial pressure of oxygen as a function of distance through the electrode, $c_{O2}(r_a, z)$ is the oxygen concentration at the film-agglomerate interface, a is the area of the film per unit volume, and $D_{O2}$ is the diffusion coefficient of the oxygen in the electrolyte:

$$R_{O_2} = \frac{dN_{O_2}}{dz} = aD_{O_2}\frac{P_{O_2}(z)c_{O_2}^{diss} - c_{O_2}(r_a,z)}{\delta} \qquad (7.19)$$

For simplicity we will assume that the reaction rate ($R_{O_2}$) depends only on the dissolved oxygen concentration although the actual rate is also first order with respect to hydrogen ion concentration; an analysis containing this latter term can be found in the work performed by Ridge and White.[21]

The reaction rate can be expressed by the following rate law where $\varepsilon_C$ is an effectiveness factor of the electrocatalyst, which depends on the relative rates of diffusion and reaction and $K_e$ is the rate constant:

$$R_{O_2} = \varepsilon_c K_e c_{O_2}(r_a, z) \qquad (7.20)$$

The numerical value of the rate constant, $K_e$, is obtained by assuming that activation is the only overpotential loss and then using the current density calculated from the Butler-Volmer equation. By considering only the cathodic term, the Butler-Volmer equation is reduced to the following equation where $b$ is the Tafel slope or $2.3RT/(1-\alpha)nF$.

$$i_{act} = i\exp\frac{2.3(V_r - V)}{b} \qquad (7.21)$$

The rate constant can then be evaluated using the equation:

$$K_e = \frac{i_{act}(V)}{4Fz_1 P_{O_2} C_{O_2}^{diss}} \qquad (7.22)$$

where $z_1$ is the thickness of the electrocatalyst layer.

In order to obtain a theoretical half-cell potential $V$ vs. current density ($i$) plot, it is necessary to relate the potential to the current density. This is established using Ohm's law. It is important to consider the relationship in both the electrocatalyst layer as well as in the electrolyte as expressed in the following equations:

$$\frac{dV_{e^-}}{dz} = \frac{i_{e^-}(z)}{\sigma_{catalyst}} \qquad (7.23)$$

$$\frac{dV_{H^+}}{dz} = \frac{i_{H^+}(z)}{\sigma_{catalyst}} \qquad (7.24)$$

where $V_i$, $i_i$, and $\sigma_s$ represent the potential, current density, and effective conductivity, respectively (the subscript $i$ represents the electrons in the electrode, $H^+$ the protons in the electrolyte, and s depict the catalyst in the electrolyte).

The electronic current density at a particular position is dependent on the flux of oxygen at that point. This is expressed in the equation:

$$i_{e^-}(z) = -4FN_{O_2}(z) \tag{7.25}$$

The current density due to the hydrogen ion transport is the difference between the current density at the outer edge of the electrode and the current density at z as shown in the equation:

$$i_{H^+}(z) = i_{e^-}(0) - i_{e^-}(z) \tag{7.26}$$

The rate of oxygen consumption is dependent on the difference between the potentials of the electrode and electrolyte. Therefore, Eqs. (7.23) and (7.24) can be simplified to the following equation:

$$\frac{dV}{dz} = \frac{i_{e^-}(0)}{\sigma_{electrolyte}} + 4F\left(\frac{1}{\sigma_{catalyst}} + \frac{1}{\sigma_{electrolyte}}\right)N_{O_2} \tag{7.27}$$

The solution of this equation along with the diffusion equations numerically will yield the half-cell potential ($V$) versus current density ($i$) plot, which can be compared to experimental data.

The above analysis includes the effects of gas diffusion, dissolution of oxygen in the electrolyte, and ohmic resistance of the electrocatalyst layer. As mentioned earlier, the effect of hydrogen-ion transport on the rate of reaction at the cathode has been considered by Ridge and White.[21] In addition to PEMFCs and PAFCs, AFCs, MCFCs, and DMFCs have also been treated in a similar manner.

Also, the effects of structural parameters such as polymer volume fraction, electrocatalyst layer thickness and platinum loading in the electrocatalyst layer have been analyzed.[23-25] The results of the simulations show that in case of poor impregnation of the electrolyte into the active layer, the proton conductivity is the limiting electrode process and the current is only generated by a thin layer close to the membrane layer; the higher the exchange current density of the electrochemical reaction, the thinner is the region where the charge transfer process occurs. The simulation indicates that there is an optimal value of polymer-volume fraction ($\varepsilon$), which depends on the thickness of the electrocatalyst layer. The optimal value of $\varepsilon$ is usually in the range of 0.4 to 0.5. The optimal value is dependent on a trade-off between gas diffusion as a controlling factor and ohmic overpotential losses for thicker electrocatalyst layers. In an analogous manner, the thickness of the active layer has to be chosen on the basis of trade-off considerations. An increase of the active layer thickness provides a larger active area for the electrochemical reaction, but at the same time, mass-transport problems will be encountered at high current densities. Due to the fact that electrical current is only generated by a thin section of the active layer, an increase of the platinum load in the active layer beyond a certain

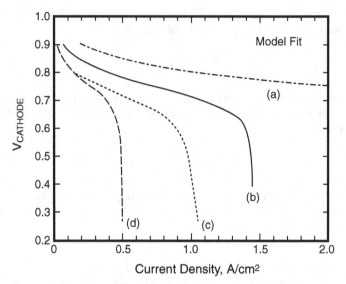

**Figure 7.13.** Comparison of simulation and experimental results for a 5 cm$^2$ PEMFC cell. Curve (a) is for 5 atm $O_2$, (b) is for 5 atm air, (c) is for a 2 atm $O_2/N_2$ mixture with 13.5% $O_2$, and (d) is for a 5 atm $O_2/N_2$ mixture with 5.2% $O_2$. Reproduced from Reference 26, Copyright (1991), with permission from The Electrochemical Society, Inc.

amount (20 wt% Pt/C) would increase the cost, but only slightly improve the performance.

The agreement between theory and experimental results was demonstrated by a number of authors. Figure 7.13 represents the results from one of these studies.[19-26] In particular, one study has shown that the agglomerate model (i.e., the active layer

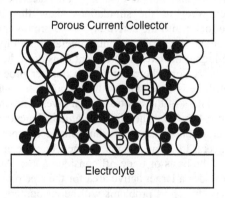

**Figure 7.14.** Schematic of a typical SOFC electrode.

contains intersecting macropores through which the gases diffuse before entering the microchannels of the carbon agglomerates) is more appropriate than the macro-homogeneous one (i.e., the active layer is considered as a homogeneous domain, in which the various transport media are simply superimposed to each other) because the former provides a better explanation for the mechanism of diffusion of the reactants in the active layer.[23]

### 7.2.3.3. Application to 2-Phase Electrodes (SOFC): Simulation of Composite Electrodes.

As mentioned in the previous sections, a key issue for 2-phase electrodes (e.g., in SOFC electrodes) is that the three-dimensional reaction zone is extended widely throughout the electrode. The electrodes are usually composed of a mixture of electron and ionic conducting particles. The optimal composition of this mixture is of fundamental importance in order for the particles of the same type (electronic conductor or ionic conductor) to be in contact with each other so that a network is formed through the electrode. It is under such conditions that a high conductivity is reached. In Figure 7.14 an electrode is shown, and three different clusters are formed by the particles for the ionic conductor. The clusters labeled A present many ramifications and connect with each other to form a network. The B type clusters are shorter chains connected only to the electrolyte; they transport electrons or ions only within a limited thickness of the electrode. In the case where many B clusters are present, the local ionic conductivity depends very strongly on the position along the electrode thickness. Finally, the C clusters are completely insulated from the ends of the electrode; their presence only represents a loss of ionic or electronic conductivity. There is a threshold[27,28] in the volume concentration of the electronic or ionic conducting particles, i.e., the percolation threshold, within which the particles form only B and C clusters in the electrode. On the other hand, a network of A clusters is formed above the threshold, even if the electrode still contains a few B and C clusters. It has to be pointed out that the percolation thresholds strictly depend on the ratio between the dimensions of the two different types of particles. Moreover, if the ratio between the particle dimensions and the electrode thickness is not small enough, the percolation limits are significantly altered.[29,30] Conductivity measurements[29,30] around the electronic percolation threshold reveal a steep variation from typical values for the ionic conducting phase to the typical values for the electronic conducting phase.

The simulation model of a cermet SOFC electrode[31] refers to the electrode structure represented in Figure 7.14. In this case, due to the very high operating temperature of SOFCs, some simplifications can be made to the generalized treatment, described in Section 7.2.3.1. It can easily be demonstrated that diffusion limitations are negligible,[31] unless the electrodes are composed of extremely small particles. A further simplification arises from the fact that, at high temperatures and when no diffusion limitations occur, the exponential terms of the Butler-Volmer equation can be linearized. With these simplifications, the equations can be integrated to yield analytical solutions and the results can be expressed on the basis of the total electrode resistance $R$, which includes both ohmic and activation overpotential effects:

$$\frac{1}{R} = \frac{\Gamma \sinh(\Gamma)}{a\left(\rho_{io}^{eff} + \rho_{el}^{eff}\right)\left\{\cosh(\Gamma) + \Omega[2 + \Gamma\sinh(\Gamma) - 2\cosh(\Gamma)]\right\}} \qquad (7.28)$$

where $a$ is the electrode thickness and $\Gamma$ is the ratio between the electrode thickness and the characteristic thickness, where the electrochemical reaction takes place, and is expressed by:

$$\Gamma = \frac{a}{\sqrt{\dfrac{R_g T}{i_0 A F (\rho_{io}^{eff} + \rho_{el}^{eff})}}} \qquad (7.29)$$

In Eq. (7.28) the term $\Omega$ is given by:

$$\Omega = \frac{\rho_{io}^{eff}\, \rho_{el}^{eff}}{\left(\rho_{io}^{eff} + \rho_{el}^{eff}\right)^2} \qquad (7.30)$$

Also in this case, several of the parameters appearing in the model are effective parameters, which strongly depend on the electrode composition and structure. In the case of SOFCs, it is very interesting to evaluate the value of these parameters as a function of the electrode composition, assuming that the particles that form the electrode are spherical, and all particles of the same type (ionic conductor or electronic conductor) have the same diameter. Using this assumption and the percolation theory,[20] the effective resistivity $\rho^{eff}{}_{io}$ or $\rho^{eff}{}_{el}$ is expressed as:

$$\frac{1}{\rho^{eff}} = \sigma^{eff} = \gamma\sigma_0\left(\frac{n - n_c}{1 - n_c}\right)^{\mu} \qquad (7.31)$$

where $\sigma_0$ is the bulk conductivity, $n$ is the fraction number of particles of the same type, and $\mu$ is an exponent typical of the percolation theory, whose value is 2 in three dimensional systems. Finally, $\gamma$ is a proportionality constant, which is an adjustable parameter in the model, and $n_c$ is the critical fraction number, i.e., the percolation threshold. Indeed, Eq. (7.31) is valid only above the percolation thresholds (below this, the effective conductivity is zero). The parameter $n_c$ is evaluated by means of the theory of particle coordination number in a random packing of bimodal spheres,[32,33] which is based upon the following relationships:

$$Z_{io} = 3 + \frac{Z - 3}{n_{io} + (1 - n_{io})P^2} \qquad (7.32)$$

$$Z_{el} = 3 + \frac{(Z-3)P^2}{n_{io} + (1 - n_{io})P^2} \qquad (7.33)$$

where the subscripts io and el identify the ionic and electronic conductors, respectively. $Z_{io}$ and $Z_{el}$ are the coordination numbers, i.e. the average number of contacts of the ionic and electronic conducing particles, respectively. The parameter $P$ is defined by the relationship $P = r_{el}/r_{io}$, and $Z$ is the overall average coordination number, that is six in a binary random packing of spheres. The average number of contacts between electronic and ionic conducting particles is then given by:

$$Z_{io-el} = n_{el} \frac{Z_{io} Z_{el}}{Z} \qquad (7.34)$$

Equation (7.34) is used in the model to evaluate (i) the percolation thresholds, which are obtained when $Z_{io-io} = 2$ or $Z_{el-el} = 2$, and (ii) the overall number of contacts between particles of different types within the electrode, and thus the overall active area $A$. For the latter calculation, the circular area around each contact (evaluated through the adjustable parameter $\kappa = \sin^2\theta$, where $\theta$ is the contact angle between two particles of different types) is assumed to be electrochemically active. The fact that small particles ($d = 0.3$ to $0.4$ μm) have been considered in this work supports the latter hypothesis. The critical fraction number, $n_c$, that determines the percolation threshold is only a function of $P$, i.e., the ratio between the dimensions of the particles that form the electrode.

Results of the modeling analysis are presented in Figures 7.15 and 7.16 for a cermet electrode composed of Pt and EDB (erbia-doped bismuth oxide) $(Bi_2O_3)_{0.7}(Er_2O_3)_{0.3}$ at an operating temperature of 900 °C. The reciprocal electrode resistivity, $1/R$, has been calculated for an EDB/Pt cathode on the basis of the model previously discussed. The exchange current density ($i_0$) and the electrical conductivity of Pt were assumed to be 400 A/m$^2$ and $2.2 \times 10^6$ S/m.[28] Literature data[28] indicate that $(Bi_2O_3)_{0.8}(Er_2O_3)_{0.2}$ has a conductivity of 37 S/m at 700 °C. The EDB conductivity was given a value of 20 S/m at 900 °C. The diameter of both the EDB and Pt particles was assumed to be 0.1 μm; it was also assumed $\gamma = 0.5$ and $\theta = 15° = \pi/12$ rad.

In Figure 7.15, $1/R$ is plotted as a function of the electrode composition. The percolation thresholds, which in the case of particles having the same diameter in identical positions, are at $\varphi = 0.294$ for each of the phases present in the structure. Maximum performance is achieved when $\varphi_{el}^{max} = 0.37$, for an electrode thickness of 90 μm. The $1/R$ value is about $8.2 \times 10^4$ S/m$^2$, corresponding to an overpotential of

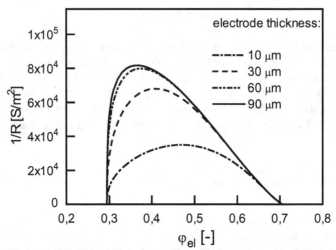

**Figure 7.15.** Reciprocal electrode resistance R as a function of the electrode composition for EDB/Pt cathodes at 900 °C. Reprinted from Reference 31, Copyright (1998), with permission from Elsevier.

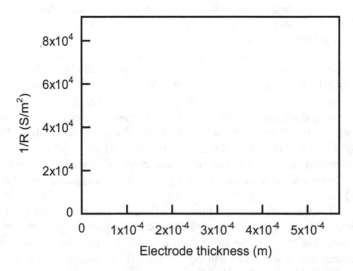

**Figure 7.16.** Reciprocal electrode resistance as a function of the electrode thickness for EDB/Pt cathodes at 900 °C. Reprinted from Reference 31, Copyright (1998), with permission from Elsevier.

0.037 V at a current density 3,000 A/m$^2$. The maximum value of $1/R$ is reached near the percolation threshold of the electronic conducting phase for a thick electrode (a $\approx$ 90 µm), and for $\varphi_{el} \approx 0.5$ for a thin electrode (a $\approx$ 10 µm). This can be explained by considering Eq. (7.28). As a first approximation the term $\Omega[2 + \Gamma \sinh(\Gamma) - 2\cosh(\Gamma)]$ can be neglected ($\Omega$ is approximately zero at compositions far from the percolation threshold of the electronic conducting phase). Thus, Eq. (7.28) can be rewritten as follows:

$$\frac{1}{R} = \frac{\Gamma}{a(\rho_{io}^{eff} + \rho_{el}^{eff})} \tanh(\Gamma) \tag{7.35}$$

For an electrode thickness $a \rightarrow 0$ and also $\Gamma \rightarrow 0$:

$$\frac{1}{R}_{\text{limit}\,\Gamma\rightarrow 0} = \frac{\Gamma^2}{a(\rho_{io}^{eff} + \rho_{el}^{eff})} = \frac{i_0 a F}{R_g T} A \tag{7.36}$$

where $A$ refers to the ratio of active area to unit volume.

It must be noted that the number of B clusters is negligible and the continuous model is applicable. The only composition-dependent parameter in the preceding expression is $A$ and the inverse relationship of the electrode resistance as a function of the electrode composition is maximum in correspondence to the composition that determines the maximum contact area between the two phases ($\varphi_{el} = 0.5$).

For a large electrode thickness, $a \rightarrow \infty$, and $\Gamma \rightarrow \infty$. Thus,

$$\frac{1}{R}_{\text{limit}\,\Gamma\rightarrow\infty} = \frac{\Gamma}{a(\rho_{io}^{eff} + \rho_{el}^{eff})} = \sqrt{\frac{i_0 F}{R_g T} \frac{A}{(\rho_{io}^{eff} + \rho_{el}^{eff})}} \tag{7.37}$$

The maximum performance is obtained when the ratio $A/(\rho_{io}^{eff} + \rho_{el}^{eff})$ attains a maximum value, i.e. where a compromise is reached between a maximum value of $A$ and a minimum value of the sum of the resistivities. Thus, the reciprocal electrode resistance reaches a maximum near the percolation threshold of the electronic conductor phase; in addition, the minimum value of the sum of the resistivities is reached near that point.

In the region outside the thresholds, the model gives the approximate result where $1/R = 0$, while a value slightly above 0 is found experimentally. For example, Kenjo et al.[13] reported that $1/R \approx 0.4 \times 10^4$ S/m$^2$ for pure platinum electrodes. They found that the performance outside the thresholds is much lower than the values obtained within the thresholds and the model result of $1/R = 0$, even if affected by a degree of uncertainty, is satisfactory in the sense that it indicates an electrode with such characteristics as not to be very interesting from an application point of view.

In Figure 7.16, $1/R$ is plotted versus the electrode thickness for various electrode compositions and electrode thickness less than 500 μm. For $0.31 < \varphi_{el} < 0.706$, $1/R$ increases up to a maximum value by increasing the electrode thickness, and then remains almost constant for an electrode thickness less than 500 μm for these compositions. The electrical conductivity of the electron conducting phase of the electrode is very small, and thus the charges can flow in the electron conducting phase with extremely small ohmic losses. Under such conditions, it can be demonstrated that the electrochemical reaction takes place within a thin area in the vicinity of the electrode/bulk electrolyte interface. By increasing the electrode thickness, the electrode resistance does not increase since the ohmic overpotential is negligible; when $\varphi_{el}$ is between 0.294 to 0.31 there is a maximum of $1/R$ that represents a compromise between a sufficiently large active area for the electrochemical reaction and acceptable ohmic losses. In this range, ohmic losses throughout both the electronic and the ionic conducting phases are significant, i.e., $\Omega$ is greater than 0.

For two-phase electrodes, model validation[31,34] has shown good results, as shown in Figures 7.17 and 7.18, for the performance of the electrode as a function of both the electrode thickness and the electrode composition.

## 7.3.  ELECTROLYTE OVERPOTENTIALS: LIMITS OF APPLICABILITY OF OHM'S LAW

### 7.3.1. General Treatment of Electrolyte Overpotentials

The electrolyte losses are predominantly of an ohmic nature. However, the mechanism of ion conduction is quite different from one type of fuel cell to another, due to the different characteristics of their electrolytes. The ion-conduction mechanisms in electrolytes is a complex phenomenon; a simplified scheme of the basic conduction mechanisms occurring in different type of fuel cell electrolytes is shown in Table 7.2.

In addition to the fact that the conduction mechanisms are different, it is also necessary to take into account that Ohm's law must be coupled to other phenomenological laws which deal with the change of conductivity, $\sigma$, as a function of the physicochemical parameters. One of the most important effects is due to temperature, and this is usually expressed by the equation:

$$\sigma = A \exp\left(-\frac{E_a}{RT}\right) \tag{7.38}$$

Also, in the case of liquid electrolytes, the molar conductivity, $\Lambda$, depends on the electrolyte concentration in a complicated way. Under conditions of high dilution, the Kohlrausch law applies:

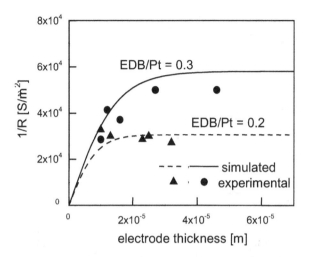

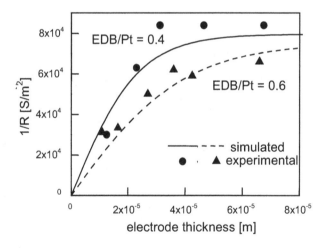

**Figure 7.17.** Comparison of simulation results with experimental data for different ratios of EDB/Pt in cathodes at 900 °C. Reprinted from Reference 31, Copyright (1998), with permission from Elsevier.

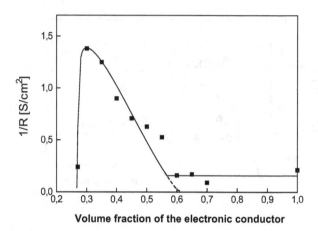

**Figure 7.18.** Simulation results (lines) compared to literature experimental data (points) for YSZ/Pt cathodes at 900 °C. Reprinted from Reference 34, Copyright (2002), with permission from Elsevier.

**TABLE 7.2**
**Mechanism of Ion Conduction in Different Types of Fuel Cells**

| Fuel cell type | Electrolyte | Ion | Mechanism of conduction |
|---|---|---|---|
| AFC | KOH | $OH^-$ | Ion migration |
| PEMFC | Nafion | $H^+$ | Proton hopping from one $SO_3^-$ site to a neighboring one (Grotthus mechanism), via $H_2O$ molecules. |
| DMFC | Nafion | $H^+$ | Proton hopping from one $SO_3^-$ site to a neighboring one (Grotthus mechanism). |
| PAFC | $H_3PO_4/H_4P_2O_7$ | $H^+$ | Proton hopping from $H_2PO_4^-$ or $H_3P_2O_7^-$ to neighboring ones via $H_3PO_4$ or $H_4P_2O_7$ (Grotthus mechanism). |
| MCFC | $Li_2CO_3 + K_2CO_3$ | $CO_3^{2-}$ | Ion migration. |
| SOFC | YSZ | $O^{2-}$ | Oxygen ion vacancies are formed in a defective lattice structure and act as ionic charge carriers. |

$$\Lambda = \Lambda^0 - K\sqrt{c} \tag{7.39}$$

where $\Lambda^0$ and $K$ are phenomenological parameters. The relationship between conductivity and molar conductivity is given by the following relationship:

$$\Lambda = \sigma^{-1}c^{-1} \tag{7.40}$$

In addition, in the two types of fuel cell types employing a Nafion electrolyte membrane (PEMFC and DMFC), the situation is even more complicated, since the water content of the membrane strongly influences the local proton conductivity, as discussed in the next subsection.

### 7.3.1.1. *Nafion Electrolyte: Water Transport Problems and Their Effect on Electrical Conductivity.* Figure 7.19 shows the water transport phenomena occurring in a PEMFC under fuel cell operating conditions.

Simulation studies of these phenomena have been carried out using different approaches. In all of these, the membrane region was considered as a homogeneous domain under steady-state conditions. The first study was reported by Bernardi and Verbrugge.[17,18] Their treatment is based on the dilute-solution theory (Nernst-Plank equation):

$$\mathbf{N}_i = -z_i \frac{F}{RT} D_i c_i \nabla \Phi - D_i \nabla c_i + c_i \mathbf{v} \tag{7.41}$$

According to this equation, a species dissolved in the fluid in the pore of a membrane is transported from one electrode to the other via the proton conducting membrane under the effects of the electrical potential gradient, diffusion, and convection. The parameter $\mathbf{N}_i$ represents the flux of the species i, $D_i$ is the diffusion, coefficient, $z_i$ and $c_i$ are the charge number and the concentration of the mobile species within the membrane, $\mathbf{v}$ is the velocity, and $\Phi$ the electrical potential. This equation is coupled to the Schögl's equation, which provides a description of the fluid dynamics in the membrane pores:

$$\mathbf{v} = \frac{k_\phi}{\mu} z_f c_f F \nabla \Phi - \frac{k_p}{\mu} \nabla p \tag{7.42}$$

where $k_\phi$ and $k_p$ are the electrokinetic permeability and the membrane hydraulic permeability, respectively, $\mu$ is the viscosity, $\nabla P$ is the pressure gradient, and $z_f$ and $c_f$ represent the charge number and the concentration of the fixed species within the membrane. The above equation states that the electric potential and pressure gradients generate convection within the pores of the ion-exchange membrane. These equations are coupled to:

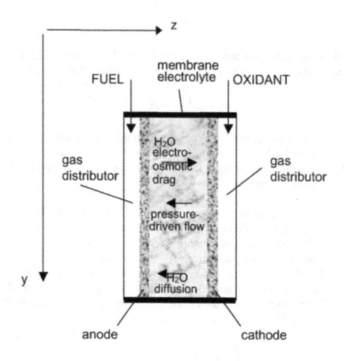

**Figure 7.19.** Water transport phenomena in a PEMFC membrane.

(a) the necessary current conservation condition:

$$\nabla \cdot \mathbf{i} = 0 \qquad (7.43)$$

(b) the steady-state material balance:

$$\nabla \cdot \mathbf{N}_i = 0 \qquad (7.44)$$

(c) the equation of continuity for incompressible fluid flow:

$$\nabla \cdot \mathbf{v} = 0 \qquad (7.45)$$

(d) the condition of electroneutrality within the membrane:

$$z_f c_f + \sum_i z_i c_i = 0 \qquad (7.46)$$

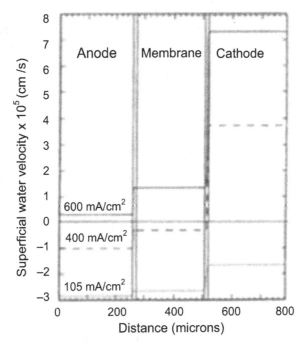

**Figure 7.20.** Model calculations of superficial water velocity throughout the fuel cell for different values of current density. See text for operating conditions. Reprinted from Reference 17, Copyright (1991) AIChE, with permission from The American Institute of Chemical Engineers. All rights reserved.

In this model, the membrane and the reactant gases are assumed to be fully humidified. Hence, the water content of the membrane and the dehydration effects are not taken into account in evaluating the membrane overpotential. Instead, the flux of water through the membrane plays a role. The results of the calculations of water velocity through the electrolyte are shown in Figure 7.20 for different values of cell current density. In the case study, the cell is considered to be at a temperature of 80 °C, with air and hydrogen as the reactants on the cathodic and anodic side, respectively; pressures are 5 atm and 3 atm on the cathodic and anodic sides, respectively. According to Figure 7.20, the pressure driven flow is the most important effect at low current densities (100 mA/cm$^2$), which causes a flux of water throughout the cell from the anodic to the cathodic side, as evidenced by the negative value of water velocity at the cathode diffuser. Under these conditions, even though water is produced by the electrochemical reaction, liquid water must be supplied on the gas side of the cathode (which can be done by humidifying the cathodic reactant gas at a temperature higher than the fuel cell temperature). The reverse situation occurs when the current density is 600 mA/cm². Under such a condition, the electro-osmotic drag is the prevailing phenomenon, which causes the electrochemically produced water to generate a build-up of hydrostatic pressure sustained by the hydrophobic nature of the cathode backing. Thus, the anode, instead of the cathode, needs to be supplied with water in this case. The intermediate case (current density 400 mA/cm²) shows the most interesting situation, where the water produced in the active electrocatalyst layer flows out from both sides of the fuel cell (about 75% of it flows out the cathode and the remaining flows out of the anode). For this type of flow pattern, it may not be necessary to supply liquid water to the cell. This situation occurs approximately between 150 and 550 mA/cm².

The effect of the water content of the membrane on the proton conductivity and on the overpotential losses of the MEA were analyzed.[26] In this case, the water content of the membrane, which varied along the thickness of the membrane, was determined by a local water balance according to the equation:

$$\chi \frac{I}{n_e F} = \zeta \lambda \frac{I}{n_e F} - \frac{\rho_{memb}}{M} D_\lambda \frac{\partial \lambda}{\partial z} \qquad (7.47)$$

where the term on the left hand side is the net flux of water through the membrane ($\chi$ is the ratio between the net water flux through the membrane and the amount of water produced by the electrochemical reaction), and the first and second terms on the right hand side represent the electro-osmotic drag (water transport from the anodic to the cathodic side due to the solvated proton flow) and the water diffusion from high to low concentration regions (usually from the cathodic to the anodic side of the membrane, see Figure 7.19), respectively. The parameter $\zeta$ is the electro-osmotic drag coefficient, $\lambda$ is the local number of water molecules per fixed $SO_3^-$ charges in the membrane, $\rho_{memb}$ is the density of the dry membrane, and $D_\lambda$ is the diffusion coefficient of water in the membrane pores, which is related to the local humidity content through the following experimental relationship:

$$D_\lambda = \left( \varphi_1 + \varphi_2 \lambda + \varphi_3 \lambda^2 + \varphi_4 \lambda^3 \right) \exp \left[ \varphi_5 \left( \frac{1}{\varphi_6} - \frac{1}{T} \right) \right] \qquad (7.48)$$

where $T$ is the temperature and $\varphi_1$-$\varphi_5$ are phenomenological coefficients. The integration of the above equations, with the appropriate boundary conditions, leads to the evaluation of the net water flux through the membrane and of the profile of water content, $\lambda$, along the thickness of the membrane. In turn, the proton conductivity of the membrane is variable along the membrane thickness, and is a function of the local hydration, $\lambda$:

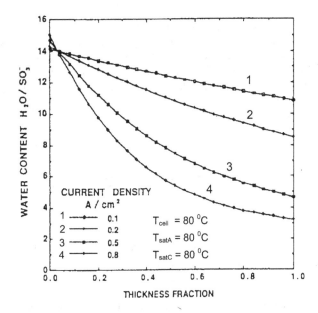

**Figure 7.21.** Simulated water profiles in a $H_2$/air PEMFC, 80°C , 3 atm,. Cathode and anode on the left and right hand side, respectively. Reprinted from Reference 26, Copyright (1991), with permission from The Electrochemical Society Inc.

$$\sigma_{memb} = (\gamma_1 \lambda - \gamma_2) \exp\left[ \gamma_3 \left( \frac{1}{\gamma_4} - \frac{1}{T} \right) \right] \qquad (7.49)$$

where $\gamma_1$ to $\gamma_4$ are phenomenological coefficients, evaluated via experimental measurements.

Results of a modeling analysis[26] (Figure 7.21) shows that the water content of the membrane is higher at the cathodic side than at the anodic side, and that severe dehydration can occur at the anodic side at high current densities, when the electro-osmotic water removal is fast and the back diffusion phenomenon is not efficient enough to re-equilibrate the humidity level. This results in a significant increase of the membrane resistance at high current densities, causing departures from linearity of the $E$ vs. $i$ plot of the fuel cell (Figure 7.22). An important remark that the author makes about their model is that, while the model does not predict the need to humidify the cathode feed stream at any appreciable current density with fuel cells based on Nafion 117 membranes, the highest performance is obtained only with well-humidified cathodic flow streams. The author suggests that the reason for this discrepancy is due, again, to significant evaporation effects (not included in the model), which causes an excessive water loss from the membrane that cannot be compensated by the water produced at the cathode. Moreover, another effect that the MEA models do not take into account is the presence of an excess of liquid water in the cathode, which could contribute to a high degree of hydration of the membrane, but at the same time may lead to high resistances for diffusion of oxygen. A solution to compromise the two conflicting effects of excess liquid water at the cathode is of fundamental importance for the optimized operation of PEMFCs.

## 7.4.   ADDITIONAL EFFICIENCY LOSSES AT SINGLE-CELL LEVEL

### 7.4.1.   Overview

When evaluating the cell potential vs. current density relationship for the single cell, one must take into account that the overall cell performance is the result of an average of the local performance, obtained point by point on the cell surface. In principle, the performance is not uniform over the cell plane due to the non-uniformity of reactant composition and temperature. Thus, for a correct evaluation of the cell potential vs. current density relationship, the local values of reactant

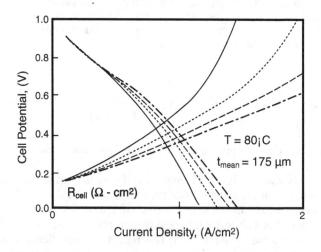

**Figure 7.22.** Simulated cell potential and membrane resistance vs. current density in a $H_2$/air PEMFC, 80 °C, 3 atm. Membrane thickness: (-) 50 μm, (-×-) 100μm, (...) 140 μm, and (---) 175 μm. Reprinted from Reference 26, Copyright (1991), with permission from The Electrochemical Society Inc.

composition, and temperatures must be evaluated, in order to correctly assess the current distribution over the cell surface. The evaluation of the cell performance is based on the solution of mass and energy balances, coupled to the local kinetics; the latter is the local current vs. voltage relationship, as described in Section 7.2.3.1. The equations of mass and energy balances will be discussed in Sections 7.4.2 and 7.4.3.

All the treatments presented in this section rely on the assumption that the reactant gases flow uni-directionally along gas channels of various geometries. This holds, for example, in planar cells where the gas channels usually consist of small grooves machined into the interconnectors (Figure 7.23). In an analogous fashion, in tubular cells (typical SOFCs), the reactant gases follow a one-directional path along suitably shaped gas chambers.

Under these conditions,[35] the heat and mass transfer in the gas channels can be treated as being one-dimensional without considerable errors. Furthermore, the streamwise diffusion terms can be neglected since the convective mass and heat fluxes are the prevailing contributions. Thus, the mass and energy balances of the gas phase can be written in a form used for chemical reactors of the plug-flow type.

## 7.4.2. Mass Balances

The streamwise molar flux gradient, $\partial n_i / \partial x$, depends on the conversion rate, $r_i$. Thus the mass balance can be written as:

$$\frac{\partial n_i}{\partial x} = r_i \tag{7.50}$$

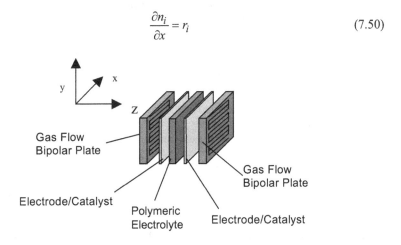

**Figure 7.23.** Components of a single cell in a PEMFC cell stack.

where i denotes the particular gas component converted. The reaction rate $r_i$ can be calculated using Faraday's law:

$$r_i = v_i \frac{I}{a \cdot n \cdot F} \tag{7.51}$$

where $n$ is the number of electrons transferred in the electrochemical reaction, $a$ is the activity coefficient of the reactant of product, and $v_i$ is the stoichiometric factor of the reacting component, and it is negative for reactants and positive for products. For example, in the case of a cell fed with hydrogen, $ne = 2$, corresponding to the following reactions in an SOFC:

Cathodic reaction:           $\frac{1}{2} O_2 + 2e^- \rightarrow O^{2-}$                          (7.52)

Anodic reaction:            $H_2 + O^{2-} \rightarrow H_2O + 2e^-$                          (7.53)

---

Overall reaction:           $H_2 + \frac{1}{2} O_2 \rightarrow H_2O$                          (7.54)

### 7.4.3.  Energy Balance in Gaseous Phase

The streamwise temperature gradients in the gas flows are assumed to depend only on the convection due to both heat and mass transfer from the channel walls to the bulk of gases, as expressed by:

$$\sum_i n_i \cdot c_{p,i} \cdot \frac{\partial T_g}{\partial x} + \sum_i c_{p,i} \cdot \frac{\partial n_i}{\partial x} \cdot \left(T_g - T_s\right) + h_g \cdot B_g \cdot \frac{1}{a} \cdot \left(T_g - T_s\right) = 0 \tag{7.55}$$

where $B_g$ is the ratio between the gas-solid heat exchange area to the cell area, $c_{p,i}$ is the specific heat at constant pressure of species $i$, $T_g$ is the temperature of the gas, $T_s$ is the temperature of the solid, and $h_g$ is the heat-transfer coefficient calculated from the Nusselt number, $Nu$:

$$h_g = Nu \cdot \frac{k}{d_h} \tag{7.56}$$

The value of the $Nu$ number generally depends on the channel geometry, the Reynolds number, $Re$, and the Prandtl number, $Pr$. For laminar flow, however, which holds for the fuel cell channels, the Nusselt number is a constant.

## 7.4.4.   Energy Balance of Solid Structure

The energy balance of the solid material describes the unsteady heat conduction in the quasi-homogeneous solid structure of the stack. The convective heat, transferred from the gas flow to the solid, and the reaction enthalpies occur as source terms, in the following equation:

$$\frac{1}{s}\sum_g h_g B_g \cdot \left(T_g - T_s\right) - \frac{1}{s}\left(\frac{\Delta H}{ne \cdot F} + V\right)I + K_x \frac{\partial^2 T_s}{\partial x^2} = 0 \qquad (7.57)$$

where $K_x$ is the effective thermal conductivity of the solid structure, accounting for the parallel heat conduction through the h parallel layers:

$$K_x = \frac{\sum_h K_h \cdot \delta_h}{\sum_h \delta_h} \qquad (7.58)$$

## 7.4.5.   Numerical Integration and Boundary Conditions

The mass and energy balance equations are solved with appropriate boundary conditions, which are (i) the inlet flow rates, compositions, and temperatures of the reactant gases, and (ii) the thermal conditions of the cell, which can be operated either adiabatically or with thermal exchange with the surroundings. The integration of the system of equations, reported above, can be performed by applying finite differencing, coupled to a successive overrelaxation method for the solution of the thermal balance.[38] For solving the equations concerned, the following steps can be followed:

(a)   specify the operating conditions in the input file and set up an initial solid temperature distribution;
(b)   calculate the current density and solve the mass balance;
(c)   solve the energy balance of the gaseous phase to determine the temperature distribution along the gas channels; and
(d)   solve the energy balance of the solid to find a new temperature field.

Steps (b) through (d) are repeated until the steady state solution is reached within a given threshold of accuracy.

## 7.4.6.   Application to PEMFCs

The main aspect of the conduction of PEMFCs is related to the management of water.[36] Indeed, when the level of humidity within the membrane is too low, membrane dehydration occurs; on the other hand, if it is too high, flooding takes

**TABLE 7.3**
**Operating Conditions for the PEMFC Experiments Carried Out by De Nora**
**(Simulation Results are Presented in Figure 7.24)[36]**

| Case | Fuel | Oxidant | P $10^5$ Pa fuel | P $10^5$ Pa oxidant | Fuel flow rate mol s$^{-1}$ | Oxidant flow rate mol s$^{-1}$ |
|------|------|---------|-----------------|---------------------|----------------------------|-------------------------------|
| 1 | $H_2$ | $O_2$ | 1.35 | 1.55 | dead end | dead end |
| 2 | $H_2$ | $O_2$ | 3.5 | 4 | dead end | dead end |
| 3 | $H_2$ | Air | 1.35 | 1.55 | dead end | $3 \times 10^{-3}$ |
| 4 | $H_2$ | Air | 3.5 | 4 | dead end | $4.5 \times 10^{-3}$ |

place, which is enhanced by the production of water from the electrochemical reaction. A further critical aspect in the conduction of PEMFCs is the distribution of temperature on the cell plane. Local temperatures that peak above 130 °C (glass transition temperature of Nafion) have to be carefully avoided, as they irreversibly damage the membrane locally.

A simulation study has been made to gain a better understanding of the problems of (i) temperature peaks, (ii) membrane dehydration, and (iii) flooding. For this purpose, the simulation framework, discussed in the introductory chapters has been applied, and it has been modified in order to include a specific electrode

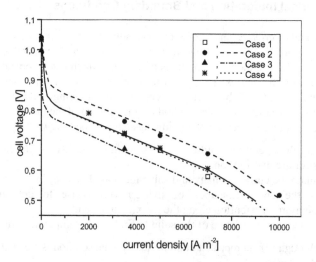

**Figure 7.24.** Comparison between simulation results (lines) and experimental data (symbols). Operating conditions reported in Table 7.3. Reprinted from Reference 36, Copyright (2001), with permission from Elsevier

kinetic model for the simulation of the membrane and electrode overpotentials as a function of the hydration level. Furthermore, specific PEMFC phenomena such as water condensation in the gas channels have been accounted for in the energy, mass and momentum balances.

The first step in the evaluation of the model results is to compare the simulation results with the experimental data provided by De Nora (presently Nuvera), Italy. The experimental cell had an area of 15 cm × 15 cm, and had a co-flow geometry; the reactant gases were fed at the top of the cell and exited from the bottom. The experimental data were taken under the operating conditions reported in Table 7.3; moreover, the cell was kept at a constant temperature (70 °C) using a thermostatic bath, and the temperature of both the anodic and cathodic feeding gases was maintained at the same value as well. According to Table 7.3, the reactant gases were fed and were dead-ended by closing the outlet duct, thus the reactant flow rate was stoichiometric. In addition, reactant gases were not humidified. For normal operating conditions, the reactants were humidified at 70 °C. Figure 7.24 shows the experimental and simulated cell potential-current density plots for the PEMFC described above. The model shows a good agreement with the experimental data for all the operating conditions applied.

The performance of a PEMFC has also been analyzed on the basis of the validated model. In this case, the temperature of the cell is not uniform as in the case discussed above, but it is determined by the energy transport over the cell plane. The operating conditions are reported in Table 7.4, where the reactant compositions correspond to full humidification at 70 °C, and the reactant flow rates correspond to a utilization factor of 90%. The results of the simulation are shown in Figure 7.25 and reported in Table 7.4. Figure 7.25a shows the distribution of the temperature on the fuel-cell plane, where a decrease along the y coordinate is determined by the direction of the flux of the cooling water. The temperatures of the gases are assumed to be the same as the cell temperature. In spite of the small dimensions of the cell and of the high coolant flowrate, the temperature gradient is quite high (Figure 7.25a). The possibility of hot spots appearing on the cell plane is demonstrated in Figure 7.25a. Analogous simulations, assuming a flow rate of cooling water of 0.3 cm$^3$/s (instead of 0.8 cm$^3$/s as in the case study), show that the maximum temperature rises as high as 110 °C. This causes severe evaporation problems and is quite close to the threshold imposed by the glass transition temperature of Nafion (130 °C).

Also, in the case study, high temperatures are reached on the edge of the cell where the reactant gases are fed in. This is the reason for the low values of the relative humidity of the gases at that point (the gases being humidified at a lower temperature, i.e., 70 °C) (Figure 7.25b and 7.25c.) Thus, membrane dehydration occurs in that region (Figure 7.25d) and, as a consequence, the membrane resistance is high (about 0.5 Ω cm$^2$). On the other hand, in the region near the edge where the reactants exit the cell, the cell temperature is low and the fuel streams tend to become saturated with water (Figure 7.25b and 7.25c). This is particularly faster at the cathodic side where the electro-osmotic flux across the membrane and the water

**TABLE 7.4**
**Operating Conditions and Simulation Results for the Simulated PEMFC (see Figure 7.25)[36]**

| Operating Conditions | |
|---|---|
| Fuel composition | 77% $H_2$, 23% $H_2O$ |
| Oxidant composition | 80% $O_2$, 20% $H_2O$ |
| Fuel pressure | $1.35 \times 10^5$ Pa |
| Oxidant pressure | $1.55 \times 10^5$ Pa |
| Cooling-water pressure | $3 \times 10^5$ Pa |
| Fuel flow rate | $1.2 \times 10^{-3}$ mol/s |
| Oxidant flow rate | $5.6 \times 10^{-4}$ mol/s |
| Cooling-water flow rate | $0.8$ cm$^3$/s |
| Fuel, oxidant, and cooling water inlet temperature | 343 K |
| Ambient temperature | 300 K |
| Overall electrical current | 160 A |
| Simulation results | |
| Cell voltage | 0.67 V |
| Exhaust fuel temperature | 349 K |
| Exhaust oxidant temperature | 349 K |
| Cooling-water outlet temperature | 358 K |
| Flow rate of liquid water in the exhaust fuel | 0 |
| Flow rate of liquid water in the exhaust oxidant | $2.1 \times 10^{-3}$ cm$^3$/s |
| Cell voltage | 0.67 V |
| Exhaust fuel temperature | 349 K |
| Exhaust oxidant temperature | 349 K |
| Cooling water outlet temperature | 358 K |
| Flow rate of liquid water in the exhaust fuel | 0 |

production from the electrochemical reaction contribute to the increase of the partial pressure of water vapor.

Thus, even if the performance of the fuel cell appears good (cell potential of 0.67 V at a current density of 700 mA/cm$^2$), the results of the simulation show that two opposing problems occur on the cell plane: membrane dehydration and water condensation within the cathodic gaseous stream. Both these effects are expected to be even more critical in cells of larger area, operated at higher current densities. The problem of water condensation is of particular concern because it leads to the formation of water droplets, causing blockages, which can lead to variable performances of the fuel cell. This behaviour has often been found in the experimental operation of PEMFCs, and it is one of the critical aspects of the conduction of this type of cell. Thus, in such a case the modeling study does not take into account a detailed investigation of this serious malfunctioning, but it also offers a starting point for further optimization studies.

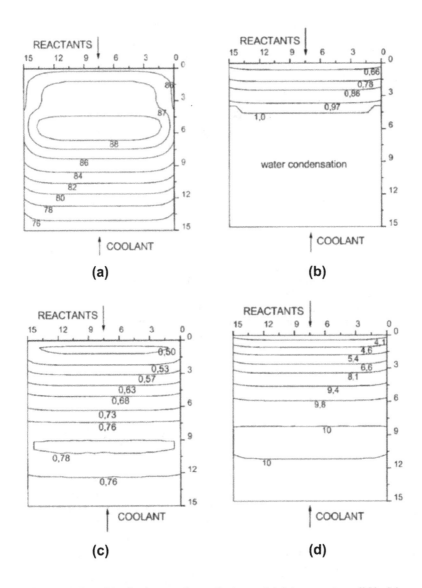

**Figure 7.25.** Distribution on the cell plane of (a) temperature (°C), (b) cathodic-relative humidity, (c) anodic-relative humidity, and (d) membrane hydration. Operating conditions and other simulation results are shown in Table 7.4. Reprinted from Reference 36, Copyright (2001), with permission from Elsevier.

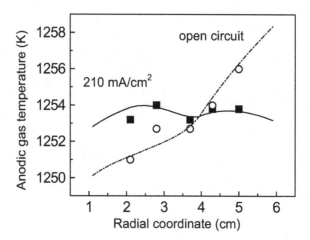

**Figure 7.26.** Experimental data (symbols) and simulation results (lines) for the anodic gas temperature along the cell radius. Operating conditions: shell temperature 1193 K, hydrogen 2.5 g/h, air 300 g/h. Reprinted from Reference 37, Copyright (1998), with permission from The Electrochemical Society, Inc.

## 7.4.7. Application to SOFC

Simulation results for the case of a Sulzer Hexis fuel cell are reported in this section.[37] In particular, results for fuel temperature, being almost identical to the SOFC temperature, are reported in Figure 7.26, together with single-cell experimental data obtained from a 5-cell stack operated in an enclosed container, maintained at a uniform and controlled temperature. The geometrical parameters and the materials of the experimental and the simulated cells are presented in Table 7.5. The temperature of the shell as well as the gas feeding flowrates are reported in the figure captions. Figure 7.26 shows the distribution of fuel temperature along the cell radius. It can be observed that, under open circuit conditions, an increase in temperature occurs at the outer rim of the cell, while under electrical load there is a decrease in this region. Moreover, the inner cell temperature increases with increasing current density. In fact, at low current densities, fuel is mostly burnt by a post-combustion reaction occurring at the cell outer rim, while at a higher current density most of the hydrogen is consumed by the electrochemical reaction in the inner part of the cell causing heat dissipation due to irreversibilities of the electrochemical reaction. The results show that temperature

## TABLE 7.5
### Design Parameters for the Experimental and the Simulated HEXIS SOFC

| Design characteristics | |
|---|---|
| Inner and outer cell radii, m | 0.01, 0.06 |
| Inner active area radius, m | 0.02 |
| Outer active area radius, m | 0.06 |
| Anode thickness, m | $10^{-4}$ |
| Cathode thickness, m | $5 \times 10^{-5}$ |
| Electrolyte thickness, m | $1.8 \times 10^{-4}$ |
| Interconnector thickness, m | $4.8 \times 10^{-3}$ |
| Cell component materials | |
| Interconnector Plansee alloy | $Cr_5FeY_2O_3$ |
| Electrolyte | $(ZrO_2)_{0.92} (Y_2O_3)_{0.08}$ |
| Anode | $NiO/(ZrO_2)_{0.92} (Y_2O_3)_{0.08}$ |
| Cathode | $La_{1-x}Sr_x MnO_3$ |

distribution along the cell radius is almost even (temperature gradient $\nabla T < 2$ K/cm), demonstrating the benefit of the SOFC with the integrated air pre-heater concept. This result is also confirmed by a comparison of experimental and simulation data conducted under different operating conditions; in all cases the average error of the calculated results is of the same order of magnitude as the experimental one.

The comparison of the experimental data and model predictions for the cell potential vs. current density plots shows good agreement as well. Some simulation results are reported in Figure 7.27 together with the experimental data. Characteristic curves are evaluated by keeping constant feeding flowrates, so that different points along the plots correspond to different utilization factors (the utilization factor being defined as the ratio between the flowrate consumed by the electrochemical reaction and the flowrate initially fed to the cell). Also in this case, a close agreement is found between numerical and experimental results (average error ~2%), validating the reliability of the simulation model for the prediction of the electrochemical performance of the cell. Other results, obtained under different operating conditions and not reported here, support this conclusion.

After ascertaining the reliability of the simulation model, predictions for the temperature distribution obtained under operating conditions of high fuel utilization have been made. These conditions are very interesting from an application point of view, since no recycling of the unutilized fuel is necessary, but are potentially serious from the point of view of temperature gradients arising within the cell. Figure 7.28 shows that temperature gradients in the solid electrolyte are still acceptable, while traditional cells would present temperature gradients, one order of magnitude higher when operated under the same conditions.[38] In addition, it has also been reported that an optimization of the thickness of the integrated heat

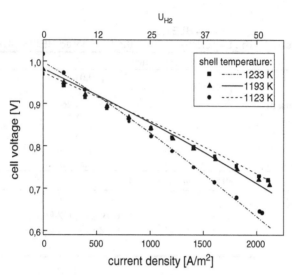

**Figure 7.27.** Experimental (symbols) and simulated (lines) characteristic curves. Operating conditions: hydrogen 1.5 g/h, air 250 g/h. Varied parameter: shell temperature. Reprinted from Reference 37, Copyright (1998), with permission from The Electrochemical Society, Inc.

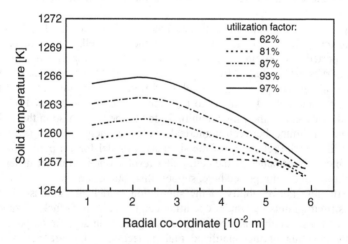

**Figure 7.28.** Simulated results of temperature distribution along the radial coordinate at high-fuel utilization. Operating conditions: hydrogen 1.5 g/h, air 250 g/h, shell temperature 1233 K. Reprinted from Reference 37, Copyright (1998), with permission from The Electrochemical Society, Inc.

exchanger would allow a further improvement of temperature distribution in the solid structure.[38] Such considerations have led to the conclusion that solid-oxide fuel cells, with an integrated air pre-heater, offer an interesting option of operating at high-hydrogen utilization factors as they do not present serious problems of mechanical stresses in the ceramic materials.

## 7.5.    FURTHER EFFICIENCY LOSSES IN CELL STACKS

### 7.5.1    Overview

In the experimental studies, differences have been found between the performance of individual or stacked fuel cells. In some cases, the causes are specific for the particular type of fuel cell under consideration. For example, in MCFCs, problems related to electrolyte migration from one cell to another in the stack occur, while in PEMFCs heat and water transport cause significantly different effects depending on the cell area and number of cells in stacks in the fuel cell system. In this chapter, we will only focus on one effect, common to all the types of planar fuel cell stacks, i.e., the non-homogeneous distribution of reactant-feed flow rate among stacked cells which is due to fluid-dynamic effects (Figure 7.29). Such problems can cause significant consequences in planar stacks—as for instance the cell with the smallest flow rate determines the current. In particular, it is important to forecast and avoid working conditions, which depend on flow geometry, stack design parameters, and overall inlet flow rates and where some of the cells attain their feed rates from the outlet manifold of the stack and discharge the exhaust gases into the inlet manifold of the stack.

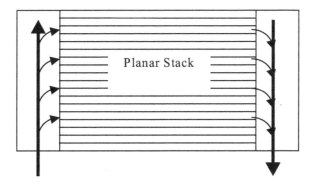

**Figure 7.29**. A sketch of the flow pattern in the bipolar plate of a planar fuel cell stack, inlet and outlet manifolds.

## 7.5.2. Fluid Dynamics of Fuel-Cell Stacks

The theoretical treatment of this phenomenon[39] is based on momentum equations, which correlate the flow rate of a channel to the pressure drop occurring along the stream length due to gravitational, viscous, and inertial effects.

Figure 7.30 represents a schematic for a duct with lateral inlet- and outlet-flow rates. The projection of the momentum balance in the z-direction, applied to the control volume formed by a portion of the pipe between sections 1 and 2, has the following form:

$$N_2 \rho S v_2^2 + \rho Q_{out} v_{out} + p_2 S + \rho g z_2 S = N_1 \rho S v_1^2 + \rho Q_{in} v_{in} + p_1 S + \rho g z_1 S + \Sigma F \quad (7.59)$$

where it is assumed that the density of the gas and the area of the cross-section is a constant; $\Sigma F$ is the sum of the components of the external forces exerted by the duct walls on the fluid; $N$ is a coefficient which accounts for the non-uniform velocity distribution across the duct cross-section; the terms $\rho Q_{out} v_{out}$ and $\rho Q_{in} v_{in}$ represent the momentum components of lateral effluxes and influxes, respectively, $\rho$ refers to density, $S$ to cross-sectional area of manifolds, $Q$ to volumetric rate of concentrated lateral flows, $v$ to velocity, $g$ to the gravitational constant, and $z$ to the vertical coordinate. It has been suggested on the basis of experimental tests that lateral inlet flow rates have velocities exactly perpendicular to the manifold and that their

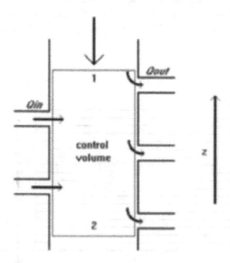

**Figure 7.30.** Duct with inlet and outlet flow rates, showing a control volume for momentum balance.

contribution to the momentum balance equation is zero.[39] On the contrary, in the same publication by considering small lateral effluxes, the authors assume that the velocity at the outlet of the control volume has a component along the main duct direction, which is equal to the velocity of the main flow. Thus, considering outflows and inflows as continuously distributed and expressing the $\Sigma F$ term—which represents the friction forces—as a function of the friction factor $\lambda$, the equation was written in the following differential form:

$$\frac{dp}{dz} = -2N\rho v\frac{dv}{dz} - N\frac{\rho}{S}vq_{out} - \rho g - \frac{\lambda}{R_h}\frac{1}{2}\rho v^2 \qquad (7.60)$$

where $q$ refers to the volumetric rate of distributed lateral flows.

The friction factor can be expressed by the formulae for the established flow through ducts. The Reynolds number varies over a wide range from the laminar to the turbulent flow regime. However, the problems associated with the transition phenomena have not been taken into account. The following equations have been applied:

- laminar flow: $\qquad \lambda = 8K'\left(\frac{R_h}{s}\right)^2 \frac{1}{Re} \qquad (7.61)$

- turbulent flow: $\qquad \lambda = 0.0868 \cdot Re_{mod}^{-\frac{1}{4}} \qquad (7.62)$

where $Re = 4\rho vR_h/\mu$; $Re_{mod} = f_1Re$ and $R_h$ is the hydraulic radius defined as the ratio of the cross section of the pipe to its wetted perimeter; $f_1$, as well as $K'$, depends on the geometrical dimension of the duct.

Equation (7.60) has been used to model the feed and exhaust manifolds of the stack, while the modified form of the same equation has been applied to the flow channels of each cell. Indeed, in the cell channels the gravitational term is absent, and it has been demonstrated that inertial effects are negligibly low compared to friction forces.[39] Therefore, the pressure drop in the rectangular channels under laminar flow conditions can be determined by the well-known relationship:[40]

$$\Delta p_{friction} = K\left(\frac{\mu l}{ab^3}\right)Q_{channel} \qquad (7.63)$$

where $K$ is a shape parameter dependent on the ratio $a/b$ of both sides of the rectangular cross section of channels.

### 7.5.3. Impact of Fluid-Dynamic Effects on Stack Performance

The equations in Section 7.5.2, when integrated, allow an evaluation of the distribution of pressure and flow rate along the manifolds of the stack.[39,41] The theoretical results are in agreement with experimental data, measured in a simulated fuel cell. Such an experimental setup was composed of a series of channeled plexiglass plates, superimposed to each other and fed through manifolds, similar to the structure illustrated in Figure 7.29. No electrochemical current was generated in such an experimental set-up; instead the fluid dynamics of the manifolds and the channels were able to mimic that in a real planar fuel-cell stack. The results, presented in Figure 7.31 and Figure 7.32, show very good agreement between theoretical calculations and experimental data. Figure 7.31 shows data of pressure distribution along the manifolds, where the cell number zero denoted the top of the stack, and the cell number 100 indicated the bottom; the experimental data are indicated by the points while the continuous lines represent the simulation. The decrease of pressure along the inlet manifold is caused by the deceleration of the flow and demonstrates that the inertial forces play a predominant role. The results for flow-rate distribution (Figure 7.32), for which the experimental data were obtained using a hot wire anemometry technique, illustrate the good agreement between the experimental data and theoretical results.

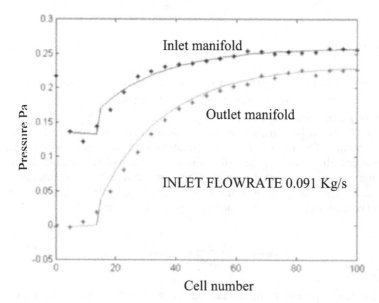

**Figure 7.31.** Distribution of pressure: (*) experimental data; (—) simulation results. Reprinted from Reference 39, Copyright (1994), with permission from Elsevier.

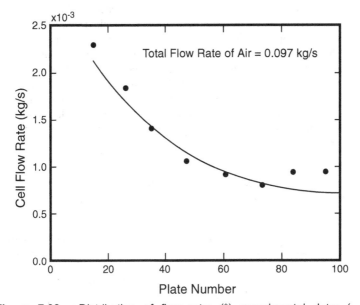

**Figure 7.32.**   Distribution of flow rate: (*) experimental data; (—) simulation results. Reprinted from Reference 39, Copyright (1994), with permission from Elsevier.

The results, presented above, are useful to predict the behavior of fuel-cell stacks and monolithic reactors by the integration of the fluid-dynamic equations as described in Section 7.5.2. For example, for an SOFC stack, consisting of 120 cells, this type of calculation has demonstrated that the flow distribution among the different cells in the stack varies by of 7% to 20% from the average value for the anodic and the cathodic side, respectively.[41] It is possible to avoid this problem by means of a change of the flow geometry of the cell, and this has been demonstrated on the basis of simulation results. Other options to avoid the problems of non-uniform distribution of reactants among the different cells of a stack rely on a proper choice of the dimensions of the manifolds. In such a case, simulation studies have proven to be a useful optimization tool.[41]

## 7.6.   MODELING OF FUEL-CELL POWER PLANTS

### 7.6.1.   Overview

The design of fuel-cell power plants depends on the type of fuel cell and on the application. However, in all the cases, the system simulation is based on a modular design, in the sense that each component of the plant is modeled separately and then the modules are connected to each other. The model for each module is based on relevant mass and thermal balances, and the results are checked with experimental performance data. Appropriate numerical tools are used for coupling the different modules to each other and to carry out the overall numerical calculation, using Matlab Simulink,[42] Aspen,[43] and IpsePRO[44]. In this section, we deal with the type of model that is developed for the plant simulation with particular regard to the stack model. Indeed, it would be appropriate to include, in the plant, a detailed stack model similar to the one described in Section 7.5. However, in order to facilitate the convergence of the plant simulation tool, simplified stack models are often used. Two types of simplified models have been proposed:

- a model based on semi-empirical equations,[43,44] and
- a model where the microscopic balance equations reported in Sections 7.5 are extended to macroscopic balances between the inlet and the outlet of the fuel cell module.[42]

The first approach has the drawback of being not fully reliable under operating conditions significantly different from the operating conditions used to set the adjustable parameters of the semi-empirical equations. The second one is reliable, and gives results very close to those obtained from the detailed model over a reasonably wide range of operating conditions.

### 7.6.2. Example: Simulation of a 300 kW$_e$ SOFC/Gas Turbine Hybrid System

The SOFCs integrated with gas turbines of small size (less than 1 MW) are attracting wide interest[45-48] as these systems are able to simultaneously solve some of the key problems of small gas turbines (low efficiency and NO$_x$ emissions due to the combustor) and SOFCs (predicted cost is $1000-1500/kW). The pressurized SOFC module substitutes the combustor in the regenerated turbine plant, and the clean effluent has a temperature of about 900 °C, which well matches the requirements of the inlet temperature of first generation microturbines. Plant efficiencies are predicted to be close to 60%, and thus the cost of the resulting energy should be lower than that of a gas turbine plant of the same size (taking into account both capital and variable costs). Integrated SOFC/microturbines (Hybrid Systems) are at an early stage of testing at the moment,[45] and techno-economic optimization is an issue currently being addressed by many developers.

Figure 7.33 shows the simplified layout of a hybrid system, which has been obtained by integration of the SOFC to the combustor of a typical Micro-Gas Turbine (MGT) plant. According to the typical features of combined SOFC/MGT plants,[14] the MGT has a size of about 50 kW$_e$, and the SOFC stack has an overall active area of 95 m$^2$, which produces about 240 kW of power at the design point (the cell specific power is about 2.5 kW/m$^2$). The SOFC includes the reformer, the SOFC stack, and the mixer. The mixer is an ejector, where the fuel and the recycled

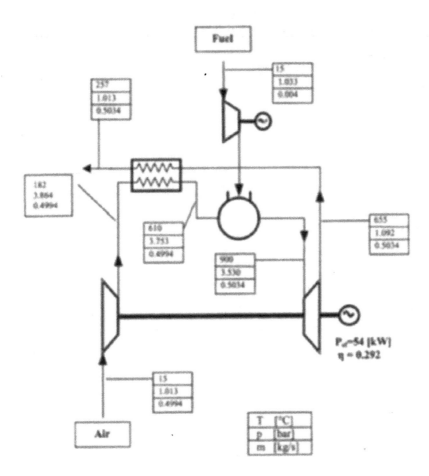

**Figure 7.33.** Schematic of an SOFC/Micro-Gas Turbine hybrid plant. Reprinted from Reference 42, Copyright (2002), with permission from Elsevier.

flow rate mix prior to entering the reformer. The reformer is a typical catalytic reactor, where the following reactions take place:

Reforming:              $CH_4 + H_2O \leftrightarrow CO + 3\,H_2$                    (7.64)

Shifting:               $CO + H_2O \leftrightarrow CO_2 + H_2$                     (7.65)

The sensible heat of the reactants provides the energy necessary for the reforming reaction to occur. The reactants are methane and water, which are

## TABLE 7.6
### Basic Equations for the Modeling of the SOFC Group

| | Mixer | |
|---|---|---|

| Mass balance | $F_i^{out} = \sum_j F_{i,j}^{in}$ | (7.66) |

| Energy balance | $T^{out} = \dfrac{\sum_i C_{p,i} \sum_j F_{i,j} T_j^{in}}{\sum_i F_i^{out} C_{p,i}}$ | (7.67) |

| | Reformer | |
|---|---|---|

| Equilibrium chemical reactions | $K_{p,ref} = \dfrac{P_{H_2}^3 P_{CO}}{P_{CH_4} P_{H_2O}}$ | reforming | (7.68) |

| | $K_{p,shift} = \dfrac{P_{CO_2} P_{H_2}}{P_{CO} P_{H_2O}}$ | shift conversion | (7.69) |

| Mass balance | $F_i^{out} = F_i^{in} + \sum_k \nu_{i,k} R_k$ | (7.70) |

| Energy balance | $T^{out} = \dfrac{\sum_i F_i^{in} C_{p,i} T^{in} + \sum_k R_k (-\Delta H_k)}{\sum_i F_i^{out} C_{p,i}}$ | (7.71) |

| | Fuel cell stack | |
|---|---|---|

| Equilibrium chemical reactions | $K_{p,ref} = \dfrac{P_{H_2}^3 P_{CO}}{P_{CH_4} P_{H_2O}}$ | reforming | (7.72) |

| | $K_{p,shift} = \dfrac{P_{CO_2} P_{H_2}}{P_{CO} P_{H_2O}}$ | shift conversion | (7.73) |

| Electrochemical reaction | $V_{oc} = \dfrac{-\Delta G}{2F} = \dfrac{-\Delta G^0}{2F} + \dfrac{R_g T}{2F} \ln \dfrac{P_{H_2} \cdot P_{O_2}^{1/2}}{P_{H_2O}}$ | (7.74) |

| | $\rho_{el} = A_{el} \exp\left(-\dfrac{B_{el}}{T}\right)$ | (7.75) |

| | $\rho_a = A_a \exp\left(-\dfrac{B_a}{T}\right)$ | (7.76) |

| | $\rho_c = A_c \exp\left(-\dfrac{B_c}{T}\right)$ | (7.77) |

**TABLE 7.6 (Continuation)**

Fuel cell stack

| | | |
|---|---|---|
| Electrochemical reaction | $\dfrac{1}{\Omega_{pol,a}} = D_1 \dfrac{2F}{R_g T} \left( \dfrac{P_{H2}}{p^o} \right)^{m_1} e^{-E_a/RT}$ | (7.78) |
| | $\dfrac{1}{\Omega_{pol,c}} = D_2 \dfrac{4F}{R_g T} \left( \dfrac{P_{O2}}{p^o} \right)^{m_2} e^{-E_c/RT}$ | (7.79) |
| | $\Omega_{tot} = \sum \Omega_{ohm} + \sum \Omega_{pol}$ | (7.80) |
| | $V_p = \Omega_{tot} I$ | (7.81) |
| | $V = V_{oc} - V_p$ | (7.82) |
| | $W_{el} = VI$ | (7.83) |
| Mass balance | $F_i^{out} = F_i^{in} + \sum_k v_{i,k} R_k$ | (7.84) |
| Energy balance | $T^{out} = \dfrac{\sum_i F_i^{in} C_{p,i} T^{in} + \sum_k R_k(-\Delta H_k) - W_{el}}{\sum_i F_i^{out} C_{p,i}}$ | (7.85) |

supplied to the reformer through the fuel and the recycling of the water-rich anodic-exhaust of the SOFC. The recycled flow rate is usually chosen in such a way to meet the condition that the steam-to-carbon ratio is about 2.4. This condition ensures that the problem of carbon deposition is avoided in the SOFC stack. As far as the SOFC is concerned, no specific details of the geometry are taken into consideration in the simulation. After the recycled flow rate has been drawn from the anodic side, the SOFC exhausts mix and burn in the outlet part of the SOFC stack so that their temperature rises to the level required by the downstream MGT group.

The equations used for the simulation studies of the various elements of the SOFC group are presented in Table 7.6. The equations of the mixer simply express the conservation of mass and energy through the mixing process. The equations of mass and energy balance for the reformer (Eqs. 7.70, 7.71) include, in addition to the input and output contributions, the generation term due to the chemical reactions. The reaction rates, $R_k$ (where $k = 1$ for the reforming reaction, and $k = 2$ for the shift reaction) are evaluated through Eqs. (7.68) and (7.69), which express an equilibrium condition. The assumption of thermodynamic equilibrium at an operating temperature of about 550 °C is fully justified in the presence of a suitable catalyst in the reforming reactor. In the SOFC stack, the calculation of the electrical current-voltage relationship is made through the evaluation of the thermodynamic voltage (Eq. 7.74), which in SOFCs is equal to the voltage of the stack under open-circuit conditions. The current-voltage behavior of the stack is evaluated by subtracting the overall voltage losses from the thermodynamic potential for each value of the electrical current. Ohmic losses (Eqs. 7.75-7.77) and activation losses (Eqs. 7.78 and 7.79) have been considered, since concentration overpotentials are usually negligible under the operating conditions of an SOFC.

The balance equations of the SOFC stack account for the process of conversion of chemical energy to electrical energy. Indeed, in this study, the balance equations are written as macroscopic balances, in the form of finite equations. These equations simply express a balance between inlet and outlet flows of mass and energy in each component of the group. In the mass balance (Eq. 7.84), both chemical end electrochemical reactions have been included, and the $k$ index varies between 1 and 3, where $k = 1$ is for the reforming, $k = 2$ is for the shift-conversion reaction, and $k = 3$ is for the electrochemical reaction. In this case, the shift-conversion and reforming reaction rates are also evaluated from the constants of thermodynamic equilibrium (Eqs. 7.72 and 7.74); the electrode kinetics $R_3$ is correlated to the electrical current $I$ supplied by the stack:

$$R_3 = \frac{I}{2F} \tag{7.66}$$

### TABLE 7.7
### Parameters Used in the Equations Reported in Table 7.6

| Parameter | Parameter value |
|---|---|
| Coefficient in Eq. (7.75) | $A_{el} = 0.00294\,[\text{ohm}^{-1}]$ |
| Coefficient in Eq. (7.75) | $B_{el} = -10350\,[\text{K}]$ |
| Coefficient in Eq. (7.76) | $A_a = 0.00298\,[\text{ohm}^{-1}]$ |
| Coefficient in Eq. (7.76) | $B_a = 1392\,[\text{K}]$ |
| Coefficient in Eq. (7.77) | $A_c = 0.008114\,[\text{ohm}^{-1}]$ |
| Coefficient in Eq. (7.77) | $B_c = -600\,[\text{K}]$ |
| Coefficient in Eq. (7.78) | $D_1 = 2.13*10^8\,[\text{A}/\text{m}^2]$ |
| Coefficient in Eq. (7.79) | $D_2 = 1.49*10^8\,[\text{A}/\text{m}^2]$ |
| Anodic activation energy | $E_a = 110000\,[\text{J Kmol}^{-1}]$ |
| Cathodic activation energy | $E_c = 110000\,[\text{J Kmol}^{-1}]$ |
| Contact resistance | $R_c = 0.15\,[\text{ohm}]$ |
| Cathode thickness | $T_c = 0.035\,[\text{cm}]$ |
| Electrolyte thickness | $T_c = 0.017\,[\text{cm}]$ |
| Anode thickness | $T_c = 0.030\,[\text{cm}]$ |

The energy balance (Eq. 7.85) includes the electrical power, $W_{el,}$ and the enthalpy changes of the chemical and electrochemical reactions, and allows the evaluation of the outlet temperature of the gases, which is equal to the average temperature of the stack. As the latter temperature is an input value for the calculation of the stack current-potential relationship ($T$ appears in Eqs. 7.74-7.79 as discussed above), the system of Eqs. (7.66)-(7.85) was solved through a numerical method, starting from a first-attempt value for the stack temperature and calculating a new temperature at each iteration until convergence was reached. The values of all the parameters appearing in the equations have been taken from the literature[49] and are stated in Table 7.7.

The simulation of the compressor module is usually based on the experimental plots (efficiency and pressure ratio versus non-dimensional flow rate).[50] Air inlet pressure and temperature, rotational speed, and a first guess pressure ratio are the input data for the compressor module. The value of the operating pressure ratio value of MGT is arrived at using a matching technique between the compressor and the downstream components.[51] In the combustion chamber the complete combustion reaction hypothesis is usually applied, using the outlet compressor data and the fuel mass flow rate as input. The simulation of the expander is based on the turbine non-dimensional plots,[52] and uses the data at the combustion chamber outlet and the regenerator downstream pressure losses as inputs. Some of the results obtained using this type of approach are shown in Figure 7.33, where the simulation results for temperature, pressure, and mass flow are reported in each section of the plant under design-point operating conditions. This type of approach can be applied also in order to simulate off-design operating conditions[42] and in optimization studies.

### *Suggested Reading:*

1.   J. T. Pukrushpan, A. G. Stefanopoulou, and H. Peng, *Control of Fuel Cell Power Systems: Principles, Modeling, Analysis and Feedback Design,(Advances in Industrial Control).* Springer, Berlin, 2004.

2.   R. E. White, M. W. Verbrugge and J. F. Stockel (Eds.), *Proceedings of the Symposium on Modeling of Batteries and Fuel Cells*, The Electrochemical Society, Inc., Pennington, NJ, 1991.

3.   J. Larminie, A. Dicks, *Fuel Cell Systems Explained*, John Wiley & Sons, England, 2003.

### Cited References

1. L. G. Austin, in *Handbook of Fuel Cell Technology*, edited by C. Berger, (Prentice Hall, Inc., Englewood Cliffs, New Jersey, USA, 1968), Chapter 1.

2. S. Srinivasan, H. D. Hurwitz and J. O'M. Bockris *J. Chem. Phys.* **46**, 3108 (1967).

3. F. G. Will, *J. Electrochem. Soc.* **110** 152 (1963).

4. S. Srinivasan and H. D. Hurwitz , *Electrochim Acta* **12**, 495 (1967).

5. J. O'M. Bockris and B. D. Cahan, *J. Chem. Phys.* **50**, 1307 (1969).
6. R. Kh. Burstein, V. S. Markin, A. G. Pshenichnikov, Yu. A. Chizmadzhev, and Yu. G. Chirkov, *Electrochim Acta* **9**, 173 (1964).
7. J. Giner and C. Hunter, *J. Electrochem. Soc.* **116**, 1124 (1969).
8. D. T. Wasan, T. Schmidt, and B. S. Baker, *Chem. Eng. Progress Symposium Series*, **63**, 77 (1967).
9. P. Costa and E. Arato, *Chem. Biochem. Eng.* **4**, 9 (1990).
10. S. C. Yang, M. B. Cutlip, and P. Stonehart, *Electrochim. Acta*, **35**, 869 (1990).
11. M. B. Cutlip, S. C.Yang, and P. Stonehart, *Electrochim. Acta* **36**, 547 (1991).
12. S. Sunde, *J. Electrochem. Soc.* **143**, 1930 (1996).
13. T. Kenjo, S. Osawa and K. Fujiwaka, *J. Electrochem. Soc.* **138**, 349 (1991).
14. M. Viitanen and M. J. Lampinen, *J. Power Sources* **32**, 207 (1990).
15. S. Sunde, *J. Electrochem. Soc.* **143**, 1123 (1996).
16. J. O'M. Bockris and S. Srinivasan, *Fuel Cells: Their Electrochemistry*, (McGraw Hill Publishing Company, New York, NY, 1969).
17. D. M. Bernardi and M. W. Verbugge, *AIChE J.* **37**, 1151 (1991).
18. D. M. Bernardi and M. W. Verbrugge, *J. Electrochem. Soc.* **139**, 2477 (1992).
19. T. E. Springer, M. S. Wilson, and S. Gottesfeld, *J. Electrochem. Soc.* **140**, 3513 (1993).
20. S. Gottesfeld and T. A. Zawodzinski, in *Advances in Electrochemical Science and Engineering*, edited by R. C Alkire, H. Gerisher, D. M. Kolb and C. W. Tobias (Wiley-VCH, Weinheim, Germany, 1997), Vol. 5, p. 195.
21. S. J. Ridge and R. E. White, *J. Electrochem. Soc.* **136**, 1902 (1989).
22. R. P. Iczkowski and M. B. Cutlip, *J. Electrochem. Soc.* **137**, 1433 (1980).
23. F. Gloaguen and R. Durand, *J. Appl. Electrochem.* **27**, 1029 (1997).
24. J. -T. Wang and R. F. Savinell, *Electrochim. Acta* **37**, 2737 (1992).
25. Y. W. Rho, S. Srinivasan, and Y. T. Kho, *J. Electrochem. Soc.* **141**, 2089 (1994).
26. T. E. Springer, T. A. Zawodzinski, and S. Gottesfeld, *J. Electrochem. Soc.* **138**, 2334 (1991).
27. D. Stauffer, *Introduction to Percolation Theory* (Taylor & Francis, London, 1985).
28. M. Sahimi, *Applications of Percolation Theory* (Taylor & Francis, London, 1994).
29. D. W. Dees, T. D. Claar, T. E. Easler, D. C. Fee, and F. C. Mrazek, *J. Electrochem. Soc.* **134**, 2141, (1987).
30. T. Kawada, N. Sakai, H. Yokokawa, M. Dokiya, M. Mori, and T. Iwata, *J. Electrochem. Soc.* **137**, 3042, (1990).
31. P. Costamagna, P. Costa, and V. Antonucci, *Electrochim. Acta* **43**, 375 (1998).
32. D. Bouvard and F. F. Lange, *Acta Metall. Mater.* **39**, 3083 (1991).
33. C. H. Kuo and P. K. Gupta, *Acta Metall. Mater.* **43**, 397 (1995).
34. P. Costamagna, M. Panizza, G. Cerisola, and A. Barbucci, *Electrochimica Acta*, **47**, 1079 (2002).

35. E. Achenbach, *J. Power Sources* **49**, 333 (1994).
36. P. Costamagna, *Chemical Engineering Science* **56**, 323 (2001).
37. P. Costamagna and K. Honegger, *J. Electrochem. Soc.* **145**, 3995 (1998).
38. P. Costamagna, *Journal of Power Sources* **69**, 1 (1997).
39. P. Costamagna, E. Arato, E. Achenbach, and U. Reus, *J. Power Sources* **52**, 243 (1994).
40. R. H. Perry and D. W. Green, *Perry's Chemical Engineers' Handbook*, (McGraw Hill, New York, NY, 1984).
41. P. Costamagna, PhD Thesis, University of Genoa, Italy, February 1997.
42. P. Costamagna, L. Magistri and A. F. Massardo, *J. Power Sources* **96**, 352 (2002).
43. S. Campanari, *Journal of Engineering for Gas Turbines and Power. Transactions of the ASME* **122**, 239 (2000).
44. J. Palsson, A. Selimovic, and L. Sjunneson, *J. Power Sources* **86**, 442 (2000).
45. S. E. Veyo, L. A. Shockling, J. T. Dederer, J. E. Gillett, and W. L. Lundberg, ASME Paper 2000-GT-0550.
46. F. Massardo and F. Lubelli, *Journal of Engineering for Gas Turbines and Power* **122**, 27 (2000).
47. J. Palsson, A. Selimovich, and L. Sjunnesson, *J. Power Sources* **86**, 442 (1999).
48. K. Hassmann, W. K Heidug, and S. Veyo, *Brennstoff Warme Kraft*, **11/12**, 40 (1999).
49. U. Bossel, *Facts and Figures* (IEA Task Report, Berne, April 1992).
50. H. Uchida, M. Shiraki, A. Bessho, and Y. Yagi, ASME Paper 94- GT-73.
51. P. Pilidis, Digital Simulation of Gas Turbine Performance, PhD Thesis, University of Glasow, 1983.
52. K. R. Pullen, N. C. Baines, and S. H. Hill, ASME Paper 92-GT-93.

## PROBLEMS

1. Evaluate the assumption of negligible water in the vapor phase in the 1-D steady state PEM model presented in this chapter. Is this a good assumption? How does the solution change if evaporation of product water and condensation in the pores are included?

2. Explain why the agglomerate model is most appropriate for the case of PEMFCs. What effects are apparent in this model that are lumped or neglected in other models?

3. Evaluate the electrical resistance of the PEN configuration represented in Figure P.7.1 (PEN strip with diagonal terminals). The arrows show the electron paths through the PEN structure, which involve in-plane conduction in the electrodes and cross-plane conduction through the electrolyte. The black stripes represent the interconnections. You will find out that, having defined $t$ as the thickness of each layer, $\rho$ the resistivity, and the parameters $L, E, B, C$ and $J$ as follows:

$$L = \sqrt{\dfrac{\rho_{el}t_{el}}{\left(\dfrac{t_2}{\rho_2}\right)^{-1} + \left(\dfrac{t_1}{\rho_1}\right)^{-1}}}$$

$$E = \left(\dfrac{t_2}{\rho_2}\right)^{-1}\left(\dfrac{t_1}{\rho_1}\right)$$

$$C = \rho_1 t_1 + \rho_{el}t_{el} + \rho_2 t_2$$

$$B = \dfrac{E}{\left(1+E\right)^2}$$

$$J = \dfrac{X}{L}$$

then, the resistance of the PEN structure in ohms is given by:

$$RA = CJ\left\{\coth(J) + B\left[J - 2\tanh\left(J/2\right)\right]\right\}$$

where $A$ is the active electrolyte area in m$^2$.

4.  In different parts of this chapter, the Butler-Volmer equation has been simplified into the Tafel equation or into a linear equation. Demonstrate how these two simplified forms of the Butler-Vomer equation are obtained under different simplifying hypotheses, and discuss their range of validity.

5.  Figure 7.28 shows a typical intersection among V-I curves collected at different operating temperatures. Can you explain this effect?

6.  Let us consider an SOFC anode fed with a hydrogen/water mixture. By assuming that the electrochemical reaction occurs only at the interface between electrode and electrolyte, demonstrate that diffusion limitations occurring in the electrode cause a departure of the Nernst from the value calculated on the basis of the hydrogen/water concentrations in the feeding mixture that is described by the following equation:

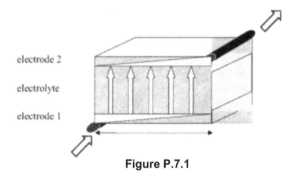

**Figure P.7.1**

$$\eta_a^{conc} = -\frac{R_g T}{n_e F} \ln \left( \frac{1 - \dfrac{i R_g T t_a}{n_e F D_{eff,a} P y_{H_2}^0}}{1 + \dfrac{i R_g T t_a}{n_e F D_{eff,a} P y_{H_2O}^0}} \right)$$

where $t_a$ is the anode thickness, $P$ is pressure, $T$ is temperature, $R_g$ is the gas constant, $i$ is the current density, and $y_{H2}^0$ and $y_{H2O}^0$ are the hydrogen and water concentrations in the anodic feeding mixture, respectively. The equation above is obtained by using the Fick's law (Eq. 7.13) to evaluate the hydrogen and water concentrations at the interface between the electrode and the electrolyte, and then evaluating the difference between (i) the Nernst potential evaluated on the basis of the concentrations in the feeding mixture, and (ii) the Nernst potential evaluated on the basis of the concentrations at the interface between the electrode and the electrolyte. How would this change for an MCFC anode?

7. Demonstrate how Eq. (7.28) is obtained by integrating the generalized treatment, described in Section 7.2.3.1., considering diffusion limitations as negligible and linearizing the Butler-Volmer equation. Discuss also the reasons for the similarity with the equation obtained in Problem 1.

8. Stack simulation. Write a simple software (using a programming language of your choice) solving the system of equations of Table 7.6. This will be the basis of your own simulation tool for an SOFC hybrid plant. What modifications should be made in order to simulate an MCFC hybrid plant?

# CHAPTER 8

# FUELS: PROCESSING, STORAGE, TRANSMISSION, DISTRIBUTION, AND SAFETY

## 8.1    HYDROGEN: THE IDEAL FUEL

### 8.1.1.    Hydrogen: The Most Electroactive and Environmentally Clean Fuel for All Types of Fuel Cells

Since the invention of the fuel cell by Sir William Grove in 1839, it has been well recognized that hydrogen is the ideal fuel for all types of fuel cells. The major accomplishments in fuel cell R&D during the 19th and 20th centuries, were:

- the proposition by the eminent physical chemist Friedrich Wilhelm Ostwald that the direct conversion of chemical energy to electricity would have a great effect on energy conservation and on minimizing environmental pollution;
- the development of a 5 kW alkaline fuel cell, using hydrogen and oxygen, by Francis T. Bacon, which was started in 1932 and completed in 1952;
- the discovery of the proton-exchange membrane fuel cell (referred to as the solid polymer electrolyte fuel cell, from its birth in the 1960s) and its use in NASA's Gemini space flights as an auxiliary power plant in the 1960s; and
- following Bacon's discovery, the development of highly efficient hydrogen/oxygen advanced alkaline fuel cells initially by Pratt and

---

This chapter was written by S. Srinivasan and J. Ogden.

Whitney and subsequently by International Fuel Cells (Division of United Technology Corporation) and their applications as an auxiliary power source for NASA's Apollo and space shuttle flights from the late 1960s until the present time.

The Chapters on Basic Electrochemistry (Chapters 1 to 3) and the Fundamental Aspects of Fuel Cells (Chapters 4 and 5) clearly demonstrate the reasons for pure hydrogen being the preferred fuel for all types of fuel cells. The most important reason is that it is the most electroactive fuel in a fuel cell i.e., it has exchange current densities for the anodic reaction that are at least three to five orders of magnitude higher than that of any other fuel at low to intermediate temperatures. In the molten-carbonate and solid-oxide fuel cells, carbon monoxide, a product of the steam-reforming reaction, is also quite electro-active. However, here too one could interpret that the anodic reaction occurs via hydrogen because on the anode electrocatalyst, the shift conversion reaction could produce hydrogen. Methanol is reasonably active in low temperature fuel cells and perhaps, in the future, this fuel and ethanol may have some prospects for direct use in MCFCs and SOFCs. The hydrocarbon fuels are considerably less active than alcohols. Further, the use of pure hydrogen might facilitate the design of fuel cell systems. In fuel cell vehicles, for example, the vehicle is likely to be less complex, more energy efficient, have better performance, and cost less if hydrogen is used directly on board.

There are other potential societal benefits of using hydrogen as a future energy carrier. Roughly two thirds of current global greenhouse gas emissions are associated with direct combustion of fuels for transportation and heating (the remainder of emissions come from electric power generation). There is a rising international concern about the potential effects of global climate change, associated with emissions of $CO_2$ from fossil fuel use, and increasing levels of $CO_2$ in the atmosphere. Many countries have signed the Kyoto conference agreement to reduce the level of $CO_2$ emissions by 10%, from 1990 levels, by the year 2010, and there are continuing efforts by various countries to limit $CO_2$ emissions. In addition, a significant fraction of air pollutant emissions and about two thirds of primary fossil energy use are associated with the direct use of fuels. A variety of alternative transportation fuels and of efficient, low-polluting vehicle technologies have been proposed to help address these challenges. Of these, hydrogen, used in fuel cells, offers the greatest potential for simultaneously reducing emissions of greenhouse gases and air pollutants.

Hydrogen, like electricity, is an *energy currency* or an *energy carrier* that must be produced from other primary resources such as fossil fuels, renewables, or nuclear power. Hydrogen can be made from a variety of primary resources including fossil fuels (natural gas or coal), renewables (biomass, hydropower, wind, and solar), and nuclear power. If hydrogen is made from renewable energy sources or from fossil fuels with capture and sequestration of $CO_2$, it would be possible to produce and use energy with near zero emissions of greenhouse gases or air

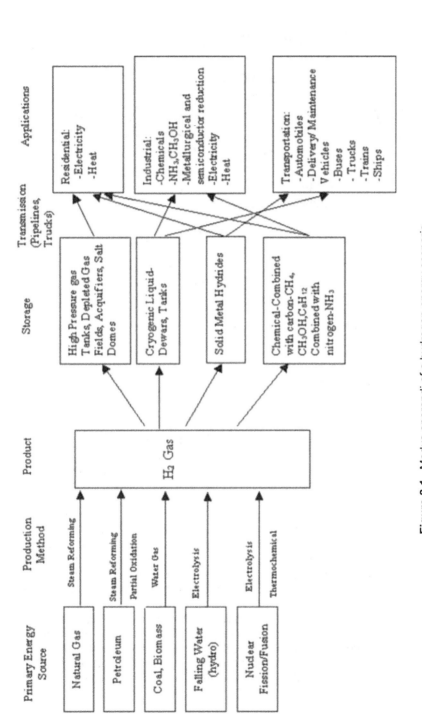

**Figure 8.1.** Modus operandi of a hydrogen-energy scenario.

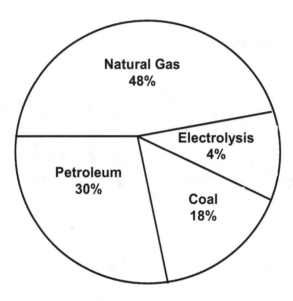

**Figure 8.2.** Global production sources of hydrogen.

pollutants. The following Section describes possible scenarios for a *hydrogen-energy system*.

### 8.1.2.   A Modus Operandi for a Hydrogen Energy Scenario

In a review article published about 18 years ago,[1] a clear illustration was presented to demonstrate the modus operandi of a hydrogen-energy system (Figure 8.1), which takes into consideration that hydrogen is an energy currency/energy carrier. First, the hydrogen has to be produced from a primary energy source. Second, one must consider the most efficient and economic method for its production. Third, hydrogen is produced in a gaseous state. The method of its storage will depend on the application and the economy of scale. Fourth, the hydrogen will have to be transmitted to the site of application. Fifth, there is a multitude of applications. In Figure 8.1, these are classified under residential, industrial, and transportation applications.

Hydrogen is produced on a large scale today for chemical applications such as ammonia and methanol synthesis and oil refining. The total world hydrogen production is about 500 billion $Nm^3$ per year. This is equivalent to about 2% of the world's primary energy consumption. Globally, about 48% of hydrogen is made from natural gas, about 30% from oil, 18% from coal, and 4% from water electrolysis (Figure 8.2). Of this amount, about 50% is used in     ammonia production

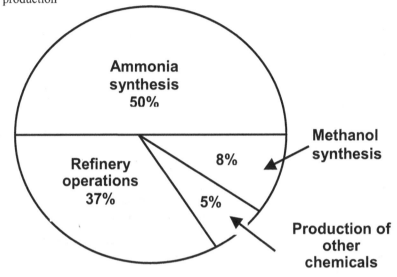

**Figure 8.3.** Global consumption of hydrogen.

for fertilizer, 37% for petroleum refining, 8% for methanol synthesis, 4% for other chemical uses, and 1% for aerospace (Figure 8.3). About 90 billion $Nm^3$ per year are produced in the US today, primarily for chemical and refining applications.[2] This is equivalent to about 1% of the US primary energy use. In the United States, about 95% of the hydrogen is currently produced from steam-reforming of natural gas, and the remainder via recovery as by-product from chemical operations (such as chlor-alkali plants), partial oxidation of hydrocarbons (such as coal or petroleum residuals at refineries), or by electrolysis of water. If in the future hydrogen is widely used as an energy carrier, its production would increase greatly beyond the current industrial levels.

Historically, there has been a trend toward *decarbonization* of the energy system. This was shown in studies carried out by Marchetti et al.,[3] and Nakicenovic et al.[4] It was shown that there is a semi-logarithmic relation for the hydrogen to carbon (H/C) ratio in the world's fuel mix as a function of time (Figure 8.4). The relationship is striking, and we are presently in an era where the H/C ratio is 4, as evidenced by the relatively high utilization of natural gas as a fuel. This decarbonization of the energy supply would have to continue at several times the historical rate to assure that atmospheric concentrations of $CO_2$ remain at levels that are twice the pre-industrial level or less. As noted above, presently fuels contribute

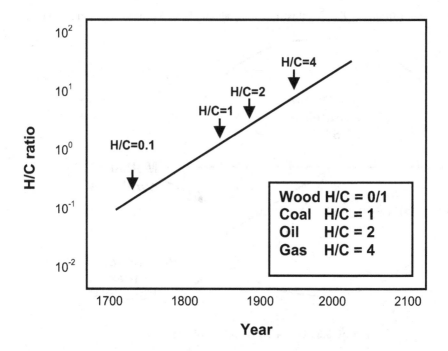

**Figure 8.4.** Global evolution of H/C ratio in the world's fuel mix.

to over half of the global greenhouse-gas emissions. Even if the electricity production sector is completely decarbonized, emissions from the fuel sector will have to be decreased by a factor of 3 to 5 by the end of the 21$^{st}$ century to achieve stable levels of atmospheric $CO_2$, i.e., 450-550 ppm.

The methods for production of hydrogen from the primary energy and renewable energy sources are dealt with in Sections 8.1.3 to 8.3.6. Section 8.4 summarizes methods of production of other fuels for direct or indirect utilization in fuel cells. Fuel storage, transmission, distribution, and safety are discussed in Sections 8.5 to 8.7.

### 8.1.3. Fossil Fuels: The Main Source of Hydrogen for the Foreseeable Future

Figure 8.2 clearly illustrates that the fossil fuels, natural gas, petroleum, and coal are the primary energy sources contributing to the production of more than 95% of hydrogen for the chemical and petroleum refining industry today. In recent years, there has been a trend by the petroleum industry to produce hydrogen from natural gas rather than from residual fuel oil for the hydrocracking of the heavier fractions of petroleum; the purpose was to enrich the production of gasoline and diesel fuel. The main reason for the use of natural gas to produce hydrogen' is that fuel processing with natural gas is cleaner and more economical. In some countries (China, India, Germany, and South Africa), the extent of utilization of coal as a primary energy source for the production of hydrogen is higher than in the USA. Another interesting fact is that during World War II, the Germans and the Japanese developed the technology for production of.coal-derived gasoline, using hydrogen that was also produced from coal. In view of the relative abundance of natural gas and coal, these primary energy sources will continue to contribute significantly to the production of hydrogen, even in the era of a h*ydrogen energy scenario*, in which the demands for hydrogen could increase many times over that for hydrogen required by the chemical industry.

Pure hydrogen can be produced from fossil fuels for storage and subsequent use in a fuel cell, or a hydrogen rich gas can be produced in a fuel processor, integrated with a fuel cell. Methods for pure hydrogen production from natural gas and coal are described in Sections 8.1.4 to 8.1.6. The topic of fuel processors for hydrogen production for integration with fuel cells will be dealt with in some detail in Section 8.2. Methods of hydrogen production from renewable and nuclear sources are addressed in Section 8.3.

### 8.1.4.   Natural Gas: The Most Promising Fuel for All Types of Fuel Cells

Since the 1990s, a stronger emphasis has been made on the utilization of natural gas as a primary energy source for all types of fuel cells.[5] This approach was further reinforced by the increased attraction for its use in fuel cell/gas turbine hybrid systems. Further, since natural gas is a considerably cleaner fuel for fuel processors than petroleum or coal and the hydrogen content is higher for the former than for the latter two fuels, the main goals of the major worldwide fuel cell programs are to use natural gas or natural-gas derived fuels. Figure 8.5 provides a vision of such an approach for all types of fuel cells. This figure also illustrates the applications of these types of fuel cells. For the low temperature, acid (PEMFC) and alkaline electrolyte (AFC) fuel cells, pure hydrogen is the desired fuel. The fuel processing involves steam reforming, shift conversion, preferential oxidation, and pressure swing absorption (see Section 8.2.1). Natural gas is also abundantly used for the production of methanol. Methanol, a liquid fuel, which has values of specific energy and energy density equal to about half those for gasoline, is an ideal fuel to be carried on board for transportation vehicles. The production of methanol from natural gas involves the combination of products of the steam-reformer reaction ($H_2$ and CO). The methanol can be used directly in a PEMFC or subjected to further fuel processing to produce hydrogen, as shown in Figure 8.5. The PAFC and PEMFC will be the most appropriate fuel cells for use of methanol or hydrogen produced by this route and the applications are multifold: power generation, electric vehicles, stand-by power, and remote power. An advantage of the PAFC is that its operating

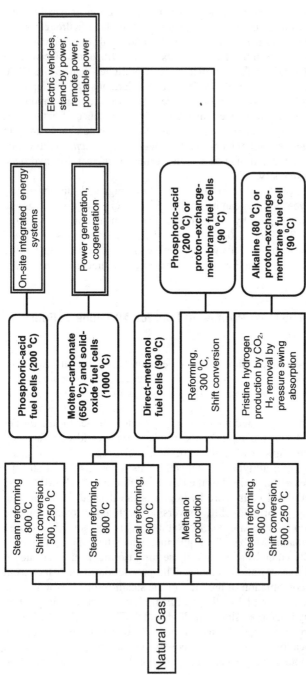

**Figure 8.5.** Natural gas: the ideal fuel for all types of fuel cells. Demonstrated and potential applications of fuel cells are marked on bold. Reprinted from Reference 5, Copyright (2001) with permission from Plenum Press.

temperature is about 200 °C, and at this temperature, a 1 to 2% level of CO in the $H_2$ produced by steam-reforming/shift-conversion of natural gas is not a poison. Thus, the step of preferential oxidation is not required. The PAFC technology is well established with this fuel-producing technology, and PAFC systems ranging in power level from 40 kW to 10 MW have been demonstrated. The major advance has been with UTC-fuel cells commercializing the 200 kW natural gas fuelled power plant (the PC-25) for cogeneration applications. The advantages of the high temperature MCFC and SOFC systems are that the steam-reforming shift conversion in the electrochemical stack is more than adequate for the anodic oxidation of not only hydrogen, but also of carbon monoxide. Such types of direct fuel cells are being developed for power generation and cogeneration applications (see Chapter 9 for details).

### 8.1.5. Coal Gasifiers for Hydrogen Production

Due to the relative abundance of coal in several countries in the world (an estimate is that it can satisfy the energy requirements of mankind for the next 500 years), there has been enthusiasm during the past 50 years for using gasified coal products in fuel cells. Basically, there are three types of coal gasifiers: moving beds, fluidized beds, and entrained beds. Steam, oxygen, or air is used to partially oxidize the coal. The moving bed gasifier mainly produces gaseous hydrocarbons, methane and ethane, and liquid fuels such as naphthalene, tars, oils, and phenolics at a temperature in the range 425 to 650 °C. The entrained bed gasifier operates at a high temperature (above 1250 °C) and the products are mainly $H_2$, CO, and $CO_2$. The fluidized bed gasifier has an intermediate temperature for operation (925 to 1040 °C), and the products are composed of a mixture of gases, as generated in other types of gasifiers. The heat needed for the reactions, in all three cases, is supplied by the exothermic reaction, i.e., the partial oxidation of coal. Impurities include $H_2S$, COS, $NH_3$, HCN, particulates, tars, oils, and phenols. These are removed by hot gas or cold gas clean-up technologies. The latter methods consist of uses of cyclones and particulate scrubbers, COS hydrolysis reactors, liquid scrubbers for $NH_3$ and $H_2S$, and sulfur recovery (Claus process). All these requirements make the systems complex and expensive, and there is also the need for heat exchangers and coolers. For the various types of oxygen-blown gasifiers the yield of hydrogen varies from about 15 to 30% and that of CO from 5 to 60%. However, if relatively pure hydrogen is the desired product, a shift converter, an acid-gas removal system, and a methanator could be incorporated with the coal gasifier system, as used in the Koppers-Totzek atmospheric pressure and the pressurized Texaco and Shell processes. The hydrogen content from these gasifiers is about 95%.

## 8.2    FUEL PROCESSORS FOR INTEGRATION WITH FUEL CELLS

### 8.2.1.    Steam-Reforming, Shift-Conversion, Pressure Swing, and Adsorption of Natural Gas for Ultra-Pure H₂ Production for AFCs and PEMFCs

The ideal fuel for a PEMFC or an AFC is ultra pure hydrogen. When natural gas is used as the primary fuel for pure hydrogen production, several chemical steps are performed: sulfur removal, steam-reforming, shift-conversion, and hydrogen purification (via pressure swing adsorption).

It is essential to first remove the sulfur from natural gas by passing the feed-stream at 290-370 °C over a Co-Mo catalyst in the presence of 5% $H_2$ in order to remove the sulfur as $H_2S$. The out-flowing gas is then cooled and scrubbed with a mono-ethanol amine solution, followed by its absorption over a ZnO catalyst at 340 to 370 °C to reduce the sulfur content to about 0.5 ppm.

The next step is to pass this gas through the steam-reformer where the following reactions take place:

$$CH_4 + H_2O \rightarrow CO + H_2O \tag{8.1}$$

$$CO + H_2O \rightarrow CO_2 + H_2 \tag{8.2}$$

The first reaction (methane plus steam-reforming reaction) is endothermic, while the second (water gas shift-reaction) is exothermic. It must be noted that the water gas shift-conversion reaction does not take place completely. The catalyst for the reformer reaction is nickel oxide supported on calcium aluminate, alumina, or calcium aluminum titanate. The operating temperature is 650 to 700 °C and the pressure is between 22 and 24 atm. The outlet gas temperature is 870-885 °C. The product composition is 76% $H_2$, 12% CO, 10% $CO_2$, and 1.3% $CH_4$.

The subsequent step is to further carry out the water gas-conversion shift reaction for a nearly complete conversion of the CO to $CO_2$. This is done in two stages: (i) in a high temperature shift reactor at 340-350 °C using chromium promoted iron oxide catalyst; and (ii) in a low temperature shift reactor at 200-300 °C over a Cu-Zn catalyst, supported on alumina catalyst. The output mixture of gases is cooled by a regenerative scrubbing process to produce 98% $H_2$, 0.3% CO, 0.019% $CO_2$, and 1.5% $CH_4$. The residual gases are converted to $CH_4$ (methanation reaction) by passing the gases reheated to 315 °C over a nickel oxide catalyst. These intermediate steps are schematically illustrated in Figure 8.6. The hydrogen can be further purified by a cryogenic procedure.

If there is a need for ultra pure hydrogen, the use of the pressure swing absorption (PSA) method for hydrogen purification is attractive. The process diagrams for conventional scrubbing and pressure swing absorption are represented

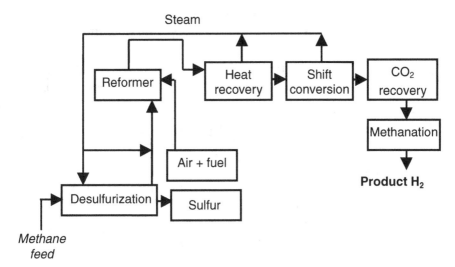

**Figure 8.6.** Flowchart for ultra-pure hydrogen production by conventional steam-methane-reforming/shift-conversion of natural gas.

in Figure 8.7. The advantage of the PSA method is that it eliminates the low temperature shift, $CO_2$ scrubbing, methanation, and cryogenic steps. The PSA method involves passing the product gases from the high temperature shift reactor into an absorbent bed, which removes all the constituents except $H_2$. To regenerate or reactivate the bed, the reactor is depressurized, purged with hydrogen, and repressurized. During this process, the feed is switched to another bed that is in parallel with the bed being reactivated. The main advantages of the PSA method are:

- lower maintenance costs,
- higher reliability,
- efficient heat recovery, and
- higher efficiency than a conventional plant, 84.6% vs. 83.2%, respectively.

The hydrogen produced by the conventional and PSA methods is therefore ideally suited for use in PEMFCs and AFCs.

### 8.2.2.  Steam Reforming, Shift Conversion, and Preferential Oxidation of Natural Gas, Gasoline or Methanol for PEMFCs

The PEMFC has an acid electrolyte environment. Thus, unlike in the case of an AFC, it is not necessary to remove $CO_2$ from the product gas stream after the steam-reforming and water-gas shift reactions. However, since the product gas may contain

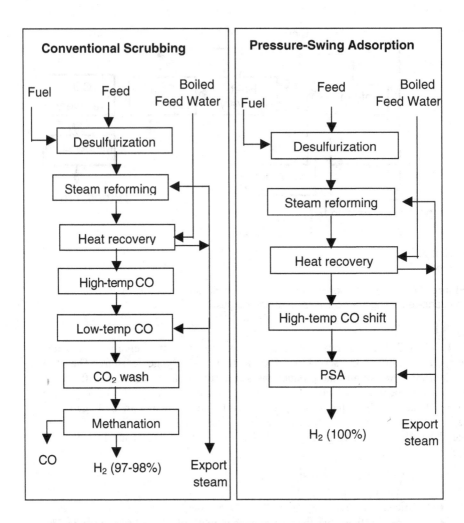

**Figure 8.7.** Flow charts for production of ultra-pure hydrogen using the conventional scrubbing and pressure swing absorption (PSA) methods.

0.5-1.0% of CO—a deadly poison for the anode in the PEMFC—an additional fuel-processing step is necessary, i.e., the preferential oxidation of this gas in a separate chamber after the low-temperature shift reactor (Figure 8.8). The exit gases from this reactor are passed through a platinum-on-alumina catalyst into the preferential oxidizer reactor along with 2% air. The CO is first selectively adsorbed on this catalyst and then oxidized by the air to $CO_2$ . Until recently, the quoted value for the

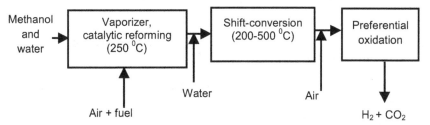

**Figure 8.8.** Typical steam-reforming/shift-conversion/preferential-oxidation of natural gas, gasoline, or methanol for PEMFCs. The flow chart shown is for hydrogen production from methanol.

reduction in CO level was 500 ppm, when hydrocarbons were used as fuels and 50 ppm when methanol was used as the fuel. However, more recently Epyx Corporation (now Nuvera Fuel Cells) claimed that they could reduce the CO level to 10 ppm when gasoline was the fuel to be processed.[6] The PEMFC can tolerate this level of CO at 80 °C without any substantial anode poisoning.

There has been great interest in the use of this type of fuel processor to be carried on board a PEMFC-powered-transportation vehicle. The advantage of methanol is that it is a partially oxidized fuel (as compared with hydrocarbons) and the first step of its steam-reforming requires a considerably lower temperature (280 to 300 °C) than that for hydrocarbons. Further, the product gas from the entire fuel processing has less than 50 ppm CO, whereas the CO levels are higher with hydrocarbons. However, there has been more emphasis on gasoline fuel processing by the US Department of Energy. The reason for this is the large availability of petroleum resources (gasoline) for at least the next 20 to 30 years as well as the higher energy density for gasoline as compared to methanol (about twice, on both, weight and volumetric bases).

Methanol is produced from the primary fuel—i.e., natural gas—on a large scale by the plastics industry for the production of formaldehyde. The products of the steam-reforming reaction (Eq. 8.1) are passed through a reactor using a CuO-ZnO catalyst for the formation of methanol:

$$CO + 2H_2 \rightarrow CH_3OH \tag{8.3}$$

Coal is another abundant primary energy source for methanol production. In this case, the products of the coal-gasification reaction, i.e.,

$$C + H_2O \rightarrow CO + H_2 \tag{8.4}$$

are combined in the same manner, as expressed by Eq. (8.3), to form methanol. In developing countries, there is also interest in methanol production from biomass (see Section 8.3.5).

### 8.2.3. Steam-Reforming and Shift-Conversion of Natural Gas or Methanol for $H_2$ Production for PAFCs

An attractive feature of the PAFC is that it operates at a temperature of about 200 °C and at this temperature the tolerance level of CO by the anode Pt electrocatalyst is about 1%. Thus, the heavy and bulky preferential oxidation reactor is not required. Further, the waste heat from the PAFC is of reasonably high quality, and it can be used for the production of steam required for the reformer reaction.

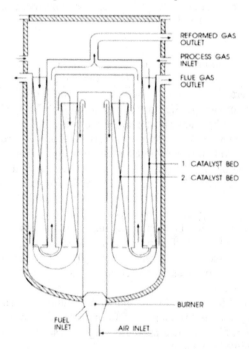

**Figure 8.9.** Design of a UTC fuel-cell regenerative-tube reformer for hydrogen production from natural gas and use in PAFCs. Reprinted from K. Kordesch and G. Simader, *Fuel Cellls and their Applications*, VCH, Weinheim, Copyright (1996), with permission from Wiley, VCH.

The heat utilization is even more effective when methanol rather than natural gas is used as the fuel because the reformer reaction, required for the former is about 300 °C, while for the latter it is about 650–700 °C.

There are three types of reforming furnaces, used for hydrogen production: (i) a top fired with co-current flow of the process and synthesis gases; (ii) a bottom fired with counter-current flow of gases; and (iii) a side-fired reactor with a homogenous heat flux along the entire length of the tube. UTC Fuel Cells, the leading PAFC developer in the world, uses a regenerative-tube reformer.[7] The design of this reformer uses a combination of counter-current and co-current flows of the reactant gases and effluents (Figure 8.9). In this design, the radiant heat transfer occurs from the flu gas to the catalyst-filled annulus on one side, and heat transfer to the other side takes place via the process gas. The reactor is compact and the heat savings are significant. It was originally designed by Haldor Topsoe and was further developed by UTC Fuel Cells for fuel cell applications.

## 8.2.4. Partial Oxidation and Shift Conversion of Hydrocarbons and Alcohols for PEMFCs

Hydrogen and hydrogen-rich synthesis gas can be produced by non-catalytic partial oxidation of hydrocarbons (e.g., refinery, residual oil) using the Texaco or Shell processes. The overall reaction is:

$$C_nH_m + \frac{n}{2}O_2 \rightarrow nCO + \frac{m}{2}H_2 \tag{8.5}$$

The process diagram for the complete conversion to hydrogen, with acid gas removal and with sulphur separation is shown in Fig. 8.10. Partial oxidation of the hydrocarbon occurs with less than the stoichiometric amount of oxygen at 1300 to 1400 °C. The high temperature and low temperature shift reactions are then carried out. The heat recovered from the shift reaction is utilized for steam production. The hydrogen produced from this process is approximately 97.5% pure (i.e., after $CO_2$ removal, as in the steam-reforming-shift-conversion process).

EPYX Corporation, Nuvera, and IFC have been actively engaged in the development of fuel flexible fuel processors of this type to be integrated with 50 kW PEMFC fuel cell stacks from IFC and Plug Power in the PNGV program. The fuels include gasoline and methanol. One of the problems encountered with this type of fuel cell processor for automotive applications is that the intermediate steps of shift conversion, preferential oxidation, and air bleeding are necessary. As a consequence, the start-up time from ambient temperature is about 5 to 10 minutes. There is also a significant energy penalty, and the CO content is enhanced during transients.

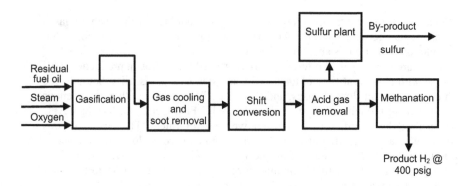

**Figure 8.10.** Flowchart for ultra-pure hydrogen production from hydrocarbons by partial-oxidation→shift-conversion→ acid-gas-removal→methanation.

### 8.2.5. Autothermal Reforming of Methanol and Gasoline for PEMFC Powered Vehicles

This novel method is being developed at Argonne National Laboratory (ANL).[8] Basically, three methods of fuel processing have been considered and/or developed for on-board fuel processing in PEMFC powered vehicles: steam reforming, partial-oxidation reforming, and autothermal reforming. The steam-reforming fuel processor has the disadvantage of reactor design with inefficient heat transfer from the shift-reactor to the steam-reformer. Thus, it is not effective enough for rapid start-up and transient response in automobiles. Partial oxidation fuel processing involves a very high temperature (over 1000 °C) for the first step, which is exothermic:

$$C_nH_m + Air \rightarrow nCO_2 + \frac{m}{2}H_2 + N_2 \qquad (8.6)$$

By injecting an appropriate amount of steam into the system, the steam reforming reaction can be carried out. The partial oxidation step can occur with or without a catalyst. The auto-thermal reforming process is a hybrid of the steam-reforming and partial oxidation fuel processors. When methanol is used as the fuel, the autothermal reforming process is represented by:

$$CH_3OH + H_2O \rightarrow CO_2 + 3H_2 \qquad (8.7)$$

$$CH_3OH + x(O_2 + 3.76N_2) + (1 - 2x)H_2O \rightarrow CO_2 + (3 - 2x)H_2 + 3.76N_2 \qquad (8.8)$$

in which $x$ is less than 2.3%. This type of fuel processor is being developed by Johnson Mathey and by Argonne National Laboratory. By carrying out the reactions in the presence of catalysts, the operating temperature can be lowered, thus favoring the water-gas shift reaction. The lower temperature also favors reduction in fuel consumption. To inhibit coke formation, the oxygen/carbon ratio in the inlet stream to the reactor is optimized. Steam is used as a source of oxygen to lower the temperature, which is necessary to avoid coke formation. Autothermal reformers are also designed for efficient heat transfer from the exothermic partial oxidation intermediate step to the steam-reforming of the fuel in the same reactor. A laboratory scale design of the reactor used by Argonne National Laboratory is shown in Figure 8.11. The fuel, water, and air are introduced via a nozzle into the reactor containing a zinc oxide/copper oxide catalyst. A small electric coil is used to ignite the mixture. The product gases exit at the bottom and are periodically analyzed by an on-line gas chromatograph. Catalyst materials are also evaluated in the packed bed micro-reactor (Figure. 8.11). The temperature is controlled using a surrounding furnace. Using methanol as the fuel at a feed rate of 54 ml/min, 35-50 ml/min of air, and 10-20 ml of water, the products are 50% $H_2$, 20% $CO_2$, and 1%

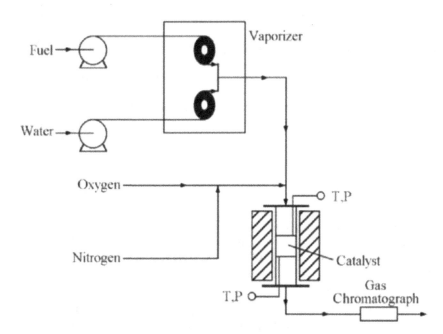

**Figure 8.11.** Laboratory-scale design of an autothermal reformer for hydrogen production from methanol or gasoline, evaluation of catalyst materials, and on-line chromatography product analysis by Argonne National Laboratory. Reprinted from Reference 9, Copyright (1997) with permission from the authors.

CO. It must be noted that it is assumed in Eq. (8.7) that air and oxygen are both used for the partial oxidation intermediate step. The parameter, x, is the oxygen to methanol molar ratio ($0 < x < 0.5$). The enthalpy change for the reaction, $\Delta H$, is given by:

$$\Delta H = 131.572 \quad \text{kJ/mol} \quad CH_3OH \qquad (8.9)$$

This assumes that all species are in their standard states at 25 °C. The reaction is thermoneutral at $x = 0.23$.

The conversion of gasoline (e.g., octane) to $H_2$ may be expressed by:

$$C_8H_{18} + xO_2 + 3.76N_2 + (16 - 2x)H_2O_1 \rightarrow$$
$$8CO_2 + (16 - 2x + 9)H_2 + 3.76N_2 \qquad (8.10)$$

where $x$ is the oxygen/fuel molar ratio ($0 < x < 4$). The enthalpy change for the reaction is:

$$\Delta H = 1685.572 \quad \text{kJ/mol} \quad C_8H_{18} \qquad (8.11)$$

The reaction is thermoneutral at $x = 2.94$. With a proprietary catalyst used by ANL, the hydrogen content in the exit gas of the reactor was over 50%. With $O_2$ instead of air for the partial oxidation step, the hydrogen content could be increased to over 60%.

Figure 8.12 shows a comparative performance analysis of the three pathways for hydrogen production from hydrocarbon and alcohol fuels.[9,10] The steam reformer/shift conversion pathway presents the challenge of separating water from the combustion products and its subsequent feed for the reformer reaction. Because of this, heat is only extracted from the combustion process and the products are vented into the atmosphere. The partial oxidation and autothermal reforming processors are more energy-efficient, compact, and lightweight; these reactions are thus more attractive for on-board fuel processing. Such types of fuel processors produce hydrogen at considerably lower levels than in large industrial-scale hydrogen generation plants using steam-reforming/shift-conversion and coal gasification processes. The former type of fuel processors must meet some stringent requirements:

- frequent cyclability, i.e., start-up and shut down;
- ability to provide hydrogen at transient rates for start-up and acceleration (this may cause problems of enhancing CO levels);
- high rate of manufacturability to achieve low capital cost (equivalent to about $50/kW); and
- high reliability and lifetime of at least 10 years.

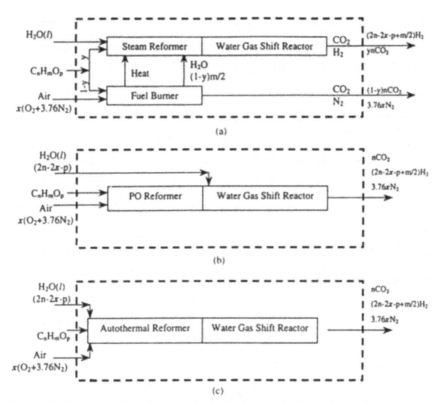

**Figure 8.12.** Comparative pathways for hydrogen production using hydrocarbon and alcohol fuels. Reprinted from Reference 10, Copyright (2001), with permission from Elsevier.

One other important aspect is that coke formation must be prevented by operating at a high temperature, with high oxygen to carbon atomic ratios. The oxygen is generally fed as water into the fuel processors for this purpose.

Scaled-up versions of such types of fuel processors have been designed, developed, and tested by ANL, EPYX/A.D Little, Nuvera, and UTC-IFC (present name UTC Fuel Cells).[7] The ANL unit has a $H_2$ generation capacity of 6 kW and includes the catalytic autothermal reformer, a zinc oxide bed for sulphur removal, and a water gas shift reactor. The catalysts were developed at ANL. This reactor produced hydrogen at a composition of 40% with less than 4% CO. The EPYX/A.D Little fuel processor was built to be integrated with a 50 kW PEMFC and claims were made that the CO level from the shift reactor could be reduced to 10 ppm. The Nuvera unit was also built for coupling with a 50 kW PEMFC from Plug Power; UTC-Fuel Cells developed their own fuel processors for integrating with their 50 kW PEMFC. The Nuvera unit is pressurized; the UTC-Fuel Cells unit operates at ambient pressure. The projected specific power and power density for integrated fuel processor/fuel cell system are 140 W/kg and 140 W/l; these values were about half those targeted for the year 2004 in the PNGV program.[6] The start-up time is currently about 10 minutes, while the target in 2004 was considerably less than a minute.

### 8.2.6.   Steam Reforming of Natural Gas for MCFC or SOFC: Gas Turbine Hybrids

Fuel processing is greatly simplified for the high-temperature MCFCs and SOFCs as compared with that for the low and intermediate temperature fuel cells. As illustrated in Figure 8.5, the process only involves the steam-reforming of natural gas to produce $H_2$ and CO. Carbon monoxide like $H_2$, is an anodic fuel in MCFCs and SOFCs and not a poison, as in the low and intermediate temperature PEMFCs and PAFCs. Though it is very possible that the oxidation of the CO occurs electrochemically on the nickel-based electrocatalysts in the MCFCs and SOFCs, it is equally probable that the shift conversion reaction occurs first on this electrocatalyst to produce hydrogen, which is then electro-oxidized.

Fuel processors for these types of fuel cells are of two types:

- the fuel processor sub-system is a unit separated from the electrochemical cell stack; (i.e., the external reformer); and
- the fuel processor is integrated with the electrochemical cell stack (i.e., the internal reformer).[11]

The latter is the most favored processor and it was developed by Fuel Cell Energy. Its advantage is that the heat required for the endothermic reforming reaction ($\Delta H = 225.18$ kJ/mol) is directly transferred from the heat generated in the exothermic fuel cell reaction in the electrochemical stack. Thus, the need for an auxiliary heat exchanger is eliminated.

The internal reformer may be further divided into two types: the indirect internal reformer (IIR) and the direct internal reformer (DIR) (see Figure 8.13). In the IIR, the reformer catalyst is separated from the fuel cell electrocatalyst, whereas in the DIR the fuel cell electrocatalyst serves also as the reformer catalyst and because of the hydrogen being instantly consumed in the fuel cell, the methane reformation reaction is accelerated. In both these types, the heat transfer is efficient, without a separate heat exchanger and the product steam, formed by the fuel cell reaction, accelerates the reformer and shift conversion reactions to produce more hydrogen. The extent of methane conversion increased with the amount of fuel utilization, i.e., in a DIR MCFC at 650 °C and 1 atm pressure with the steam/carbon ratio being maintained at 2, therefore there was more than 99% of methane conversion when the fuel utilization in the MCFC was greater the 65%. In the IIR, nickel supported on MgO or $LiAlO_2$ was used as the electrocatalyst. Internal

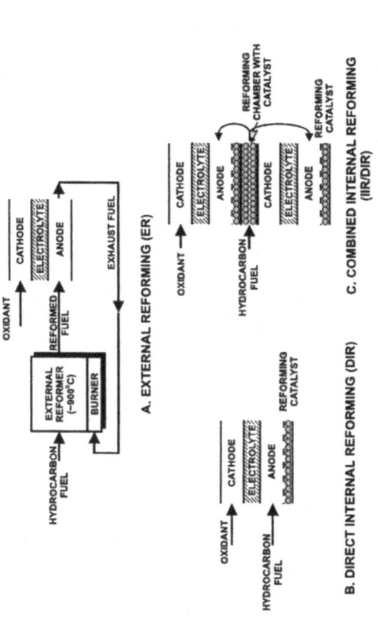

**Figure 8.13.** Process diagram for external, internal, and combined internal reforming of natural gas. Reprinted from Reference 10, Copyright (2001) with permission from Elsevier.

reforming in stacks, ranging in power levels from a few kW to 250 kW, has been demonstrated by Fuel Cell Energy. In some recent work at this company, both the IIR and DIR have been coupled, in an optimized manner, to attain the highest level of methane conversion, at a fuel utilization rate of over 70% in long-term lifetime studies.

Ansaldo Ricerche in Italy is in the process of developing an alternate configuration for the steam reformer, the *sensible-heat reformer* (SHR), to be closely integrated with the electrochemical cell stack.[12] The attractive features of the design are:

- good thermal management,
- low heat losses,
- low gas volumes, and
- minimal differential pressure between the internal/external sides of components.

Performance evaluations have been carried out at part-load and during load-following. The multifuel compatibility is also being tested.

SOFCs, like MCFCs, have also been developed for use with external and/or internal reformers. An advantage of the SOFC over the MCFC is that $CO_2$ recovery from the exit anode stream for feeding into the oxidant stream is not necessary. The SOFC only needs oxygen as the cathodic reactant, while the MCFC needs both $O_2$ and $CO_2$ to form the carbonate ion that transports the ionic current from the cathode to the anode. Most of the SOFC power plants developed to date are coupled with external steam reformers using natural gas. An interesting concept, invented by Siemens, was to use a pre-reformer at a relatively low temperature in the range of 250-500 °C to convert high molecular weight hydrocarbons, present in natural gas, to hydrogen.[13] The exit gases methane, steam, and small amounts of hydrogen, $CO_2$, and CO from the pre-reformer were delivered to the anode chamber in the SOFC where internal reforming converted the methane to $H_2$ and CO. One advantage of this route was that the small amount of hydrogen from the pre-reformer maintained the anode electrocatalyst in a reduced state.

Siemens-Westinghouse have designed a natural-gas fuelled-pressurized SOFC system in order to attain high power densities and efficiencies. As illustrated in the schematic for this system (Figure 8.14), there are two stages for internal fuel-reforming and utilization, one operating at high pressures and the other at low pressures. Reforming occurs in the anode chamber and the heat required for the reaction is supplied by the anodic reaction. Steam is provided for internal reforming, as well as for preventing soot formation via a gas circulation loop. The $NO_x$ emission level is only about 4 ppm. The fuel utilization mainly occurs at the anode. The overall efficiency of the hybrid system (fuel cell/gas turbine) to convert the chemical energy of the fuel to electrical energy is estimated at 67%, which is based on the lower heating value of the fuel.

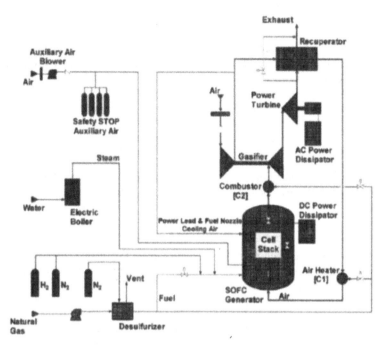

**Figure 8.14.** Schematic of a Seimens-Westinghouse hybrid-solid-oxide-fuel cell/gas-turbine-generator with internal reforming of natural gas. Reprinted from Reference 13.

### 8.2.7.  Coal Gasifiers for SOFC/Gas Turbine Hybrids, with H₂ Separation and CO₂ Sequestration

In recent years, several authors (Hendrik, Doctor, Simbeck, Spath, Williams, Kreutz et al.)[14] have assessed the prospects for producing hydrogen and carbon monoxide from coal via gasification and for their utilization in gas turbine/SOFC hybrid power plants. It is technically feasible to capture $CO_2$ during hydrogen production. The $CO_2$ can then be compressed to supercritical pressures (> 10 MPa), transported by pipeline, and injected into underground storage sites (such as depleted oil or gas reservoirs or deep saline aquifers) for permanent sequestration. This would allow production of hydrogen from fossil fuels, with greatly reduced emissions of $CO_2$ to the atmosphere (Herzog et al.,[15] Williams[16,17]).

Another method of fuel processing involves the use of the *coal-steam-iron cycle* which is one of the oldest methods for hydrogen production. The reaction involved is:

$$3Fe + 4H_2O \rightarrow Fe_3O_4 + 4H_2 \qquad (8.12)$$

Hydrogen, produced by this process is ultra-pure. Syngas, produced by the coal gasification reaction (Eq. 8.4) is utilized to regenerate the iron from the iron oxide, according to the equations:

$$Fe_3O_4 + 4CO \rightarrow 3Fe + 4CO_2 \qquad (8.13)$$

$$Fe_3O_4 + 4H_2 \rightarrow 3Fe + 4H_2O \qquad (8.14)$$

The reader may be puzzled by the overall reaction, as written above. The overall reaction is the sum of the reactions (8.4), (8.12) and (8.13) and is expressed as:

$$C + 2H_2O \rightarrow CO_2 + H_2 \qquad (8.15)$$

i.e., the production of hydrogen from coal using steam as the second reactant. The use of the steam-iron reaction is one route to overcome the problem of impurities formed by the complete direct oxidation of coal. Instead, by use of the steam produced by reaction (8.4), the steam-iron reaction occurs. Reactions (8.13) and (8.14) produce iron in a spongy form with a high surface area. The coal-steam-iron cycle occurs at a temperature of 700-900 $^0$C. One can also have a pressurized reactor. The fuel-cell company H Power proposed the steam-iron reaction for pure hydrogen generation and its utilization in a proton exchange membrane fuel cell. It was also suggested that the iron oxide could be generated at central sites, using syngas, produced from coal. The advantage of this method is that there will be no $CO_2$ emission from the fuel cell vehicle. The $CO_2$ removal at the central site could be sequestered to minimize atmospheric pollution.

For power generation, Steinberg and Cheng proposed a hybrid system,[18] as illustrated in Figure 8.15. This scheme involves using a fraction of the hydrogen and carbon monoxide for combustion and the use of heat/steam in a gas-turbine/steam-turbine hybrid electric power generator to produce electric power. In this scheme, the steam-iron reaction is used to generate electricity from a fuel cell power plant.

## 8.3.  HYDROGEN PRODUCTION FROM NUCLEAR AND RENEWABLE ENERGY RESOURCES

### 8.3.1.  Role of Nuclear and Renewable Energy Resources

There is little doubt that the fossil fuels will be the major source of hydrogen in the near to intermediate term and also possibly in the long term, particularly if the technologies for carbon capture and sequestration are successfully implemented.

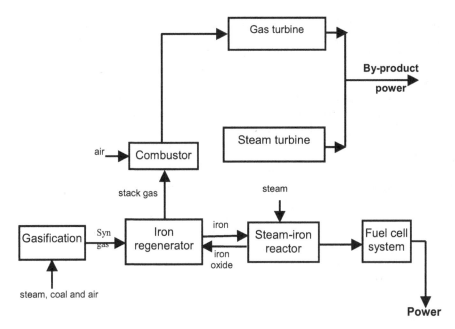

**Figure 8.15.** Coal-steam-iron cycle for ultra-pure hydrogen production and its utilization in fuel cells and for use of stack gas combustion for utilization in gas turbine/steam turbine. Reprinted from Reference 18, Copyright (1989) with permission from International Journal of Hydrogen Energy.

**TABLE 8.1**
**Global Energy Resources[a]**

| Source | Reserves plus resources, (thousand exajoules) |
|---|---|
| *Primary Energy Resources[b]* | |
| Coal | 199.7 |
| Oil | 32.4 |
| Natural gas | 49.8 |
| Uranium | 32.0 |
| *Renewable Energy Resources[c]* | |
| Hydro | 0.05 |
| Biomass | 0.28 |
| Solar | 1.57 |
| Wind | 0.64 |
| Geothermal | 5.00 |

[a] Extracted Reference 19
[b] Data for primary-energy sources is for reserves plus resource base
[c] Data for renewable-energy resources in thousands of exajoules/g

The main energy alternatives in the long term, will be to focus on renewable energy resources (see Table. 8.1). It is also necessary to examine the potential contribution of nuclear-energy resources, as shown in Table 8.1. This table clearly shows that nuclear and solar energies could be the winners to displace fossil fuels. Even though at the present time there are countries like France and Belgium that are strongly dependent on nuclear power, the strong antinuclear lobby and safety considerations have recently retarded the growth of nuclear power. Solar power can solve the world's energy problems but due to the requirement of large areas of land, the high cost of solar power and the need for extensively coupled energy storage systems, the growth rate has been slow. The other renewable energy sources—hydro, wind, geothermal, tidal, and biomass—are site specific in respect to their availability, but will still have some role in contributing to the energy sector.

With all the renewable energy resources, the main focus has been on electricity generation. Another important aspect is the vital need of a liquid or gaseous fuel for applications such as transportation and portable power. The ideal fuel, which will meet this demand and provide a solution to the global warming problem, is hydrogen. In the following Sections, we shall deal with the methods of the production of hydrogen using the renewable energy resources.

## 8.3.2.  Water Electrolysis

8.3.2.1.  *Thermodynamic and Electrode Kinetic Aspects.* Water electrolysis is one the simplest and best known technologies for hydrogen production, whether it is on a small or large scale. The overall reaction is the decomposition of water into its components:

$$2H_2O \rightarrow 2H_2 + O_2 \tag{8.16}$$

It is exactly the reverse of this reaction which occurs in a fuel cell. The standard free energy change and enthalpy change have values of 236.83 kJ/mol and 285.58 kJ/mol, respectively, for the case of liquid water being the reactant and the products being gases at a pressure of 1 atm and temperature of 25 °C. The decomposition of water is an energy driven reaction, while the fuel cell reaction is a spontaneous one. During water electrolysis, electrical energy is supplied from a power source for this reaction. As stated in Chapters 1 and 2, the thermodynamic reversible potential (at 25 °C and 1 atm pressure of reactant gases) for the reaction is 1.229 V. Thus, this is the minimum voltage that can drive a water electrolysis cell. However, since the entropy change ($\Delta S$) for the reaction is negative, the electrolysis cell will absorb heat from the surroundings if water electrolysis is carried out at a potential between the thermodynamic reversible potential and thermoneutral potential ($E_t$) of 1.484 V (this potential corresponds to the enthalpy change of the reaction $\Delta H^0 = -nFE_t$.) Other factors, which have to be taken into consideration, are the activation overpotential losses at the anode and cathode and also the ohmic overpotential losses, mainly in the electrolyte. Mass transport overpotential losses are minimal during water electrolysis. Thus, the cell potential $(E)$ vs. current density $(i)$ relation for water electrolysis can be expressed by the equation:

$$E = E_r + b_a \ln \frac{i}{i_{o,a}} + b_c \ln \frac{i}{i_{o,c}} + iR \qquad (8.17)$$

where $E_r$ is the reversible potential, $i_0$'s are the exchange current densities (the suffixes $a$ and $c$ denote anodic and cathodic reactions, respectively), and $R$ is the internal resistance of the water electrolysis cell. As in the case of fuel cells, the slow kinetics of the oxygen electrode reaction causes high activation overpotential losses at the desired current densities.

Taking into consideration the expressions for the variation of the reversible potential with temperature and pressure (see Chapter 1), from a thermodynamic point of view it is more favorable to operate water electrolysis cells at high temperatures and low pressures. However from a technological point of view, high temperatures and high pressures are most beneficial in (i) enhancing the electrode kinetics of the reactions, (ii) reducing ohmic overpotential, and (iii) minimizing energy requirements for compression and storage of product gases.

The efficiency of a water electrolyzer $(\varepsilon)$ may be defined by the equation:

$$\varepsilon = \frac{\text{Chemical Energy Output}}{\text{Electrical Energy Output}} = \frac{\Delta H}{nFE} \qquad (8.18)$$

The assumption made in this equation is that the coulombic-current efficiency for water electrolysis is unity. This is practically always the case. If water electrolysis is carried out at a cell potential of 1.48 V, the thermoneutral potential, the efficiency is 100%. Invariably, the activation overpotential at the anode and the ohmic overpotential in the cell raise the $E$ values to more than 1.6 V (typically about 1.8 V) and thus, the practical efficiencies are in the range of 75 to 90%.

Another important factor has to be taken into consideration in ascertaining the efficiency of hydrogen production using a water electrolyzer, i.e., the efficiency for electric-power generation from the primary energy source. This efficiency is about 30 to 35% for a thermal power plant, when using coal or oil, and may be as high as 50% for a combined cycle gas turbine power plant using natural gas. Using these values and assuming the efficiency of the water electrolysis plant to be about 80%, the overall efficiency for hydrogen production from the fossil fuels via the water electrolysis route will be in the range of 35 to 45%. It is for this reason that the chemical route of fuel processing is preferred for the large-scale production of hydrogen (Section 8.2). For a smaller scale production—as for instance in the semiconductor industry where ultra-pure hydrogen is required—the water electrolysis route is more economical.

Entering the non-fossil renewable energy resource era, hydrogen production by water electrolysis will be, most probably, the most techno-economically feasible route. Efficiencies for hydrogen generation will be considerably higher with the hydroelectric and wind-generator power plants than with the nuclear or photovoltaic power plants. The costs of such power plants may play an overriding role in determining the cost of hydrogen generation.

8.3.2.2. *Types of Water Electrolyzers and Status of Technologies.* Three types of water electrolyzers have been developed on the type of electrolyte used: alkaline, proton-exchange membranes, and solid-oxide electrolytes.

The most commercialized one is that with an alkaline (6–8 N KOH) electrolyte. In this case too, there is a further classification: monopolar (tank-type cells) and bipolar (filter-press type cells). Figure 8.16 illustrates the fundamental difference in design of the two types of alkaline water electrolyzers.[20] In the monopolar design, the cells in a stack are connected externally in a series/parallel arrangement, while in the bipolar design, a bipolar plate serves as a current collector for the anode on one side and for the cathode on the other (the latter is just as in the case of all types of fuel cells except that for the solid oxide water electrolyzer). The advantages of the bipolar design is that electron flow during electrolysis from one cell to the next is across the whole bipolar plate and ohmic losses are thus minimal. In the unipolar design the current flow is in the longitudinal direction in the electrodes, and thus, the ohmic losses are higher. In the former case, it is possible to attain higher current densities at the same cell voltage.

Significant advances have been made in optimizing the structure and composition of the electrode to minimize activation and ohmic overpotentials in alkaline-water electrolyzers. Noble metals or their alloy electrocatalysts are not required in the alkaline environment. Nickel-based electrocatalysts are mostly used for the oxygen evolution reaction. By use of lithium-doped nickel oxide, nickel cobalt spinel ($NiCo_2O_4$), and perovskite ($Ni_{0.2}Co_{0.8}O_3$), anodic overpotentials are decreased and more importantly, the degradation in performance with time, caused by oxidation of $Ni^{3+}$ to $Ni^{4+}$, has been minimized. For the hydrogen electrode, high surface area nickel or stainless steel are used as electrode materials. The activation overpotential at this electrode is reduced by use of NiB, Ni-S, Ni-Al, Ni-Mo, or $NiCo_2S_4$ thiospinel electrocatalysts. Further, by the use of these compound materials, degradation due to hydrogen entry into nickel, which causes embrittlement problems, are minimized. Even though in the early times of technology the unipolar electrolyzers were the more common ones, in more recent times the transition has been to bipolar electrolyzers. Alkaline water electrolyzers are generally operated at about 70 to 80 °C. Operation under pressure is desirable to minimize energy requirements. Such a large scale electrolyzer of this type, which has been commercialized, was manufactured by Lurgi.

Table 8.2 presents the operating characteristics of the well-developed water electrolyzers. It also includes the operating characteristics of the novel proton-exchange-membrane water electrolyzer (PEMWE), formerly referred to as the solid-

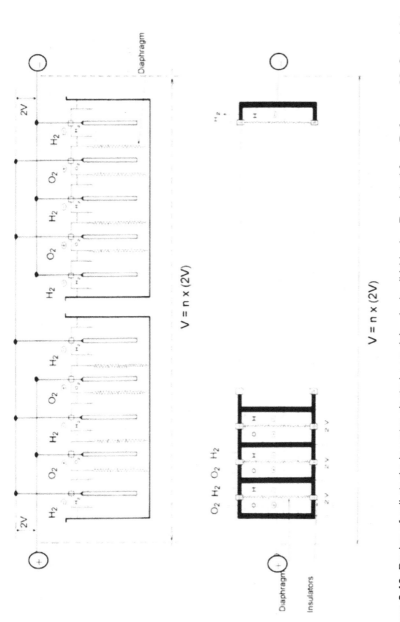

**Figure 8.16.** Design of cell stacks in water electrolyzers (a) unipolar (b) bipolar. Reprinted from Reference 20, Copyright (1981), with permission from Plenum.

**TABLE 8.2**
**Operating and Performance Characteristics of Water Electrolyzers[a]**

| Manufacturer | Brown Boveri and Cie | Brown Boveri Denora SPA | Lurgi GmBH | Norsk Hydro A.S | The Electrolyzer Corp. Ltd. | Krebsokosmo | Teledyne Energy Systems | G.E's SPE |
|---|---|---|---|---|---|---|---|---|
| Cell type | Bipolar filter press | Bipolar filter press | Bipolar filter press | Bipolar filter press | Monopolar tank | Bipolar filter press | Bipolar filter press | Bipolar filter press |
| Operating pressure (atm) | Ambient | Ambient | 32 atm | Ambient | Ambient | Ambient | 2.4 atm | 3.9 atm |
| Operating temperature ($^{0}$C) | 80 | 80 | 90 | 80 | 70 | 80 | 82 | 80 |
| Electrolyte | 25% KOH | 29% KOH | 25% KOH | 25% KOH | 28% KOH | 28% KOH | 35% KOH | DuPont Nafion-1200 EW |
| Current density (A/m$^2$) | 2000 | 1500 | 2000 | 1750 | 1340 | 3000 | 2000 | 5000 |
| Cell voltage (V) | 2.04 | 1.85 (increases to 1.95 after 2 y) | 1.86 | 1.75 (after 1 yr. operation) | 1.90 | 1.90 | 1.90 | 1.70 |
| Current Efficiency (%) | >99.90 | ~98.50 | 98.75 | >98.00 | >99.90 | >99.9 | - | - |
| Oxygen Purity (%) | >99.60 | 99.60 | 99.30-99.50 | 99.30-99.70 | 99.70 | 99.50 | >96.00 | >98.00 |
| Hydrogen Purity (%) | >99.80 | 99.9 | 99.80-99.90 | 98.80-99.90 | 99.9 | 99.9 | 99.99 | >99.00 |
| Power Consumption (DC-kWh per normal m$^3$ H$_2$) | 4.90 | 4.60 | 4.50 | 4.30 | 4.60 | 4.50 | 6.00 | 4.10 |

[a]Reprinted from Reference 1, Copyright (1985) with permission from John Wiley and Sons.

polymer-electrolyte water electrolyzer. This technology was a spin-off of General Electric Company's proton exchange membrane fuel cell technology. A schematic of the single cell in a PEMWE is illustrated in Figure 8.17. The compact design with a thin electrolyte (Nafion) layer and an electrode structure permits the ready escape of evolved gases through the electrodes to the flow channels in the bipolar plate. Ohmic overpotential problems due to gas bubble formation (commonly encountered in alkaline-water electrolyzers) are thus minimized. This makes it possible to attain current densities two to three times higher at the same cell voltage in PEMWEs than those in AWEs and in solid-oxide electrolyte water electrolyzers (see Figure 8.18). The PEMWEs are capable of generation rates of 1 A/cm$^2$ for hydrogen and oxygen, at a cell potential of 1.8 V. Since the electrolyte in the PEMWE is acidic, it is necessary to use noble metal-based electrocatalysts. Platinum is the best electrocatalyst for the hydrogen evolution reaction while a mixed

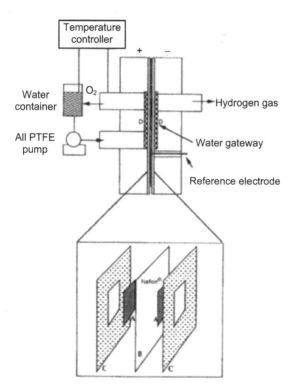

**Figure 8.17.** A design for a single cell in a proton exchange membrane water electrolyzer. Reprinted from Reference 23, Copyright (1993) with permission from Elsevier.

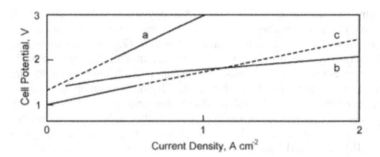

**Figure 8.18.** Typical cell potential vs. current density plots for (a) alkaline, (b) proton exchange membrane, and (c) solid oxide electrolyte water electrolyzers.

oxide electrocatalyst with the composition of $RuIr_{0.5}Ta_{0.5}O_x$ with $x = 2$ was found to be the best one for oxygen evolution.[21,22] These electrocatalysts are electrodeposited or vapor deposited on a metallic substrate. Fine titanium mesh was used for this purpose. Just as in the case of PEMFCs, the electrodes were impregnated with a proton conductor (e.g., Nafion) to extend the three-dimensioned reaction zone.

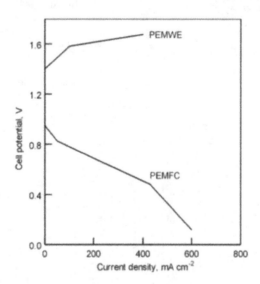

**Figure 8.19.** Cell potential vs. current density plots for a regenerative proton exchange membrane water electrolyzer (PEMWE)/fuel cell (PEMFL).

An added attraction of the PEMWE is that it can also be used in the reverse mode as a fuel cell, i.e., the system would function as a regenerative fuel cell. The cell potential vs. current density performances in both modes, as obtained in a recent research investigation,[23] is shown in Figure 8.19. In the fuel cell mode, it was better to use the RuIrTaO$_x$ electrocatalyst for the hydrogen oxidation reaction and the Pt electrocatalyst for oxygen reduction. Since the electrode structures were not optimized, the performance in the fuel cell mode was considerably less satisfactory than in the state-of-the-art PEMFCs. Giner Inc. has reported[24] the development of a unique, highly-efficient, bifunctional oxygen electrode structure that shows similar performance to the typical, efficient discrete PEMWE and PEMFC oxygen catalysts. Round trip electrical efficiency (PEMFC voltage/PEMWE voltage) for a single unit regenerative fuel cell with this structure is approximately 46% at 500 mA/cm$^2$, 80 $^{\circ}$C, ambient pressure. The General Electric Technology was sold to the Hamilton Standards Division of United Technologies Corporation in the 1980s. Since then, the latter company has had contracts from NASA to develop PEMWEs and Regenerative fuel cells.

The third type of water electrolyzer with a solid oxide electrolyte, was developed by Brown, Boveri (presently ABB)[25] and Dornier.[26] Just as in SOFCs, the electrolyte is ZrO$_2$-Y$_2$O$_3$. Nickel was used for the cathode electrocatalyst and LaMnO$_3$ doped with Sr for the anode electrocatalysts. Advantages of operation at 1000 $^{\circ}$C are: (i) the thermodynamic reversible potential is about 200 mV less than at below 100 $^{\circ}$C; (ii) the temperature effect on the kinetics of the electrochemical reactions greatly reduces the activation overpotentials; and (iii) since the system is a two phase (solid-vapor system), the same cell can be used in both modes (fuel cell/water electrolysis). The net result is that the overall efficiency for the regenerative system can be as high as 70%, just as in the case of rechargeable batteries. Water vapor electrolysis appears attractive because of the considerable decrease in the thermodynamic reversible potential by about 0.2 to 0.3 V at 1000 $^{\circ}$C, as compared with that at 80 $^{\circ}$C, as well as for the significantly higher exchange current densities for hydrogen and oxygen evolution. However, it presents several drawbacks. First, it needs process heat to evaporate the water and maintain the cell temperature because of the positive value of the entropy change, and if the cell potential is less than the thermoneutral potential, the cell will cool. Also, there is a requirement of thermally stable and compatible materials for the cells and cell stacks. Finally, it needs the use of sophisticated and expensive fabrication techniques for the cell stacks, as in the case of SOFCs (see Chapter 4).

### 8.3.3. Photoelectrolysis

The first demonstration of this method of hydrogen production was by Fujushima and Honda, who used a single crystal TiO$_2$ anode electrocatalyst and a platinum cathode electrocatalyst.[27] In a photoelectrochemical cell, an interface of a n-type semiconductor with an electrolyte was illuminated with light energy above

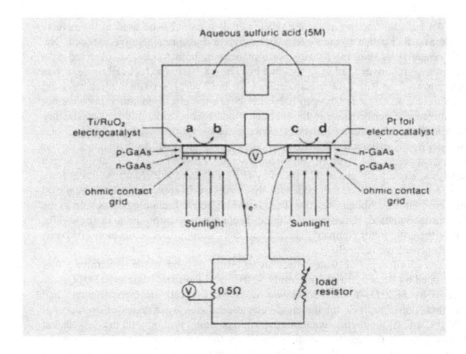

**Figure 8.20.** Schematic of a photoelectrochemical cell for production of hydrogen.

the band-gap of the material (Figure 8.20). This caused the transport of electrons toward the bulk of the semiconductor and holes towards the surface (the photoanode). The opposite behavior is observed at the photocathode. Thus, light promotes photoreduction at p-type semiconductors and photooxidation at n-type materials. One factor, which must be taken into consideration, is that the splitting of water requires an energy input of 285.58 kJ/mol of $H_2$ (2.96 eV). Thus, it is necessary to use a semiconductor material with a band-gap of at least 3 eV at one of the electrodes (e.g., $TiO_2$) to produce two moles of $H_2$ and one mole of oxygen. But at this band gap the material can capture only the ultraviolet light from the solar spectrum, which corresponds to about 2% of the solar energy. Thus, attempts were made to achieve a higher response to solar radiation by the use of sensitizers and dye molecules (e.g., ruthenium 2-2'-bipyridine). These molecules capture the incident sunlight, producing excited states that inject charge carriers into the conduction band of the semiconductor.

A stand-alone photoelectrolysis cell with a reasonably good efficiency was first demonstrated by Kainthala et al.[28] (see Chapter 1 and Figure 1.5). The semiconductor materials used for the electrodes were a Pt-catalyzed indium-phosphide cathode and a $MnO_2$ coated gallium-arsenide anode (the $MnO_2$ coating was to prevent electrooxidation). An overall efficiency of about 8% was obtained. More recently, values of 16% have been claimed, making this approach potentially better than the dual approach of using photovoltaic/electrolysis cells to decompose water. However, problems in photoelectrochemical cells are most challenging because of the needs of very large area electrodes, circulation of the electrolytes, and auxiliary requirements for thermal management. On the contrary, the technology for manufacturing photovoltaic cells, producing kW to MW electric power has been demonstrated, and these systems can be readily coupled with large water electrolyzers, which are already available (see Section 8.3.2). Thus, there is only an academic interest in methods to enhance the efficiencies of photoelectrochemical reactions using (i) dye sensitizers, (ii) electrocatalysts for recombination of the H and/or O atoms produced in the electrochemical reactions, and (iii) alternate reactions with redox couples to lower the band-gap requirements for photosplitting of the reactant molecules (e.g., HBr or HI).

## 8.3.4.  Thermochemical Decomposition of Water

In the 1970s, there was great interest in hydrogen production using thermal energy, particularly utilizing the high-grade waste heat energy from thermal power plants as well as from chemical plants. It was stated in the preceding Section that the dissociation energy for the decomposition of abundant liquid water into its constituents is 285.58 kJ/mol $H_2$. At any temperature, water is in equilibrium with hydrogen and oxygen according to equation 8.12. The equilibrium constant for this reaction is expressed by:

$$K = \frac{P_{H_2} P_{O_2}^{1/2}}{P_{H_2O}} \qquad (8.19)$$

The extent of the thermal dissociation of water as a function of temperature is shown in Figure 8.21. For a complete dissociation in a stand-alone system, a temperature of about 4000 K is required. Further, it will be necessary to separate the product gases, for instance by using some selective membranes.

The concept of a thermochemical cycle for water decomposition was proposed so as to employ different schemes for the intermediate steps. Those having positive entropies are driven at higher temperatures and the intermediate steps having negative free energies at lower temperatures. Thus, if the first reaction has a positive entropy change, the second reaction may occur with a more favorable reaction rate at the same or preferably lower temperatures. Several thermochemical cycles have been researched for hydrogen production. As an example, a route to hydrogen production is:

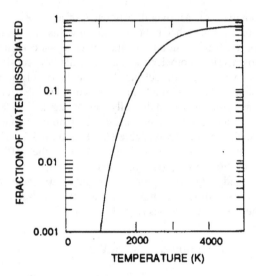

**Figure 8.21.** Dissociation of water as a function of temperature.

$$SO_2 + H_2O + MO \xrightarrow{\quad 500 \text{ K} \quad} MSO_4 + H_2 \qquad (8.20)$$

$$MSO_4 \xrightarrow{\quad 1100 \text{ K} \quad} MO + SO_2 + \frac{1}{2}O_2 \qquad (8.21)$$

$$SO_3 \rightarrow SO_2 + \frac{1}{2}O_2 \qquad (8.22)$$

where M is a metal, like iron. The theoretical efficiency of a thermochemical cycle is limited by the Carnot efficiency. However, because at least one of the intermediate steps is carried out at a high temperature, high values for efficiency could be predicted. Further cost of chemicals used in several thermochemical cycles (e.g., $Fe_2O_3/FeSO_4$, $FeCl_2/Fe_3O_4$, $HI/SO_2$, $CaO/I_2$, $CaBr_2$) are quite low, and there is a considerable use of inexpensive heat energy rather than expensive electrical energy as needed for water electrolysis for hydrogen production. Thus, there were many research programs in the USA, Russia, Germany, and Japan in the 1970s and early 1980s to develop such systems but the prospects of using this technology were found to be bleak due to the following reasons:

(a)  Carnot limitations makes it difficult to attain more than two thirds of theoretical efficiency;
(b)  high efficiencies, on the basis of thermodynamics, could not be attained in practical systems, because of kinetic constraints;
(c)  several of the intermediate chemical reactants or products rapidly corrode reaction vessels at the high temperatures required for at least one of the steps in the overall cycle;
(d)  energy requirements for pumping and product gas separation are high; and
(e)  loss of chemicals because of lack of complete cyclicity.

An alternate route that appeared more promising is the thermochemical-electrochemical hybrid cycle, extensively investigated by the Westinghouse Electric Corporation. It has the following intermediate steps:

$$\text{Anode:} \qquad SO_2 + 2H_2O \rightarrow H_2SO_4 + 2H^+ + 2e_0^- \qquad (8.23)$$

$$\text{Cathode:} \qquad 2H^+ + 2e_0^- \rightarrow H_2 \qquad (8.24)$$

$$\text{Thermal:} \qquad H_2SO_4 \rightarrow SO_2 + H_2O + \frac{1}{2}O_2 \qquad (8.25)$$

In the electrochemical cell, akin to a water electrolysis cell, hydrogen is evolved at the cathode and $SO_2$ is oxidized to $H_2SO_4$. The latter reaction occurs instead of the oxygen evolution reaction, which requires a considerably higher voltage (about 1 V). Lowering the cell potential by using $SO_2$ instead of water as the anodic reactant is thus possible because of the significant difference in the thermodynamic reversible potentials for the $SO_2$ oxidation and oxygen evolution reaction. Sulfuric acid is produced at a concentration of about 80%. It is then vaporized to form $SO_3$ and $H_2O$, and $SO_3$ is decomposed catalytically to $SO_2$ and $O_2$ in a thermal reactor. Even though there is some overpotential associated with the $SO_2$ oxidation in the electrochemical cell, Westinghouse estimated a decrease in electric energy consumption by 40% using this hybrid cycle instead of water electrolysis.

### 8.3.5.  Biomass Fuel Production and Conversion to Hydrogen

Biomass energy resources are derived from plants, trees, and crops, as well as from agricultural and forest residues and organic waste-streams. According to the World Energy Assessment (WEA) in the United Nations Development Program (UNDP), the contribution of biomass to the world's total energy supply (electricity, heat, and fuels) is about 9.13% (45 ± 10 exajoules per year). The main biomass energyconversion routes,as described in UNDP's WEA, are illustrated in Table 8.3.

**TABLE 8.3**
**Main Biomass Energy Conversion Routes**[a]

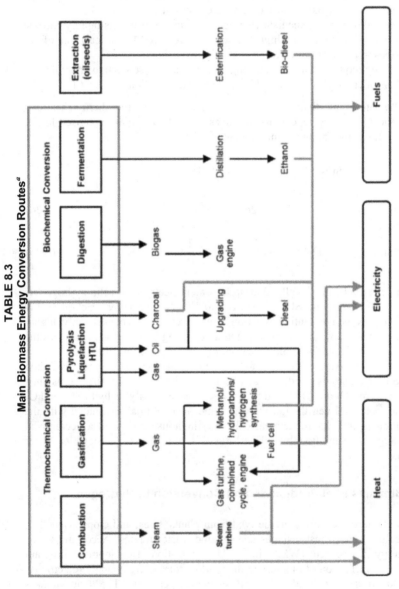

[a]Reprinted from Reference 19.

The focus of this Section will be on the production of $H_2$ and of $H_2$ plus CO for utilization in PAFCs, MCFCs, and SOFCs.

Table 8.3 presents several routes, thermal gasification, pyrolysis, liquefaction, anaerobic digestion, fermentation, or extraction of oil seeds for production of gaseous-liquid hydrocarbon or alcohol fuels. Most of these fuels can be processed to produce hydrogen or hydrogen plus carbon monoxide for subsequent utilization in fuel cells. Alternatively, the fuels could be used in thermal engines (gas turbines, IC engines, diesel engines). In some countries (Scandinavian countries, India, China) where biomass derived fuels are extensively used for combined electricity and heat production, the most efficient method is the biomass-integrated-gasifier/combined-cycle gas turbine. For the fuel cell application, $H_2$ or $H_2$ plus CO production involves the sequence of steam reforming and the two-stage shift conversion (see Section 8.2.3). If pure hydrogen (purity of 99.99%) is required, as for PEMFCs or AFCs, the pressure-swing absorption method (see Section 8.2.1) can be used.

Even though there are several routes (Tables 8.3) for production of biomass derived liquid or gaseous fuels, the fuel cell community has concentrated on the processing of biomass derived gaseous or liquid fuels produced either by biomass gasification or anaerobic digestion. However, the fuels produced by the other methods shown in Table 8.3 could also be similarly processed. Biomass gasification produces syngas. Fluidized bed gasifiers are generally used for large-scale syngas production. It is preferable to use oxygen rather than air for the gasification, because the latter method produces a high amount of nitrogen. The product gas must be cleaned up before use for hydrogen or hydrogen plus carbon monoxide production. Particulate matter is removed using a cyclone separator, and water quenching removes small particles, condensable hydrocarbons, alkali compounds, and ammonia. Fabric filters can be used for eliminating sub-micron particles. A zinc oxide bed can capture sulphur compounds.

## 8.3.6.  Biological/Biochemical Production of Hydrogen

Several methods have been proposed and investigated for the biological/biochemical production of hydrogen. These will be briefly summarized as follows:

(a)  *Use of photosynthetic catalysts.* In this approach, hydrogen is generated from water and sunlight using photosynthetic catalysts such as heterocystous blue-green algae. The algae contain the enzyme hydrogenase, which catalyzes the photochemical decomposition of water to produce hydrogen. This reaction is inhibited by the formation of oxygen in the reactor. Efficiencies of only 3% have been achieved. Since the biochemical production also occurs at low temperatures, this method may be useful for only low rates of hydrogen production.

(b)  Anaerobic Digestion. This method can produce a variety of biochemicals from vegetable and animal wastes using fermentative and acetogenic bacteria. The fermentative bacteria can cause a breakdown of several biochemicals, as for example polysaccharides into sugars; proteins into peptides and amino acids; fats into glycerine and fatty acids; and nucleic acids into nitrogen heterocyclics ribose and inorganic phosphates. The acetogenic bacteria cause a further breakdown of the sugars, alcohols, and higher acids into acetic acid, hydrogen, and water. Typical reactions may be represented by:

Fermentative bacteria:

$$C_6H_{10}O_5 + H_2O \rightarrow (CH_2O)_6 \qquad \Delta H^0 = -17.7 \text{ kJ/mol} \qquad (8.26)$$

$$(CH_2O)_6 + 3H_2O \rightarrow CH_3CH_2OH + 4CO_2 + 6H_2$$
$$\Delta H^0 = -192 \text{ kJ/mol} \qquad (8.27)$$

Acetogenic bacteria:

$$(CH_2O)_6 + 4H_2O \rightarrow 2CH_3COOH + 4CO_2 + 6H_2$$
$$\Delta H^0 = -21.6 \text{ kJ/mol} \qquad (8.28)$$

$$CH_3CH_2COOH + 2H_2O \rightarrow CH_3COOH + CO_2 + 3H_2$$
$$\Delta H^0 = -71 \text{ kJ/mol} \qquad (8.29)$$

These reactions are not inhibited by the presence of oxygen. Using this approach, the hydrogen generation occurs at low efficiencies and at low rates. This approach for hydrogen production is far fetched because (i) it requires large land requirements, (ii) the hydrogen generation rates are low, and (iii) capital costs are prohibitive. There may be some applicability in remote rural areas where the power requirements will be low.

## 8.4.  OTHER FUELS FOR DIRECT OR INDIRECT UTILIZATION IN FUEL CELLS

### 8.4.1.  Partially-Oxygenated Carbonaceous Fuels

8.4.1.1.  *Methanol and Ethanol.* Hydrogen is the most electroactive fuel for low and intermediate temperature fuel cells (PEMFC, AFC, PAFC), while $H_2$ or $H_2$ plus CO are the most electroactive for the high temperature fuel cells. Even in the latter case, it is very possible that CO, along with steam, undergoes the water gas shift reaction on the anode electocatalyst and the half-cell reaction at this electrode is the electro-oxidation of hydrogen. During the 1960s, there was great interest in the direct utilization of saturated and unsaturated aliphatic hydrocarbons in, predominantly, phosphoric acid fuel cells. However, in practically all these cases, the exchange current densities for their electro-oxidation were found to be extremely low, even considerably lower than for the cathodic reduction of oxygen. The reason for this is the difficulty of breaking up the C-H, and more so the C-C bonds, in these hydrocarbons. Since the 1970s, progress has been made at the Energy Research Corporation (recently with a name change to Fuel Cell Energy) in the direct utilization of natural gas in MCFCs and more recently this approach has been investigated in SOFCs. In these cases, the anode electrocatalyst (or sometimes a second catalyst behind this one in the electrode) first performs the steam-reforming reaction and perhaps the shift-conversion reaction and then the $H_2$ or $H_2$ plus CO are electro-oxidized at the anode.

The alternate approach in the 1960s was to investigate partially oxidized organic fuels for direct utilization in fuel cells. From that period until the present time, methanol has been found to be the most electroactive organic fuel. It has been found to be relatively easy to dissociatively adsorb methanol on a platinum electrocatalyst but somewhat challenging to further electrochemically oxidize these adsorbed species to carbon dioxide and water. The details of the electrode kinetics and electrocatalysis of electrooxidation of methanol are presented in Chapter 4, Section 4.3. Since the early 1990s, there was more success in developing DMFCs, with perfluorosulfonic acid membranes, rather than with sulphuric acid, as the electrolyte as had been previously used from the 1960s (see Chapter 4). Another partially oxidized hydrocarbon fuel of great interest for fuel cell application is ethanol. Both methanol and ethanol are liquid fuels, like gasoline, and have about 50% to 60% of the energy density of the latter fuel. Thus, they are appealing fuels for the transportation application. The difficulty of the direct utilization of ethanol in fuel cells is because even though it dissociatively adsorbs on the platinum electrocatalyst, it is difficult to break up the C-C bond. Further, in low temperatures fuel cells, it is only partially oxidized to acetic acid. However, according to a recent study,[29] it was shown that at about 140 to 150 °C, there is complete electro-oxidation of ethanol to $CO_2$ and $H_2O$ in a direct-ethanol fuel cell with a Nafion/silicon oxide-composite proton-exchange membrane (see Chapter 4). Another partially oxidized fuel of some interest was glucose for biomedical applications, e.g., as a fuel-cell power-source for pacemakers or artificial hearts. The electroactivity of this fuel was far less than that of methanol or ethanol. In    the remainder of this Section, we shall make some brief comments about the production of methanol and of ethanol, (partially-oxidized organic fuels) to be used directly or indirectly in fuel cells.

The largest application of methanol is in the plastics industry and on a large scale it is produced predominantly in the USA from natural gas by first steam reforming it to CO and $H_2$ (Eq. 8.1), and then, by the reaction of these two gases on a catalyst to produce methanol (Eq. 8.3). For this reaction, the pressure is 60-80 atm, the catalyst is CuO-ZnO, and the desired operating temperature is 250-280 °C. During the steam reforming of methanol, $CO_2$ is also formed, and this gas can also be converted to methanol. An effective catalyst for methanol production by this route is:

$$CO_2 + 3H_2 \rightarrow CH_3OH + H_2O \tag{8.30}$$

This reaction increases the yield of methanol production.

Another largely available primary energy source for methanol production is coal. In this case, syngas is first produced by the reaction represented by Eq. (8.4), and this reaction is then followed by the one expressed by Eq. (8.3). Since the reaction producing methanol is exothermic, the reactors are designed to have efficient cooling by water circulation. In some cases, cooling is affected by feeding cold synthesis gas and having heat exchangers.

A third approach for methanol production is from biomass. Biomass is a renewable, clean feedstock with less environmental problems than coal. The major problem is their scattered availability. Liquid fuels can be obtained from biomass by pyrolysis, direct liquefaction, or by converting biomass, first to syngas and then to methanol (Eq. 8.3).

In respect to economics, the cost of methanol production on a large scale is lowest when natural gas is used as the primary energy source. It is somewhat higher from coal and still higher from biomass. The lowest cost of methanol production is about twice that of gasoline, based on an energy equivalent basis.

Ethanol is mainly produced on a large scale by fermenting sugars in sugar cane, maize, and corn. The countries noted for large-scale production of ethanol are Brazil, France, the USA, and Zimbawe. The total amount of ethanol produced worldwide is about 20 billion liters ($\approx 450$ petajoules). The fermentation process involves the breakdown of the sugars in very much the same way as for the production of wines, using organisms such as yeast and bacteria. The biocatalysts for the reactions are acids or enzymes, commonly referred to as amylases or xylanases. The advantages of biological processes are their high specificities for the production of desired products from the biomass. Further, these processes occur at ambient temperatures and pressures. One problem is that the costs of production of such fuels are higher by about a factor of two, compared to the costs of production of gasoline and diesel fuel from petroleum resources. This is the reason for the gradual decrease in the manufacture of alcohol-fueled automobiles in Brazil, which was at the highest level in the 1980s. Nevertheless, with the great concerns of dependence on foreign oil, the environmental pollution problems from automobiles, and the steadily decreasing economic situation of farm workers, there is still interest in ethanol as a fuel for the transportation application. Further, because of the increasing prospects of developing direct-ethanol fuel cells (DEFCs) operating at intermediate temperatures, there is some hope for utilization of this potentially abundant fuel directly in fuel cells.

8.4.1.2.   *Dimethyl Ether (DME).*   During the last 4 to 6 years, there has been interest in using dimethyl ether as a fuel directly in diesel engines for transportation application and in fuel cells. DME, like methanol or ethanol, is a clean fuel; since it is also a partially oxidized fuel, it undergoes complete combustion to carbon dioxide in a thermal engine. In a compression-ignition diesel-injection (CIDI) engine, DME has a high cetane number and the tail-pipe emissions are extremely low, unlike in the case when diesel fuel is used. It has also been recently demonstrated that in a proton-exchange-membrane fuel cell, DME has about the same level of electrochemical activity and yields $CO_2$ and water as its main production of electro-oxidation.[30] Since DME is a gas at room temperature (its boiling point is $-25$ °C), it has to be stored at a pressure of about 10 bar, i.e., somewhat the same as for liquid propane gas.

Just as in the case of methanol, DME can be produced from natural gas or syngas. The reactor is practically the same as that for methanol production except that it also contains a solid dehydration catalyst for the reaction:

$$2CH_3OH \rightarrow CH_3OCH_3 + H_2O \qquad (8.31)$$

By selecting the operating conditions, the extent of coproduction of dimethyl ether and methanol can be varied over a wide range (5 to 95%).

At the present time, DME is mainly used as an alternative to chlorofluorohydrocarbons (CFCs) in aerosol sprays. Unlike in the case of CFCs, DME does not affect the ozone layer. It is also non-toxic and non-carcinogenic. Other uses of DME are in the synthesis of oxygenated hydrocarbons (e.g., methyl acetate) and higher hydrocarbon ethers. DME is also used as an intermediate in the Mobil MTG process for the production of gasoline from methanol. The companies, Haldor Topsoe and Amoco, estimate that if low cost natural gas is available, DME can be produced at costs comparable to that of diesel fuel.

### 8.4.2. Nitrogenous Fuels

8.4.2.1. *General Comments.*   The nitrogenous fuels ammonia and hydrazine have hydrogen contents comparable to gaseous, liquid, or solid hydrogen fuels (see Section 8.5.1). Ammonia is widely used by the agricultural industry for the production of fertilizers. It has also been used as the fuel for farm vehicles powered by internal combustion engines. It is difficult to electrooxidize ammonia completely to $N_2$ and water in fuel cells. A partial oxidation product is hydroxyl amine. However, there has been interest in using ammonia as a storage medium for fuel cells. The hydrocracking of ammonia can be carried out in a relatively simple fuel processor at a temperature of 450 °C and at a pressure of 10 bar. The product hydrogen can then be used in a fuel cell.

Hydrazine is a rocket fuel. It is electrochemically active in a fuel cell. The accepted view is that in an alkaline fuel cell, hydrazine is decomposed to nitrogen and hydrogen and the hydrogen is then electrooxidized at the anode. Since hydrazine is a liquid with a relatively high solubility in water, it is necessary to have a separator in the fuel cell to prevent its diffusion to the cathode, where it causes depolarization. The following Sections will deal briefly with the synthesis of ammonia and of hydrazine.

8.4.2.2. *Ammonia.* Ammonia is synthesized on a large scale for the agricultural industry by the Haber process. In the USA, the primary reactants are natural gas and nitrogen. Natural gas is first converted to hydrogen by the steam reforming and shift conversion reactions, followed by the methanation or pressures swing absorption methanol for its purification (see Section 8.1). Hydrogen and nitrogen (separated from air by liquefaction) are reacted using iron as the catalyst, in the well-known Haber process for the large scale production of ammonia, according to the following reaction:

$$N_2 + 3H_2 \rightarrow 2NH_3 \qquad \Delta H^0 = -92 \text{ kJ/mol} \qquad (8.32)$$

The number of moles in the reactant mixture is reduced during product formation and thus the reaction is exothermic. According to Le Chatelier's principle, higher pressures and lower temperatures accelerate the reaction rate. Industrial production of ammonia is carried out in the pressure range 200 to 1000 atm and at a temperature of about 450 °C. Apart from iron, several other metals have been tested as catalysts (e.g., Mo, W, Mn, Ru, etc). By using only the metals as catalysts, there is some degradation in activity. However, by using catalytic promoters, degradation is minimized. The generally accepted mechanism for ammonia synthesis follows the sequence:

$$N_2 + \frac{3}{2}H_2 \rightarrow N_{ads} + 3H_{ads} \qquad (8.33)$$

$$N_{ads} + H_{ads} \rightarrow NH_{ads} \qquad (8.34)$$

$$NH_{ads} + H_{ads} \rightarrow NH_{2ads} \qquad (8.35)$$

$$NH_{2ads} + H_{ads} \rightarrow NH_3 \qquad (8.36)$$

Since the adsorption behavior of atomic nitrogen and of hydrogen are of the Temkin type (see Chapter 1), there is a decrease in the activation energy for adsorption with increase of coverage of the intermediate species. The nitrogen-adsorption intermediate step (Eq. 8.29) is the rate-limiting one. The quantity of ammonia production in the USA exceeds 30 million tons/year, and the bulk of it is utilized by the fertilizer industry.

8.4.2.3. *Hydrazine.* The industrial production of hydrazine uses the Raschig process. It involves the partial oxidation of ammonia using sodium hypochlorite in the presence of gelatin or glue at a temperature in the range of 160 to 180 °C and elevated pressure. The product is a dilute liquor at a concentration of only 2%. It is concentrated by fractional distillation to a product 85 to 100% hydrazine hydrate ($N_2H_4 \cdot H_2O$). If anhydrous hydrazine is required, this product is dehydrated using barium oxide or sodium hydroxide. Anhydrous hydrazine can also be obtained from hydrazine sulphate, by its reaction with ammonia.

## 8.5    Fuel Storage

### 8.5.1.  Hydrogen

*8.5.1.1.   Multifold Options but Many Challenges.* Hydrogen is the *ideal* fuel for all types of fuel cells and it is the *pristine* one for PEMFCs and AFCs. Table 8.4 summarizes the investigated methods for hydrogen storage and their performance characteristics. Hydrogen can be stored as a compressed gas, liquid, solid, or combined with chemicals. Each method has some advantages and disadvantages. Technical and economic aspects of these storage methods will be summarized in the following Sections. It is assumed in these Sections that hydrogen is produced on a commercial scale by one or more of the methods described in the preceding Sections. Thus, an energy analysis should deal with the entry of hydrogen into the storage system, any conversion in the storage system and the release of hydrogen for input into the fuel cell. The economic aspects should also be taken into consideration.

Section 8.5.3 considers a techno-economic analysis of hydrogen storage versus alternate fuels, mainly hydrocarbons, alcohols, and ethers. But in these cases, it must be remembered that for the low and intermediate temperature fuel cells, these fuels will have to be converted to hydrogen (see Section 8.2) and for the high temperature fuel cells, to hydrogen plus carbon monoxide. The exceptions are methanol, ethanol, and dimethyl ether that may be used directly in proton-exchange-membrane fuel cells. Another exception, as described in Section 8.2.6, is the case where a fuel, like natural gas, can be fed directly to a MCFC or SOFC and the anode electrocatalyst (or a separate catalyst) converts methane into hydrogen and carbon monoxide, which are then electrooxidized. Energy efficiencies and costs of hydrogen entering the fuel cell will also be taken into consideration in Section 8.5.3.

*8.5.1.2.  Compressed Hydrogen Stored Underground for Power Generation.* Large quantities of hydrogen can be stored underground in natural gas fields, aquifiers, and salt domes. The capacities can be as large as 1 billion $Nm^3$ in aquifiers or gas fields and several million $Nm^3$ in caverns. The pressure for storage is about 60 atm. Only a fraction of this amount (one to two thirds) is available for the storage cycle, mainly to maintain a cushion gas pressure. These systems can provide 1 to 10 M $Nm^3$ of hydrogen. The energy density for hydrogen

## TABLE 8.4
### Techno-Economic Assessments of Hydrogen Storage Systems

| Storage Method | Hydrogen weight (%) | Content (vol g/l) | Energy density (kJ/kg) | Higher heating value (kJ/l) | Relative cost[a] | Demonstrated/ potential applications |
|---|---|---|---|---|---|---|
| Gas | | | | | | |
| Steel cylinder (60 kg/50 l/200 atm) | 1.5 | 18 | 2,132 | 1,003 | 2 | Lab chemical, transportation |
| Alumina composite (75 kg/125 l/200 atm) | 2.6 | 17 | 3,700 | 1,739 | 3 | Transportation, portable power |
| Glass microspheres | 6 | 6 | 8,527 | 853 | 5 | Scientific curiosity |
| Zeolites | 0.8 | 6 | 1,128 | 814 | 5 | Residential industrial fuel |
| Liquid | | | | | | |
| Cryogenic 300 m³ semitraiter | 12.5 | 71 | 17,765 | 9,919 | 1 | Rocket fuel, space fuel cells |
| Solid | | | | | | |
| $FeTiH_2$ | 1.6 | 96 | 2,278 | 13,564 | 3 | |
| $LaNi_5H_6$ | 1.4 | 89 | 1,990 | 12,749 | 3 | Portable power, transportation |
| $Mg_2NiH_4$ | 3.2 | 81 | 4,514 | 11,474 | 3 | |
| Combined with Chemicals | | | | | | |
| n-Octane | 15.8 | 11 | 47,652 | 33,524 | 1 | Transportation |
| Methanol | 12.5 | 150 | 22,321 | 17,665 | 1 | Transportation, portable power |
| Ammonia | 17.6 | 136 | 22,363 | 17,222 | 1 | Farm vehicle, portable power |

[a]Code for Relative Cost: 1 → Least expensive capital cost (~ $10/kg of $H_2$); 5 → Most expensive

is only about one third that for natural gas. Therefore, it is more costly to store hydrogen by this method, not only for this reason, but also due to the production costs of hydrogen from primary energy sources raising the cost of hydrogen. The energy efficiency for this method of storage should take into consideration the electrical energy for compression of the gas. According to some analysis, fuel processing of $CH_4$ to $H_2$ is carried out under pressure. Therefore, for further compression of $H_2$, the energy requirement for energy storage is reduced to about 10% of the fuel energy. Some of this energy may be recovered by using a compressor-expander cycle for the entry of hydrogen into the storage system and its delivery to the fuel cell.

*8.5.1.3. Compressed Hydrogen Storage in Cylinders Tanks for Transportation and Portable Power.* The most common method for storage of hydrogen on a relatively small scale for automobiles and buses is as a compressed gas in stainless steel cylinders at a pressure of about 130 to 150 bar. The cylinder weight is about 10 kg and its volume is about 40 liters. The hydrogen content is only 0.5 kg. Thus, the specific energy and energy density are very low (see Table 8.4). Considerable efforts have been made to substantially increase the hydrogen content by using light-weight high-strength materials, such as fiber-glass reinforced aluminum and carbon composites, as well as using higher pressures (up to about 350 bar). As seen from Table 8.4, the specific energy and energy density can be improved by about a factor of at least five by using the advanced materials for construction of the cylinders. The electric energy penalty for compression of the hydrogen is about 10% of that in the fuel. Since the lifetime of a cylinder can exceed 5 to 10 years, there is only a marginal cost, which can be added to the cost of the hydrogen.

On a somewhat larger scale and more so for hydrogen distribution stations, one can have above-ground compressed gas storage in pressure tanks. The pressures can be in the range of 100 to 500 bar. These can contain 175,000 to 250,000 liters of hydrogen. Since tank storage of this type is modular, there is only a linear economy for scale-up. Energy penalty and cost for such a method of storage (mainly due to electricity costs for compression and a minor amount due to capital costs) are about 7-10% of the fuel energy and $3,000-$5,000/GJ, respectively.

Since the early 1970s, there was some interest in storing hydrogen in glass bead microspheres. The idea was that the glass microspheres could withstand hydrogen pressures up to 700 bar, while the pressure in the container was much less. Thus, from a safety point of view, this method could be considerably better. However, because of the slow rate of breakage of the glass beads with long term repeated cycling with hydrogen (i.e., hydrogen in, hydrogen out), this approach has virtually been abandoned.

*8.5.1.4. Liquid Hydrogen Storage for Space, Military, and Transportation Applications.* Hydrogen is the lightest fuel in terms of specific energy (kWh/kg) and oxygen is the lightest oxidant. It is for this reason that NASA chose these two chemicals in the cryogenic form for space flights, which last for two weeks or longer. The largest storage vessel used by NASA has a capacity of 3 million liters, corresponding to about 10 million kWh. In the long term space program (e.g., the space station and Mars), it is proposed to use photovoltaic power generators to generate electricity for utilization during the *light* periods, as well as for electrolyzing water (produced from fuel cells) to regenerate hydrogen and oxygen and then liquefy these two gases. The military is also interested in the storage of hydrogen and oxygen cryogenically for fuel cell powered submarines. On the terrestrial front, when large quantities of hydrogen are required and need to be transported, the cryogenic method seems to be the ideal one.

The liquefaction of hydrogen is a highly energy intensive process, theoretically requiring about 3.3 kWh/kg. However, during liquefaction, the ortho-para conversion of $H_2$ occurs, and this increases the theoretical energy to about 3.8 kWh/kg. The practical energy requirements are, however, considerably higher, by a factor of about 6 for small liquefiers and 3 for large ones. The net result is that at least 35% of the hydrogen energy is required for the liquefaction process.

The technology for liquefaction of hydrogen typically involves a flow process, as represented in Figure 8.22. Reciprocating compressors had been traditionally used but since centrifugal compressors are more compact and less expensive, there is interest in the use of the latter in more advanced plants. Hydrogen liquefaction involves the use of an admixture of a high molecular weight gas (propane, see

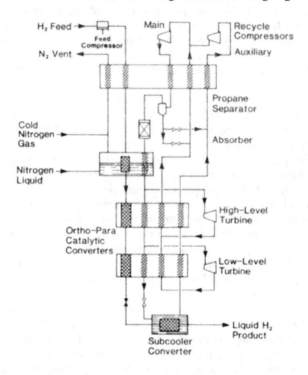

**Figure 8.22.** Process diagram for liquefaction of hydrogen. Reprinted from Reference 1, Copyright (1985) with permission from John Wiley and Sons.

Figure 8.22) to use the centrifugal compressors more efficiently. The propane gas is subsequently separated from $H_2$ by liquefaction. The condensate is blended with the hydrogen and fed back to the compressors. Cold nitrogen is introduced at this stage for cooling, and the input of liquid nitrogen provides additional cooling and promotes the ortho to para conversion of hydrogen. The latter is an important step during liquefaction of hydrogen. Since the ortho to para conversion step is exothermic, efficient heat removal becomes necessary. Use of catalysts can minimize the heat generation and boil-off loss can be minimized. The energy required for the ortho to para conversion (up to 95%) is about 18% of the total energy for the liquefaction process. Four difficulties are associated with the liquefaction process:

(a) liquefaction of hydrogen occurs at an extremely low temperature, 20.4 K. (cf., BP of methane is 109 K and of $N_2$ is 78 K). The fall in temperature of gas introduced into a cold tank simulates a cryogenic pump and thus cause introduction of impurities;

(b) the Joule-Thomson (J-T) inversion temperature for hydrogen (204 K) is below room temperature. An auxiliary liquefaction plant for another gas (e.g., $N_2$) is thus required to lower the temperature of hydrogen below its J-T inversion temperature. This increases energy costs and energy penalties;

(c) the refrigeration process is Carnot limited and this efficiency reduces to zero at the absolute zero temperature. The liquefaction temperature is only 20 °C higher; and

(d) at room temperature the ratio of ortho to para hydrogen is 3:1. The para form has the lower energy. At the temperature of liquefaction, the percentage composition of para is 99.7%. Even though this reaction (ortho to para conversion) is exothermic, it only occurs as the temperature is substantially lowered and thus the energy requirements for liquefaction are high.

Dewar flasks are used to store liquid hydrogen ($LH_2$). The flask is then immersed in a second one containing liquid nitrogen ($LN_2$) to prevent heat absorption by thermal radiation.[*] By having the outer flask containing $LN_2$, $\Delta T$ is significantly smaller. This method is good only for small quantities of $LH_2$ ($10^5$ to $10^6$ liters). Using thermal radiation shields positioned within the multilayer insulation, additional heat loss is prevented. The largest $LH_2$ storage vessels at NASA's Kennedy Space Center are two 3–4 million liter spherical containers. The material for construction of the inner shell (20 m diameter) is austenitic steel. The carbon outer shell has a diameter of 23 m. The operating pressure for the delivery of hydrogen is 6 atm and the boil-off rate is about 0.02%/d, when the tank is full. The capital cost for liquefaction decreases with scale-up, but reaches a steady value at about 225,000 kg/d. $LH_2$ tanks have also been developed in Germany for delivery of the gas to fuel cells. BMW has demonstrated IC engine powered vehicles with $LH_2$. The storage tanks in the vehicles are of the Dewar flask type. The walls are 3 cm thick and contain aluminum foil interlaced with fiberglass matting. More recently, Daimler Benz has also exhibited a fuel-cell automobile, powered with $LH_2$. The main difference between the two types of fuel feed to the two vehicles is that in the

---

[*] Thermal heat radiation is proportional to $\Delta T^4$, where $\Delta T$ is the temperature difference between $LH_2$ in the flask and the outside temperature.

former case, hydrogen is delivered in the liquid form, while in the latter, it is in the gaseous form and also heated to the desired temperature. The waste heat from the fuel cells may be efficiently used for this purpose.

In respect to economics, the cost of $LH_2$ is about \$12/GT for large-scale use. The costs are higher by a factor of 5 to 10 for small-scale use in vehicles. It may be more appropriate to consider $LH_2$ for use in fleet vehicles, ships, and trains. Apart from the use in the NASA program, $LH_2$ is used on a large scale by the petroleum refining and ammonia production industries.

8.5.1.5.  *Solid Storage of Hydrogen as Metal Hydrides.*  The preceding Sections dealt with the storage of pure hydrogen as a compressed gas and as a liquid. Storing pure hydrogen as a solid is considerably more energy intensive than storing it cryogenically. However, there is one alternate approach in storing hydrogen as a solid, i.e., as a metal hydride. Researchers at Brookhaven National Laboratory in New York, USA and at the Philips Research Laboratory in Eindhoven, Netherlands, both of which started in the 1960s, carried out the pioneering work in this area. Quantum leaps have been made in metal hydride technology, within the last 10 years, mainly for the development and commercialization of nickel/metal hydride batteries. The use of metal hydrides as a source of hydrogen for fuel cells has been the subject of great interest since the 1970s. There has been a multitude of demonstrations of the integration of the metal hydride-storage systems with fuel cells. Other applications of such hydrogen-storage systems include their use in heat pumps and compressors. Metal hydrides consist of three types:

- ionic, e.g., magnesium hydride,
- covalent, e.g., hydrides of Be and of the Group 3 metals, and
- intermetallic, hydrides of the transition and rare earth metals.

For the hydrogen storage application, the ideal type of hydrides is the metals/alloy, which reversibly absorb and desorb hydrogen. An example is given by the Fe-Ti alloy:

$$FeTi + H_2 \rightarrow FeTiH_2 \qquad (8.37)$$

The hydrogen absorption/desorption characteristics of a metal or alloy are best represented by the pressure–composition–temperature ($P$-$C$-$T$) isotherms, as illustrated in Figure 8.23. The extent of absorption depends on the equilibrium pressure. The hydrogen molecule dissociatively adsorbs on the metal or alloy. At relatively low pressures, the adsorbed hydrogen diffuses into the bulk and occupies interstitial sites in the metallic lattice and the isotherm ascends steeply as hydrogen absorbs into the metal, forming a solid solution or the α-phase. In Figure 8.23, region A-B signifies a limited amount of hydrogen in the metal or alloy. The region

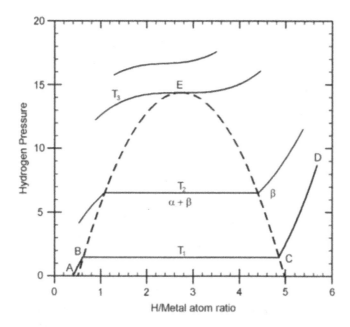

**Figure 8.23.** Typical pressure-composition-temperature (P-C-T) plots for hydrogen absorption by materials and alloys at three temperatures ($T_1 < T_2 < T_3$).

B-C represents the appearance of a second phase, i.e., the β-phase, which is in equilibrium with the solid solution or α phase. The β-phase is the one in which there is the highest amount of hydrogen absorption into the metal or alloy. According to the Gibbs phase rule, the pressure must be constant in this region, i.e., shown in the Figure as a plateau pressure. After the β-phase region absorbs a saturation amount of hydrogen, it can be further absorbed in the C-D region, but in this region the equilibrium pressure sharply increases with the amount of hydrogen absorbed. With the increase of temperature, the plateau pressure region becomes smaller (see Figure 8.23), accounting for a lower amount of hydrogen absorption. The parabolic curve represents the β-phase region in which hydrogen absorption/desorption occurs as a function of temperature and pressure.

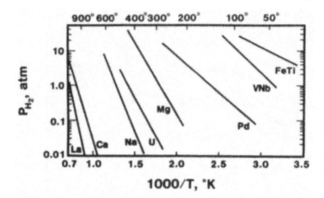

**Figure 8.24.** Equilibrium pressure for hydrogen absorption of some metals and alloys as a function of temperature. Reprinted from Reference 1, with permission from John Wiley and Sons.

Another plot of importance in selecting the metal hydride systems is that of the equilibrium pressure as a function of temperature (Figure 8.24). The range of pressures that is of interest for a hydrogen storage system for fuel cell applications is about 1 to 5 atm. Also, it is necessary for the hydrogen desorption to occur at below 100 °C. Thus, from Figure 8.24, it can be seen that Fe, Ti, and VNb alloys are potential candidates. Not shown in this graph are the $AB_5$ and $AB_2$ alloys, which show considerably better *P-C-T* characteristics. Typical examples of these alloys are $LaNi_5$ and $FeTi_2$. The $AB_5$ alloys are predominantly used for the hydride electrodes in the rechargeable nickel/metal hydride batteries because these have the ideal characteristics of maximum absorption of hydrogen (one atom of hydrogen for each atom of the metal). The alloys are modified with partial substitution of the parent components by other elements (e.g., cerium, copper, tin) which have effects of lowering the plateau pressure, inhibiting the oxidation of the parental elements, and increasing the lattice parameters. The first step in the absorption of hydrogen by the alloy is the dissociative adsorption of hydrogen on the surface of the alloy. This is a catalytic process and nickel, in the alloy $LaNi_5$, fulfills this role. The oxidation of nickel is inhibited by the presence of small amounts of Ce in the alloy.

As seen from Table 8.4 on a volumetric basis, the hydrogen content in a metal hydride is somewhat higher than for $LH_2$ and somewhat lower than for hydrogen combined with carbon or nitrogen. However, on a weight % basis, the hydrogen content is lower by a factor of ten, when making such a comparison. Magnesium hydride ($MgH_2$) has a hydrogen content of 12 %, but its plateau pressure (dissociation pressure) is too low at room temperature, and it will have to be heated up to a temperature of about 400 °C to increase its dissociation pressure to about 1 atm. Again, alloying magnesium with nickel facilitates the dissociative adsorption of hydrogen, but even in this case, it is necessary to increase the temperature up to about 280 °C to achieve a 1 atm equilibrium pressure of hydrogen. Table 8.4 shows that alloying magnesium with nickel reduces the hydrogen content in the hydride by a factor of about four. To overcome the weight penalty of the $FeTiH_2$ hydride, Daimler-Benz designed an ingenious dual metal hydride system for storage on board in an IC engine powered vehicle. Only a small quantity of $FeTiH_2$ was carried for start-up and acceleration. Soon after the exhaust temperature became hot enough, the source of hydrogen was a Mg-Ni hydride, which was the main source of the hydrogen fuel for cruising. Toyota, Honda, and Mazda have demonstrated fuel cell powered vehicles, with hydrogen stored as metal hydrides. In all these cases, the $AB_5$ alloys were used for the hydrides. Due to the relatively low hydrogen content, the ranges of the vehicles were about 200 km or less with a fully charged storage system.

From a safety point of view, the metal hydride storage system is attractive (see Section 8.7) but its disadvantages are the following:

- though the hydrogen content on a volumetric basis is good, on a weight percent basis it is low. Thus, it is not sufficiently competitive with fuel processing to produce hydrogen. In the latter case, one must also take into account the weight and volume of the fuel-processor;
- during hydriding/dehydriding, there is particle attrition. Even though there is some advantage of fine particles for the kinetics of hydriding/dehydriding reactions, from a safety point of view, it is essential that the particles do not reduce to submicron sizes;
- thermal management is critical. Hydriding of a metal is an exothermic reaction and hence, the storage unit needs cooling. Dehydriding, for supply of hydrogen, requires heating. In general, the thermal energies required for these reactions are small with the desired hydrides;
- the kinetics hydriding/dehydriding reactions are very sensitive to impurities (e.g., CO, $O_2$, and sulphur compounds). The oxygen, present at a critical level, is also a fire hazard. Hydrides in ultra–low particle sizes are pyrophoric; and
- from an economic point of view, hydrogen storage using metals or alloys is quite expensive (of the order of a $1000/kg of metal hydride).

An alternative approach for application of a metal hydride storage system is to use ionic hydrides such as calcium hydrides ($CaH_2$), lithium aluminum hydride ($LiAlH_4$), sodium borohydride ($NaBH_4$), and a combination of these with water, on site or on board, to generate hydrogen, as for example:

**TABLE 8.5**
**Physical Properties of Hydrogen, Methane, and Gasoline**

|  | Hydrogen | Methane | Gasoline |
|---|---|---|---|
| Molecular weight (g/mole) | 2.016 | 16.04 | ~ 110 ($C_n H_{1.87n}$) |
| Mass density (kg/Nm³) at standard conditions, i.e., $P = 1$ atm, $T = 0$ °C | 0.09 | 0.72 | 720-780 (liquid) |
| Mass density of liquid $H_2$ at -253 °C (kg/Nm³) | 70.9 | – | – |
| Boiling point (K) | 20.2 | 111.6 | 310–478 |
| Higher heating value (MJ/kg) | 142.0 | 55.5 | 47.3 |
| Lower heating value (MJ/kg) | 120.0 | 50.0 | 44.0 |
| Flammability limits (% volume) | 4.0–75.0 | 5.3–15.0 | 1.0–7.6 |
| Detonability limits (% volume) | 18.3–59.0 | 6.3–13.5 | 1.1–3.3 |
| Diffusion velocity in air (m/s) | 2.0 | 0.51 | 0.17 |
| Buoyant velocity in air (m/s) | 1.2–9.0 | 0.8–6.0 | non-buoyant |
| Ignition energy at stoichiometric mixture (mJ) | 0.02 | 0.29 | 0.24 |
| Ignition energy at lower flammability limit (mJ) | 10 | 20 | n.a. |
| Flame velocity in air (cm/s) | 265–325 | 37–45 | 37–43 |

$$NaBH_4 + 2H_2O \rightarrow NaBO_2 + 4H_2 \qquad\qquad (8.38)$$

These reactions are quite exothermic and have to be carried out in a highly controlled manner. In the case of $NaBH_4$, control of the pH and use of a catalyst are essential. Such systems have been proposed for transportation applications. There are several disadvantages for such an approach:

- these hydrides are highly stable and the regeneration of the hydrides from the product of hydrogen generation (e.g., see reverse of Eq. 34, $NaBH_4$ from $NaBO_2$) is highly energy intensive and expensive (cost of $NaBH_4$ is $45/kg);
- carrying a high quantity of water is essential for the reaction. The weight and volume of the water have to be taken into account in ascertaining the energy density and specific energy of the storage system;

- as stated above, there are safety issues in hydrogen generation from these hydrides when combining them with water. The reactions have to be carried in optimally designed reactors; and
- the hydroxide and oxide products are highly corrosive. This makes the selection of materials for the containers quite challenging.

There is, however, interest in such types of hydrogen storage/hydrogen generation systems for low-power fuel-cell systems (a few W to a few kW).

### 8.5.2. Hydrogen Storage in Combination with Other Elements or Compounds

The ideal element for hydrogen storage is in combination with carbon as hydrocarbon or alcohols. This topic has been extensively dealt with in Sections 8.1 to 8.3. One interesting carbon-hydrogen system that was not discussed, is the benzene-cyclohexane system. The second most interesting element is nitrogen (see Section 8.4.2). The other possible interesting elements are Be, B, and Si. The hydrides of these elements are quite toxic. In practically all these cases the hydrides are of the covalent type.

### 8.5.3. Techno-economic Analysis of Hydrogen Storage

Development of a safe, compact, lightweight, and low cost storage for hydrogen is an important issue, particularly for hydrogen vehicles. Also, it is desirable that onboard hydrogen storage vehicles could be rapidly refuelled (< 5 minutes) with a relatively small expenditure of energy, for example, for compression. Recently, the United States Department of Energy set goals for hydrogen storage onboard vehicles (US DOE 2002). Current physical (compressed gas) storage at about 350 bar already satisfies most of these goals, with the exception of volumetric energy density. The simplicity of compressed-gas storage makes it attractive, especially for vehicles like buses, where space constraints are not as stringent as on cars. However, the volumetric energy density is lower for compressed gas than for liquid hydrogen or metal hydrides. For very efficient hydrogen vehicles, the amount of hydrogen energy storage required is acceptably small for a 500-km travel range, and the volume may not be as large a constraint.[31]

The cost and performance of onboard hydrogen-storage systems for fuel-cell vehicles has been estimated by Directed Technologies, Inc. and the Ford Motor Company,[31,32] assuming that these systems are mass produced. Mass produced costs of $1000 per car are estimated for advanced compressed gas cylinders holding hydrogen at about 350 bar. Costs for liquid-hydrogen systems are projected to be about $500 per car. The costs of hydride systems are less well characterized, as these systems are further from commercialization, but are projected to be several thousand dollars per car.

## 8.6.    FUEL TRANSMISSION AND DISTRIBUTION: HYDROGEN VERSUS OTHER ALTERNATE FUELS

Most studies indicate that compressed hydrogen gas is the simplest, near-term option to store hydrogen onboard vehicles. A number of possibilities for producing and delivering compressed gaseous hydrogen transportation fuel to vehicles can be considered, which employ commercial or near commercial technologies for hydrogen production, storage, and distribution. These include:

- hydrogen produced from natural gas in a large, centralized steam-reforming plant and truck delivered as a liquid to refuelling stations;
- hydrogen produced in a large, centralized steam-reforming plant, and delivered via small scale hydrogen gas pipeline to refuelling stations;
- hydrogen from chemical industry sources, e.g., excess capacity in refineries which have recently upgraded their hydrogen production capacity, etc., with pipeline delivery to a refuelling station;
- hydrogen produced at the refuelling station via small scale steam-reforming of natural gas, in either a conventional or an advanced steam-reformer of the type developed as part of fuel cell cogeneration systems; and
- hydrogen produced via small scale water electrolysis at the refuelling station or in residential homes.

In the longer term, other centralized methods of hydrogen production might be used including gasification of biomass, coal or municipal solid waste, or electrolysis powered by wind, solar, or nuclear power. Thermochemical hydrogen production systems might include capture and sequestration of byproduct $CO_2$.

The capital cost of developing an extensive gaseous-hydrogen refuelling infrastructure for hydrogen fuel-cell vehicles has been estimated by several authors[31,33-35] to be in the range of several hundred to several thousand dollars per car, depending on the hydrogen supply pathway and level of demand. For the first few demonstration projects, the cost of hydrogen refuelling stations will be considerably higher than this. However, after building several hundreds of large-size hydrogen refuelling stations (serving fleets totaling perhaps several hundred thousand vehicles), refuelling station capital costs should decrease.

The best hydrogen supply option is site-specific, depending on local energy prices, the size of the demand, and the distance from a nearby source of hydrogen. Starting with centrally refuelled fleet vehicle markets would defer the need to build a widespread public hydrogen infrastructure until the technology had been proved and costs had been reduced in fleet use.

The cost of building a hydrogen infrastructure has been compared to the cost of implementing a new infrastructure for other alternative fuels.[33-36] Techno-economic analysts have found that the capital cost of hydrogen infrastructure per car, based on various near to mid-term hydrogen supply options, is comparable to that for a methanol infrastructure, assuming high levels of implementation and highly-efficient hydrogen vehicles.[34,37] Maintaining the current gasoline infrastructure would be more costly than implementing a new infrastructure based on hydrogen production via onsite reforming of natural gas. A hydrogen infrastructure based on coal with $CO_2$ sequestration and hydrogen delivery is projected to be more costly than one based on natural gas[38] and renewable hydrogen routes are also projected to be more capital intensive.[16]

## 8.7.     Fuel Safety: Hydrogen versus Alternate Fuels

When hydrogen is proposed as a future fuel, the average person may ask the question about the Hindenburg, the Challenger, or even the hydrogen bomb disasters. Clearly, consumers will not accept hydrogen or any new fuel unless it is as safe as our current fuels. Table 8.5 shows some safety related physicochemical properties of hydrogen as compared to two commonly accepted fuels natural gas and gasoline.[39] In some ways, hydrogen is clearly safer than gasoline. For example, it is very buoyant and disperses quickly from a leak. This contrasts with gasoline, which puddles rather than disperses, and where fumes can build up and persist even outside. Hydrogen is non-toxic, which is also an advantage.

Other aspects of hydrogen are potential safety concerns. Hydrogen is a small molecule and is more likely to leak than other gaseous fuels. Leak prevention, which can be accomplished through proper equipment design and maintenance as well as reliable leak-detection, are key safety issues for hydrogen. Hydrogen can cause embrittlement of certain steels, resulting in cracks, leaks, and failure. However, with properly selected materials, the possibility of embrittlement can be avoided.

Hydrogen has a wide range of flammability and detonability limits, e.g., a wide range of mixtures of hydrogen in air will support a flame or an explosion. In practice, the lower flammability limit is most important. For example, if the hydrogen concentration builds up in a closed space through a leak, problems might be expected when the lower flammability limit is reached. The lower flammability limit is comparable for hydrogen and natural gas.

The ignition energy (e.g., energy required in a spark or thermal source to ignite a flammable mixture of fuel in air) is low for all three fuels compared to real sources such as electrostatic sparks. The ignition energy is about an order of magnitude lower for hydrogen than for methane or gasoline under stoichiometric conditions. But at the lower flammability limit, the point where problems are likely to begin, the ignition energy is about the same for methane and hydrogen. If hydrogen leaks in a closed space, a large volume of flammable mixtures can occur, increasing the likelihood of encountering an ignition source. The flame velocity is high in hydrogen-air mixtures, carrying the risk of a fire transitioning to an explosion in a confined space. For this reason, it is recommended that hydrogen refuelling and storage be done outdoors, whenever feasible, or in well-ventilated indoor areas.

During vehicle refuelling and maintenance, it is important to avoid producing flammable mixtures, by excluding air from storage tanks, refuelling lines, etc. This can be done with double locks on lines, and by maintaining a positive pressure in hydrogen tanks and lines to exclude air. If necessary, lines and tanks should be purged with nitrogen prior to filling with hydrogen. Hydrogen burns with a nearly invisible flame, and radiates little heat, making fire detection difficult in the daytime. However, infrared detectors or special heat sensitive paints on hydrogen equipment allow rapid detection.

Safe handling of large quantities of hydrogen is routine in the chemical industry. Proposed use of hydrogen in vehicles has raised the question of whether this experience can be translated into robust, safe hydrogen vehicles and refuelling systems for the consumer, a topic that has been addressed in several recent papers. Safety engineers at Air Products and Chemicals, Inc., a large producer of chemical hydrogen,[40,41] delineated procedures for safe operation in hydrogen vehicle refuelling. According to a 1994 hydrogen vehicle safety study by researchers at Sandia National Laboratories[42]

"...there is abundant evidence that hydrogen can be handled safely, if its unique properties—sometimes better, sometimes worse and sometimes just different from other fuels—are respected."

A 1997 report on hydrogen safety by Ford Motor Company[43] concluded that the safety of a hydrogen fuel cell vehicle would be potentially better than that of a gasoline or propane vehicle, with proper engineering.

To assure that safe practices for using hydrogen fuel are employed and standardized, there has been a considerable effort by industry and government groups within the US and several other countries in recent years to develop codes and standards for hydrogen and fuel-cell systems. Development of low cost, reliable hydrogen sensors is an ongoing area of research.

## Suggested Reading

1.  H. Audus, O. Karstaad, and M. Kowal, Decarbonization of Fossil Fuels: Hydrogen as an Energy Carrier, in *Hydrogen Energy Progress XI*, Proceedings of the 11th World Hydrogen Energy Conference, held June 23-28 in Stuttgart Germany (1996), pp. 525–534.
2.  J. O'M. Bockris, A.K.N. Reddy and M. Gamboa-Aldeco, *Modern Electrochemistry: Fundamentals of Electrodics* (Kluwer Academic/Plenum Publishers, NY, 2000), Vol. 2A.
3.  A. V. Da Rosa, *Fundamentals of Energy Processes*, Stanford University Publications, Palo Alto, CA, 1997, Vol. 2A.
4.  S. Denn, Hydrogen Futures: Towards a Sustainable Energy System, *Int. J. Hydrogen Energy,* **27**, 235 (2002).
5.  Directed Technologies, Inc., Air Products and Chemicals, BOC Gases, The Electrolyser Corp., and Praxair, Inc., *Hydrogen Infrastructure Report*, report prepared for Ford Motor Company under USDOE Contract No. DE-AC02-94CE50389, July 1997.

6.  R. Doctor, et al., Hydrogen Production and $CO_2$ Recovery, Transport and Use from a KRW Oxygen Blown Gasification Combined Cycle System (ANL, May 1999), Table 7.3.

7.  J. Dybkjayer ans S. W. Madsden, *Advanced Reforming Technologies for Hydrogen Production* (Haldo Topsoe Publications, Lyngby, Denmark, 1988).

8.  E. G. and G. Services Parsons, Inc., and Science Applications International Cooperation, *Fuel Cell Handbook,* under Contract No. DE-AM26-99FT40575, with US Department of Energy – National Energy Technology Laboratory, W. Virginia, USA, 2000, 5th Edition, Chapter 9.

9.  C. N. Hamelinck and A. P. C. Faaij, Future Prospects for Production of Methanol and Hydrogen from Biomass, *J. Power Sources* **111**, 1 (2002).

10. K. Kordesch and G. Simader, *Fuel Cells and their Applications*, VCH, Weinheim, Germany, 1996, Chapter 8.

11. T. G. Kreutz, R. H. Williams, R. H. Socolow, P. Chiesa, and G. Lozza, Production of Hydrogen and Electricity from Coal with $CO_2$ Capture, *The 6th International Meeting on Greenhouse Gas Control*, Kyoto, Japan, October 1-4, 2002.

12. D. Z. Megede, Fuel Processors for Fuel Cell Vehicles, *J. Power Sources* **106**, 35 (2002).

13. S. T. Nauman and C. Myren, Fuel Processing of Biogas for Small Fuel-Cell Powerplants, *J. Power Sources* **56**, 45 (1995).

14. C. E. G. Padro and V. Putsche, Survey of the Economics of Hydrogen Technologies, NREL/TP-570-27079, September, 1999.

15. J. R. Rostrup-Nielsen, Advanced Reforming Technologies for Hydrogen Production, *Phys. Chem. Chem. Phys.* **3**, 283 (2001).

16. D. Simbeck, A Portfolio Selection Approach for Power Plant $CO_2$ Capture, Separation, and R&D options, in *Proceedings of the Fourth International conference on Carbon Dioxide Removal* (Pergamon Press, 1999), pp. 119–124.

17. P. L. Spath and W. A. Amos, Technoeconomic Analysis of Hydrogen Production From low BTU western coal with $CO_2$ sequestration and coalbed methane recovery including delivered $H_2$ costs, Milestone Report to the US DOE $H_2$ Program, 1999.

18. S. Srinivasan, R. Mosdale, P. Stevens, and C. Yang, Fuel cells for the 21st century: reaching the era of clean and efficient power generation, *Ann. Rev. Energy Environ* **24**, 281 (1999).

19. W. P. Teagan, J. Bentley, and B. Barnet, Cost Reductions of Fuel Cells for Transportation applications: fuel processing options, *J. Power Sources* **71**, 80 (1998).

20. United States Department of Energy, Hydrogen storage workshop, Argonne National Laboratory, Argonne, IL, http://www.cartech.doe.gov/publications/2002hydrogen.html, August 14-15, 2002.

## Cited References

1. B. V. Tilak and S. Srinivasan, in *Handbook of Energy Systems Engineering – Production and Utilization*, edited by L. C. Wilbur (John Wiley and Sons, NY, USA, 1985), Chapter 15.
2. United States Department of Energy, Office of Transportation Technologies, http://www.eere.energy.gov/hydrogenandfuelcells/hydrogen.
3. C. Marchetti, *Chemical Economic Engineering Review* **5**, 7 (1973).
4. Global Energy Perspectives, in *Institute for Applied Systems Analysis and the World Energy Council*, Ed. by N. Nakicenovic, A. Grubler, and A. McDonald, Cambridge University Press, Cambridge, England, 1998.
5. C. Lamy, J-M. Leger, and S. Srinivasan, in *Modern Aspects of Electrochemistry*, edited by J. O'M. Bockris, B. E. Conway, and R. E. White, Kluwer Academic/PlenumPubvlishers, New York, USA, 2001, Chapter 3.
6. National Research Council, *Review of the Research Program of the Partnership for a New Generation of Vehicles, Seventh Report,* National Academy Press, Washington, DC, USA, 2001.
7. K. Kordesch and G. Simader, *Fuel Cells and their Applications*, VCH, Weinheim, Germany, 1996, 305.
8. E. D. Doss, R. Kumar, R. K. Ahluwalia, and M. Krumpelt, *J. Power Sources* **102**, 1 (2001).
9. S. Ahmed, R. Doshi, S. H. D. Lee, R. Kumar, and M. Krumpelt, *Proceedings of the $32^{nd}$ Intersociety Energy Conversion Engineering Conference,* 2, American Institute of Chemical Engineers, New York, NY, USA, 1997, Paper No. 97081.
10. S. Ahmed and M. Krumpelt, *Int. J. Hydrogen Energy* **26**, 291 (2001).
11. B. S. Baker and H. C. Maru, Carbonate Fuel Cells: A Decade of Progress, *$191^{st}$ Meeting of Electrochemical Society*, May 1997.
12. http://www.ansaldofuelcells.com.
13. R. A. George, A. C. Casanova, and S. E. Veya, *Fuel Cell Seminar Abstracts*, Palm Springs, CA, Nov. 18-21 (Courtesy Associates, Washington DC, USA, 2002), No. 2002977.
14. C. Hendrik, Carbon Dioxide Removal from Coal-Fired Power Plants, Ph. D. Thesis, Department of Science, Technology and Society, Utrecht University, Utrecht, Netherlands, 1994.
15. H. J. Herzog, E. Drake, and E. Adams, *$CO_2$ Capture, Re-Use and Storage Technologies for Mitigating Global Climate Change*, Final Report for USDOE, Contract No. DE-AF22-96PC01257, January 1997.
16. R. H. Williams, Toward Zero Emissions for Transportation Using Fossil Fuels, in *VII Biennial Conference on Transportation, Energy and Environmental Policy (Managing Transitions in the Transportation Sector: How Fast and How Far),* Ed. by K. S. Kurani and D. Sperling, Transportation Research Board, Washington, DC, 2002.
17. R. H. Williams, Decarbonized Fossil Energy Carriers and Their Energy Technological Competitors, in *IPCC Workshop on Carbon Capture and Storage*, Regina, Saskatchewan, Canada, 2002.

18. M. Steinberg and H. C. Chen, *Int. J. Hydrogen Energy* **14**, 797 (1989).
19. United Nations Department of Economics and Social Affairs, *World Energy Assessment—Energy and the Challenge of Sustainability*, United Nations Development Program, World Energy Council, New York, NY, 2000.
20. B. V. Tilak, P. W. T. Lu, J. E. Colman, and S. Srinivasan, in *Modern Aspects of Electrochemistry*, edited by J. O'M. Bockris, B. E. Conway, E. Yeager, and R. E. White, Plenum Press, New York, NY, 1981, Vol. 2, Chap. 1.
21. R. S. Yeo, J. Orehotsky, W. Visscher, and S. Srinivasan, *J. Electrochem. Soc.* **128**, 1900 (1981).
22. R. M. Dempsey, A. R. Fregala, A. B. LaConti, and T. F. Enos, Device for Evolution of Oxygen with Ternary Electrocatalysts Containing Valve Metals, U.S. Patent 4,311,569 (January 19, 1982).
23. K. Petrov, K. Xiao, E. R. Gonzalez, S. Srinivasan, A. J. Appleby, and O. J. Murphy, *Int. J. Hydrogen Energy* **18**, 907 (1993).
24. L. L. Sweete, N. D. Kackley, and A. B. LaConti, *Proceedings of the 27th IECEC*, San Diego, CA, August 3-7, 1992, Vol. 1, p. 1—101.
25. F. J. Rohr, *Proceedings of the Workshop on High Temperature Fuel Cells, May 5-6, 1977* (Brookhaven National Laboratory, Upton, NY, USA, 1978), p.122.
26. W. Doenitz, W. Schmidberger, and E. Steinhell, in *Proceedings of the Symposium on Industrial Water Electrolysis*, **78-4**, edited by S. Srinivasan, F. J. Salzano, and A. R. Landgrebe, The Electrochemical Society, Princeton, New Jersey, USA, 1978, p. 266.
27. A. Fujishima and K. Honda, *Nature* **238**, 37 (1972).
28. R. Kainthla, S. U. M. Khan and J.O'M. Bockris, *Int. J. Hydrogen Energy* **12**, 381 (1987).
29. A. S. Arico, S. Srinivasan, and V. Antonucci, *Fuel Cells* **1**, 133 (2001).
30. M. M. Mench, H. M. Chance and C. Y. Wang, *J. Electrochem. Soc.* **151**, A144 (2004).
31. C. E. Thomas, B. D. James, F. D. Lomax, and I. F. Kuhn, Integrated Analysis of Hydrogen Passenger Vehicle Transportation Pathways, Final Report to the National Renewable Energy Laboratory, U.S. Department of Energy, Golden, CO, Under Subcontract No. AXE-6-16685-01, 1998.
32. B. D. James, G. N. Baum, F. D. Lomax, C. E. Thomas, and I. F. Kuhn, Comparison of Hydrogen Storage for Fuel Cell Vehicles, Final Report Task 4.2, United States Department of Energy, May, 1996.
33. J. M. Ogden, M. M. Steinbugler and T. G. Kreutz, *J. Power Sources* **79**, 143 (1999).
34. C. E. Thomas, J. P. Reardon, F. D. Lomax, J. Pinyan, and I. F. Kuhn, *Distributed Hydrogen Fueling Systems Analysis*, United States Department of Energy Hydrogen Program Review, NREL/CP-570-30535, 2001.
35. M. Mintz, S. Folga, J. Gillette, and J. Molburg, Hydrogen: On the Horizon or Just a Mirage?, *Society of Automotive Engineers*, paper No. 02FCC-155, 2002.
36. C. E. Thomas, B. D. James, F. D. Lomax, and I. F. Kuhn, Societal impacts of Fuel Options for Fuel Cell Vehicles, *Society of Automotive Engineers Technical*

*Paper* No. 982496, presented at the *SAE Fall Fuels and Lubricants Meeting and Exposition*, San Francisco, CA, October 19-22, 1998.

37. J. M. Ogden, *Ann. Rev. Energy Environ.* **24**, 227 (1999).
38. J. M. Ogden, R. H. Williams and E. D. Larson, *Energy Policy* **32**, 7 (2004).
39. J. Hord, *International Journal of Hydrogen Energy* **3**, 157 (1976).
40. J. G. Hansel, G. W. Mattern, and R. N. Miller, *International Journal of Hydrogen Energy* **18**, 783 (1993).
41. R. E. Linney and J. G. Hansel, Safety Considerations in the Design of Hydrogen-Powered Vehicles. Part 2, *Proceedings of the 11th World Hydrogen Energy Conference*, Stuttgart, 1996, pp. 2159-2168.
42. J. T. Ringland, Safety Issues for Hydrogen Powered Vehicles, Technical Report, Sandia National Laboratories, Livermore, CA, March 1994.
43. Ford Motor Company, *Direct Hydrogen Fuelled Proton Exchange Membrane Fuel Cell System for Transportation Applications: Hydrogen Vehicle Safety Report*, contract No.DE-AC02-94CE50389, May 1997.

## PROBLEMS

1.  Various analysts have projected that a lightweight, mid-sized hydrogen fuel cell automobile would have an energy consumption of about 12 g of hydrogen/mile in a typical urban-highway driving.

    (a) How does this compare to the average energy consumption in today's gasoline car? Use the conversion factors below and assume that today's gasoline car has an average fuel economy of 27 miles/gal gasoline.

    (b) Compare the amount of hydrogen-energy needed by this car in one year to the amount of gasoline-energy needed by today's average US gasoline car. Assume that the vehicle is driven 11,000 miles per year (the US average).

    (c) There are about 140 million cars in the US. Find the total hydrogen energy use that would be needed for cars in the US, if all the vehicles were efficient hydrogen vehicles. How much energy would be saved in the US by switching to hydrogen vehicles?

    (d) At present, the US uses about 10 EJ of oil-energy per year for light-duty vehicles, and imports about 15 EJ of oil for all purposes. How much oil could be displaced by switching to fuel-cell vehicles using hydrogen?

    **Useful Conversion Factors**:

    1 MJ = Megajoule = $10^6$ Joules
    1 GJ = gigajoule = $10^9$ Joules
    1 EJ = exajoule = $10^{18}$ Joules
    1 kg $H_2$ = 120 MJ (lower heating value or LHV basis)
    1 gallon gasoline = 0.125 GJ (LHV basis)

2.  Hydrogen can be produced from fossil fuels such as natural gas or coal.

(a) Hydrogen can be produced from natural gas at a conversion efficiency of about 80%. The US currently uses about 22 EJ of natural gas per year. How much would natural-gas use be increased if all the cars in the US ran on hydrogen (use results from Problem 8.1)?

(b) Hydrogen can be produced from coal at a conversion efficiency of about 65%. The US currently uses 35 EJ of coal per year. How much would coal use be increased if all cars ran on hydrogen?

(c) When hydrogen is produced from coal, $CO_2$ can be captured and sequestered underground. Typically, about 16 kg of $CO_2$ are captured per kg of $H_2$ produced. How many tons of $CO_2$ should be processed per day if $H_2$ is made from coal with a $CO_2$ sequestration from 10% of US cars? If one large $CO_2$ injection well will take 2500-tons $CO_2$/day, how many wells would be needed to handle the $CO_2$ from 10% of US cars?

3. Hydrogen can also be produced from a variety of renewable sources. Consider three renewable options for hydrogen production: solar-photovoltaic-powered electrolysis, wind-powered electrolysis, and biomass gasification. Use the following data to find the land requirements in $km^2$ to produce enough hydrogen for 100,000 $H_2$ fuel cell cars:

> Biomass energy production: 10 dry tons per hectare per year
> 1-dry ton of biomass = 20 GJ
> Conversion efficiency of biomass to hydrogen = 60%
> 1 hectare = $10^4 \, m^2$
> Solar PV conversion efficiency to electricity = 15%
> Electrolysis efficiency = 80%
> PV System losses (wiring, PV-electrolyzer coupling) = 10%
> Average annual insolation = 200 Watts/$m^2$
> Required land area = twice the solar PV array size (to avoid self-shading by adjacent arrays)
> Wind power peak production = 1 MW for a wind turbine with a 50 meter rotor diameter.
> Spacing for wind power systems = 5 D x 10 D, where D = wind turbine diameter
> Wind power capacity factor = 35% (e.g., annual average production = 35% of peak power production)

How much land is needed per car? How does this compare to a typical roof area (100 $m^2$) or lot size for a suburban house (1/4 hectare)?

4. For a gas pipeline of length L, and inlet and outlet pressures of $P_1$ and $P_2$, the volumetric flow rate $Q$, and the pipeline diameter $D$, are related as follows:

$$Q = \frac{\pi}{8} \frac{T_b}{P_b} \left(\frac{1}{f}\right)^{0.5} \left(\frac{R}{W_a GTLZ}\right)^{0.5} \left(P_1^2 - P_2^2\right)^{0.5} D^{2.5}$$

where:

$Q$  = flow rate $(mN^3/s)$
$L$  = length of pipeline (m)
$R$  = universal gas constant = 8314.34 J/(kg mol K)
$P_1$ = inlet pressure $(N/m^{2)}$
$P_2$ = outlet pressure $(N/m^2)$
$W_a$ = molecular weight of air = 28.97
$G$  = dimensionless gas specific gravity
     = 0.0696 for $H_2$, 0.553 for $CH_4$, 1.0 for air
$T$  = gas temperature (K)
$T_b$ = reference temperature = 298 K
$D$  = pipe diameter (m)
$P_b$ = reference pressure = 101325 $N/m^2$ (1 atm)
$f$  = dimensionless friction factor (depends on the flow regime)
$Z$  = compressibility = 1

For a fully turbulent flow, the *rough pipe* formula can be used:

$$\left(\frac{1}{f}\right)^{0.5} = 4\log_{10}(3.7D/k) + 2.273$$

where $k$ = roughness factor = 0.0007, and $D$ is given in inches.

The hydrogen output of a large steam-methane reformer is 3 million $Nm^3$/day. (i) Find the diameter of a 50-km pipeline with an inlet pressure of 7.3 MPa, and an outlet pressure of 1.4 MPa, carrying this amount of hydrogen. (ii) What is the hydrogen-energy flow rate through the pipeline (expressed in MW), given that the heating value of $H_2$ is about 0.012 kJ/ $Nm^3$. (iii) How many efficient $H_2$ fuel-cell cars could this pipeline supply with fuel? (iv) If it is assumed that the pipeline installed capital costs are given by the maximum of \$155,000/km and \$10,000/[km x (cm of pipeline diameter)], how much does the pipeline cost? What is the capital cost per car to build the pipeline?

# PART IV:

# APPLICATIONS, TECHNO-ECONOMIC ASSESSMENTS, AND PROGNOSIS OF FUEL CELLS

# CHAPTER 9

# STATUS OF FUEL CELL TECHNOLOGIES

## 9.1.  SCOPE OF CHAPTER

This book is divided into three parts so that the reader (particularly a graduate student following a course on fuel cells or a new researcher entering the field) can have a perspective of the diverse fields of science, engineering, and economics needed for research, technology development, and commercialization of fuel cells. This chapter, the first one in Part C on the Technology Development of Fuel Cells, focuses on the experimental work involved in the transition from single cells to electrochemical cell stacks for the leading types of fuel cell technologies (PEMFC, DMFC, AFC, PAFC, MCFC, and SOFC). There are several organizations worldwide (universities, national laboratories, and industries) heavily involved in R&D activities on these fuel cells. Considerable progress has been made and some of these technologies have reached the power plant/power source stage of demonstration for power generation/cogeneration, transportation, and portable power applications (see Chapter 10). The topic of the development and demonstration of power plants/power sources, including the balance of plants (BOP), is presented in Chapter 10. It is not possible in this chapter to deal with all the advances made in the leading fuel cell technologies from the early 1960s to the present time. There have been thousands of publications on fuel cell technologies (books, chapters in books, review articles in journals, original articles in journals and proceedings of conferences, etc.). Selected references are in the list of Suggested Reading at the end of this Chapter. Also, since most of the work prior to the mid 1990s has been extensively covered in the literature, our focus is on the

---

This chapter was written by S. Srinivasan and B. Kirby.

advances made from the mid 1990s to the present time. The reader will find that the PEMFC technology is covered extensively than the other leading five fuel cell technologies. The reason for this is that more than 80% of the R&D efforts since the mid 1990s have been on this technology, which shows great promise for the three major applications noted above.

Apart from the leading fuel cell technologies, there are other types of fuels cells that have been investigated (see Figure 4.2). Some of these have even reached the technology development/demonstration level. But at the present time, the prospects for these types to reach the applications level, even in the intermediate to long term, are fairly small except for regenerative fuel cells for electric energy storage. Thus, only a synopsis of these types of fuel cells is made in Section 9.8.


## 9.2    PROTON EXCHANGE MEMBRANE FUEL CELLS (PEMFC)

### 9.2.1.  Evolutionary Aspects

The proton exchange membrane fuel cell (PEMFC) is one of the most elegant types of fuel cells in terms of its design and mode of operation. It consists of a solid polymeric proton conducting membrane (the electrolyte), which is sandwiched between two platinum catalyzed porous gas diffusion electrodes in a single cell. As mentioned in Chapter 4, the proton exchange membrane fuel cell was the first type of fuel cell to find an application as an auxiliary power source in NASA's Gemini Space Vehicles in the early 1960s. Because of the problems of relatively low power densities ($< 100$ mW/cm$^2$) and short lifetimes ($< 2$ weeks) caused by the low performing proton conducting membrane, the PEMFC power source was replaced by an alkaline fuel cell (AFC) system for NASA's Apollo and Space Shuttle flights from the mid 1960s until the present time. PEMFC R&D programs were thereafter quite dormant until the early 1970s. There was then a revolutionary step, when the General Electric Company/Dupont Chemical Company invented the replacement of the originally used polystyrene sulfonic acid (PSSA) proton conducting membrane, a perfluorosulfonic acid (PFSA) proton conducting membrane called Nafion. There was a "quantum jump" with this revolutionary step in the ability of PEMFCs to attain high power densities ($> 400$ mW/cm$^2$) and lifetimes (over several tens of thousand of hours).

Since the PEMFC and AFC applications in the Gemini and Apollo Space flights, the first impetus for R&D on PEMFCs was the energy crisis, which resulted from the oil embargo in the Middle East in 1973. It was then followed by legislation in California and other states to reduce the level of environmental pollutants from transportation vehicles. An added incentive for the development of PEMFCs was that the PEMFC was one of the three systems to be chosen in the Partnership for Next Generation of Vehicles (PNGV), initiated by the Clinton/Gore administration, for the development of ultra-low emission vehicles by the year 2004. The other two types of power plants chosen for this program were Compression Ignition Direct Injection (CIDI) and Gas Turbine engines. Batteries were chosen as the hybrid

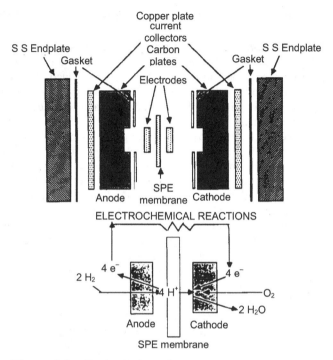

**Figure 9.1.** Basic design of a single cell in a PEMFC. Reprinted from Reference 7, Copyright (1990) with permission from Elsevier.

power source for all the three power plants, in order to capture the energy during braking of the vehicle and use it for start up and acceleration in urban type drive vehicles.

PEMFCs have attained the highest performance levels and longest lifetimes of all types of fuel cells. The PEMFC system is also unique in that it has the technical capabilities of covering a wide range of power levels. Since the late 1980s, the momentum has been rising exponentially to develop PEMFCs and a multitude of advancements have been made in PEMFC R&D, resulting in several thousand publications. Within the space limitations for this chapter, it is possible only to present a brief description of the significant advances made in this technology. The reader is referred to the publications in the lists of Suggested Reading and Cited References for more details. An overview of these advances is in the review articles by Costamagna and Srinivasan.[1]

## 9.2.2. Research and Development of High Performance Single Cells

### 9.2.2.1. *Design, Component Materials, and Assembly.* A basic

design of a single cell is shown in Figure 9.1. Its components are the porous gas diffusion electrodes for the anode and cathode, the proton conducting membrane electrolyte, and the current collectors with the flow fields. The active layer with the electrocatalyst is generally deposited on the electrode but can also be deposited on the proton conducting membrane. In the former case, the electrode consists of three layers – a substrate layer (Teflonized carbon cloth or paper—about 30% Teflon), a diffusion layer (with high surface area carbon and Teflon, again about 30% Teflon), and the active layer (with the nano-size Pt or Pt alloy particles, 200 to 400 nm, supported on high surface area carbon, e.g., Vulcan XC 72). The active layers of the electrode are impregnated with the proton conductor (Nafion) to enhance the ionic conductivity in the three-dimensional reaction zone. The electrodes are hot-pressed on to the Nafion membrane at a temperature of about 130°C and at a pressure of about 140 atm. The membrane and electrode assembly (MEA) is then placed in the appropriate position between the current collectors. The external surface area of the Nafion (i.e., without the electrode) serves as a gasket to prevent any leakage of the reactant gases from the anode side to the cathode side or vice-versa. The current collector contains the flow fields for the gases.

In respect to electrocatalysts, Pt or Pt alloys are the best ones to date. The reason is that the electrolyte environment is quite acidic (as strong as 2N sulfuric acid). Hence, it is not possible to use transition metals or their alloys. Other electrocatalysts have been investigated, but to date, these have not yielded the desired performance characteristics in respect to power density or lifetime.

The discovery of the perfluorosulfonic acid membrane by Dupont, Inc. made it possible to enhance the power density of a PEMFC by five to ten times. It is interesting to analyze the *quantum jumps* made in the development of proton conducting membranes from 1959 to 1980 (see Table 9.1). There are two reasons for the dominance of PEMFC technology with Nafion as compared with phenol sulfonic and polystyrene sulfonic acid membranes used previously. PFSAs, like

**TABLE 9.1**

**Quantum Jumps in the Development of Proton Conductive Membranes**

| Time | Membrane | Power density $(kW/m^2)$ | Lifetime (thousands of hours) |
|------|----------|--------------------------|-------------------------------|
| 1959-1961 | Phenol sulfonic | 0.05–0.1 | 0.3–1 |
| 1962-1965 | Polystyrene sulfonic | 0.4–0.6 | 0.3–2 |
| 1966-1967 | Polytrifluorostyrene sulfonic | 0.75–0.8 | 1–10 |
| 1968-1970 | Nafion (experimental) | 0.8–1 | 1–100 |
| 1971-1980 | Nafion (production) | 6–8 | 10–100 |

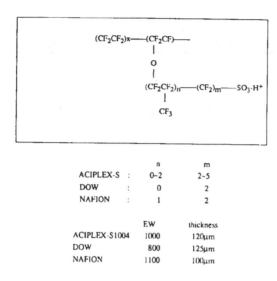

| | n | m |
|---|---|---|
| ACIPLEX-S : | 0-2 | 2-5 |
| DOW : | 0 | 2 |
| NAFION : | 1 | 2 |

| | EW | thickness |
|---|---|---|
| ACIPLEX-S1004 | 1000 | 120µm |
| DOW | 800 | 125µm |
| NAFION | 1100 | 100µm |

**Figure 9.2.** Basic structure and composition of some perfluorosulfonic acid membranes of DuPont, Dow, and Asahi Chemical. Reprinted from Reference 2, Copyright (1995) with permission from Elsevier.

Nafion, have a Teflon-like backbone except for side chains with ether like linkages, followed by $CF_2$ groups prior to the sulfonic acid group (Figure 9.2). The high electronegativity of the fluorine atom bonded to the sulfonic acid enhances the acidity of Nafion to that of a superacid like trifluoromethane sulfonic acid ($CF_3SO_3H$). C-F bonds are also highly electrochemically stable compared to C-H bonds (present in PSSAs and other C-H based polymeric acids), particularly at potentials in the range of operation of the oxygen electrode, enhancing the lifetime by about four orders of magnitude (see Table 9.1). High proton conductivity and high mechanical/chemical stabilities are pre-requisites for proton conducting membranes for PEMFCs. Efforts were made successfully by Dow Chemical Company and Asahi Chemical Company to enhance proton conductivities in PFSAs by some modification in the structures, i.e., altering the m and n parameters in the structure of the PFSA (Figure 9.2). Using this approach, the equivalent weights of these acids were reduced and hence the acidity increased, by increasing the number of sulfonic acid groups. This approach markedly reduced ohmic overpotential in the single cell (see Section 9.2.2.2). Another approach, used by W. L. Gore and Associates, was to develop ultra-thin supported membranes. The Nafion was impregnated into a supporting Teflon mesh, another major advance in PEMFC technology.

Since the late 1990s, there have been several research programs to find alternatives to PFSAs and modify PFSA membranes with composite materials.

**TABLE 9.2**
**Short List of Recently Investigated Proton Exchange Membranes (PEM) for PEMFCs and Some of their Physicochemical Characteristics**

| PEM | Ionic conductivity (S/cm$^2$) | Water uptake (moles/SO$_3$H) | Thermal stability (°C) | H$_2$/O$_2$ fuel cell and other test results |
|---|---|---|---|---|
| Perfluorosulfonic acid composite membranes | | | | |
| Nafion/SiO$_x$ | | | | 0.6 A/cm$^2$ at 0.6 V, 120 °C and 3 atm |
| Nafion/TiO$_2$ | | | | 0.6 A/cm$^2$ at 0.6 V, 120 °C and 3 atm |
| Nafion/Zirconium-hydrogen-phosphate (ZHP) | 10$^{-1}$ at 100% RH | 12 at 100% RH | 150 | 0.8 A/cm$^2$ at 0.6 V, 120 °C and 3 atm |
| Alternatives to Nafion membrane | | | | |
| Sulfonated poly-sulfone | 10$^{-2}$ for 48% sulfonation | 18 at 100% RH | 180 | IEC = 1.23 meq/g 0.5 A/cm$^2$, 0.5 V at 80 °C and 3 atm |
| Polybenzimidasole (PBI) | 1.5x10$^{-2}$ at 80% RH and 25–160 °C | 20 wt.%, equiv. to 3.5 moles/ PBI unit | Decreases with increase of SO$_3$H content | 0.6 A/cm$^2$ at 0.6 V, 80°C, 1 atm |
| Polyphenylene-benzo-bis-oxazole, (PBO) | 0.18 at 95% RH and 80 °C | 2.5 at 100% RH | 250 | |
| Sulfonated poly-ether-ethyl-ketone | 10$^{-2}$ at < 100 °C | 2.5 at 100% RH | 350 | No fuel cell tests |

These activities were promoted to produce low cost membranes and to operate PEMFCs at temperatures above 100 °C. A short list of these efforts is presented in Table 9.2, and the results of these studies are discussed in Section 9.2.2.2. W. L. Gore Associates deposit the active layer, consisting of the supported electrocatalysts and the PFSA, directly on the supported membrane. The thickness of the supported membrane is about 20 μm and that of the active layer is about half this value.

### 9.2.2.2. *Optimization of Structure of Membrane and Electrode Assembly (MEA).* The membrane and electrode assembly (MEA) is the heart of

the PEMFC. Its structure and composition are of vital importance to minimize all forms of overpotential losses, to minimize the noble metal loading (and thus the cost per kW of the PEMFC) in the gas diffusion electrode by high utilization of the surface area of the nano-sized particles of the electrocatalyst, for effective thermal and water management including the latter operation of PEMFCs with minimal or no humidification), and to attain lifetimes of PEMFCs needed for power generation, portable power, and transportation applications. It is in these areas of science and technology that major *quantum jumps* were made:

(a)   The first invention, which was by Raistric[3] at Los Alamos National Laboratory, was the impregnation of the proton conductor into the active layer of an E-TEK electrode containing 0.4 mg Pt/cm$^2$ to enhance the three-dimensional reaction zone, just as in the case of a liquid electrolyte entering these pores in the active layer by capillary action. It was demonstrated that the electrode kinetics of the oxygen reduction reaction increased by about two orders of magnitude.

(b)   This invention was followed by the work of Srinivasan, Ticianelli, Derouin, and Redondo,[4] who used such electrodes in a PEMFC and demonstrated (Figure 9.3) the first cell potential versus current density plot for a PEMFC using a low Pt loaded electrode (0.4 mg/cm$^2$).

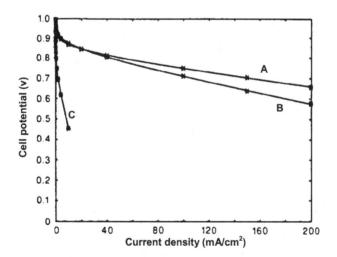

**Figure 9.3.** First demonstration of achieving a single cell performance of a PEMFC with a low platinum loading electrode. A and C, Nafion-impregnated and as-received electrodes (0.4 mg/cm$^2$) compared to that with high platinum loading (B, 4 mg/cm$^2$, non-impregnated). Reprinted from Reference 4, Copyright (1988) with permission from Elsevier.

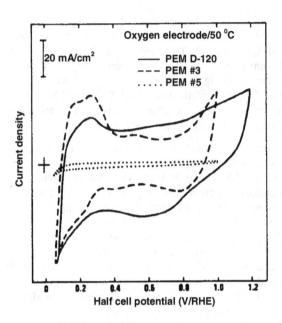

**Figure 9.4.** Cyclic voltammograms on the three types of electrodes from the PEMFC results shown in Fig 7.3 (PEM 0-120, A; PEM #3, B; PEM #5, C). Note the similarity between the elecrochemically active area of the low platinum loaded electrode impregnated with solubilized Nafion and the high platinum loaded electrode, and the very low electrochemically active surface area of the non-impregnated low platinum loaded electrode. Reprinted from Reference 4, Copyright (1988) with permission from Elsevier.

This PEMFC showed a better behavior than a PEMFC with a high Pt loaded electrode (4 mg/cm$^2$) from General Electric-United Technologies. In the latter, the active layer containing unsupported Pt black was pressed onto the Nafion membrane. It is worthwhile noting that if the E-Tek electrodes were not impregnated with Nafion solution, the cell performance was very low (see Figure 9.3). The cyclic voltammograms on the three types of electrodes in the PEMFCs (Figure 9.4) illustrate the high electrochemically active surface area of the Nafion-impregnated, low Pt loaded E-Tek electrode compared to that of the GE-UTC electrode and the very low electrochemically active surface area of the E-Tek without impregnation of Nafion into the active layer.

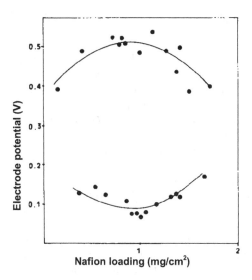

**Figure 9.5.** Effect of amount of Nafion impregnated into the active layer on the hydrogen and oxygen electrode potentials at a current density of 200 mA/cm$^2$. Reprinted from Reference 9, Copyright (1992) with permission from Elsevier.

(c)   The above work led to further enhancements in performances of PEMFCs by optimizing the amount of Nafion impregnated into the electrodes. This work was first carried out at LANL,[5,6] then at Texas A&M University,[7,8] and subsequently by researchers in other institutions (universities, national and industrial laboratories). When the amount of Nafion impregnated was too low, there was an increase of activation and ohmic overpotentials. When the amount of Nafion impregnated was too high, activation, and ohmic overpotentials were reduced while mass transport limitations arose at lower current densities. This phenomenon is illustrated in Figure 9.5.

(d)   In order to minimize ohmic and mass transport overpotentials in the active layer, the percentage of the amount of Pt in the high surface area of the carbon was increased from 10% in the state-of-the art E-Tek electrodes (as used in phosphoric acid fuel cells) to higher values. This study was conducted in the range 20 to 60% Pt/C. It was found that a value of 20% Pt/C gave the highest performance. The reason for this is that at higher percentage of Pt/C, the particle size of the Pt increased, thus reducing its BET surface area. By use of 20% Pt/C instead of 10% Pt/C, the thickness of the active layer was reduced from about 100 μm to 50 μm. This considerably reduced the activation and ohmic overpotentials of the electrodes and enhanced the electrochemical utilization of the platinum surface area.

(e) A further increase in the electrochemical utilization of the platinum surface area was made by Dhar.[10] He was the first to use only the supported electrocatalysts and solubilized Nafion in the active layer. This made it possible to have thinner active layers and even lower quantities of Pt in the electrode (0.1 to 0.25 mg/cm$^2$). This procedure was later followed by the investigators at W.L. Gore and Associates, LANL, TAMU, and several other laboratories.

(f) Further developments in optimizing the structures of membrane and electrode assemblies have been made in several laboratories (e.g., Princeton University[11,12], University of Connecticut[13]) using Nafion composites as proton conducting membranes for higher temperature operation of PEMFCs. The most effective composite materials for Nafion are oxides of silicon and titanium. Zirconium hydrogen phosphate also has been used as a constituent of the Nafion composite. These materials are hydrophilic and also enhance the glass-transition temperature of the Nafion. This makes it possible to operate the fuel cells at temperatures of about 140 °C.

9.2.2.3. *Mode of Operation / Operating Conditions.* Even though the PEMFC is the most promising type of fuel cell to attain the highest levels of energy efficiency, power density, and specific power, it has also faced major technical challenges with respect to water management, thermal management, and minimizing CO poisoning. In order to have efficient water management, a relative humidity level for the $H_2$ and $O_2$ reactant gases of about 90% is needed; otherwise, there is water loss by evaporation and the proton conductivity is markedly decreased. Thus, humidification of the gases to this level is necessary. Furthermore, the operating temperature range at low pressures ($\leq 3$ atm) is limited to about 80°C because there is a rapid increase of water vapor pressure with temperature. According to the work at LANL, TAMU, and several other laboratories, the ideal operating conditions are a temperature of 80 °C and a pressure of about 3 atm. These studies have also shown that the ideal conditions for humidification of the reactant gases are hydrogen at a temperature of 10 °C higher than the cell temperature and oxygen at a temperature of 5 °C higher than the cell temperature. The topic of water transport in PEMFCs is dealt with in Section 7.3.

The PEMFC, unlike the intermediate temperature PAFC and the high temperature MCFC and SOFC, operates at a temperature of less than 100 °C. It is also the system capable of operation at the highest power densities. If a PEMFC operates at a power density of about 0.6 W/cm$^2$ (generating electricity) the heat generation rate will be about 0.9 W/cm$^2$ (about as high as in a toaster). This heat has to be instantly removed. In the case of the intermediate and high temperature fuel cells, the temperature differential between the cell and the surroundings is considerably high, while in the case of the PEMFC it is quite low. Thus, for the former types of fuel cells, air cooling is quite effective, while for the PEMFC it is essential to use liquid cooling.

The other challenge with PEMFC is to minimize poisoning of the anode by carbon monoxide, which is present in reformed fuels. CO poisoning occurs because of competitive adsorption of $H_2$ and of CO on the electroactive Pt sites. In the hydrogen gas produced by steam-reforming and shift-conversion of natural gas, the amount of CO contained in the fuel is about 1 to 2%. This level of CO can be tolerated by a PAFC. But under the operating conditions of a PEMFC, the CO level will have to be decreased to less than 10 ppm for a Pt catalyzed anode. By using preferential oxidation of the CO in the reformed gas as an additional step in the hydrogen production process, it is possible to reduce the CO level to about 50 ppm. This level results in little loss of PEMFC performance when using Pt-Ru for the anode electrocatalyst. Air bleeding has also been proposed for reducing the CO level in reformed fuels, but this could present safety problems.

In order to minimize the water management, thermal management, and CO poisoning problems, efforts have been undertaken since 1998 to operate PEMFC at temperatures above 100 °C using PFSA composites with hydrophilic metal oxides or salts such as zirconium hydrogen phosphate. However, using this approach presents tough engineering challenges. For instance, higher pressures of operation are required because at an operating temperature of 130 °C, the vapor pressure of water is about 2.5 atm requiring at least a 3 atm total pressure of each reactant stream to attain a reasonable level of performance. There are on-going studies to operate PEMFCs at 120 to 130 °C using reactant gases at 50% relative humidity. This could reduce the pressure requirements. If the reactant gases are used at close to 100% RH, the engineering challenge is to effectively recycle the water from the exit and use it for continuous humidification. But this will involve a significant energy penalty.

An interesting aspect of higher temperature (about 130 °C) PEMFC operation is minimizing CO poisoning. Figure 9.6 shows the adsorption behavior $H_2$ and CO on the Pt electrode as a function of temperature.[12] At a temperature of 130 °C, the hydrogen coverage is reasonable enough to have a fast oxidation rate, even at a CO level of 500 ppm. This was demonstrated in a PEMFC by Adjemian[14] in a recent study using Nafion/SiO$_x$ and Nafion/TiO$_2$ composite membranes; up to 500 ppm CO was acceptable with minimal CO poisoning (Figure 9.7).

9.2.2.4. *Physicochemical Characterization of Single Cell and its Components.* Prior to setting up a single cell and examining its performance characteristics as a fuel cell, it is essential to have knowledge of the physicochemical characteristics of the cell and its components. A synopsis of the results of these studies are presented below:

(a) *Porous gas-diffusion electrodes.* Some essential physicochemical characteristics of the electrodes are the non-wetting nature of its substrate

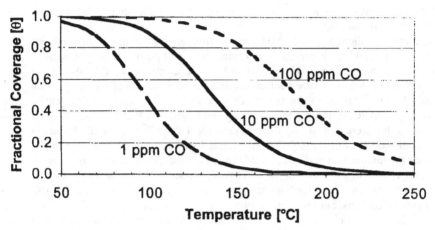

**Figure 9.6.** Extent of adsorption of hydrogen and CO on platinum as a function of temperature. Reprinted with permission.[16]

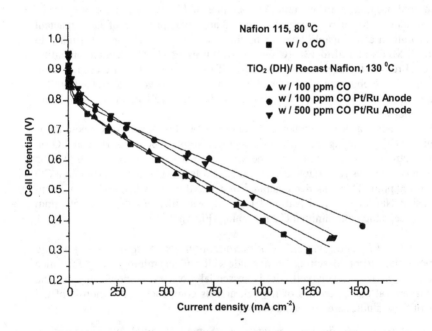

**Figure 9.7.** Cell potential vs. current density behavior of PEMFC operating at 130 °C with Nafion/TiO$_2$ composite membrane illustrating CO tolerance level of 500 ppm. DH stands for Degussa Hüls. Reprinted with permission.[14]

and diffusion layers and an adequate size of the macropores of the substrate. Detailed experimental studies have shown that Teflonized carbon paper or carbon cloth substrates (about 30% Teflon) have the desired characteristics. The wetting characteristics can be ascertained from contact angle measurements or by measuring bubble pressures of gases. Knowledge of the pore sizes and pore size distribution is also necessary. It is desirable to have larger pores (1–5 μm) in the substrate layer and smaller pores (0.1 to 1 μm) in the gas diffusion layer. The reason for this is to facilitate the exit of unreacted or inert (e.g., $N_2$ in air) gases from the substrate layers and for the even dispersion of the reactant gas into the diffusion layer and the electrocatalytic layer. The active layer in the early work at LANL and TAMU had as its constituents the carbon supported Pt electrocatalysts and Teflon. Subsequently, Nafion was impregnated into the active layer. The Teflon content in this layer was less then 20%. Teflon serves two purposes: as a binder for this layer, and for the active layer to have a certain degree of hydrophobicity for gases to enter the active layers and then get dissolved in the electrolyte. In recent times, the active layer has been only prepared using the carbon-supported electrocatalyst particles and solubilized PFSAs. The PFSA serves as the proton conductor through the active layer as well as a binder for the electrocatalyst particles in the active layer. In such electrodes, the active layers are relatively thin (10–20 μm).

It is also desirable to know the noble metal content in the active layer, even though one has an estimate of it from the amount of supported electrocatalyst used during preparation of the electrode. The easiest method of measuring the platinum content is by burning off the carbon and Teflon in a large known area of the fuel cell electrode and weighing the residual metallic content. Since the noble metal content is small, an alternative is to dissolve the noble metal or alloy in aqua regia and use an atomic absorption method for determining its content. In general, it is better to prepare large surface area electrodes since the wastage of the active components will be less when one uses the spraying or brushing methods.

Other techniques that have proven to be useful to characterize the electrodes are x-ray diffraction, scanning electron microscopy plus EDAX, electron microscopy, and transmission electron microscopy. An illustration of the structure of a membrane and electrode assembly from a SEM-EDAX examination of a cross-section is shown in Figure 9.8. Figure 9.9 shows the structure of a supported electrocatalyst from a TEM examination.

(b)  *Proton-conducting membranes.* This is the most important component in a PEMFC single cell. The first essential characteristic is that the PEM has to be mechanically stable. This is determined from stress versus strain measurements. A second important property is that its stability is in the oxidizing environment of the anode, as well as in the region of the

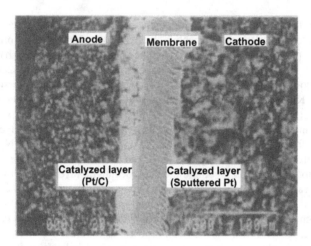

**Figure 9.8.** Structure of a membrane and electrode assembly from an SEM-EDAX examination of a cross-section. Reprinted from Reference 21, Copyright (1993) with permission from Elsevier.

operating potential of the oxygen reduction reaction at the cathode. A vigorous test of this property is determining the stability of the membrane in Fenton's reagent. PFSAs, subjected to this test, were found to be very stable. Water uptake is also important and can be determined by the difference in weight of a water saturated membrane and the dry membrane. During water uptake, the membrane gets swollen and

**TABLE 9.3**
**Saturation Water Uptake From Liquid Water at 25 °C and Water Vapor at 80 °C**

| Membrane treatment | Liquid water uptake (25 °C) | | Water vapor uptake (80 °C) | |
|---|---|---|---|---|
| | Weight % | $\lambda^a$ | Weight % | $\lambda^a$ |
| Nafion 115 | 41 | 25 | 18 | 11 |
| Recast Nafion | 43 | 26 | 21 | 13 |
| Nafion 115 ZP (25%)$^b$ | 33 | 25 | 25 | 19 |
| Nafion 115 ZP (12%)$^b$ | 34 | 24 | 22 | 15 |
| Lightly swollen Nafion 115 | 18 | 11 | 15 | 10 |
| Lightly swollen Nafion/ZP (25%) | 30 | 23 | 24 | 18 |
| Glycerol treated Nafion (150°C) | 81 | 48 | 34 | 21 |
| Glycerol treated Nafion (180°C) | 142 | 87 | — | — |

$^a\lambda$ = mole ratio of $H_2O$ to $SO_3H$
$^b$ZP = zirconium phosphate

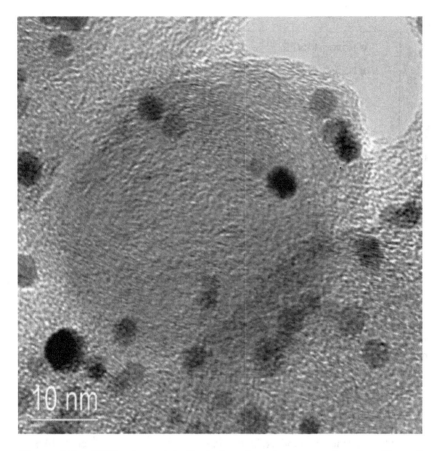

**Figure 9.9.** TEM micrograph illustrating the structure and composition of nanocrystalline Pt supported on high surface area carbon.[15]

measurements can be made of this volume expansion. PFSAs undergo a volume expansion of about 20–30%. Yang et al.[12] recently measured the number of water molecules per sulfonic acid group as a function of relative humidity. Simultaneously, measurements were also made of the specific conductivity of the membrane. Results of these studies are shown in Figure 9.10 and Table 9.3. The maximum amount of water absorbed by a PFSA membrane is about 25 moles of water per sulfonic acid group. The lower the equivalent weight of the membrane, the higher is the amount of water uptake. However, if the equivalent weight of the PFSA

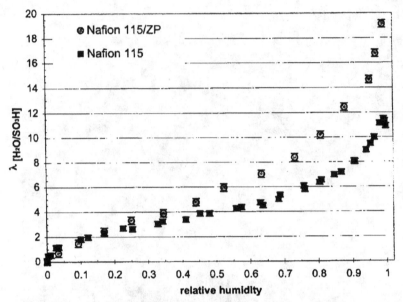

**Figure 9.10.** Water uptake by a Nafion membrane, expressed as the mole ratio of $H_2O$ to $SO_3H$, by a Nafion membrane at 80 °C as a function of relative humidity. Reprinted with permission from Reference 16.

is less than about 600 g/eq, the membrane tends to dissolve in water. A typical value of the specific conductivity of a PFSA as measured by the AC impedance method is in the range $10^{-2}$ to $10^{-1}$ ohm$^{-1}$ cm$^{-1}$. Even though this value is comparable to that of 1N $H_2SO_4$, in order to reduce the ohmic resistance in a single cell to less than 0.1 ohm/cm$^2$, it is most advantageous to use thin membranes (50 µm or less).

A novel micro-electrode method was developed by Parthasarathy et al.[17] at TAMU to determine the solubility of oxygen and its diffusion coefficient in the membrane as well as its specific conductivity and capacitance. The work of Parthasarathy et al.[17] and Ogumi et al.[18] was followed by several others using more refined micro-electrode techniques. By determining the diffusion coefficients and solubilities, one can determine the permeation rates ($k_p$, moles/s) of the reactant gases across the membrane in a single cell:

$$k_p = \frac{Dc}{\delta}$$

$$\text{(9.1)}$$

where $D$ is the diffusion coefficient, $c$ is the solubility of oxygen, and $\delta$ is the thickness of the membrane. $k_p$ can easily be converted to a crossover current density ($i_{cr}$) in a PEMFC single cell by using the formula:

$$i_{cr} = \frac{DnFc}{\delta} \tag{9.2}$$

Other methods to determine physicochemical characteristics of the membrane include x-ray diffraction, small angle x-ray scattering, and Fourier Transform Infra Red (FTIR) Spectroscopy. X-ray studies have recently been used to reveal the crystallinity of the membranes and the structures of the different phases in Nafion composite membranes (Figure 9.11)[16] and FTIR for the identification of the functional groups in the Nafion membranes[19] and in polystyrenesulfonic acid (PSSA) cross linked membranes.[20]

(c)  *Membrane and electrode assemblies.*  Optimizing the structure and composition of the MEA is a prerequisite for obtaining the maximum performance in a PEMFC. In the two preceding Sections, some of the most important characteristics of the individual components have been presented. The electrodes have to be well bonded to the PEM on either

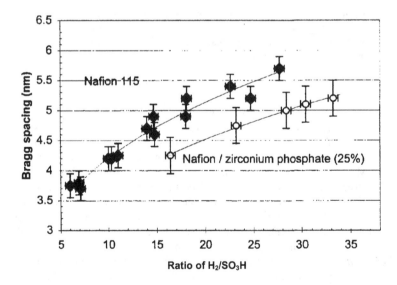

**Figure 9.11.**  Small angle x-ray scattering of Nafion (control and a Nafion/zirconium phosphate membrane). Reprinted with permission from Reference 16.

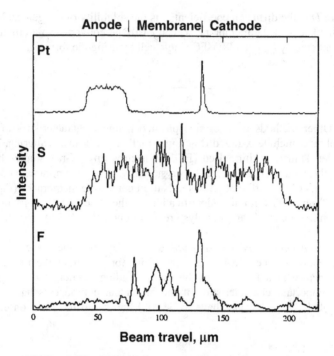

**Figure 9.12.** EPMA of a cross-section of a membrane and electrode assembly showing the locations of the Pt, S, and F atoms and their distributions. Reprinted with permission from Elsevier.[21]

side to minimize the contact resistance. The procedure used by W. L. Gore and Associates for depositing the active layers directly on both sides of the membrane minimizes the contact resistance problems and has been followed by several other laboratories. Electron probe micro-analysis (EPMA) of membrane and electrode assembly cross-sections has revealed the structures for individual layers (see Figure 9.8) in the MEA and the locations of the functional atomic elements (Figure 9.12)[21]. Examining structural changes between a control MEA and an active MEA after evaluation in a PEMFC can reveal whether there is any dissolution of Pt or its component alloying element from the electrodes.

(d) *Current collectors.* In order to avoid formation of water droplets in the channels of the current collector, Ballard Power Systems, Inc. (BPSI) uses a serpentine flow. Other flow designs (parallel flow, series parallel flow, and interdigitated flow) have also been successfully used (see Section 9.7). The current collectors should have a high electronic conductivity to minimize electronic overpotential losses, be chemically stable in the oxidizing and reducing environments, and be impermeable to the reactant gases. Graphitic plates are used as current collectors in single cells and as bipolar plates in a cell stack. The graphitic plates contain high surface area carbon and a resin (furfural) for bonding. Generally, the flow channels are machined but this is a very expensive procedure. Molded plates have been made for PAFCs but this method is expensive and time-consuming. Measurements on current collectors are relatively simple to determine the chemical stability in an oxidizing environment, electronic resistance, and gas permeability. The polymer-reinforced graphite plates have been successfully used. With metallic plates such as gold-plated aluminum or stainless steel current collectors/bipolar plates, passivation or corrosion problems have been encountered in long-term studies.

## 9.2.2.5. Performance Characteristics of Single Cells

(a) *General comments.* The PEMFC system has been the most widely investigated fuel cell system since the mid 1980s because it has the greatest potential for applications in the three major areas: power generation/cogeneration, transportation, and portable power. Great progress has been made in this technology, as a result of which several thousand publications have appeared in the scientific and engineering literature. Summarizing remarks of the most important aspects of performance analyses of single cells are made here. Water and thermal management are dealt with in Chapter 7 and are not included here.

(b) *Cyclic voltammetry.* Cyclic voltammetry is probably the only method to ascertain the electrochemically active surface areas of the electrodes and the percent utilization of platinum. Figure 9.4 illustrated the importance of Nafion impregnation of the active layer to enhance the three-dimensional reaction zone. Figure 9.13 shows the effect of increasing the percent Pt deposited on the carbon support in the E-Tek electrode, while maintaining the same Pt content in the electrode (0.4 mg Pt/cm$^2$).[5] Also shown here is the behavior of an uncatalyzed electrode. In this case, there are no hydrogen and oxygen adsorption/desorption peaks, while the catalyzed electrodes show an increase in the adsorption/desorption peak areas for hydrogen with increasing percentage of Pt; there is a significant increase in the peak area when the Pt content is increased from 10 to 20%. In the case where a very thin film of Pt is sputtered on the electrode (amount of Pt in sputtered film is 1 mg/cm$^2$), there is a further increase in the area of hydrogen adsorption/desorption peaks. The electrochemically active surface area is calculated from the hydrogen adsorption or desorption peak areas, taking into consideration that for a smooth Pt surface, the charge required to adsorb/desorb hydrogen for a fully adsorbed platinum surface is 220 µC/cm$^2$. In order to calculate the percent utilization of the platinum, we can calculate the total platinum surface area from the BET

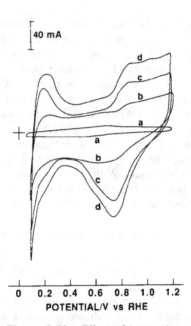

**Figure 9.13.** Effect of increasing percent Pt deposited on high surface carbon in E-TEK electrode (maintaining same absolute Pt content) on its cyclic voltammogram: (a) uncatalyzed; (b) 10 wt% Pt/C; (c) 20 wt% Pt/C; and (d) 20 wt% Pt/C plus 50 nm sputtered film of Pt. Reprinted from Reference 5, Copyright (1988), with permission from Elsevier.

surface area of the Pt (about 500 $m^2$/g) and from the amount of Pt used in the electrode (0.4 mg/$cm^2$). For the case shown, the roughness factor of the electrode is about 100 (roughness factor is the true surface area of Pt per geometric surface area of electrode). Thus, the utilization is about 20%. The cyclic voltammetric procedure has also shown that when very thin active layers are prepared with only the supported electrocatalyst and solubilized Nafion, the roughness factor can increase to about 200–250 for the electrodes and the utilization of Pt can increase to 40–50%. A second case, shown in Figure 9.14, is with Pt alloy electrocatalysts.[22] Even though these electrocatalysts were alloyed with transition metals (Pt₃M with M being Cr, Co, Ni, or Fe), the hydrogen adsorption/

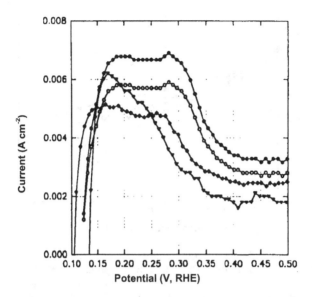

**Figure 9.14.** Cyclic voltammograms of fuel cell electrodes containing Pt and Pt alloy electrocatalysts. Roughness factor for Pt (●) 61, Pt + Ni (○) 56, Pt + Co (◆) 35, and Pt + Cr (▼) 48. Reprinted from Reference 22, Copyright (1993), with permission from Elsevier.

desorption areas are clearly visible. One normally does not use the oxygen adsorption/desorption areas for the calculation of the roughness factor, because these adsorption/desorption areas also have small contributions from the charges for oxidation/reduction of active groups on carbon such as hydroquinone/quinone.

(c) *Effects of temperature, pressure, and flow rate on the cell potential vs. current density behavior.* As in the case of chemical reactions, increasing the temperature or pressure of the reactant gases enhances the kinetics of the half-cell reactions at the anode and the cathode. The effect of these parameters is more significant for the kinetics of the electro-reduction of oxygen than for the electro-oxidation of hydrogen. Figure 9.15 illustrates the effects of temperature in the range from 50 °C to 95 °C when the reactants are $H_2/O_2$ and $H_2/Air$. The semi-exponential activation overpotential regions are less steep at higher temperatures. Two other effects are also apparent with increasing temperature: the linear regions have lesser slopes, indicating that the ohmic resistances have decreased because of the higher conductivity of the electrolyte; and mass transport limitations occur at higher current densities because of the increase of

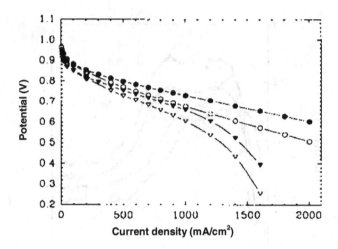

**Figure 9.15.** Effect of temperature on cell potential vs. current density plots for a PEMFC at 5 atm pressure. (●) 95 °C oxygen, (○) 50 °C oxygen, (▼) 95 °C air, (▽) 50 °C air. Reprinted from Reference 8, Copyright (1991), with permission from Elsevier.

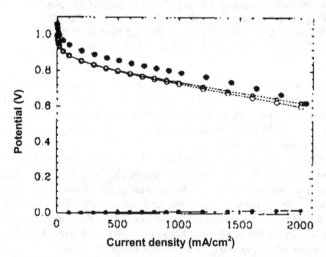

**Figure 9.16.** Half-cell potential vs. current density plot for a PEMFC. 10 mg/cm$^2$ Pt loaded electrodes, (●) $E_{cell}$; 0.45 mg/cm$^2$ Pt loaded electrodes, (○) $E_{cell}$, (□) $E_{O_2}$, (■) $E_{H_2}$. Note the semi-exponential region at low current densities for the electro-reduction of oxygen and the linear region for the electro-oxidation of hydrogen. Reprinted from Reference 8, Copyright (1991), with permission from Elsevier.

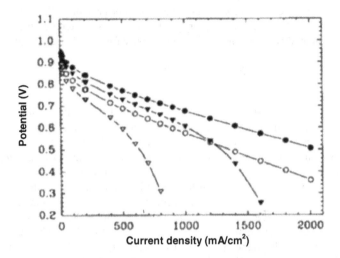

**Figure 9.17.** Effect of pressure of reactant gases on the cell potential vs. current density plots for a PEMFC at 50°C. (●) 5 atm $O_2$, (○) 1 atm $O_2$, (▼) 5 atm air, (▽) 1 atm air. Reprinted from Reference 8, Copyright (1991), with permission from Elsevier.

diffusion coefficients of the reactant with temperature. It must be noted that the results in Figure 9.15 were obtained with the reactant gases humidified to about 90 to 100%. The effect of temperature on the activation overpotential is considerably higher for the electro-reduction of oxygen than for the electro-oxidation of hydrogen because the exchange current densities are about $10^3$ to $10^4$ times lower for the former reaction than for the latter. It is for this reason that the semi-exponential region is exhibited in the half-cell potential vs. current density plot for the electro-reduction of oxygen at low current densities and a linear region is observed over the entire current density range for the electro-oxidation of hydrogen (see Figure 9.16).

There are three features related to the effect of pressure on the $E$ vs. $i$ plots, shown in Figure 9.17 for PEMFCs with $H_2/O_2$ and $H_2/Air$ as reactants: the slopes in the semi-exponential region of the E vs. i plots are steeper in $H_2/Air$ than in $H_2/O_2$ PEMFCs; the slopes of the linear region of the E vs. i have lesser values for $H_2/O_2$ than for $H_2/Air$ PEMFCs; and mass transport limitation starts at lower current densities for $H_2/Air$ than for $H_2/O_2$ PEMFCs. The reasons for these behaviors are: reaction rates increase with pressure and this decreases the activation overpotential; the slope of the linear region includes a small contribution from mass transport overpotential; and the rates of diffusion of the reactants to the active sites are enhanced by operating at higher pressures. It can be easily shown that the effect of pressure ($PO_2$), at a constant temperature on the cell potential ($E$) at a constant current density follows the equation:

$$E = E_0 + b \log P_{O_2} \qquad (9.3)$$

i.e., the slope of the line $E$ vs. log $P_{O_2}$ is the Tafel slope for the oxygen reduction reaction, $b$. Another noteworthy aspect is that the reaction order for the electro-reduction of oxygen in fuel cell electrodes in low- to intermediate-temperature fuel cells is about 0.5, while it is unity for the same reaction on a smooth platinum electrode. The reason for this is that in the former case the oxygen has to diffuse through a three dimensional reaction zone while the diffusion occurs to a planar surface in the latter.

One noticeable feature in Figures 9.16 and 9.18 is the significant difference in performances in $H_2/O_2$ and $H_2/Air$ PEMFCs. The reason for the considerably higher performances in $H_2/O_2$ PEMFCs than in $H_2/Air$ PEMFCs is the five fold higher oxygen partial pressure in pure $O_2$ than in air. Another contribution for this behavior is the *barrier layer* effect of $N_2$ at the cathode. During operation of the $H_2/Air$ PEMFC, it is necessary for the inert nitrogen in air (80%) to back-diffuse from the active layer to the back side of the substrate. There is some diffusional limitation for this and thus there is a build-up of nitrogen concentration in the pores of the diffusion and substrate layers (generally referred to as a barrier layer effect). Thus, the rate of diffusion of active sites from the substrate to the active layer is reduced. To validate this concept, an interesting study was conducted at TAMU.[23] Gas mixtures of $O_2/N_2$, $O_2/Ar$, and $O_2/He$ were introduced at varying concentrations in PEMFCs and their performances evaluated (Figure 9.18a-c). It was observed that a PEMFC with $O_2/N_2$ and $O_2/Ar$ showed similar characteristics, while the linear region was extended (indicating lower mass transport limitations) when $O_2/He$ was used. The reason for this is that the molecular weights and molecular sizes of oxygen and argon are about the same while He is a much lighter and smaller molecule than nitrogen or argon. Thus, the binary diffusion coefficient is considerably higher for $O_2/He$ than for $O_2/N_2$ or $O_2/Ar$. Also, there is a strikingly higher increase in performance due to the pressure effect when the $O_2$ concentration is doubled from 20% (as in air) to 40% than from 40% to a higher value. Thus, if there were some energy efficient method of oxygen separation from air, the PEMFC performance could be considerably improved.

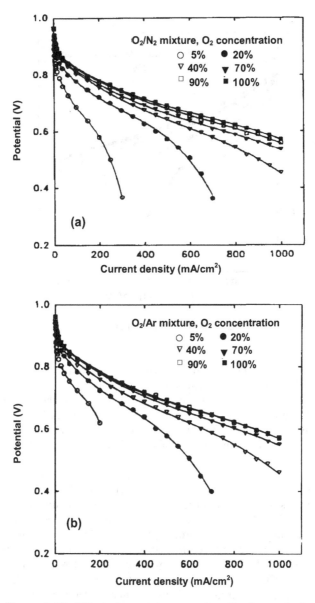

**Figure 9.18.** Effect of percent oxygen in cathodic reactant on performance of PEMFC with pure $H_2$ as anodic reactant: (a) $O_2/N_2$; (b) $O_2/Ar$; (c) $O_2/He$. Reprinted from Reference 23, Copyright (1994), with permission from The Electrochemical Society, Inc.

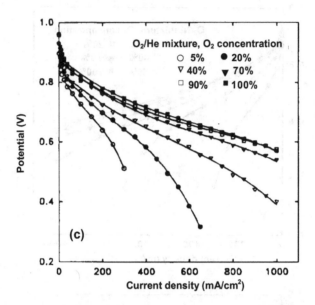

**Figure 9.18.** Continuation.

Increasing flow rates also increases the performances of $H_2/O_2$ and $H_2$/Air PEMFCs.[24] Lower utilization of the reactant gases for the fuel cell reaction leads to practically constant concentration of the reactant gases over the entire area of the electrode (from inlet to outlet). However, this is not a practical solution for fuel cells because the operating conditions in PEMFCs are so designed to have about 85% utilization of hydrogen and a 50% utilization of oxygen, either as pure oxygen or air.

d) *Effects of electrocatalysts on cell potential vs. current density behavior.* Most of these studies have been conducted in PEMFCs with Nafion membranes and Nafion-impregnated active layers. Platinum has always been and is still the best electrocatalyst for the electro-oxidation of hydrogen and the same electrocatalyst is used for the oxygen electrode. As observed in Figure 9.16, the activation overpotential at the hydrogen electrode is only 20 mV at a current density of 1 A/cm$^2$ in a PEMFC operating at 80 °C and 3 atm. On the contrary, the activation overpotential at the oxygen electrode is about 400 mV under the same operating conditions. About one half of this value is consumed even under open-circuit conditions. This is because of the very low exchange current density for the oxygen electrode (about $10^{-6}$ to $10^{-4}$ A/cm$^2$ at a porous gas diffusion electrode) and competing anodic reactions (Pt oxidation and/or oxidation of organic impurities if present in the electrolyte) setting

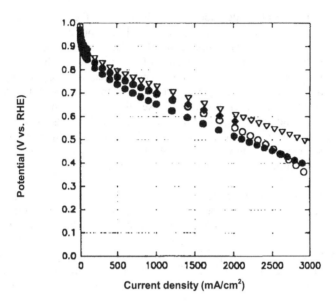

**Figure 9.19.** Effect of electrocatalyst on cell potential vs. current density plot for a PMFC. 95°C, 5 atm, Pt (●), Pt + Ni (○), Pt + Co (◆), Pt + Cr (▽). Reprinted from Reference 22, Copyright (1993), with permission from Elsevier.

up a mixed potential. The Tafel slope for oxygen reduction is $60 \pm 10$ mV/decade and is practically independent of temperature and pressure. The Tafel slope has higher values when organic impurities are present in the electrolyte and/or in the active layer of the electrode.

A *quantum jump* was made when the electrocatalyst loading was reduced from 4 mg/cm$^2$ unsupported Pt black (particle size 10 to 20 nm) to 0.4 mg/cm$^2$ for nanocrystals of Pt (particle size 2 to 4 nm), supported on high surface area carbon. In Vulcan XC, 72R has been used for the carbon support in most of the studies, but other supports such as acetylene black, black pearls, and regal have been evaluated and some showed promising results. Assume that the PEMFC cell potential is 0.7 V at a current density of 1 A/cm$^2$. The Pt loading on both electrodes is 0.8 mg/cm$^2$, and the cost of platinum is \$30/g, the cost of Pt per kW will be about \$20/kW. For an 80 kW automobile, the cost of the Pt will be prohibitive, \$1600. A reduction in Pt loading by a factor of 10 is needed to meet the target of a total Pt loading of 0.25 mg/cm$^2$ for both electrodes set by the USA Department of Energy. Another severe restriction is that the annual global production of Pt is about $2 \times 10^8$g. An 80 kW PEMFC powered automobile will require about 90 g. If 10% of the automobiles sold annually (total sales about 20 million/yr) in the USA were to be PEMFC powered, the Pt requirement would be about $2 \times 10^8$ g, equal to the total annual worldwide production of Pt. Thus, a major challenge is to reduce the Pt loading by a factor of 10 to 100 or find alternate electrocatalysts. There has been reasonable success in enhancing the performance of a PEMFC by using Pt alloy electrocatalysts (Pt-Cr, Pt-Co, Pt-Ni) for the oxygen electrode but these alloys contain about 75 atomic percent Pt and the oxygen reduction reaction still needs about 0.3 to 0.4 mg/cm$^2$ of Pt.[22] Typical $E$ vs. $i$ plots for PEMFCs with such types of electrocatalysts are shown in Figure 9.19. The plots show that by using a Pt alloy electrocatalyst, the power densities can be increased by a factor of two to three.

Alternatives to noble metal electrocatalysts have also been evaluated. Heat-treated organometallic macrocyclics (like cobalt tetraphenyl prophyrin) have proven to be as electrocatalytic as Pt for the electro-reduction of oxygen in alkaline media (as in AFCs), but in a PEMFC, even though the initial performances looked promising, there was a significant degradation of PEMFC performance with time, the main reason for the degradation being the slow corrosion rate of the transition metal.

(e)  *Effects of types of PEM on cell potential vs. current density behavior.* The Dupont Chemical Company membrane, Nafion, has been used as the proton-conducting material in most of the PEMFC power plants/power sources developed to date. The Japanese fuel cell researchers and developers have successfully used Asahi Chemical Company's Aciplex membranes and these have exhibited similar or better performances than Nafion. There is a slight structural difference between the two types of membranes, as shown in Figure 9.2. Asahi Chemical Company's membranes, in general, have a lower equivalent weight and hence a higher specific conductivity than Dupont's membrane. Another PEM, which was very promising, is the Dow membrane. This too had a somewhat slightly different structure than the Asahi Chemical Company or DuPont membrane (Figure 9.2).

In order to reduce the ohmic overpotential in the membrane, which is the predominant cause of efficiency losses in the linear region of the $E$ vs. $i$ plots, the thickness of the aforementioned membranes has been reduced from 175 to 20–50 μm in recent studies. The significant increases in performance by decreasing the thickness of the membrane are illustrated in Figure 9.20.[23] One may note that the open circuit potential of the PEMFC is lower with decreasing thickness of the membrane. This is mainly a result of increased crossover of hydrogen from the anode to the cathode, which then depolarizes the oxygen electrode. Another approach to reduce the thickness of the membrane even further was invented by W.L. Gore and Associates. In this case, solubilized Nafion was impregnated into a Teflon mesh to produce recast reinforced membranes.

The mechanical strength was improved because of the reinforcement by the Teflon mesh. Also, the ohmic resistance was significantly reduced because the recast Gore membrane containing Nafion 115 has a higher conductivity than the extruded Nafion film and the thickness of the membrane is less than in Nafion 112 (20–30 μm vs. 50 μm). Typical PEMFC performance with a PEMFC Gore membrane is also shown in Figure 9.20. Here too, the small amount of crossover of $H_2$ from the anode to the cathode causes the open-circuit potential to drop to 0.85–0.90 V from 1.0 V for a PEMFC with a Nafion or Aciplex membrane having a thickness of 100–125 μm.

The perfluorosulfonic acid membrane costs are extremely high at the present time, about \$800/m². Assuming that a PEMFC operates at a cell potential of 0.7 V and 1 A/cm², the cost of Nafion will be \$112/kW! According to the DuPont Chemical Company and the Asahi Chemical Company, the projected costs of the PFSAs will be about one third of the present cost if the total area of production per year is increased to the level required for 10% of PEMFC powered automobiles. At the present time,

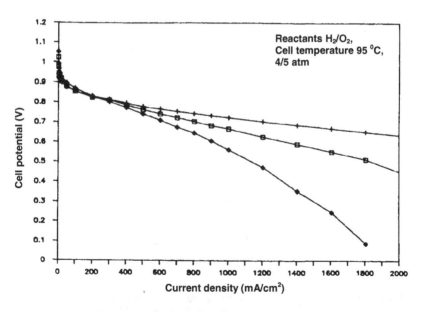

**Figure 9.20.** Effect of thickness of perfluorosulfonic acid membrane on performance of $H_2/O_2$ PEMFC operating at 95°C and 4/5 atm. Thickness: 175 □m (◆), 100 □m (□), 50 □m (+). Reprinted from Reference 7, Copyright (1990), with permission from Elsevier.

the main application of PFSAs is for the chlor-alkali plants (see Section 3.3); this niche application does not require large-scale production. Dupont has also recently announced that if PEMFCs will be developed for a greater percentage of automobiles manufactured per year and will also find applications for power generation/cogeneration and for portable power applications, the cost of Nafion could be reduced to about $80/m². In this case, the cost of Nafion will reduce to about $10/kW and this will be an acceptable cost for the automotive application.

The high cost of PFSAs is due to the expensive fluorination step involved in manufacturing the membranes. Therefore, alternate schemes are being examined to reduce the cost. Table 9.2 depicts some types of membranes investigated and their physicochemical characteristics relevant to the PEMFC application. Only a few of these have been subjected to evaluations in PEMFCs. The promising ones are the partially fluorinated polystyrene sulfonic acid membranes (by Ballard Power Systems, Inc.), radiation-grafted and/or cross-linked polystyrene sulfonic acid membranes, sulfonated polybenzimidazole membranes, and sulfonated polysulfone membranes. Performance parameters of PEMFCs

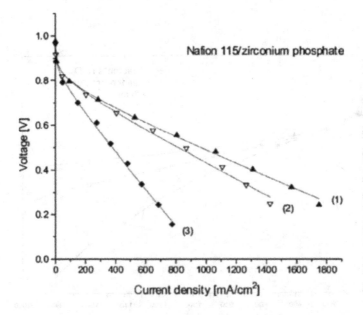

**Figure 9.21.** H₂/O₂ PEMFC performance with Nafion/Zr(HPO₄)₂ membranes at different temperatures and 3 atm: (▲) 130/120/130, (▽) 130/130/130, (◆) 130/140/130 [anode gas/cell/cathode gas, °C]. Reprinted from Reference 12, Copyright (2001), with permission from Elsevier.

with some of these membranes are at a considerably lower level than in those with the PFSA membranes. One reason for the poor performance is their lower specific conductivities. Further, the lifetimes of PEMFCs with the alternate membranes are considerably shorter than those of PEMFCs with the PFSA membranes because of chemical degradation in the oxidative environment at the oxygen cathode. A solution to this problem is to use laminated membranes, i.e., thin layers of PFSA membranes laminated on both sides of the alternate membranes. There has been some success with this approach by Chen et al.[20] and by Yu et al.[25]

Projects at Princeton University, The University of Connecticut, and several other laboratories in the USA and in foreign countries have focussed on raising the operating temperature to 120-130 °C by using Nafion/metal oxide and Nafion/zirconium hydrogen phosphate composite membranes. By optimization of the structures of the MEAs, reasonable levels of performances have been achieved (Figure 9.21). Recast composite membranes (Nafion/SiO$_x$, Nafion/TiO$_2$) have also been evaluated. The performance results at 3 atm and 130 °C are about the same as at 80 °C. It must, however, be noted that the operating pressure should be at least 3 atm because the partial vapor pressure of water at 130 °C is 2.5 atm, and it is necessary to have a reactant (H$_2$, O$_2$) gas pressure of at least 0.5 atm. When using air at this total pressure, the partial pressure of oxygen will only be 0.1 atm and still higher operating pressures may be required. Further, in the research at Princeton University, it was found that at least a 90% humidification of the reactant gases is necessary, as otherwise there will be a loss of water from the membrane causing an increase in the ohmic resistance of the PEMFC. One of the US Department of Energy goals is to operate cells at a considerably lower level of humidification ($\approx$ 50% relative humidity) and there is some success in this direction by the researchers at the University of Connecticut.[13]

(f)  *Effects of under- or non-humidification of reactant gases.* In addition to humidity requirements discussed, effects of under- or non-humidification of the reactant gases on the PEMFC performance have been extensively investigated at temperatures below 100 °C. Dhar was the first researcher to show that H$_2$/O$_2$ fuel cells can be operated at about 60 °C and 1 atm without humidification of the reactant gases.[10] His approach was to use a thin membrane, 50 μm thick, and use electrodes prepared only with the supported electrocatalysts and solubilized Nafion in the active layer of the electrode. Furthermore, these active layers were quite thin (less than 20 μm). In order to attain a reasonably high level of performance, it was necessary to feed the PEMFC with the reactant gases at low flow rates (close to stoichiometric) and to operate the single cell at 60 °C to minimize the loss of water content from the membrane by evaporation. The PEMFC performance results were considerably less satisfactory when air was used as the cathodic reactant because, in order to maintain the same stoichiometry, the flow rate had to be five times higher. Another method of operation when pure $H_2$ is the anodic reactant is to use it non-humidified in dead-ended anode chambers. The oxygen or air is humidified to nearly 100%. Even though there is transport of water with the protons from the anode to the cathode, the rate of back-diffusion of water more than compensates for the loss of water in the active layer of the anode. However, when using this method, it is necessary to periodically purge the water and inert gases (e.g.,, small amount of $N_2$ from air which crosses over from cathode to anode) accumulated during the PEMFC operation. This method is being evaluated by Plug Power and several other companies. A third method, used by Watanabe and coworkers,[26] was to disperse a relatively small number of nanocrystals of Pt particles in the proton-conducting membrane. The rationale for this method was that there is a low crossover rate for the reactant gases from anode to cathode and vice-versa, and the Pt nanocrystals chemically catalyze these gases to react and produce water. The method is more effective with thin membranes because of the higher crossover rates of these gases and hence of the production rate of water. This method has not been widely pursued because it is still not as effective as humidification of the gases, and due to the possibility of electronic short circuits and creation of hot spots in the membrane.

(g)  *Effects of CO poisoning on cell potential vs. current density behavior.* The tolerance level of CO in PEMFC at 80 °C is about 10 ppm. Ultrapure hydrogen (purity 99.9% $H_2$) is the ideal fuel for PEMFCs. This level of purity can be reached when hydrogen is produced by water electrolysis or thermal decomposition of water. However, the bulk of hydrogen production is from natural gas and to a lesser extent from coal and higher hydrocarbons (see Chapter 8). When hydrogen is produced from these fuels, the level of CO can be as high as 2%. But this can be reduced to 50–100 ppm with subsequent purification steps in fuel processing. The tolerance level of CO in the state-of-the-art PEMFCs (i.e., 80 °C, 3 atm) can be increased to about 50 ppm by using Pt alloy electrocatalysts. Pt-Ru (50/50 atomic %) is the best, but it is still necessary to increase the Pt loading at least by a factor of 3 to 5 to attain a satisfactory and stable performance in the PEMFC. Figure 9.22 shows a comparison of the performances of PEMFCs with Pt and Pt-Ru alloys electrocatalyst for the hydrogen electrode with the hydrogen gas containing different levels of CO.[27] With CO levels of 25 ppm and above in the $H_2$ reactant, the PEMFCs with the Pt electrocatalyst show a significant degradation in performance, while the CO tolerance is significantly enhanced with the Pt-Ru alloy electrocatalyst.

A second method of increasing CO tolerance is by oxygen-bleeding (0.4 to 2%) into the hydrogen stream.[27] In a study at Los Alamos National Laboratory, it was shown that this method produced PEMFC

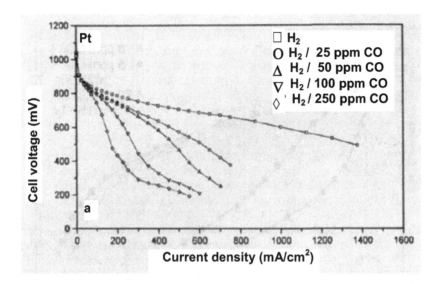

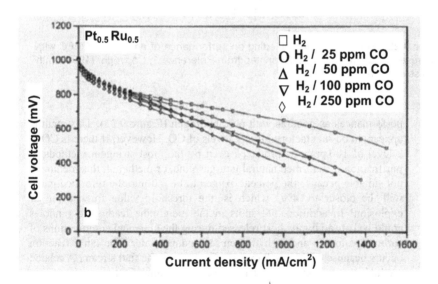

**Figure 9.22.** Effect of CO concentration in hydrogen on Performance of $H_2/O_2$ PEMFC with (a) pure Pt and (b) $Pt_{0.5}Ru_{0.5}$ (4 mg/cm$^2$) anode electrocatalyst at 80°C, 2 atm. Reprinted from Reference 1, Copyright (1999), with permission from Elsevier.

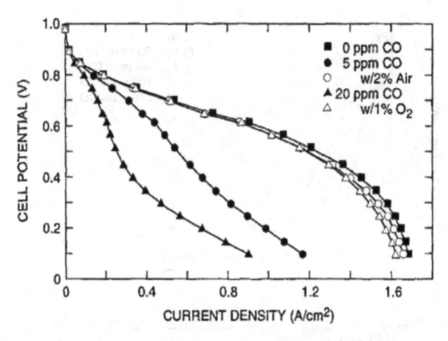

**Figure 9.23.** Effect of oxygen bleeding on performance of an $H_2/O_2$ PEMFC with CO present in the fuel stream. Reprinted from Reference 1, Copyright (1994), with permission from Elsevier.

performances as good as with pure hydrogen (Figure 9.23). This method appears to be satisfactory for low levels of CO. However, if there is CO at a level of 100 ppm, as produced even by the most stringent methods of purification of reformed natural gas, gasoline, or methanol, this method is not suitable because the percent oxygen to be fed into the reformed gases will be closer to 4%, which is the threshold value for causing an explosion! In addition, hot spots in the membrane leading to pinholes could be caused by the heat released during the chemical combinations of carbon monoxide and of hydrogen to produce water. The latter reaction occurs because of the small excess of oxygen in the fuel stream. A related method was to add 1–5% of $H_2O_2$ into the humidification system. This caused the oxidative removal of the adsorbed CO from the Pt surface.

A *quantum jump* in enhancing the level of CO tolerance was made by the researchers at LANL and at Princeton University by raising the operating temperature to about 130 °C. As stated previously, the best method of minimizing water loss from the membrane is by using a Nafion composite membranes (Nafion/SiO$_x$ or Nafion/TiO$_2$) and operating the PEMFC at a pressure of 3 atm. In the work at Princeton University, it was shown that the PEMFCs fed with $H_2$ containing 500 ppm of CO showed a minimum loss in performance (Figure 9.7). It was shown that the PEMFC with the Nafion/$TiO_2$ composite exhibited a superior behavior to that with Nafion/$SiO_x$, probably because the $TiO_2$ had some catalytic effect on the removal of CO adsorbed on the Pt-Ru electrocatalyst.

Another source of CO in the fuel stream is $CO_2$ from the reformed gases, up to 20% concentration. The reverse water gas shift reaction can produce a high level of CO causing poisoning. It was shown by Vickers[28] that another source of $CO_2$ in the anode compartment is created by the diffusion of $CO_2$ (present in the air stream at 350 ppm) from the cathode to the anode where the reverse of the water gas shift reaction occurs. The CO levels thus produced could be as high as 40 ppm.

(h) *Life-testing and performance degradation.* PEMFC power plants/power sources are potential candidates for power generation/cogeneration, transportation and portable power applications. Targets for lifetime requirements of the electrochemical cell stacks are 40,000 h (continuous operation) for power generation, 3000–5000 operating hours over a 5-year period for transportation, and 1000 h (intermittent operation) for portable power. Lifetime studies have been conducted in several laboratories worldwide, first in single cells, then in electrochemical cell stacks, and finally in PEMFC power plants/power sources. In this Section, we shall mainly consider the results of lifetimes at the single cell level. One of the most impressive results, obtained by General Electric Company in the 1970s, was a lifetime of 100,000 h in a single cell soon after Nafion replaced the polystyrene sulfonic acid proton conducting membrane. During this period, the PEMFC was operated at a relatively low current density. Since the mid 1980s, PEMFC single cells have been subjected to lifetime studies under different operating conditions. These results and explanations for their degradations in performance are presented below:

- In PEMFCs with the DuPont Nafion or Asahi-Chemical proton-conducting membranes (thickness 100–125 μm), the performance is reasonably stable for a period of about 3000 h when $H_2$ and $O_2$ are the reactants. There is some decrease in the cell potential initially for about 50 h, after which the cell potential is constant. The current density was maintained constant at about 300–500 mA/$cm^2$. After about 3000 h, there is a slow degradation in the performance, 5 to 10 mV over a 1000 h period. There are two reasons for this decay in performance. There could be some metallic impurities (e.g., Fe, Ni, Cu) in the humidified gas streams, probably from the humidification sub-systems, which exchange with the proton in the PFSA causing an increase in the ohmic overpotential.[29] Metallic impurities may also diffuse into the proton conducting membrane when metallic current collectors are used (e.g., stainless steel, even if gold-plated). The second reason is that small amounts of hydrogen peroxide can form at the cathode during electro-reduction of oxygen and/or at the anode, effected by the slow diffusion of oxygen from cathode to anode, where it is reduced to hydrogen peroxide or a peroxide radical. These peroxide species cause local degradation of the membranes, releasing small amounts of hydrofluoric acid in the effluent gas streams, as shown by researchers at the General Electric Company[30] and UTC-Fuel Cells.[31]

- When $H_2$/Air are the reactants, one has to consider the effects of the small amount of $CO_2$ in air (350ppm), which diffuses from the cathode to the anode and gets converted to CO by the water-gas shift reaction. The CO is instantaneously adsorbed on the platinum causing performance degradation. This problem can be partially overcome by using a Pt-Ru anode electrocatalyst. CO poisoning is more severe when a PEMFC is operated on reformed gases. The CO level has to be reduced to ultra low levels like 10 ppm. The reformed fuel also contains at least 20 to 30% $CO_2$ and this concentration is much higher than the amount of $CO_2$ from air which diffuses from the cathode to the anode (see previous sub-section). Surprisingly, there is hardly any data on loss in performance in PEMFCs using reformed $H_2$, probably because most lifetime data were obtained with pure $H_2$ as the fuel.

- It has been repeatedly mentioned in the PEMFC literature that water management in the MEA is one of the toughest challenges. The optimal condition for an MEA is when the proton conducting membrane, as well as that impregnated into the active layer, is practically 100% hydrated. Water loss causes an increase of the cell resistance and hence of the ohmic overpotential in the PEM. In several lifetime studies, it has been found that the main cause of performance degradation is due to evaporative losses of water from the PEMs. This is particularly so when the PEMFCs are operated at temperatures above 100 °C, even if the PEMs contain a hydrophilic composite material (e.g., $SiO_x$ or zirconium hydrogen phosphate).

## 9.2.3.  Technology Development of Cell Stazcks

9.2.3.1. *Scale-up of Single Cells.*  In the basic and applied research investigations on the development of PEMFC single cells, the bulk of the work at universities and national laboratories has been conducted with MEAs having an active area of 5 cm$^2$. However, there are several cases where the active areas have been scaled up to 10, 25, or 50 cm$^2$ depending on the type of R&D investigation. For electrochemical stack development, the industrial laboratories focus on the potential application of the fuel cell and the desired power output from the electrochemical cell stack. An advantage of the PEMFC is that its performance is extremely good at high current densities, such as 1 A/cm$^2$, at a cell potential of about 0.65 V. This current density is 3 to 5 times higher than in the other types of fuel cells (see Sections 9.3 to 9.7). Thus, for the desired total current generated, the total surface area (geometrical) of the cell could be significantly reduced compared to other types of fuel cells. Power output from the cells is higher than 250 kW for power generation and 50 kW for transportation applications. The most advanced fuel cell companies in the world have been developing electrochemical cell stacks with an active area of the electrode of 200–300 cm$^2$. This area of electrode is also being used for PEMFC at lower power levels, about 5 kW.

One problem with PEMFCs, which is not encountered in AFCs, PAFCs, MCFCs, or SOFCs, is that there is some loss of efficiency and power density with scale-up of the area of the electrode. The main reason is that removal of water becomes more difficult from larger area cells. The water could originate from that used for humidification of the cells and/or from the product water. Water tends to condense in regions of the flow channels or form droplets in the substrate and/or diffusion layer of the fuel cell electrodes. This causes problems of diffusion of the reactants to the active sites and of removal of the products away from the active sites to the back of the substrate layer. The net result is an increase of mass transport overpotential. Several solutions have been proposed to resolve the water management problem. Cyclic and rapid ejection of excess water can be done through purge valves, though this method is neither very practical nor quite effective. Ballard Power Systems, Inc. (BPSI) have used a partial removal of water from the anode side by under humidifying the hydrogen and thus setting up a concentration gradient for water to be transported from the cathode to the anode. BPSI also uses a serpentine flow of the reactant gases to ensure that there is no blockages in the channels of the bipolar plates. Büchi and Srinivasan[32] have investigated a counter-flow of unhumidified reactant gases in the cell. The idea was that there would be a water balance in the cell caused by electro-osmotic transport of water from the anode to the cathode and of diffusion of product in the reverse direction.

### 9.2.3.2. *Bipolar Plates and Flow Fields.*

In order to attain the desired voltage and power output of a fuel cell power plant/power source it is necessary to scale up the size of the electrode and have cells connected in series. Section 9.2.3.1 addressed the subject of scale-up of electrode area. In this section, we deal with the approaches to attain the desired cell-stack potential. This potential depends on the application. For power generation, the power is delivered at 120 V in the USA and Canada, but 240 V in practically all other countries. For the transportation application, the desired voltage is 120 or 240 V, and for portable power, it is about 12 V. Customarily, one uses a power conditioning system (DC-DC or DC-AC) in practical applications.

The electrochemical cells have to be stacked in series, and for this purpose bipolar plates with the appropriate flow fields are used. The most extensively used material for a bipolar plate is graphitic carbon molded with a polymeric binder at a high temperature and pressure. Furfural is the best binder known to date. The bipolar plate is one of the most expensive components in a fuel cell stack. This is because in the state-of-the art PEMFCs, the flow fields in it are machined. If one uses a graphite polymer complex, which is molded to form the bipolar plate, the cost will be considerably less. Use of metallic plates is an alternative approach. Gold-plated stainless steel or aluminum sheets, which are stamped to produce the flow fields, have shown good initial performances, but over a period of operation of PEMFCs there is a degradation of performance due to corrosion and/or passivation problems. Delphi is also investigating aluminum-carbon composites. In this method, layers of graphitic carbon with a polymeric resin are bonded onto the aluminum plates.

In respect to flow fields for the delivery of the reactant gases to the cells, the designs consist of series, parallel, or series-parallel flow. With the series flow arrangement (a serpentine flow), water blockages in the flow channel can be avoided. An alternative approach is to use recessed plates with porous carbon or a stainless steel mesh to provide the flow fields (UTC Fuel Cells, Nuvera). One other method, which initially appeared attractive, was to use an interdigitated flow. In this case, the reactant gas is forced in through the substrate to the active sites and out the exit path. The inlet and outlet paths are not otherwise connected, except by this passage through the substrate. This involves using a higher pressure.

### 9.2.3.3. *Assembly of Electrochemical Cell Stacks.*

There is very little information in the published literature on this topic, mainly because the technology development work is carried out by industries that consider this information proprietary. Thus, only the principles of electrochemical cell stack assembly are covered in this section. Basically, electrochemical cell stacks consist of single cell connected via bipolar plates. The advantages of the bipolar design over the unipolar are considerably less weight and volume, and hardly any ohmic losses because the electron flow from the anode to the cathode is via the entire area of the bipolar plate. In the case of the unipolar construction, the electronic current is drawn from one edge of the current collector of one cell and transferred to the edge in the next cell. One advantage of the unipolar design is that if one cell in a cell stack becomes defective, it can be easily replaced.

In most of the cell stacks developed to date, the bipolar plate is machined with the flow channels on both sides for feeding the reactant gases into the electrode. But there are some technologies in which these flow fields are only on one side with the backside containing flow channels for the cooling liquid. The bipolar plate thus consists of two half-plates, adhered to each backside for the flow of the coolant. These are generally thicker than when the bipolar plates have the flow channels for the reactant gases on either side.

In order to have a compact electrochemical cell stack, it is necessary to pack many cells within the desired volume. The MEAs have a thickness of about 500 to 600 μm. But considerably more space is required for the bipolar plates. The reason for this is that the depth of the channels for the flow of reactant gases is about 0.5 mm with similar width. In order to have a high mechanical stability of the bipolar plate and to make the plate impervious to the gases, the center of the flat plate region will have to be about 1 mm. Thus, the minimum thickness of the plates will be about 2 mm, and therefore the thickness of a single cell in a stack will be roughly 2 to 3 mm. However, in a stack, coolant plates are necessary and if a PEMFC operates at relatively high power densities, one needs to have at least one coolant plate for every two cells. The coolant plates will also have about the same thickness as the bipolar plates. The net result is that in a cell stack there will be 2 to 3 cells/cm. If metallic bipolar plates (coated stainless steel or aluminum) are used, the thickness of the cells and coolant plates can be reduced by at least 50% and one or two more cells can be packed per cm thickness of the cell stack.

The next hurdle to overcome is to have gas-tight seals in the stack, as otherwise there could be a crossover of the gases in the stack. The proton conducting membrane, which extends the active area on all sides by at least 2 cm serves as an adequate sealant under compression of the stack to prevent any gas crossover from the anode to the cathode. However, to reinforce the sealing, O-ring seals on the edges have been extensively used. Another aspect in the assembly of cell stacks is for the provision of reactant gas inlets and outlets to the individual cells using manifolds to distribute the gases.

The active cell areas in the cell stacks are determined by the power requirements for the projected application. At the present time, for power generation/cogeneration applications, the active area of the electrode is about 250 cm$^2$ in a 250 kW power plant. For other applications (transportation, portable power in the range 5 to 80 kW), too, cell stacks have been developed with the same active area. At a low power level (< 100 W), cell stacks have been developed with active areas as low as 10 cm$^2$.

The number of cells in a stack depends on the needed stack power output and the voltage requirements. The cell stacks have also been designed to be modular units with lower power outputs, as for example in the BPSI power plant for buses there are four modules, each generating 32 kW, which are connected in series and/or parallel. The highest power level electrochemical cell stack assembled to date is by BPSI for a 250 kW power plant for power generation/cogeneration.

### 9.2.3.4. *Water Management.*

Water management in PEMFC single cells and cell stacks is a major challenge, and an adequate balancing of a fully water-saturated proton conducting membrane and removal of excess water (including PEMFC product water) is essential. A simulation study of water management for a PEMFC stack for automotive drive cycles has recently been carried out by Haraldsson, Markel, and Wipke.[33] This study dealt with the effects of condenser size and cathode inlet relative humidity on system water balance under drive cycles in a hybrid PEMFC/battery automobile. This study aimed at maintaining a zero water balance during the simulated drive cycle of the vehicle. The water balance depends on the operating parameters: temperature, pressure, relative humidity, gas flow rates, and system parameters such as condenser size. The simulation studies were carried out at The National Renewable Energy Laboratory and at the Virginia Institute of Technology. A schematic of the PEMFC model used in this study is depicted in Figure 9.24.

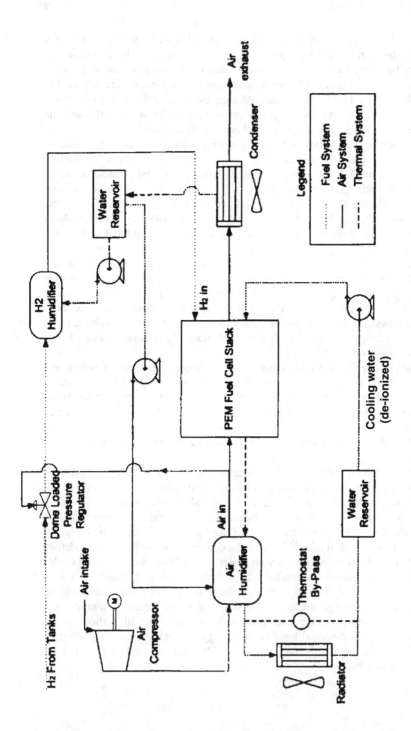

**Figure 9.24.** Schematic of the Virginia Tech fuel cell system model for water management in PEMFC. Reprinted from Reference 33, Copyright (2003), with permission from ASME.

A brief analysis of the approaches for water management and the results obtained and/or projected is presented here:

(a)   Ideally, the reactant gases would be humidified to at least 90% RH to prevent evaporative losses of water from the proton conducting membrane. One method involves use of external humidifiers in which case it was found that the ideal conditions are to have the hydrogen and oxygen/air humidified at 10 °C and 5 °C higher than the cell temperature, respectively. The flow rates of the gases in a multicell stack are customarily 1.2 times stoichiometric for the anodic reactants and two times stoichiometric for the cathodic reactants. A second method uses internal humidifiers in the multicell stack. In one design, the gases enter the cell stack in humidification chambers (similar to a single cell but with only the proton conducting membrane, instead of the membrane and electrode assembly in the center). The reactant gases enter one side of the membrane and water is circulated on the opposite side. The water diffuses through the membrane and humidifies the gases. The number of such cells required depends on the number of electrochemical cells in the multicell stack. BPSI and formerly The General Electric Company used this approach.

(b)   BPSI has also under-humidified hydrogen in order for some water transport to occur from the cathode side to the anode side in order to prevent flooding of the cathode side. Their results illustrate that the stoichiometric flow rate of air could be reduced considerably.

(c)   Another potential method of water management used by Plug Power, Inc., and other companies is to have only the oxygen/air humidified and have the hydrogen gas delivered to dead-ended anodic chambers. In this method, diffusion of water from the wet cathode chamber to the anode chamber provides adequate saturation of the active layer and of the proton conducting membrane and thus minimizes ohmic overpotential losses.

(d)   Attempts have been made to operate PEMFCs without external humidification. One of these was to have the cell design with counterflow of the reactant gases (as mentioned in Section 9.2.3.1) and the second is to use thin proton conducting membranes. These methods have shown promising results when $H_2$ and $O_2$ are used as reactants, the operating temperature is below 60 °C, and flow rate of gases is low, closer to stoichiometric even for oxygen. It does not work satisfactorily with air because the flow rate will have to be increased by 5 times to maintain the same stoichiometry.

(e)   Since 1998, there have been attempts to use PFSA composite membranes to retain water in the membranes (see sub Section 9.2.1). Though the results are somewhat promising up to operating temperatures of 130 °C, PEMFC stacks with such membranes have not been developed to date.

9.2.3.5.   *Thermal Management.*   Thermal management is as challenging a problem as water management. These two problems are inter-related because it is vital that the cells in an electrochemical stack are maintained at practically the same temperature. The main reasons as to why thermal management is extremely difficult in a PEMFC are that the PEMFC is the only fuel cell system that can operate at high power densities (> 0.5 W/cm$^2$), and the operating temperature in the state-of-the-art PEMFCs is below 100 °C. Thus, the temperature differential between the cell and the environment is relatively small, making it difficult to remove the 1.3 W of heat produced for every watt of electricity generated.

In Chapter 7, results of some simulation studies are presented. For more detailed analysis of simulation, it is suggested that the reader refers to the Suggested Reading List. Let us consider the possible cooling methods for PEMFCs that have been investigated[34] and make some comments on their suitability. The emphasis will be on high power applications.

(a) *Passive cooling.* In this method, the aluminum or anisotropic graphitic fins are extended from the stacks at room temperature to reject the heat to the environment. This method may be acceptable for low power level PEMFCs. A 25 kW small cell stack (4 cells each with area 12 cm$^2$) with aluminum fins (5 fins of 14 cm$^2$ and 6 mm thick), rejected only 6 kW of heat.

Heat pipes can be more effectively used. In this method, a fluid is evaporated at one end and cooled at the other. Plug Power has developed a system where the heat pipes have a design similar to that of bipolar plates. Evaporative cooling methods have also been proposed by Los Alamos National Laboratory and Texas A&M University.

(b) *Active cooling.* Air cooling works effectively for the intermediate and high temperature fuel cells. It has also been evaluated for PEMFCs, but the flow rates of air will have to be excessively high in the cooling plates, even when the fuel cell operates at low power densities. The need for very high flow rates also requires large pressure drops causing parasitic efficiency losses.

(c) *Liquid cooling.* At the present time, thermal management is effectively carried out with this method of cooling by practically all fuel cell developers. Cooling plates are located between neighboring cells or every two or three cells in a stack. The most common coolant is water. UTC-Fuel Cells has developed an innovative cooling method, using porous bipolar plates filled with water. This prevents the reactant gases from being transported from one side to the other. The coolant flows through the bipolar plate. There is a slight pressure difference between the anode and cathode chambers, which transports the fluid medium. Parasitic energy consumption of pumps is reduced by depressurizing the coolant liquid. A schematic of the concept is depicted in Figure 9.25. Alternatives to water for coolants are anti-freezing compounds (e.g., glycerol).

Using liquid coolants also has some problems such as leaks, freezing, and metallic impurities that cause short circuits. One must also overcome freeze-thaw cycles when using liquid coolants. One possibility is to

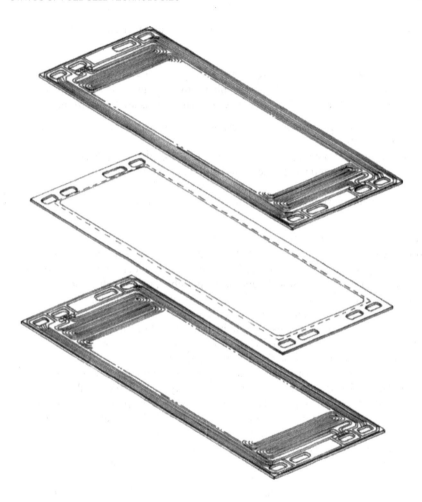

**Figure 9.25.** Schematic of liquid cooling subsystem for a PEMFC. Coolant water flows around active area while fuel and oxygen are supplied to the active area. Reprinted from Reference 35.

maintain the cell stack at a temperature of about 10 °C higher than the coolant liquid by operating the cell stack at a low power density. However, there will be some loss in efficiency for energy conversion. The biggest challenge is for transportation applications, particularly in the winter season.

(d) *Heat exchangers.* A heat exchanger is also an essential auxiliary for a cooling subsystem. For example, for a transportation PEMFC system, one could have an external radiator which will cool and recirculate the exiting liquid coolant from the stack.

9.2.3.6. *Performance Evaluation.* Before incorporating an electro-chemical cell stack in a PEMFC power plant/power source, its performance is determined in the laboratory. The main types of experiments are to determine:

(a)  the cell stack potential as a function of current density using galvanostatic experiments;
(b)  individual cell potential in the stack as a function of current density;
(c)  ohmic overpotentials in the individual cells in the stacks, using AC impedance methods;
(d)  temperature variations from cell to cell in a stack; and
(e)  lifetimes.

Lifetime is most difficult because the lifetime goals are about 5 y for power generation/cogeneration applications, 3000 operating hours over a 5 y lifetime for transportation applications (studies need be conducted under simulated drive cycles), and 100 to several thousand hours for the portable power applications.

Ideally, the individual cell potential vs. current density behavior in a stack should be the same as in a single cell, but with PEMFCs, it has generally been found that the individual cells in a stack exhibit slightly lower performances than in single cells, unlike PAFCs, MCFCs and SOFCs. The main reasons for this are the problems connected with water and thermal management.

Hardly any results of variation in stack potential with operating time are available in the published literature. For practically all types of fuel cells, the goal is to have only a decrease of cell potential by 1 mV over a 1000 h period. To date, the variations have been in the range of 5 to 10 mV per 1000 h. Degradation can occur by slow oxidation of the proton conducting membrane at the anode/membrane and cathode/membrane interfaces by peroxide or free radical oxygen species. This has been detected by measuring the HF concentration in the effluents of fuel cells. Organic impurities can poison the hydrogen electrode. Metallic impurities in the cell components and/or water in the humidified gases increase ohmic overpotential. Flooding of the electrodes causes mass transport problems and drying in the active layer and/or the membrane resulting in loss of electrochemically active surface area of the electrode and/or in ohmic losses.

## 9.3    DIRECT METHANOL FUEL CELLS (DMFC)

### 9.3.1.  Evolutionary Aspects

Since the early 1950s, methanol was found to be the most electroactive fuel for direct electrochemical oxidation in fuel cells and since then it has been a *fuel cell researcher's dream* to develop direct methanol fuel cells (DMFCs). Pavela[36] was the first scientist to make this discovery in 1954. It was found that the kinetics for electro-oxidation was faster in alkaline media (KOH or NaOH). Justi and Winsel[57] used DSK electrodes (porous nickel for the anode and porous nickel-silver for the cathode) to design and build a DMFC system. It was followed by the work of Wynn[38] who assembled a system with a platinized porous carbon anode and a porous carbon cathode impregnated with silver-cobalt-aluminum mixed oxides. With air as the cathodic reactant, this system had a cell potential of 0.35 V at a current density of 8 $\mu A/cm^2$. An attempt was made by Hunger[39] to develop a PEMFC with an anion exchange membrane on both sides of which were pressed catalytic electrodes. The performance was quite low.

During the early 1960s, the most impressive work was carried out by Murray and Grimes[40] at Allis Chalmers Manufacturing Company, by developing a 40-cell DMFC stack operating at 50 °C. The anode consisted of a nickel sheet on which was deposited a Pt-Pd electrocatalyst; a porous nickel sheet impregnated with silver served as the cathode. The electrolyte was 5 M KOH. The stack had 40 cells, the open circuit potential was 22 V, and the maximum power output was 750 W at 9 V and 83 A (180 $\mu A/cm^2$ at 0.22 V/cell). The power density and specific power at the maximum power density were 21 W/L and 11 W/kg, respectively.

One of the major disadvantages of DMFCs with alkaline electrolytes is the carbonation of the electrolyte by one of the products of the methanol oxidation reaction, namely $CO_2$. In addition, solid potassium carbonate deposits form in the pores of the active layer of the electrode. As a result, the electrolyte conductivity decreased significantly and the electrochemically active surface area markedly was lowered. Mass transport limitations were also encountered in the transport of the reactants to the active layer and removal of products from the active layer to the back of the electrode.

Thus, since the late 1960s to the early 1980s, the trend was to use $CO_2$ rejecting electrolytes, the most common during this period being concentrated sulfuric acid. The most advanced DMFC systems using this electrolyte were designed, constructed, and demonstrated by the Shell Research Center in England.[41] Surprisingly, the Pt-Ru anode electrocatalyst used in these systems is still the best electrocatalyst in the most advanced DMFC developed to date! The cathode electrocatalyst was an iridium chelate in the Shell DMFC and a carbon-supported platinum electrocatalyst in the Hitachi DMFC.[42] The platinum loadings were high (about 10 $mg/cm^2$). This resulted in Pt loadings being as high as 400 g/kW and Pt costs as high as $6000/kW! The rated power output in the Hitachi DMFC system was 5 kW and its operating temperature was 60 °C. The maximum power density in both of these systems was about 25 $mW/cm^2$. One problem when using sulfuric acid as an electrolyte in fuel cells is that the acid can be reduced first to sulfurous acid ($H_2SO_3$) and finally to $H_2S$. The sulfide ion adsorbs on the platinum electrocatalyst and poisons the methanol oxidation reaction. This is more important for the $H_2/O_2$ fuel cells than for DMFCs. But even in the latter case, partial degradation in DMFCs was due to the reduction of $H_2SO_4$. In order to overcome this problem, the Shell Research Center had used trifluoromethane sulfonic acid ($CF_3SO_3H$) in some of their research activities. This acid was found to be considerably more stable with less degradation in performance of the DMFC.

It was in the early 1990s, that R&D studies on DMFCs were initiated with proton conducting membranes as the electrolyte.[43] In practically all these investigations, Nafion is the electrolyte. As in the case of PEMFCs, the perfluorosulfonic acid electrolyte has the attractive physicochemical characteristics of high ionic conductivity, high chemical stability in oxidizing and reducing environments, high mechanical integrity, and moderate to low crossover rate of reactant gases. It also served as an excellent gasket material, surrounding the area of the membrane and electrode assembly. There has been intense activity in developing and demonstrating such types of DMFC systems. Due to the relatively low electrochemical activity of methanol, as compared with hydrogen (the exchange current density for the electro-oxidation of methanol to carbon dioxide and water is at least three orders of magnitude less than that for electro-oxidation of hydrogen – $10^{-6}$ A/cm$^2$ vs. $10^{-3}$ A/cm$^2$ on a smooth Pt electrode), the magnitude of the activation overpotential losses are considerably higher in DMFCs than in $H_2/O_2$ or $H_2/Air$ PEMFCs. The activation overpotential at the methanol electrode is about the same as that at the oxygen electrode. Even so, the progress in DMFC technology has been quite significant. At the present time, power densities of about 250 mW/cm$^2$ have been attained at energy conversion efficiencies of 30-35%.

The highlights of the advances in DMFC single cells and cell stacks are presented here. Due to the lower power densities of DMFCs, as compared with those of $H_2/O_2$ or $H_2/Air$ PEMFCs, most of the R&D efforts are on ultra-low and low power systems at the 10 W to 5 kW level.

### 9.3.2. Research and Development of High Performance Single Cells

9.3.2.1. *Cell Design and Assembly.* A schematic of a DMFC with an acid electrolyte is represented in Figure 9.26. Also indicated in the figure are the half-cell reactions that occur at the electrode. The overall reaction, which occurs in a DMFC, is

$$CH_3OH + 3/2\ O_2 \rightarrow CO_2 + 2H_2O \tag{9.4}$$

This overall reaction occurs at 100% coulombic efficiency with the best electrocatalysts: Pt-Ru for the anode and Pt for the cathode. With several other investigated electrocatalysts, partially oxidized products (formaldehyde, formic acid) are also produced in a DMFC. With a proton exchange membrane electrolyte (e.g., Nafion), the structure of the MEA is very similar to that in a PEMFC, as described in Section 9.2. The structure and composition of the oxygen electrode in a DMFC is practically the same as that in a PEMFC. The methanol electrode requires a noble metal electrocatalyst loading five to ten times higher than for a hydrogen electrode because of the poisoning of the electrode by intermediate species,

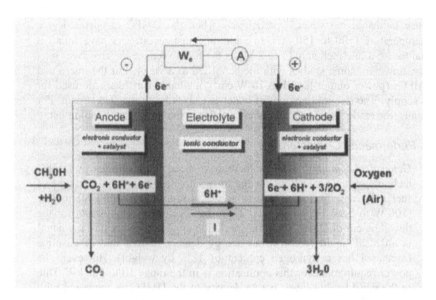

**Figure 9.26.** Schematic of a single cell in a Direct Methanol Fuel Cell (DMFC). Reprinted from Reference 23, Copyright (2001), with permission from Kluwer Academic.

predominantly CO. The Teflon contents in the substrate and diffusion layers are lower in the methanol electrode than in a hydrogen electrode in order to facilitate the transport of the liquid fuel methanol to the active layer of the electrode.

One of the serious problems with DMFCs is the high crossover rate of methanol from the anode to the cathode. This is due to the high solubility of methanol in water over the full range of compositions. There are several projects underway to modify membranes or synthesize alternatives to perfluorosulfonic acid membranes with the hopes of reducing the crossover rate of methanol. These include sulfonated polyether ethyl ketones (PEEK), polyether sulfone, polyvinylidine fluoride, grafted polystyrene, zeolite gel films (e.g., tin mordenite), and membranes doped with heteropolyanions. Most of these membranes do not have the required chemical stabilities and conductivities for the desired performances of DMFCs. A possible approach to overcome these problems is by casting the membrane and then cross-linking the polymer with compounds such as cyclic diamines (e.g., diazobicyclooctane or aminopyridine). Alternatives to these membranes and the perfluorosulfonic acids are acid doped polyacrylamide and polybenzimidazole. But one problem encountered with such membranes is the slow leaching out of the acid when the DMFC is fed with a hot methanol/water mixture.

An alternate route is to use composite membranes—e.g., Nafion with a metal oxide ($SiO_x$, $TiO_2$) or zirconium hydrogen phosphate. These types of membranes can reduce methanol crossover,[44] particularly when the DMFC is operated at a higher temperature (100 to 150 °C). In addition, most researchers have found it beneficial to use dilute methanol (1 to 2 M) as the fuel to reduce the crossover rate. There are, however, some studies with methanol fed as a vapor. For the micro and mini DMFCs (power output less than 10 W/cm$^2$), methanol cartridges are used for the fuel supply. The required amount of water is also fed into the fuel stream. For more details, the reader is referred to review articles in the Suggested Reading list.

### 9.3.2.2. Performance Characteristics

(a) *General comments.* Though there have been major strides in DMFC technology, the projections for the foreseeable future are that this type of fuel cells will find applications at the ultra-low (1 to 100 W) to low (100 W to 5 kW) power levels. There has been great enthusiasm for the development of DMFCs for transportation applications because methanol is an ideal liquid fuel with an energy density of half that of gasoline (methanol has a hydrogen content of 12% by weight). However, the power requirement for this application is in the range 10 to 100 kW. Due to the considerably lower power density of the DMFC as compared with that of a PEMFC and the high rate of methanol crossover, the interest in this area has decreased. Thus, in the following analysis of the performance characteristics of DMFCs, we shall consider these characteristics only for low and ultra-low power levels.

(b) *Cell potential, efficiency, and power density as functions of current density.* Significant improvements were made in the performance of DMFCs in the early 1990s. To illustrate this statement, a comparison is made of the $E$ vs. $i$ plots (Figure 9.27) as obtained in the 1960s by Exxon researchers[45] and in the 1990s by the LANL researchers.[46] There is practically a six-fold increase in the power density. With increase of operating temperature, these cell performances can be considerably improved. Thus, at 120 °C, and at a cell potential of 0.4 V, the current density is 400 mA/cm$^2$. By increasing the cell temperature, the extent of poisoning of the Pt-Ru anode by CO is reduced, the electrode kinetics of the methanol oxidation and the oxygen reduction reactions are enhanced, and the ohmic resistance of the electrolyte is reduced.

Another approach to enhance the performance of DMFCs is to use composite perfluorosulfonic acid/metal oxide or perfluorosulfonic acid/zirconium hydrogen phosphate (ZHP) membranes. In an international collaboration between researchers at Princeton University in the USA and CNR-TAE in Italy, Nafion/SiO$_x$ and Nafion/ZHP membranes were used in DMFCs and the performances evaluated.[47] The electrocatalysts were 60% Pt-Ru/C for electro-oxidation of methanol (noble metal loading 4 mg/cm$^2$) and 20% Pt/C for the electro-reduction of oxygen (noble metal loading 0.4 mg/cm$^2$). A comparison of the performances of the DMFCs, with composites of Nafion with SiO$_2$, TiO$_2$,

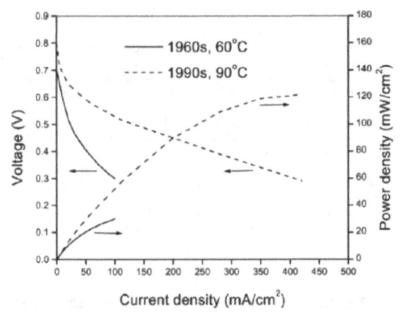

**Figure 9.27.** Comparison of DMFC (single cells) performance in the 1960s by Exxon and by Los Alamos National Laboratory in the 1990s. Reprinted from Reference 1, Copyright (1999), with permission from Elsevier.

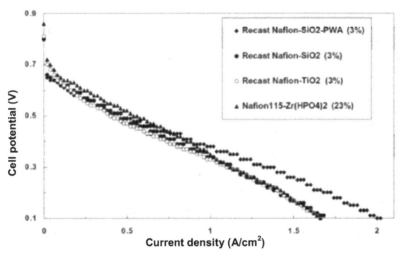

**Figure 9.28.** Comparison of DMFC (single cells) performance with composites of Nafion and $SiO_2$ (PWA = phosphotungstic acid), $TiO_2$, and $Zr(HPO_4)_2$ at 145 °C. Reprinted from Reference 48, Copyright (2004), with permission from Elsevier.

and ZHP at 145 °C is made in Figure 9.28. The impregnation of a small amount of phosphotungstic acid in the Nafion/SiO$_2$ membrane improved the performance of the DMFC.[49] In this investigation, it was also shown that the performance of the DMFC did not need humidification of the oxygen gas. The water transport from the anode to the cathode compartment inhibits the drying out of the membrane and the active layer in the cathode.

Researchers at Argonne National Laboratory have carried out investigations of DMFCs at temperatures in the range 450 to 550 °C.[50] A ceramic electrolyte ($Ce_{0.9}Gd_{0.1}O_{1.95}$) was used. The advantages of operation at such high temperatures are that CO formed as an intermediate is not a poison, methanol crossover is eliminated, and the products exit the cell in the vapor phase, simplifying the water management problem encountered in PEMFCs. Conversely, the disadvantages are long start-up times because of the high operating

## TABLE 9.4
### Operating Conditions and Performance Characteristics of DMFCS Single Cells

| ORGANIZATION | Methanol concentration | Oxidant | Temperature (°C) | Power Density (mW/cm$^2$) |
|---|---|---|---|---|
| UTC-Fuel Cells | 1 M | O$_2$/Air | 80 | 70 |
| Seimens/IRF-AS/ Johnson Mathey | 0–7M | O$_2$ | 110 | 100 |
| Los Alamos National Laboratory | 1M | Air | 60 and 90 | 80 |
| Thales-Thompson/ CNR-ITAE/ Nuvera | | | 110 | 140 |
| Jet Propulsion Laboratory | 1M | Air | 90 | 80 |
| Motorola | 1M | Air | 20 | 20 |
| Jet Propulsion Laboratory | 1M | Air | 25 | 25 |
| Forschungszentrum Julich ,GmbH | 1M | O$_2$ at 3.5 atm | 50-70 | 45–55 |
| Samsung | 2–5M | Air | 25 | 10–50 |
| Korea Institute of Science and Energy | 2–5M | O$_2$ | 25–50 | 150 |
| More Energy Ltd. | 30–45% | Air | 25 | 80 |

temperature, complex and expensive fabrication techniques for the cell components, and difficulties in developing a bipolar configuration due to thermal compatibility requirements of edge-sealing materials.

Table 9.4 summarizes the operating conditions and performance characteristics of DMFC single cells. The main goal is to achieve a stable performance level of at least 200 mW/cm$^2$ at a cell potential of about 0.5 V. This power level is adequate for portable power sources, but not so for the transportation applications.

For the ultra-low power applications (laptop computers, cell phones), several international organizations are carrying out R&D activities in developing mini and micro fuel cells (power range 1 W to 100 W). The essential feature of these cells is that they are air-breathing, or in some cases air-blowing, for supply of the oxygen reactant. Practically all of these types of fuel cells have a unipolar design for the stacks. Methanol is stored in cartridges. Its concentration is relatively high and in some cases pure methanol is used with an additional water-injection facility for the anodic reaction. More details of the mini/micro fuel cells are dealt with in Section 9.3.3.

(c)   *Methanol crossover problem and possible solutions.*   Apart from the problems of low activities of electrocatalysts and poisoning of the methanol electrode, another challenging problem is the crossover of methanol from the anode to the cathode. This has two effects: coulombic efficiencies of DMFCs can be considerably less than unity, about 60 to 70% and the crossover of methanol depolarizes the oxygen electrode. The crossover rate of methanol at the open circuit potential in a typical DMFC operating at 60 °C is about 100 mA/cm$^2$. This causes the open circuit potential (commonly referred to as open circuit voltage, OCV) of the oxygen electrode to drop to about 0.8 V/RHE from a value of 1.0 V/RHE in a H$_2$/O$_2$ fuel cell. This OCV is akin to a mixed potential at the oxygen electrode, as in the case of a PEMFC, due to the slow electrode kinetics of oxygen reduction kinetics being affected by the Pt oxidation reaction. The crossover rate decreases with increasing current density drawn from the DMFC (Figure 9.29). At a current density close to the limiting current density, the crossover rate is nearly zero. But it is not practical to operate a DMFC so close to the limiting current density. The crossover rate increases with methanol concentration and thus in most   DMFCs for portable power application and    the     proposed transportation application, the methanol concentration is maintained between 1 and 2 M. This means that the energy density of the fuel will be greatly reduced. To overcome this problem, it is possible to have a methanol recirculating flow system into and out of the anode compartment, a sensor to monitor the methanol concentration, and an automatic methanol refilling system to maintain the concentration. The crossover rate can also be reduced by using alternate membranes, membrane composites, or operating at

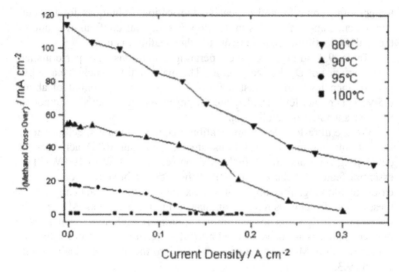

**Figure 9.29.** Crossover rate of methanol (expressed in current density) in a typical DMFC as a function of current density. Reprinted from Reference 44, Copyright (2001), with permission from Kluwer Academic.

temperatures above 100 °C. All these types of activities are still at the research stage.

### 9.3.3.  Technology Development of Cell Stacks

9.3.3.1.  *Low Power Level (100 W–5 kW).*  One of the original motivations for the development of cell stacks for DMFCs was for transportation applications and work in this direction was started in the early 1990s. The main justification was that even if the efficiency and power density are lower for a DMFC cell stack than for a PEMFC cell stack, the overall fuel cell power plant performance characteristics could be higher for a DMFC system than for a PEMFC system (in terms of efficiency, power density, specific power). The DMFC uses methanol as a fuel directly while the PEMFC must first use a fuel processor to convert methanol to hydrogen. DMFC systems at the low power level (100 W to 5 kW) could also find application for campers, residential power, and stand-by power. Summarizing remarks on stack development at these low power levels by some leading institutions follow:

(a)  UTC-Fuel Cells developed a 150 W (24 V), 600 Wh cell stack with sponsorship from the USA Defense Advanced Research Projects Agency. The intended application was as a backpack power source for soldiers. In addition, a 10-cell stack (electrode area 37 cm$^2$) was designed, built, and tested at room temperature. The hydrophilic anode contained a Pt-Ru electrocatalyst (4 mg/cm$^2$). The hydrophobic cathode had Pt black (5 mg/cm$^2$) as the electrocatalyst. The anode flow field consisted of an inlet liquid feed to the bottom of the cells and a two-phase effluent from the top of the cells back to the center of the manifold. The performance of the cell stack was in agreement with that predicted from the results of single cell studies. The stack potential was 3.6 V at a current density of 200 mA/cm$^2$. The efficiency of the cell stack was 30%.

(b) Siemens AG in Germany, IRF A/S in Denmark, and Johnson Mathey Technology Center in the UK developed a three cell stack under the auspices of the European Union Joule Program. At an operating temperature of 110 °C, a pressure of 1.5 atm, and using methanol at a concentration of 0.75 M, the cell potentials were 0.5 V and 0.48 V at current densities of 175 mA/cm$^2$ and 200 mA/cm$^2$, respectively. This work was extended to the development of a cell stack generating 0.85 kW. The cell stack had 16 cells and operated at 105 °C. The power density was 100 mW/cm$^2$ per cell.

(c) Los Alamos National Laboratory and Jet Propulsion Laboratories were also involved in cell stack development. In the former case, a 5-cell stack had an active area of the electrodes of 45 cm$^2$/cell. The cell stack was operated using 1 M methanol at 100 °C, and at an air pressure of 3 atm. The maximum power of this stack was 50 W; the power density was 1 kW/L. At 80% of the peak power the efficiency was 37%. In the work of the Jet Propulsion Laboratory in collaboration with Giner, Inc., a 5-cell stack was developed (electrode area 25 cm$^2$). The methanol concentration was 1 M and the cells operated with air at 60 and 90°C. The stack potential at 90 °C was 2 V at a cell current density of 200 mA/cm$^2$.

(d) A consortium composed of Thales-Thompson (France), Nuvera Fuel Cells (Italy), LCR (France) and Institute CNR-ITAE (Italy) has developed a 5-cell, 150 W DMFC stack, with financial support from the European Union Joule Program. Stainless steel bipolar plates were used in the stack. The electrodes had high surface area carbon Pt-Ru electrocatalysts for the anode and Pt electrocatalysts for the cathode. Other characteristics of the cell stack were: 225 cm$^2$/cell electrode area, 110 °C operating temperature, 1 M methanol concentration, and 3 atm air pressure. The power density of the stack was 140 mW/cm$^2$. The small differences in performance in a single cell and the first and fifth cells in the stack are illustrated in Figure 9.30. The difference in performance of a single cell vs. a cell in stack and cell-to-cell variation are due to variations in stack fluid dynamics.

Ballard Power Systems, Inc. (BPSI) has developed higher power level DMFC cell stacks, 3 kW and 6 kW. No details of the stack design and performance are available. According to the patent literature, the anode electrocatalyst was prepared by first oxidizing the carbon substrate

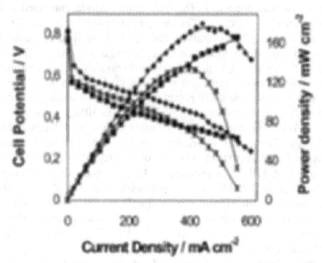

**Figure 9.30.** Performance of a DMFC single cell (◆) and cell #1(■) and #5 (*) in a 5-cell stack at 110°C. Experimental investigation conducted by consortium Thales-Thompson (France), Nuvera Fuel Cells (Italy), LCR (France), and Institute CNR-TAE (Italy) under the auspices of the European Union Joule Program. Reprinted from Reference 51, Copyright (2001), with permission from Kluwer Academic.

(carbon fiber paper or non-woven carbon fiber) electrochemically in 0.5 M $H_2SO_4$ at a potential of about 1.2 V/RHE. After oxidation of the carbon substrate, it was impregnated with the solubilized perfluoro-ulfonic acid polymer and the solvent was evaporated (dry weight of polymer 0.2 $mg/cm^2$). The electrocatalyst ink was then applied to the carbon substrate without extensive penetration into the substrate for efficient utilization of the electrocatalyst. The hydrophilic nature of the electrode and the partially oxidized carbon appear to enhance the activity for electro-oxidation of methanol. The MEA was prepared by the conventional method of hot-pressing. Oxygen and methanol flow fields were incorporated in the bipolar plates. It is unfortunate that details of the stack design are not available in the public domain.

9.3.3.2. *Ultra-Low Power Levels (1 to 100W).* Several organizations are actively involved in the development of DMFC cell stacks for the ultra-low power level applications—laptop computers, cellular phones, portable cameras, and electronic games. The rationale is that because of the high energy density of methanol and the rapid replacement of methanol cartridges for feeding the fuel into the cell stack (a refueling time of a minute or so), the DMFC power source can have advantages over even the most advanced lithium-ion batteries. It may be interesting to note that methanol has a theoretical energy density of 6000 Wh/kg. The energy density for a lithium ion battery is a factor of 10 or less. Also, the charging time for the lithium ion battery is at least 2–3 h.

A summary is presented in Table 9.5 from a recent review article[53] illustrating the characteristics of the components in the cell stacks, the operating conditions, and the power densities achieved by various organizations working on stack development. For most of the applications, the power requirements are within the range 1 to 100 W. A brief description of the stack development by some of the listed organizations is made below:

(a)    Motorola Laboratory's Solid State Research Center, in collaboration with Los Alamos National Laboratory, uses a multi-layer ceramic technology for processing and delivery of methanol and air to the fuel cell. The MEA is mounted between two ceramic plates using the upper one for passive air delivery. Four cells are externally connected in series in a planar configuration (25 cm$^2$ in area and 1 cm thick). The

## TABLE 9.5
### Characteristics of Some DMFC Stacks and Their Performance

| Organization | Electrode area (cm$^2$) | Number of cells in stack | Rated power (W) |
|---|---|---|---|
| UTC-Fuel Cells | 35 | 10 | 150 |
| Ballard Power Systems, Inc. | NA | NA | 3000–6000 |
| Los Alamos National Laboratory | 5 | 45 | 50 (300 W/L) |
| Jet Propulsion Laboratory/Giner, Inc. | 8 | 6 | 0.48 |
| IRF Fuel Cell A/S | 150 | 50 | 700 |
| Seimens/IRF-AS/ Johnson Mathey | 550 | 16 | 850 |
| Thales-Thompson/ CNR-ITAE/ Nuvera | 225 | 5 | 150 |
| Motorola | 15 | 4 | 1.5 |
| Forschungszentrum Julich ,GmbH | 100 | 40 | 100 |
| Korea Institute of Science and Energy | 50 | 6 | 50 |

cells exhibited average power densities between 15 and 22 mW/cm$^2$. Four cells are necessary for portable power applications because the operating cell potential was 0.3 V for each cell and a stack potential of over 1 V is necessary for the DC-DC converter to set up the operating voltage for the electronic devices. Ambient air is delivered to the cell at a temperature of 20 °C and 30% RH. The methanol fuel (1 M) was pumped into the anode chamber using a peristaltic pump. Motorola is also considering delivery of the methanol using small cartridges, with about the same dimensions as fountain pen cartridges.

Because of the low electrocatalytic activity for methanol oxidation, Motorola has initiated a project in collaboration with Engelhard Corporation and the University of Michigan to develop a microchannel reactor for methanol steam-reforming to generate hydrogen. The hydrogen will then be fed into the micro fuel cell and its performance will be considerably better than the DMFC fed with methanol. The power source will also be considerably smaller and provide power for a considerably longer time.

(b)  Energy Related Devices, Inc. and Manhattan Scientific, Inc. have developed miniature fuel cells with the objective of application for portable electronic devices. A low-cost sputtering method, as used in the semiconductor industry for production of microchips, is being used for deposition of electrodes on both sides of a microplastic substrate. Micropores (15–20 μm) are etched into the substrate using nuclear bombardment. Micro-fuel arrays, with external connections in series, are fabricated and have a thickness of about 1 mm. The arrays surround a fuel distribution manifold; the air distribution manifold is external to the array and is located at the rim. The reactant feed and product exit are via a wicking mechanism. The fuel delivery is through a vial penetrating the center of the array. In a prototype mini DMFC, the specific energy and energy density were 370 Wh/kg and 250 Wh/L.

(c)  Jet Propulsion Laboratory developed a flat-pack design but this type of fuel cell will have higher ohmic losses than the bipolar design. However, because of operation of these fuel cells at low power densities, ohmic losses will be reduced. The flat pack design has a back-to-back configuration. Figure 9.31 depicts a schematic of the cross-section of the fuel cell. Three twin packs connected in series are necessary for the cell phone application. Thus, there are six cells in series in a planar configuration and the power density is 25 mW/cm$^2$. The fuel cell was operating using 1 M methanol and air as the reactants. Some characteristics of the stacks and performances are presented in Table 9.5.

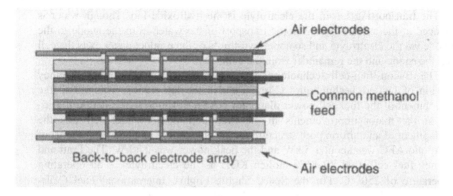

**Figure 9.31.** Schematic of JPL miniature DMFC with back-to-back multi-cell configuration. Reprinted from Reference 48, Copyright (2004), with permission from Elsevier.

## 9.4. ALKALINE FUEL CELLS (AFC)

### 9.4.1. Evolutionary Aspects

As stated in Chapter 4, the first alkaline electrolyte used in a fuel cell was molten KOH in the 19th century. The fuel was solid carbon and oxygen was the cathodic reactant. This was followed by the work of Francis Bacon, whose goal was to develop a 5 kW AFC system using pure hydrogen and oxygen as the reactants. This development and demonstration was completed over a 20-year period (1932–52). One of the main reasons for Bacon's choice of an alkaline electrolyte was to use non-noble metal electrocatalysts for the fuel cell reactions; the electrocatalysts were nickel for the hydrogen electrode and lithiated nickel oxide for the oxygen electrode. During the development of the Bacon fuel cell, Teflon had not been discovered. In order to obtain a stable three-phase zone, dual porosity electrodes were developed with the larger pores on the gas side and the smaller pores on the electrolyte side. The electrolyte in the Bacon cell was molten and the cell operating temperature and pressure were 200 °C and 50 atm, respectively.

In an alkaline fuel cell, water is produced on the anode side, according to the equation:

$$H_2 + 2OH^- \rightarrow 2H_2O + 2e^-_o \tag{9.5}$$

The cathodic reaction is:

$$\tfrac{1}{2}O_2 + H_2O + 4e^-_o \rightarrow 2OH^- \tag{9.6}$$

The transporting ion in the electrolyte is the hydroxide ion. Though water is produced at the anode, there is some transport of the water from the anode to the cathode via the electrolyte and so about two-thirds of the product water exits the cell from the anode and the remainder from the cathode.

The Bacon fuel-cell technology was inherited by the Pratt and Whitney Division of United Technologies Corporation in the late 1950s. This led to the development of the fuel cell power plants for NASA's Apollo space shuttle flights. One of the major improvements in the Pratt and Whitney fuel cell was the development of an uniform pore structure for the electrode. The rated power level of the Apollo AFC was about 1.5 kW and the peak power was 2.3 kW. The Pratt and Whitney fuel cell used 80–85% molten KOH as the electrolyte, at an operating temperature of 250 °C. For the Space Shuttle Flights, International Fuel Cells-United Technologies Corportaion (now UTC-Fuel Cells) made a revolutionary advance by developing AFCs using noble metal electrocatalysts and 35% potassium hydroxide immobilized in a reconstituted asbestos matrix as the electrolyte (maximum power greater than 1 W/cm$^2$). The performance of this fuel cell power plant is considerably better than that used in the Apollo space flights.

Several companies besides UTC-Fuel Cells have developed AFCs in the rated power range 1 to 50 kW. These companies include Allis-Chalmers, Siemens, Union Carbide, Exxon/Alsthom, Fuji, Varta, IFP, CGE, and ELENCO. All of these developments focused on terrestrial applications except the Siemens AFC, which was developed for submarine applications. A number of these systems used electrodes with low noble metal loadings. Most of this work was carried out from the early 1960s to the late 1980s. There are several books, chapters in books, and review articles in journals (see Suggested Reading list), which present in-depth descriptions of these technologies. For the next two Sections, we have chosen two types of AFCs for a brief analysis of single cell and electrochemical stack development and their performance characteristics: one developed for NASA's space shuttle flights and the other for terrestrial applications.

### 9.4.2.   Research and Development of High Performance Single Cells

9.4.2.1.*Cell Design, Assembly, and Operating Conditions.* Basically, there are two designs of single cells in AFC; one uses an immobilized electrolyte, as in the UTC Fuel Cells for NASA's Space Applications, and the other uses a circulating electrolyte, developed for terrestrial applications (Figure 9.32). An alternative material to the asbestos-matrix electrolyte is a butyl-bonded potassium titanate. A reservoir plate on the anode side replenishes the electrolyte. A porous nickel sheet or porous graphite in contact with the anode serves as the electrolyte reservoir plate. Noble metal alloys are the electrocatalysts; the anode is 80% Pt, 20% Pd (10 mg/cm$^2$); the cathode is 90% Au, 10% Pt (20 mg/cm$^2$). The electrocatalysts are pressed on to metallic substrates of silver-plated nickel screen for the anode and gold-plated nickel for the cathode. It has been proposed that gold serves as the electrocatalyst for the four electron tranfer oxygen reduction and the role of Pt

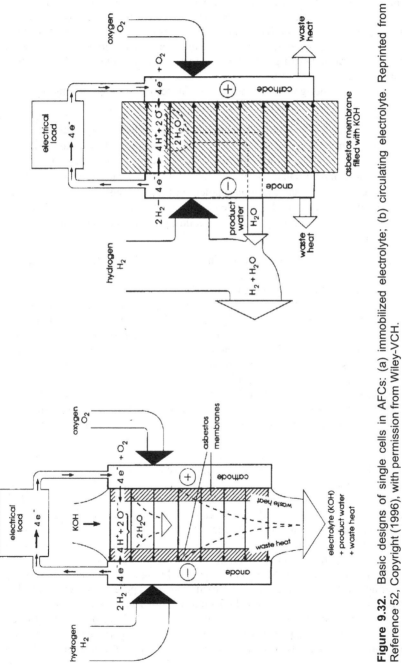

**Figure 9.32.** Basic designs of single cells in AFCs: (a) immobilized electrolyte; (b) circulating electrolyte. Reprinted from Reference 52, Copyright (1996), with permission from Wiley-VCH.

is to inhibit the sintering of the nano-sized Au particles. But in the opinion of the author, Au is a good electrocatalyst for the two electron transfer reduction of oxygen to a peroxide species ($HO_2^-$) and the Pt serves as the electrocatalyst for the subsequent two electron transfer reaction producing $OH^-$ ions. The current collectors are fabricated with gold-plated magnesium sheets and have chemically milled flow channels for the gases.

The alternative design of a single cell for terrestrial applications with a circulating electrolyte has some advantages:

- the KOH concentration is maintained constant in the single cell;
- potassium carbonate crystals formed by the reaction of small amounts of $CO_2$ present in the cathode air stream can be removed relatively easily;
- the electrolyte circulation can be used for thermal management of the cell;
- there is no $OH^-$ concentration gradient in the cell caused by the transport of the $OH^-$ ions from the cathode to the anode; and
- electrolyte circulation prevents the build up of gases (mainly inert $N_2$) in the cell.

In most of the developmental work in cells with electrolyte circulation, Pt electrocatalysts with low Pt loading electrodes (0.4 mg/cm$^2$) have been used. In an AFC technology development program in the 1970s to 1980s, Siemens in Germany used a high surface area Raney nickel electrocatalyst containing 1 to 2% Ti (Ni loading 120 mg/cm$^2$) for the anode and Raney silver (60 mg/cm$^2$, with small amounts of Ni, Bi, and Ti as additives to prevent sintering of the silver) for the cathode. An advantage of AFCs is that transition metals and their alloys and conducting oxides are stable in the concentrated KOH electrolyte. Nickel-cobalt spinels and some perovskite materials have shown high levels of performance as electrocatalysts for oxygen reduction in AFCs. One of the best electrocatalysts for oxygen reduction, as an alternative to Pt, is a heat-treated metal organic macrocyclic, e.g., heat-treated cobalt tetraphenyl porphyrin. The heat treatment is carried out at about 800°C and one could expect that most of the organic species will be removed. But there is evidence that the nitrogen atoms still remain.

One of the major technical problems in the development of AFCs has been the carbonation of the electrolyte. $CO_2$ is present at the anode at 20 to 30% concentration when reformed hydrogen is used. In the cathode, $CO_2$ present in the reactant, air (350 ppm), is also sufficient to cause the carbonation of the electrolyte. Thus, it is important to reduce the $CO_2$ concentration to near zero in both the fuel and air streams. Several methods have been proposed to reduce the $CO_2$ concentrations: soda lime or ethanol amine scrubbers, physical adsorption (selexol process), removal of $CO_2$ by diffusion of the gas through a membrane, or use of electrochemical concentration cells (e.g., oxidation of impure hydrogen at anode and its regeneration at the cathode). Economic analyses have been made for all these methods and the estimated costs for these subsystems have been found to be prohibitive for terrestrial applications of AFCs when using air as the cathodic reactant. Thus, the consensus at the present time is that, as in the case of AFCs for space applications, AFCs can find some applications when pure $H_2$ and $O_2$ are the reactants. There are prospects for such a use in the case of a regenerative fuel cell, e.g., use of photovoltaic plants to decompose water to $H_2$ and $O_2$, storage of these gases, and subsequent use in AFCs.

### 9.4.2.2.  *Performance Characteristics.*

The electro-reduction of oxygen is considerably faster in alkaline than in acid media. It is for this reason that AFCs, particularly with the immobilized electrolyte and pure $H_2/O_2$ as reactants show considerably better performance than PEMFCs. However, when $H_2$ and air are used as reactants, an AFC with a circulating electrolyte exhibits much lower performance (Figure 9.33). The inter-electrode spacing is only 50 μm in the AFC with the immobilized electrolyte, while it is at least 1 mm in the AFC with the circulating electrolyte, leading to higher ohmic overpotential. Also, the concentration of $O_2$ in the AFC with the circulating electrolyte is only one-fifth that in pure $O_2$ used in the AFC with the immobilized electrolyte.

Carbonation of the electrolyte and solid carbonate formation in the pores of the electrolyte are the predominant causes for performance degradation in AFCs. Apart from $CO_2$ present in the $H_2$ and/or air reactants, $CO_2$ is also generated in AFCs at the open circuit potentials or at low current densities because of the slow corrosion rate of the carbon support for the Pt electrocatalysts. Another problem encountered with AFCs is that there is a degradation of the Teflon used in the electrodes over a period of time. Due to surface tension effects between the Teflon and the strong KOH electrolyte, seepage occurs from the active layer to the backside of the

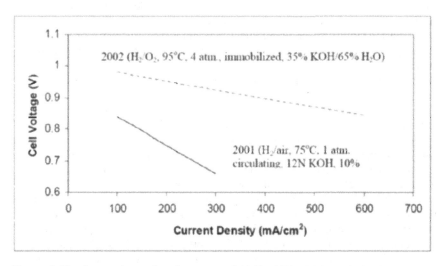

**Figure 9.33.** Comparison of performance of: $H_2/O_2$ AFC with immobilized electrolyte (UTC Fuel Cells); $H_2$/Air AFC with circulating electrolyte. Reprinted from Reference 54.

substrate layer (in contact with the current collector). This phenomenon is referred to as *electrolyte weeping* and enhances mass transport limitations for the supply of the reactant gases to the active sites of the electrode. It has been found that electrolyte weeping is less when NaOH is used as the electrolyte, but the specific conductivity of this electrolyte is less than that of KOH and hence ohmic overpotential losses will be higher. The electrolyte KOH also causes another degradation; the chemical stabilities of the gasket materials deteriorate over a period of time. This is one reason for the periodic replacement of cell stacks in the NASA Orbiter Space flight fuel cells.

9.4.2.3.   *Technology Development of Cell Stacks.*   There have been a multitude of stack development projects internationally from the 1960s until the present time. The most successful of these has been the power plant development by United Technologies Corporation. During the period from the early 1960s to the 1980s, alkaline fuel cell power plants were successfully developed and demonstrated for terrestrial applications. However, since the early 1990s, the $CO_2$ problem has decreased enthusiasm for developing AFCs for such applications. One reason for this was the rapid advances made in PEMFC technology development. In spite of this, there are a few projects on AFC R&D with the hope that the $CO_2$ problem can be resolved with sophisticated methods for its removal from hydrogen fuel.

Table 9.6 presents a list of most of the stack development projects, many of them leading to power plant projects. Also included in Table 9.6 are the operating and performance characteristics of these systems. Innovative methods have been used in several of these technologies and have been well described in previous publications. In this book, the author is focusing on UTC-Fuel Cells stack, developed for the Orbiter Space Flight Fuel Cells, and Exxon-Alsthom ELLENCOs stack development using carbon conducting plastics (electronic conduction) for current collector/bipolar plates.

As stated in the previous subsection, the state-of-the art UTC AFCs use bipolar plates made from gold-plated magnesium sheets with chemically milled flow channels for the gases. Polysulfone, and more recently polyphenylene sulfide, is used for the gaskets in the edges for the sealing of the cells. Light-weight graphite and metallic plastics are used for the electrolyte reservoir plates, which are necessary to replenish the electrolyte in the matrix. The loss of electrolyte in the matrix occurs during the fuel cell operation. The waste heat removal from the stack is via the heat-conducting foils located between every two cells in the stack. Product water is removed from the anode side using a condenser, from which it flows to a centrifugal separating device and then to a storage tank. The water thus produced is used for drinking by the astronauts and for spacecraft cooling. The rated power level is 12 kW and the peak power can be 16 kW for short periods. Each stack is 35 x 38 x 115 cm, weighs 118 kg, and contains 32 cells. The cell stack operates at about 90 °C and 0.4 MPa. The active electrode area is 465 $cm^2$. At the rated power, the stack potential is 28 V (single cell potential 0.875 V) at a current density of 0.9 A/$cm^2$. In comparison with the Apollo Space AFC, the Space Shuttle Orbiter

**TABLE 9.6**
**AFC Stack Development and Demonstration Projects and Their Performance Characteristics**

| Organization | Time Frame | Operating Temperature (°C) | Designed power (kW) | Peak power (kW) | Number of cells in stack | Electrode area (cm²) | Specific power (kW/kg) | Special features |
|---|---|---|---|---|---|---|---|---|
| Pratt and Whitney Division of UTC | 1960s and 1970s | 250 | 1.5 | 2.3 | 31 | | 20 | Molten KOH electrolyte |
| International Fuel Cells–UTC | 1970s–2000s | 60–70 | 12 | 20 | 32 | 465 | 100 | Gold-plated Mg bipolar plates |
| Exxon-Alsthom/ Occidental Chemical | 1970s | | | | 100 | 400 | | Conductive plastic bipolar plates |
| ELENCO | 1970s and 1980s | 60–80 | 10 | | | | 100 | Conductive plastic current collectors, monopolar design |
| Siemens | 1970s and 1980s | | 7 | | 70 | 340 | 80 | Unipolar design, Raney nickel anode and silver cathode electrocatalyst |
| Union Carbide, F.T. Bacon | 1930s–1950s | 200 | 5 | | | | | Porous nickel anode, lithiated nickel oxide cathode |

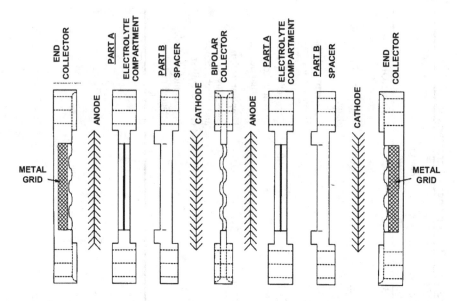

**Figure 9.34.** Components and layout of Exxon-Alsthom/Occidental AFC 2-cell stack. Reprinted from Reference 55, Copyright (1993), with permission from Plenum Press.

AFC delivers about eight times more power, is considerably lighter, and occupies a smaller volume.

Monopolar and bipolar configurations have been developed and demonstrated in AFC stacks for civilian applications. Researchers at Exxon-Alsthom were the pioneers in using electronically conducting plastics for the current collect/bipolar plates. The principle of this method is to blend small spherical particles of polypropylene or polyethylene with graphite and mold it to the desired configuration for the bipolar plate. The plate was incorporated on a 40% tac-filled polypropylene frame. Injection molding was used for the fabrication of the bipolar plate. The active area of the cell was 400 cm$^2$ with a framed total surface area of 900 cm$^2$ (Figure 9.34). One of the problems encountered with AFC cell stacks with a bipolar configuration and a circulating electrolyte was the parasitic efficiency loss due to shunt current generation from the anode of one cell in the stack to the electrodes in neighboring stacks. The shunt current could be minimized by using a small diameter path for the transport of KOH from one cell to a neighboring cell, thereby increasing the ohmic resistance. The Exxon-Alsthom group also limited the number of cells in a stack to 100 to minimize the shunt current. The electrolyte in this stack was 6 N KOH. The cell performance was reasonable: 0.72 V at a current density of 150 mA/cm$^2$.

ELENCO, a consortium of the Belgium Atomic Energy Commission, Bekaert (a Belgian Company), and Dutch State Mines used a similar approach as the Exxon-Alsthom group with conducting plastics for the current collector plates. The stack had a monopolar configuration and thus the shunt current problem was eliminated. ELENCO used compression molding for the production of the stacks. The procedure greatly reduced the *electrolyte weeping* problems encountered in most AFC development work. ELENCO electrodes contained carbon-supported Pt electrocatalysts (0.3 to 0.4 mg/cm$^2$) and were embedded in a nickel screen. The electrolyte was 6 M KOH and the operating temperature was 65 °C. There were 24 cells in a stack (300 cm$^2$ active area of electrode). The stack weighed 4 kg and the power output was 400 W using H$_2$/Air as reactants. The power density was 70 mW/cm$^2$. Modules with rated power outputs in the range 0.5-10 kW were produced and demonstrated.

## 9.5    PHOSPHORIC-ACID FUEL CELLS (PAFC)

### 9.5.1.  Evolutionary Aspects

The PAFC system was the first fuel cell technology to be commercialized for civilian applications. Interest in this system evolved in the mid 1960s because phosphoric acid could be used in a concentrated form (above 85%) in a fuel cell at intermediate temperatures of 150 to 200 °C. There was great enthusiasm in the 1960s to electro-oxidize organic fuels (natural gas, the higher hydrocarbons, methanol, and ethanol) at these higher temperatures. However, it was found that even at such temperatures, the activity of the best electrocatalyst (Pt) at high loading was minimal for the electro-oxidation of the organic fuel and for the oxygen reduction.

In a PAFC, the acid serves as the ionic electrolyte as well as the solvent. The ionization reaction of the acid is represented by the equation:

$$H_3PO_4 \rightarrow H^+ + H_2PO_4^-  \qquad (9.7)$$

This reaction is followed by the solvation of the acid according to the equation:

$$H^+ + H_3PO_4 \rightarrow H_4PO_4^+  \qquad (9.8)$$

The H$_3$PO$_4$ serves as the solvent for the Grotthus-type conducting mechanism, as in the case of aqueous acid electrolytes. The specific conductivity ($\kappa$) of H$_3$PO$_4$ is quite low at temperatures less than 100 °C. The value of $\kappa$ increases with dilution of the acid because of increasing contribution of the solvent water to proton conduction.

In order to operate a PAFC at reasonably high temperatures (about 150 °C), the acid strength needs to be about 85%. However, even at such a temperature, it was found that the PAFC performance was not satisfactory. Thus, since the late 1970s, research and development efforts were focused on higher operating temperatures and using the acid at close to 100% concentration. The acid polymerizes to pyrophosphoric acid $H_4P_2O_7$ at this temperature as:

$$2\ H_3PO_4 \rightarrow H_4P_2O_7 + H_2O \tag{9.9}$$

$H_4P_2O_7$ has a considerably higher ionization constant than $H_3PO_4$,

$$H_4P_2O_7 \rightarrow H^+ + H_3P_2O_7 \tag{9.10}$$

Activation overpotential losses were significantly lowered at the higher temperature, particularly at the oxygen electrode, because the pyrophosphoric anion is adsorbed to a considerably lesser extent than the phosphoric acid anion on the Pt electrocatalyst. The ohmic overpotential loss is also lower because of the higher conductivity of pyrophosphoric acid than phosphoric acid. An interesting basic study on the structure of the double layer was conducted by Gonzalez, Hseuh, and Srinivasan in the early 1980.[53] It was shown that the thickness of the electric double layer is greater at an electrode/concentrated phosphoric acid interface than at an electrode/aqueous acid interface, and that there is a high degree of adsorption of the phosphate ion on the electrode. These factors contribute to the low activity of the electrocatalyst for the electro-reduction of oxygen, particularly at lower temperatures.

The low activity of the electrocatalysts for the direct electrochemical oxidation of the hydrocarbons and alcohols is a result of of the high energy requirement for the breakage of C-H and C-C bonds at low and intermediate temperatures. Even if there is some oxidation of these reactions, the intermediate species during the reaction are strongly adsorbed on the Pt electrocatalysts. For the desorption of these species, it is necessary to have a neighboring site with an O or OH type species. Such species can be formed by the oxidation of water as:

$$Pt + H_2O \rightarrow PtOH + H^+ + e_o^- \tag{9.11}$$

The concentration of water is quite low and thus the extent of adsorption of such species is low. Therefore, higher overpotentials are necessary for the electro-oxidation of the organic species.

To overcome these difficulties, there was an evolutionary change in R&D of PAFC systems, e.g., to reform the organic fuels to hydrogen and use it in the PAFC. The overpotential losses were considerably lowered. The reader is referred to Chapter 8 for the details of fuel processing for the production of hydrogen. Since the early 1970s, there have been several large companies involved in technology development of phosphoric acid fuel cell systems: UTC Fuel Cells, Westinghouse Electric Corporation, Energy Research Corporation, and Engelhard in the USA; Fuji

Electric Corporation, Toshiba Corporation, and Mitsubishi Electric Corporation in Japan; and Ansaldo in Italy.

The power levels of the demonstrated PAFC systems have been in the range of 40 kW to 11 MW. The system had three major sub-systems: fuel processor, fuel cell, power conditioner. The applications were focused on stationary, distributed power plants, and on-site integrated energy systems. These fuel cell power plants were initially being considered by electric utilities as substitutes for gas turbines, which were being used to meet peak-power demands or for load-following. There were a multitude of demonstration projects of PAFC power plants by electric utilities for these applications. However, the competition was very severe from gas turbines that exhibited as good or better performance with respect to efficiency and lifetime. In addition, the cost of the PAFC systems was higher than that of a gas turbine by a factor of six to eight. However, one application that showed promise was the PAFCs as on-site integrated energy system for residential and commercial buildings at a power level of 200 kW. UTC-International Fuel Cells built a large-scale production plant and started to commercialize these plants worldwide in 1996. Presently, 200–250 plants have been sold, far fewer than the expected sales levels. The costs of these plants are prohibitive: $4000/kW vs. about $500–700/kW for gas turbines. Another negative attribute for the PAFC system was that the electrochemical cell stack has a lifetime of about 40,000 h before there is a decrease

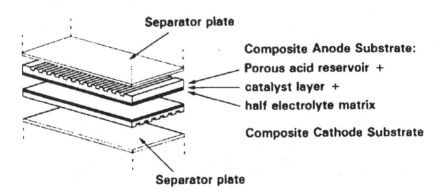

**Figure 9.35.** Components and design of the UTC Fuel Cells single cell PAFC. Reprinted from Reference 55, Copyright (1993), with permission from Plenum Press.

in the performance. However, the positive feature was that when one considers the combined heat and power generation, the overall efficiency is 80 to 85%.

At the present time, there are hardly any R&D programs on advancing the PAFC technology.

## 9.5.2. Research and Development of High-Performance Single Cells

### 9.5.2.1. *Design, Assembly and Operating Conditions.* The components of a single cell in a PAFC electrochemical cell stack are shown in Figure 9.35. The current collectors (graphite plates with the flow-channels for the reactants) are placed on the exterior sides of the carbon substrate layer. The electrolyte is held in a silicon-carbide matrix (about 50 μm thick). It is deposited on the active layer of the cathode along with Teflon. The Teflon serves as a binder. The electrode silicon-carbide composite layer is heated to dry the matrix and to sinter the Teflon. The porous-conductive substrate layer serves three purposes: gas distribution, electrolyte reservoir, and current collector. The electrolyte reservoir is necessary to replenish the small amounts of electrolyte lost from the silicon carbide matrix by evaporation during the operation of the fuel cell. The active layer in the electrode contains high surface area platinum, supported on high surface area carbon (e.g., Vulcan XC 72R). The platinum loading was about 0.4 mg/cm$^2$ on each electrode, but UTC has reduced the loading on the anode side to about 0.25 mg/cm$^2$. It is believed that the UTC-Fuel Cells uses a Pt-Co-Cr electrocatalyst for the oxygen reduction reaction. There is a reduction of activation overpotential at the cathode by about 40 mV using this electrocatalyst compared to pure Pt. The acid concentration in the most advanced PAFC systems (200 kW UTC fuel cells) is 100%. The operating temperature is 200 °C. The 200 kW PAFC system (PC 25) operates at 1 atm pressure. But there has been an 11 MW system developed by UTC Fuel Cells–Toshiba, which operates at 8 atm.

### 9.5.2.2. *Performance Characteristics*

(a) *General comments.* Research at a fundamental level in half and single cells was conducted quite extensively from about 1975 to 1985 because the PAFC system was projected to be the first of all types of fuel cells to find an application as a stationary power source. Indeed, the 200 kW PAFC power plant (PC 25 manufactured by UTC-International Fuel Cells) emerged as an on-site integrated energy system (combined heat and power) in 1996. Previous publications (see Suggested Reading) have extensively described the advances made in PAFC technology. The most notable advances, particularly showing differences between the acid based PAFC and aqueous-acid electrolyte fuel cells (e.g., PEMFC, sulfuric acid electrolyte fuel cells) are discussed here.

(b) *Cell potential vs. current density behavior.* The initial problem encountered with PAFCs was that its performance in single cells was very low compared to that of aqueous acid or alkaline electrolyte fuel cells at operating temperatures below 100 °C. Strong adsorption of the phosphate ion on the Pt electrocatalyst and low ionic conductivity of the acid created high overpotentials in the concentrated phosphoric acid (85% $H_3PO_4$) fuel cells. To overcome these problems, the temperature was initially raised to 150 °C while the electrolyte concentration was maintained at 85%. There were considerable reductions in activation and ohmic overpotentials at this temperature. As stated earlier, at about this temperature, phosphoric acid polymerizes to pyrophosphoric acid and the latter is a stronger acid with a higher ionization constant. Also, the pyrophosphate anion is considerably larger than the phosphate anion, reducing its specific adsorption considerably. In subsequent studies, further performance improvements were made by progressively increasing the concentration of phosphoric acid first to about 95%, and then to 100% in UTC-Fuel Cells PAFC power plants. The operating temperature is currently 200 °C. The benefits of such an operating temperature are better electrode kinetics of oxygen reduction, lesser ohmic overpotential losses, and higher CO tolerance. Figure 9.36 illustrates the enhancements in performance of PAFC single cells with an increase of operating temperature and pressure. The plot, representing the highest level of performance, was made from experimental data in a short stack. The following empirical equations, as ascertained from experimental data, have been proposed for the increases in cell potential, $\Delta V_T$ and $\Delta V_p$, as functions of temperature and pressure, respectively:

$$\Delta V_T = 1.15 \, (T_2 - T_1) \qquad 180 \le T \le 205 \text{ °C} \qquad (9.12)$$

$$\Delta V_p = 146 \log (P_2/P_1) \qquad 1 < P < 10 \text{ atm} \qquad (9.13)$$

(c) *Corrosion of carbon supports for the electrocatalyst and Pt dissolution.* Extensive research was carried out to determine the rates of corrosion of a wide variety of high surface area carbons as supports for the high surface area Pt electrocatalysts. Appleby reviewed this topic and presented an interesting analysis in a plot of the dependence of the corrosion current of the carbon on the electrode potential for three types of carbon.[56] Vulcan XC 72 is the most commonly used carbon in low and intermediate temperature fuel cells. Its heat treatment in an inert atmosphere up to about 900 °C reduces the rate of carbon corrosion. Acetylene black has a lower corrosion rate than Vulcan XC 72 but its specific area is considerably smaller. The corrosion rate increases with increasing electrode potential and is highest at the open circuit potential.

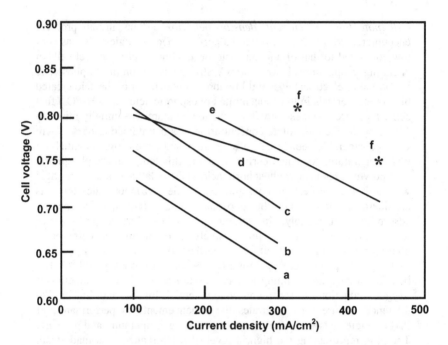

**Figure 9.36.** Enhancements in performance of H₂-Rich Fuel/Air PAFCs with increasing temperature and pressure: (a) 1977: 190°C, 3 atm, Pt loading of 0.75 mg/cm² on each electrode; (b) 1981: 190 °C, 3.4 atm, cathode Pt loading of 0.5 mg/cm²; (c) 1981: 205 °C, 6.3 atm, cathode Pt loading of 0.5 mg/cm²; (d) 1984: 205 °C, 8 atm, electrocatalyst loading not specified; (e) 1992: 205°C, 8 atm, 10 ft² short stack, 200 hrs, electrocatalyst loading not specified; and (f) 1992: 205°C, 8 atm, subscale cells, electrocatalyst loading not specified. Reprinted from Reference 54.

Another corrosion problem is that of the nano-sized particles of the Pt electrocatalysts. There is some dissolution of these particles due to Ostwald ripening, a dissolution/precipitation mechanism where the smaller Pt particles dissolve and redeposit on the larger Pt particles. This causes a loss of surface area of the Pt electrocatalyst with time. Here again, this phenomenon is more pronounced closer to the open circuit potential.

(d) *Effects of impurities in the hydrogen fuel.* The effect of CO poisoning on the overpotential at the hydrogen electrode is markedly less in PAFCs than in low temperature acid electrolyte fuel cells. Hydrogen adsorption is increasingly favored at higher temperatures while CO adsorption is increasingly disfavored. At a temperature of 200 °C, the CO tolerance level is 1–2% and thus, there is no need for a preferential oxidation step in fuel processing. However, another source of CO with reformed fuels is the $CO_2$ which is reduced to CO by the reverse of the water gas shift reaction on the anode electrocatalyst. This reaction also occurs in PEMFCs, as described in Section 9.2.2.5. Interesting plots illustrating the effects of CO and $CO_2$ on the anode potential are shown in Figure 9.37.

Hydrogen sulfide is a deadlier poison (tolerance level < 5 ppm) than CO for the anode electrocatalyst. The overpotential increase with $H_2S$ concentration (with and without CO) is illustrated in Figure 9.38. It has been proposed that the significant effect of the combined CO and $H_2S$ impurities is due to the formation of a species COS which is more strongly adsorbed on the Pt than CO or $H_2S$. During $H_2S$ adsorption, there could also be electro-oxidation of $H_2S$ to elemental sulfur, which is also strongly adsorbed on the electrocatalysts.

(e)  *Lifetime studies.* Extensive investigations have been carried out on PAFC single cells and cell stacks to determine their lifetime characteristics. The target for a cell stack has been a continuous lifetime of 40,000 h with a performance degradation of 1 mV/1000 h for stationary applications. During the preliminary phase of the research, development, and demonstration programs in the 1960s to the 1970s, the degradation rate was about 10 times the target value.Several factors contributed to the

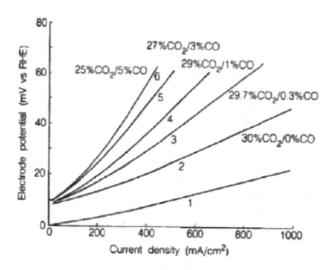

**Figure 9.37.** Effects of CO and $CO_2$ on the anode potential as a function of current density, using $H_2$ and $CO/CO_2$ gas mixtures. Curve 1, 100% $H_2$; Curves 2-6 70% $H_2$. Reprinted from Reference 54.

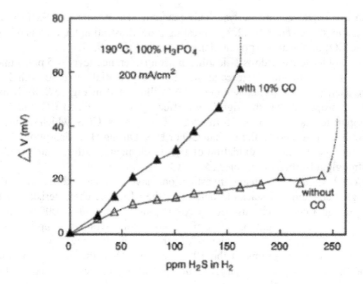

**Figure 9.38.** Increase in anode overpotential ($\Delta V$) with $H_2S$ concentration (with and without CO), using a Pt electrocatayst. Reprinted from Reference 54.

degradation, including corrosion of the carbon support and Pt dissolution/precipitation (sintering phenomenon). Performance degradation was also caused by poisoning of the Pt anode electrocatalyst by CO and $H_2S$. In the initial period, an organic polymer, Kynar, in combination with Teflon, was used as the matrix for the phosphoric acid electrolyte matrix. Kynar was not found to be stable in the hot PAFC environment and it caused significant anode poisoning. This material, however, did not affect the oxygen electrode performance. Since the late 1970s, considerable progress has been made by the PAFC system developers to identify the above causes for the degradation in lifetime, and corrective measures have been taken to minimize the performance variation with time. Maintaining the cell potential at 0.8 V or lower minimizes the rate of corrosion of the carbon support and platinum particles. An alternate method is to have the cathode in an inert atmosphere when the cell is under open-circuit conditions. CO content in the fuel is minimized. But if reformed hydrogen is used, the $CO_2$ content in this gas gets reduced to CO and raises the concentration above the tolerance level. Performance degradation can also be reduced by the complete elimination of sulfur in natural gas by hydro-desulfurization.

## 9.5.3.  Technology Development of Cell Stacks

The most advanced projects on stack development and testing have been carried out by UTC-Fuel Cells, Westinghouse Electric Corporation-Energy Research Corporation (presently Fuel Cell Energy), and Engelhard Industries Division of Engelhard Minerals and Chemicals Corporation in the USA; Toshiba, Mitsubishi, and Fuji in Japan; and Ansaldo in Italy. There have been joint ventures between UTC-Fuel Cells and Toshiba, and Engelhard and Fuji in the development and demonstration of fuel cell stacks. The most advanced PAFC cell stacks and systems in the kW to MW range have been by UTC-Fuel Cells. This company used a *ribbed substrate configuration* with a thin impervious graphite sheet for the bipolar plate. The ribbed substrates are porous and provide channels for the gases to enter the porous electrodes. Since there is a small loss of phosphoric acid by vaporization from the electrolyte layer, the ribbed substrate also serves as an electrolyte reservoir plate. In the UTC Fuel Cells cell stack, liquid cooling is used; heat is rejected by flowing water through a number of parallel, thin-walled copper tubes that run through graphite coolers in the stacks. The electrode area in the 200 kW and 670 kW stacks (in the 11 MW PAFC power plant for Tokyo Electric Power Company) is 1 m$^2$. The cell stacks were designed for operation up to a pressure of 8 atm.

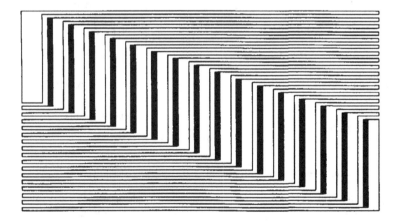

**Figure 9.39.**  Westinghouse-ERC Z pattern bipolar plate for PAFCs. Reprinted from Reference 55, Copyright (1993), with permission from Plenum Press.

Other noteworthy developments of PAFC systems include the following: Westinghouse-Energy Research Corporation used a Z-pattern bipolar plate (Figure 9.39) made by a molding process that combines graphite powder with a binder. The plate is designed so that the reactant (fuel or air) enters half of the cell and exits the cell from the opposite half. The cell was 30 x 43 cm in an air-cooled stack. Fuji based their design on the Engelhard technology and built a 50 kW stack. Mitsubishi has built and demonstrated 100 kW stacks, and Ansaldo has designed and developed modules for a 1 MW system.

With respect to operating conditions and performance, most work has been at atmospheric pressure or at pressures of 3 or 8 atm. The operating temperature was about 200 °C. The electrolyte contained in the electrolyte matrix was 100% $H_3PO_4$. The performance in a single cell was about 300 mA/cm$^2$ at a cell potential of about 0.7 V. One striking feature of PAFC stacks compared with PEMFC stacks is that there is hardly any loss in performances with scale-up of electrode area in PAFCs. The problem in PEMFCs is caused by the difficult water and thermal management problems discussed in Section 9.2. It is also worth noting that cell areas in the PAFC stacks have reached values of 1 m$^2$, whereas the maximum cell area in PEMFC stacks is 250 to 300 cm$^2$. The main advantage of the PEMFC cell stacks is that the power densities can be about two to three times higher than in PAFC stacks.

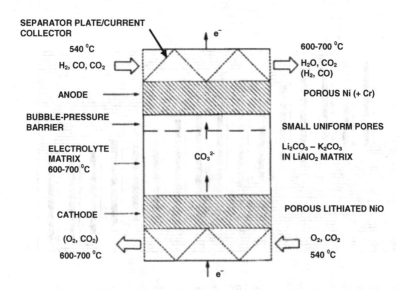

**Figure 9.40.** Schematic of single cell in an MCFC. Reprinted from Reference 57, Copyright (1993), with permission from Plenum Press.

## 9.6    MOLTEN CARBONATE FUEL CELLS (MCFC)

### 9.6.1.   Evolutionary Aspects

MCFC power plants are capable of direct utilization of natural gas and coal derived gases for electric and thermal power generation at high efficiencies. In an MCFC, the cathodic reaction consumes oxygen and carbon dioxide to produce carbonate ions, which are transported to the anode through the electrolyte. At the anode, hydrogen combines with the carbonate ion to produce water and carbon dioxide. Carbon dioxide is then shuttled back to the cathode for the cathodic reaction. A schematic of a single cell in an MCFC is shown in Figure 9.40. CO is also shown as a fuel. However, according to most investigations, CO is converted to hydrogen via the water-gas shift reaction (see Chapter 8) and hydrogen is the electroactive fuel. The state-of-art electrolyte in MCFCs is 62% $Li_2CO_3$ and 38% $K_2CO_3$. This electrolyte has a reasonable ionic conductivity only at temperatures above 600 °C. The current operating temperature is 650 °C. At temperatures above 700 °C, there is some loss of electrolyte by vaporization. One advantage of the high operating temperature of MCFCs is that activation overpotentials for the fuel cell reactions are considerably lower than those for the low and intermediate temperature fuel cells. Furthermore, expensive noble metal electrocatalysts are not required for the anodic and cathodic reactions.

Some historical development aspects of MCFCs were summarized in Chapter 4. As in the case of all other types of leading fuel cell technologies, the pioneering work on MCFCs was carried out in the latter half of the 19th century and in the first half of the 20th century. The original and most outstanding R&D work on MCFCs was carried out by Ketelaar and Broers in the 1950s in the Netherlands.[58] This work was followed by that of Baker at the Illinois Institute of Technology.[59] The energy crisis in 1973 stimulated the development of MCFCs for base load and intermediate load power generation. Several companies in the USA, Japan, Italy, and the Netherlands were actively involved in the development of multikilowatt MCFC stacks for power generation and cogeneration. This type of fuel cell was considered the second-generation fuel cell for power generation (the first being the PAFC). But MCFC technology has faced many challenges, particularly because of the relatively high corrosion rates of component materials in the electrolyte at the high operating temperatures. In spite of this problem, major advances in technology development and demonstration have been made in the last two decades mainly by Fuel Cell Energy in the USA, Ansaldo in Italy, and Ishikawajima-Harima Heavy Industries in Japan.

### 9.6.2.   Single cells

9.6.2.1. *Design, Assembly, and Operating Conditions.*   Characteristics of the state-of-the-art component materials are presented in Table 9.7. All three components (anode, cathode, and electrolyte layer) are presently manufactured by tape-casting processes. The anode contains a nickel-chromium alloy (2–10% Cr by weight); chromium prevents the sintering of the porous anode because it forms $LiCrO_2$ at the grain boundaries and minimizes diffusion of the nickel. In addition, small metal oxide particles are incorporated in the anode to inhibit mechanical creep. In addition to its electrocatalytic function, the anode also serves as an electrolyte barrier for gas crossover and a structural support for the cathode and thin electrolyte matrix. The gas crossover is minimized by a thin structured layer of fine porosity, which is filled with the molten carbonate electrolyte at the anode/electrolyte interface. This layer is referred to as the *bubble-pressure layer*. Unlike in the case of the low and intermediate temperature fuel cells, there are no non-wetting materials like Teflon for maintaining stable gas/liquid interfaces and channels for the electrolyte and the reactant gases in MCFCs. The pore-size distribution for the electrodes, electrolyte matrix, and bubble pressure layer play important roles in establishing stable gas-electrolyte interfaces in the porous electrodes because the balance in capillary pressure is used for the electrolyte distribution in the cell components. This balance in capillary pressure can be expressed by:

**TABLE 9.7**
**Characteristics of the State-of-the-Art Cell Components in MCFCs**

| Component | Current status |
|---|---|
| Anode | Ni-10 wt% Cr <br> 3-6 μm pore size <br> 50-70% porosity <br> 0.5-1.5 mm thickness <br> 0.1-1 $m^2$/g |
| Cathode | Lithiated NiO <br> 7-15 μm pore size <br> 70-80% porosity <br> 0.5-0.75 mm thickness <br> 0.5 $m^2$/g |
| Electrolyte support | $\gamma$-$LiAlO_2$ <br> 0.1-12 $m^2$/g |
| Electrolyte (mole % of alkali carbonate salt) | 62 Li-38 K <br> 50 Li-50 Na <br> 70 Li-30 K <br> ~50 wt% |
| Fabrication process | Tape cast <br> 0.5 mm thickness |

$$\frac{\gamma_a \cos\theta_a}{d_a} = \frac{\gamma_e \cos\theta_e}{d_e} = \frac{\gamma_c}{d_c} \qquad (9.14)$$

where $\gamma$ is the interfacial tension, $\theta$ is the contact angle of the electrolyte, $d$ is the pore diameter, and the subscripts $a$, $e$ and $c$ denote the anode, electrolyte, and cathode, respectively. The electrolyte pores are completely filled with electrolyte, while the electrodes are partly filled (the latter is achieved by varying the pore size distribution).

The electrolyte tile in the MCFC consists of a eutectic mixture of 62% $Li_2CO_3$ and 38% $K_2CO_3$ retained in a porous aluminate matrix. The electrolyte layer serves two purposes: to conduct carbonate ions from cathode to anode and to separate fuel and oxidant gases. Cations in the electrolyte have strong effects on the performance and endurance of MCFCs. High Li and Na contents in the electrolyte increase ionic conductivity, while high potassium content increases gas solubility. The $Li_2CO_3/Na_2CO_3$ electrolyte also increases corrosion resistance of the cathode. One problem with the electrolyte tile material is that it is susceptible to a $\gamma$- to $\alpha$-$LiAlO_2$ transformation at high temperatures over long periods of time. This occurs more at higher temperatures and in strongly basic melts.

The cathode consists of porous lithiated nickel oxide formed by in situ oxidation and lithiation of sintered nickel. Since lithiated nickel oxide is a non-stoichiometric compound ($Li_xNi_{1-x}$, where $0.02 \leq x \leq 0.04$), its electronic conductivity is several orders of magnitude higher than nickel oxide. At the present time, the thickness of the cathode layer is lower than that of the anode to minimize electronic resistance and enhance performance.

Nickel or nickel-plated steel are used as materials for current collectors. Because of the high operating temperature and a highly corrosive electrolyte, it had been a challenge to provide seals at the edges of the cell. However, the plasticity of the electrolyte tile was found to provide gas tight seals (the temperature is lower in these sites and the molten carbonate is solidified in these areas). Thus, the solid electrolyte in this area serves as a gas-tight seal. The cell is compressive and periodically tested for the extent of compression to obtain the highest level of performance.

Laboratory cells generally have an active area of 100 cm². However, as in the case of PAFCs, the MCFC cells in stacks have an electrode area of about 1 m². An MCFC operates at about 650 °C and 1 atm. Higher pressures (up to 8 atm) enhance the performance.

### 9.6.2.2. Performance Characteristics

(a)  *Cell potential vs current density behavior.* Because of the high operating temperature of MCFCs (650 °C), the exchange current density for oxygen reduction at the lithiated nickel oxide cathode is considerably higher than in low and intermediate temperature fuel cells, at least two orders of magnitude higher than in PAFCs operating at 200 °C. Thus, the initial semi-exponential region of the cell potential vs. current density plot is not observed. The reason for this is that the net current density, $i$, for oxygen reduction is expressed by the Butler-Volmer equation:

$$i = i_0 \left[ \exp\left(\frac{\alpha \eta F}{RT}\right) - \exp\left(\frac{(1-\alpha)\eta F}{RT}\right) \right]$$  (9.15)

where $\alpha$ is the transfer coefficient, $\eta$ is the overpotential, and $F$ is the Faraday constant. At room temperature, the exponential terms can be linearized when $\eta < 20$ mV (see Chapter 2), but at a temperature of 650 °C this can be done for overpotentials up to about 60 mV. There has been great improvement in the performance of single cells in MCFCs, as depicted in Figure 9.41. The cell potential vs. current density plot is linear over the entire range of current density recorded. The major contribution to the efficiency loss in MCFCs is the ohmic losses, particularly in the electrolyte layer. Ohmic losses contribute to more than half of the total overpotentials, while Nernstian losses and activation and mass transport overpotentials contribute to the remainder of losses in the cell potential. The Nernstian losses are caused by changes in concentration of the reactants during operation of the fuel cell. The reversible potential for the cell ($E_r$) is expressed by the Nernst equation:

$$E_r = E_r^0 + \frac{RT}{2F} \ln\left( \frac{P_{H_2} P_{O_2}^{1/2}}{P_{H_2O}} \frac{P_{CO_2,c}}{P_{CO_2,a}} \right)$$  (9.16)

The parameter $P$ indicates the pressures of the noted gases and the subscripts a and c denote the anode and cathode, respectively. With increasing utilization of the reactants, there is a decrease in the reversible potential of the cell, which causes the Nernstian losses. As in the case of the low and intermediate temperature fuel cells, the activation overpotential for the cathodic reduction oxygen is significantly higher than for the anodic oxidation of hydrogen. At higher current densities, mass transport overpotentials play a role, as indicated by the deviation from linearity of the $E$ vs. $i$ plot (Figure 9.42). In this figure, two plots from Fuel Cell Energy are shown: one is the experimental plot, and the other is the theoretical plot from a modeling study conducted by this organization. It can be seen from this plot that the $E$ vs. $i$ plot can be represented by:

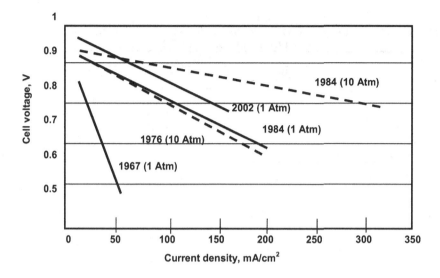

**Figure 9.41.** Progress in improving performance of MCFC single cells on reformate gas and air. Reprinted from Reference 54.

$$E = E_0 - Ri \qquad (9.17)$$

with $R$ being 1.2 $\Omega cm^2$ at current densities between 50 and 150 mA/cm$^2$ and 1.75 $\Omega cm^2$ at current densities between 150 and 200 mA/cm$^2$. It is worth noting that the magnitude of $R$ is about 5 to 6 times higher than in PEMFCs. It is for this reason that MCFCs do not attain high power densities.

(b) *Effect of temperature.* Increased operating temperature of an MCFC decreases the reversible potential of the cell, but also decreases the overpotential losses in the cell. In calculating the effect of temperature on the reversible potential of the cell, one has to take into consideration the water-gas shift and reforming reactions. Selman has calculated the effect of temperature on the gas compositions, reversible potential, and equilibrium constant for a high BTU gas mixture saturated with water vapor at room temperature (Table 9.8).[60] As expected, there is only a small effect of temperature on the reversible potential. Increasing the temperature has more significant effects of decreasing activation, mass transport, and ohmic overpotentials. A fundamental study was conducted by Baker et al. to determine the effects of temperature on the initial performance of small cells (electrode area 8.5 cm$^2$).[61] The fuel used was

**TABLE 9.8**
**Effect of Temperature on Gas Composition, Reversible Potential, and Equilibrium Constant for an MCFC**

| Component | Partial pressure (atm) | | | |
|---|---|---|---|---|
| | 298 K[a] | 800 K | 900 K | 1,000 K |
| $H_2$ | 0.775 | 0.669 | 0.649 | 0.643 |
| $CO_2$ | 0.194 | 0.088 | 0.068 | 0.053 |
| CO | | 0.106 | 0.126 | 0.141 |
| $H_2O$ | 0.031 | 0.137 | 0.157 | 0.172 |
| Cell voltage (V)[b] | | 1.155 | 1.143 | 0.133 |
| Equilibrium constant | | 0.2472 | 0.4538 | 0.7273 |

[a]$H_2O$ saturated at 1 atm
[b]vs. oxidant: $O_2$ 30%, $CO_2$ 60%, $N_2$ 10%

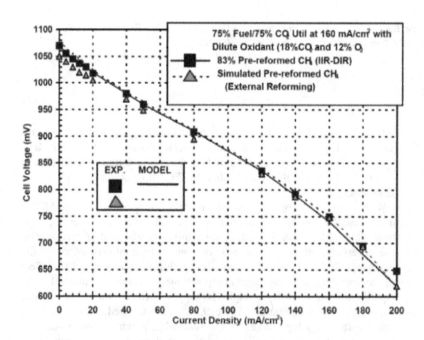

**Figure 9.42.** Experimental and theoretical potential vs. current density plots for Fuel Cell Energy's MCFC single cells. Reprinted from Reference 54.

steam-reformed natural gas and the oxidant was 70% air / 30% $CO_2$. The temperature effects of the cell potential, at a current density of 200 mA/cm$^2$, showed the following results:

$$\Delta E_t = 2.16\ (T_2 - T_1) \qquad\qquad 575 \leq T \leq 600\ °C \qquad\qquad (9.18)$$

$$\Delta E_t = 1.40\ (T_2 - T_1) \qquad\qquad 600 \leq T \leq 650\ °C \qquad\qquad (9.19)$$

$$\Delta E_t = 0.25\ (T_2 - T_1) \qquad\qquad 650 \leq T \leq 700\ °C \qquad\qquad (9.20)$$

where $\Delta E_t$ is the difference between cell potentials, in mV, at the higher ($T_2$) and lower ($T_1$) cell temperatures. The increases in cell potential are mainly due to considerably lower activation overpotential at the cathode and ohmic overpotential. From the above equations, it can be seen that there is a much smaller effect of temperature on $\Delta E_t$ between 650 and 700 °C than between 575 and 600 °C or between 600 and 650 °C. Other problems associated with operating cells above 650 °C are higher electrolyte losses by vaporization and increased corrosion rates of cell component materials.

(c)   *Effect of pressure.*   Increased pressure enhances the performance of MCFCs. Firstly, there is an increase of the reversible potential, $\Delta E_p$, of the cell, as expressed by:

$$\Delta E_p = 45.8 \log \frac{P_2}{P_1} \qquad\qquad (9.21)$$

where $\Delta E_p$, in mV, denotes the increase in the reversible potential when the cell pressure (both anodic and cathodic reactants) is increased from $P_1$ to $P_2$. In addition, there is an effect of pressure on the electrode kinetics of the fuel cell reactions. Taking both into consideration, the net pressure effect is given by:

$$\Delta E_p = 104 \log \frac{P_2}{P_1} \qquad\qquad (9.22)$$

Figure 9.43 illustrates the effect of pressure on the performance of a single cell (electrode area 70.5 cm$^2$) at a cell temperature of 650 °C. The effect of pressure on the cell performance is more significant for the cathodic reduction of oxygen than for the anodic oxidation of hydrogen. One of the reasons for this behavior is that increase of pressure enhances the gas phase reactions such as methanation:

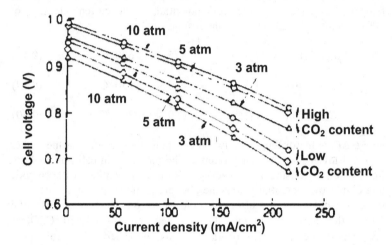

**Figure 9.43.** Effect of pressure on the performance of an MCFC single cell: electrode area, 70.5 cm$^2$; temperature, 650 °C. Reprinted from Reference 57, Copyright (1993), with permission from Plenum Press.

$$CO + 3H_2 \rightarrow CH_4 + H_2O \qquad (9.23)$$

or carbon deposition:

$$2CO \rightarrow CO_2 + C \qquad (9.24)$$

(d) *Effect of impurities.* A major advantage of the high temperature fuel cells, MCFCs and SOFCs, is that the CO present in reformed hydrogen is not a poison but a reactant. Several years ago, it was assumed that CO is electrochemically oxidized at the anode, just as in the case of hydrogen, but the recent explanation is that CO is transformed to hydrogen by the water-gas shift reaction at the anode. This reaction occurs at both the reformer catalyst and the anode electrocatalyst. The impurities which have most deleterious effects are listed in Table 9.9. These are the impurities from an air-blown coal gasifier. The amounts of impurity levels and the tolerance levels in MCFCs are also presented in Table 9.9. Hydrogen sulfide is one of the deadliest poisons and thus the tolerance level is very low. There have been extensive studies to determine the mechanism of poisoning of MCFCs by this impurity. The conclusions from these

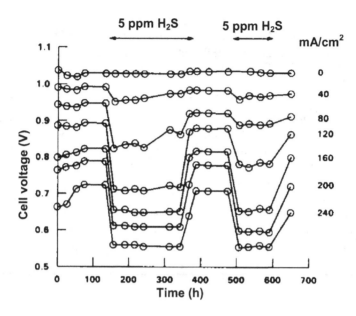

**Figure 9.44.** Effect of $H_2S$ on the performance of an MCFC single cell. 650 °C, 100 $cm^2$ cell, 5 ppm $H_2S$ added to fuel gas (10% $H_2$, 5% $CO_2$, 10% $H_2O$, 75% He) at 25% utilization. Reprinted from Reference 57, Copyright (1993), with permission from Plenum Press.

studies were that $H_2S$ causes adsorption of sulfur on the active sites of the Ni anode electrocatalyst, poisoning of the catalysts for the water-gas shift reaction by sulfur species, and its oxidation to $SO_2$ at the anode electrocatalyst, which then reacts with the electrolyte to form sodium sulfite. An interesting analysis of the effects of $H_2S$ on the performance of a single cell was carried out by Remick,[62] depicted in Figure 9.44. At the low concentration of $H_2S$ (5 ppm), there was little effect on the open circuit potential; conversely, there was an increasing effect on the cell potential with increased current density. It was also found that below a certain threshold concentration of $H_2S$, the performance degradation could be eliminated by changing the reactant gas to one with no $H_2S$.

Effects of other impurities listed in Table 9.9 are addressed below:

- *Halides.* HCl and HF react with the electrolyte to form halides, which change its composition. Thus, these impurities have to be reduced to ultra-low levels (0.5 ppm);

**TABLE 9.9**
**Most Deleterious Impurities Affecting the Performance of MCFCs and Their Tolerance Levels**

| Contaminants (typical ppm in raw coal gas) | Reaction mechanism | Qualitative tolerances | Conclusions |
|---|---|---|---|
| $H_2S$ (15,000) | $xH_2S + Ni \rightarrow NiS_x + xH_2$ | $< 0.5$ ppm $H_2S$ | Recoverable effect |
| HCl (500) | $2HCl + K_2CO_3 \rightarrow$ $2KCl(v) + H_2O + CO_2$ | $< 0.1$ ppm HCl | Long term effects possible |
| $H_2Se$ (5) | $xH_2Se + Ni \rightarrow NiSe_x + xH_2$ | $< 0.2$ ppm $H_2Se$ | Recoverable effect |
| As (10) | $AsH_3 + Ni \rightarrow$ $NiAs(s) + 3/2H_2$ | $< 0.1$ ppm As | Cumulative long term effect |
| Pb ($< 2$ ppm) | Adsorption | $< 1$ ppm Pb | Potential poisoning |
| Cd ($< 2$ ppm) | Adsorption | $< 30$ ppm Cd | Potential poisoning |
| Hg ($< 2$ ppm) | Adsorption | $< 35$ ppm Hg | Potential poisoning |
| Sn ($< 2$ ppm) | Adsorption | NA | Potential poisoning |
| $NH_3$ (2600) | NA | $< 10,000$ ppm $NH_3$ | No effect |

- *Ammonia.* No effect;
- *Arsine* ($AsH_3$). Negative effects were found at a level of 9 ppm;
- *Trace metals.* The trace metals listed in Table 9.9 can adsorb on the catalyst for the reformer reaction as well as on the anode electrode catalyst, poisoning them.

(e) *Lifetime studies.* Long term performance studies have been conducted in MCFC single cells. This type of fuel cell is mainly being developed for power generation and co-generation applications and a lifetime of at least 40,000 h for continuous operation is necessary. At the single cell level, these studies have been conducted with active electrode areas of 100 cm$^2$ and above. In the work carried out in the 1970s and 1980s, there was a considerable decrease in the performance within the first thousand hours, but subsequently the degradation in performance corresponded to about 5 mV/1000 h. A major goal of the leading MCFC developers is to reduce this rate to 2 mV/1000 h. There are several causes for the performance degradation, as listed below:

- loss of electrolyte by vaporization, which increases activation and mass transport overpotentials;
- formation of surface layers on cell components;
- electrode creep and sintering; coarsening and compression of the small particles, which again increase the activation and ohmic overpotential;

- dissolution of nickel oxide (from the cathode) in the electrolyte, followed by its reprecipitation. This is very much like an Ostwald ripening problem. In the case of thin electrolyte layers, the electron conducting lithiated nickel oxide causes short circuiting of the anode and the cathode.

Electrolyte loss by vaporization and by corrosion of the cell components is the most serious problem. Most of the electrolyte loss is due to vaporization of lithium carbonate. The amount of electrolyte loss by corrosion is greater in cells with smaller areas. In these cells there is more surface area of the hardware components of the cell exposed to the electrolyte.

When only sintered nickel was used as the anode material, sintering occurred to a great extent and caused a decrease in the surface area of the anode electrocatalyst. This problem was largely overcome by use of a Ni-Cr alloy as the anode electrocatalyst. By alloying Ni with Cr or Al, the creep resistance was also increased. Creep resistance is necessary for a low contact resistance of cell stacks. The use of nickel and Ni/Al coatings could prevent corrosion at the anode and wet seal.

## 9.6.3. Cell Stacks

9.6.3.1. *Design, Assembly, and Operating Characteristics.* Several companies worldwide have been actively involved in stack development and demonstration. The leaders include Fuel Cell Energy in the USA; Ishikawajima-Harima Heavy Industries (IHI), Hitachi, Mitsubishi Electric Company, and Toshiba Corporation in Japan; Ansaldo in Italy; ECN in the Netherlands; and Deutsche Aerospace AG in Germany. In this Section, we shall focus on the advances in stack development technology by Fuel Cell Energy because this company is at the forefront of the development of MCFC power plants for power generation/cogeneration applications.

The primary issue in the development of cell stacks is in the scale-up of the electrode area. In single cell studies, electrode areas have mostly been about $100 \text{ cm}^2$ with some results from single cells with considerably higher surface areas. MCFCs operate at a low current density ($150–160 \text{ mA/cm}^2$) requiring high surface area electrodes to produce adequate current. Fuel Cell Energy stacks have an active area of $1 \text{ m}^2$. The bipolar plates in MCFC stacks are usually fabricated using thin sheets ($\sim 300 \text{ μm}$) of an alloy (e.g., 310S or 316L stainless steel). The anode side is coated with Ni to prevent corrosion. The plate is aluminized around the edges for scaled up stacks. The cells have external manifolds and the stacks contain reforming plates to produce $H_2$ and CO from $CH_4$. Of the total reforming, 80% occurs in these plates, while the remainder occurs on the anode electrocatalyst. The concept of an *indirect internal reforming* (IIR)-*direct internal reforming* (DIR) combination in Fuel Cell Energy's stack is illustrated in Figure 9.45. A reforming catalyst is included in the anode compartment for direct hydrocarbon conversion. For each

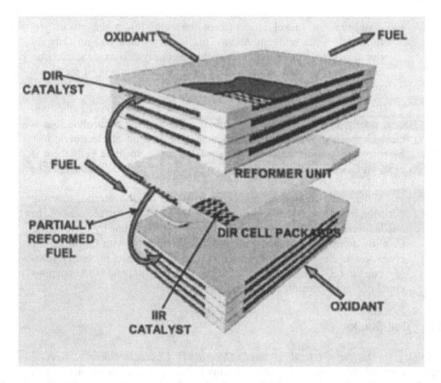

**Figure 9.45.** Indirect Internal Reforming (IIR) – Direct Internal Reforming (DIR) combination in fuel cell Energy's MCFC stack. Reprinted from Reference 63, Copyright (2003), with permission from ASME.

group of eight to ten cells in a stack, there is one indirect reforming unit. The cells are stacked together with a compression sub-system and external manifolds in a cross-flow configuration for the entry and exit of the reactant gases. Also taken into consideration are the current and temperature distribution over the entire cell area. Stacks are designed to control this distribution and prevent hot spots in the active areas of the electrodes. Thermal management is critical, but is considerably more facile than in PEMFCs. One advantage of the DIR MCFC is that part of the waste heat generated in the electrochemical energy conversion is utilized directly behind the active layer of the anode for the reforming of natural gas. The cooling is effectively carried out by the reactant gases. There is also a need to separate the $CO_2$ from the anode exit gas and then supply it to the cathode stream for the cathodic reduction of oxygen with $CO_2$ to produce the charge-carrier carbonate anion.

Fuel Cell Energy employs a stack compression system and external manifolding for the assembly of the cell stacks into a module. There are 350-400 cells per module. Interfaced with a stack is a fuel processor to remove the impurities (see Section 9.6.2.2). Air/oxygen and carbon dioxide are delivered to the system via another sub-system. In addition, there is a waste heat recovery system.

9.6.3.2. *Performance Characteristics.* The performances of the individual cells in Fuel Cell Energy stack were quite uniform. Figure 9.46 illustrates this behavior in a 360-cell stack (265 kW). Stacks have been tested for over 10,000 h. The initial cell potential in a stack was about 0.8 V at an operating current density of 160 mA/cm$^2$. During the initial period of about a 100 h, the degradation in performance was somewhat high, but over a period of about 10,000 h the degradation rate decreased to 5 mV/1000h. As stated in the Section 9.6.2.2, the target is to reduce the rate to 2 mV/1000h. The reader is referred to the articles in the Suggested Reading list for information on performance characteristics of MCFC stacks for power levels of 2.5 kW to 250 kW, not only by Fuel Cell Energy but also by other fuel cell developers. Ansaldo in Italy is also making rapid progress in stack development and testing.

Fuel Cell Energy uses Computer Assisted Design (CAD) 3-D parametric modeling in the engineering design and development of individual components and stacks in order to gain information on design verification, strength and stress analysis, and materials verification. Finite Element Analysis (FEA) methods are used for this purpose. This has been applied for models of end plates and compression hardware to ensure that engineering designs meet their requirements for flatness, uniform stack compressive load and strength under stack operating conditions. These studies have been quite fruitful from the point of view of

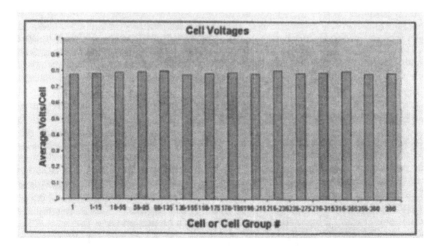

**Figure 9.46.** Performance of individual cells in Fuel Cell Energy's cell stacks, ±1.6%. Reprinted from Reference 63, Copyright (2003), with permission from ASME.

compressive load uniformity in stacked cells. In addition, Fuel Cell Energy carries out extensive engineering design failure modes and effects analysis to resolve potential design problems.

## 9.7.   SOLID-OXIDE FUEL CELLS (SOFC)

### 9.7.1.   Evolutionary Aspects

SOFC technology is unique in comparison with the five other leading fuel cell technologies in that it is an all-solid-state power system. It is a two-phase system because the fuel and oxidants are fed as gases, and the fuel cell reactions occur at the solid electrode/gas interfaces, again producing gaseous products, steam and $CO_2$. The reactant gases have ready access to the electrocatalyst particles over practically their entire surface areas. Conversely, the other five leading types of fuel cells are three-phase systems and, according to theoretical studies, the fuel cell reactions occur at the three-phase zones. Another interpretation in these cases is that the reactant gases dissolve in the electrolyte and diffuse through thin films of electrolytes to reach the electrocatalytic sites where the electrochemical reactions occur.

High operating temperatures are essential for operation of SOFCs. This is to assure high ionic (for electrolyte) and electronic (for electrodes and interconnection) conductivities. The principles of operation of a solid oxide fuel cell are illustrated in Figure 9.47. The electrolyte in the state-of-the-art SOFC is the same used in

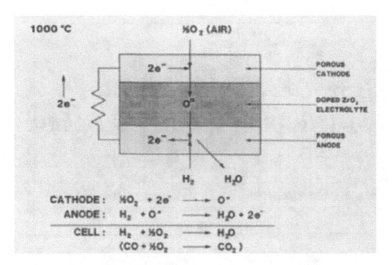

**Figure 9.47.** Mode of operation of an SOFC. Reprinted from Reference 64, Copyright (1993), with permission from Plenum Press.

the Nernst glower lamp in 1899—yttria stabilized zirconia. With this electrolyte, as well as with the cathode electrocatalyst (lanthanum manganate doped with divalent cations), and the interconnection material (lanthanum chromite doped with alkaline earth metals), the desired ionic and electronic conductivities are achieved at about 800 to 1000 °C. The conduction mechanism for the ionic electrolyte involves oxide ion transport from one vacancy to the next. The ionic conductor has a perovskite structure with some vacancies. There is mixed electronic/ionic conduction in the cathode. This is essential to enhance the three-dimensional reaction zone. The materials used for the fabrication of SOFCs have to meet the following stringent requirements: chemical stability in oxidizing and/or reducing environments, high oxide ion conductivity of the electrolyte to minimize ohmic overpotential, and thermomechanical compatibility of component materials for the anode, cathode, electrolyte, and the interconnection.

Some historical developmental aspects of SOFCs were described in Chapter 4. Most noteworthy are: the bell and spigot design by researchers at Westinghouse Electric Corporation from the 1930s to 1960s,[65] the tubular design used by Westinghouse Electric Corporation in the USA[66] and by Brown, Boveri, and Cie in Switzerland in the 1960s to the 1970s;[67] the monolithic design developed and demonstrated at Argonne National Laboratory in the 1970s and 1980s;[55,68] and the planar design by several organizations since the 1980s.[69] In the following Sections an effort is made to describe in some detail the status of the technologies for the production of the most advanced electrochemical stack for an SOFC, with the tubular design by Siemens-Westinghouse; and advances made in the last decade for the development of planar SOFCs. Several organizations are involved in the latter because performance levels as high as with PEMFCs have been demonstrated and it is projected that such systems could be developed from low to high power levels at considerably lower costs than SOFCs with the tubular design.

## 9.7.2.  Single Cells

### 9.7.2.1.  *Design, Assembly, and Operating Conditions*

(a)  *Siemens-Westinghouse tubular SOFCs.*  The design of the single cell in the Siemens-Westinghouse solid oxide fuel-cell power plant is illustrated in Figure 9.48. In the original work by Westinghouse Electric Corporation, layers of the cathode, electrolyte, and anode material were deposited on a porous calcium stablilized zirconia support tube (closed at one end) by chemical and/or electrochemical vapor deposition techniques.[71] It was subsequently realized that the zirconia tube could be replaced by a tube made of the cathode material, lanthanum manganite with alkaline, and rare earth elements. Thus, this tube served two functions: as a support tube and as the cathode. In this new design, the weight of the single cell could be significantly reduced, and mass transport of the cathodic reactant could be significantly enhanced. The

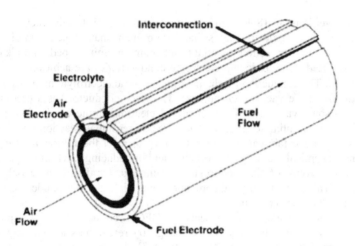

**Figure 9.48.** Schematic of single cell in Siemens-Westinghouse SOFC. Reprinted from Reference 70, Copyright (2000), with permission from Elsevier.

porous supported tube was fabricated using an extrusion method. The dense electrolyte layer is formed on the cathode tube by chemical vapor deposition (CVD) and electrochemical-vapor deposition (EVD) at 150 °C using $ZrCl_4$, $YCl_3$, $H_2$, and $H_2O$ with $O_2$ in the vapor state. The resulting chemical composition of this layer is 85–90% $ZrO_2$ and 5–10% $Y_2O_3$. The Mg doped $LaCrO_2$ layer is deposited by plasma-spraying from appropriate chloride vapors. The interconnection layer connects cells in series (see Section 9.7.2.2.). The $Ni/ZrO_2$-$Y_2O_3$ cermet anode is prepared by coating a slurry of NiO and $Y_2O_3$ stabilized $ZrO_2$, followed by EVD. In some recent studies, an alternative approach has been used to deposit the anode layer by deposition of a slurry of Ni-YSZ (yttria-stabilized zirconia) on the electrolyte followed by sintering. Use of this method instead of CVD could be quite cost-effective for the large-scale production of SOFCs. The porosities of the anode layer and the cathode tube are about 30% to facilitate the diffusion of reactants into the active layer. Oxygen/air enters the tubular cell via a concentric alumina tube and the fuel ($H_2$ or $H_2$ plus CO) flows by outside the tube.

The tube lengths have been increased from 50 cm to 150 cm. The diameter of the 150 cm tube is 2.2 cm. The dense electrolyte is 40 μm thick and the anode layer is 100 to 150 μm thick. About 90% of the weight of the single cell is that due to the cathode tube.

The cathode, lanthanum manganite doped with a divalent cation (e.g., Sr), has a perovskite type structure and exhibits reversible oxidation/reduction behavior. It is stable in air or oxygen but decomposes at oxygen partial pressures of about $10^{-14}$ atm and at a temperature of 1000 °C. It has mixed electronic and ionic conduction permitting charge transfer over the entire three-dimensional electrode structure. The mixed conduction enhances charge transfer and oxygen adsorption because of increased electrode/electrolyte interfacial areas, minimizing activation overpotential losses. The electronic conduction of strontium doped lanthanum manganite is due to hopping of an electron hole between $Mn^{3+}$ and $Mn^{4+}$. During oxygen reduction in the SOFC, the electrons released at the anode are transported through the mixed conductor. The electrons combine with the adsorbed oxygen atoms to produce oxide ions ($O^{2-}$). The ionic conduction in the mixed conductor is by migration of oxygen vacancies, between octahedral edge sites, as shown in Figure 9.49. Dense strontium-doped-lanthanum chromite has a specific conductivity of 1000 S/cm, but its effective conductivity in the cathode is much less because of the porosity and pore geometry.

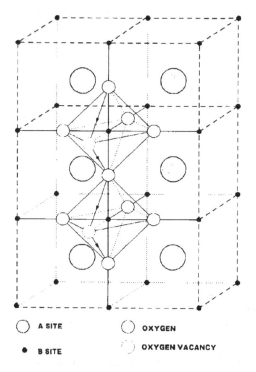

○ A SITE     ○ OXYGEN

● B SITE     OXYGEN VACANCY

**Figure 9.49.** Oxide ion conduction pathway via oxygen vacancies in perovskite structure of SOFC electrolyte. Reprinted from Reference 64, Copyright (1993), with permission from Plenum Press.

The electrolyte (YSZ) has a fluorite structure. The $Y_2O_3$ dopant stabilizes the high temperature cubic phase in zirconia and generates oxygen vacancies, according to the Kroger-Vink notation:

$$Y_2O_3 \rightarrow 2Y + 3O + V \qquad (9.25)$$

An oxygen vacancy (V) is created for every mole of the dopant. Thus an oxide ion transports from one vacancy to the other. The desired ionic conductivity (0.15 S/cm) occurs only at a temperature of 1000 °C for YSZ containing about 10% $Y_2O_3$. One important criterion for developing SOFCs is that the thermal-expansion coefficients of the component materials (anode, cathode, electrolyte, interconnection) should be nearly identical. This is the case in the materials chosen for the components by Siemens-Westinghouse. In order to reduce the fabrication costs of SOFCs, methods such as plasma-spraying or colloidal/electrophoretic deposition followed by sintering are being explored.

The anode electrocatalyst is Ni in the form of fine nickel, but its thermal expansion coefficient is much higher than that of solid nickel. Furthermore, the nickel particles can sinter forming larger particles and reducing the active surface areas at the operating temperature of 1000 °C. To overcome these problems and increase the three dimensional reaction zone, porous YSZ is formed around the nickel particles. The added advantage is that the YSZ promotes oxide ion conductivity within the active layer of the anode. Figure 9.50 exhibits the microstructure of a cross-section of the layers of the cathode/electrolyte/anode.

The interconnection connects two cells in series. In the Siemens-Westinghouse design, it is a 9 mm wide strip along the length of the tube (50 to 150 cm$^2$). The interconnection material needs to meet the following requirements:

- 100% electronic conductivity;
- stability in the oxidizing environment of the air electrode and in the reducing environment of the fuel electrode;
- low permeability of the anodic and cathodic reactants to minimize their mixing and direct chemical combination;
- thermal expansion coefficient nearly identical to the other cell component materials; and
- chemical stability with the nickel anode on one side and the strontium-doped lanthanum manganate on the other.

The material that satisfied all these requirements was found to be magnesium-doped lanthanum chromite; calcium or strontium could also be used as dopant materials. This material is a p-type semiconductor; the electronic conduction is via polaron hopping. The divalent cations

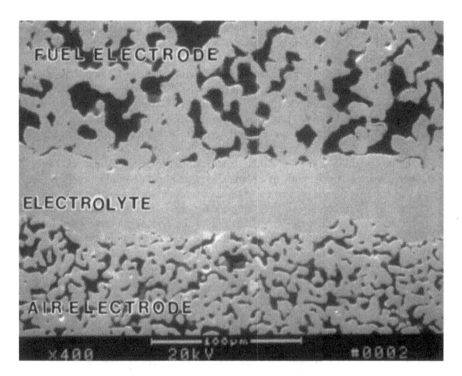

**Figure 9.50.** Microstructure of cross-section of layers of cathode/electrolyte/anode. Reprinted from Reference 70, Copyright (2000), with permission from Elsevier.

enhance the electronic conductivity. The thickness of this layer is 85 μm and it is deposited by plasma-spraying.

(b) *Planar design.* The planar design for SOFCs, which is similar to the designs of the other five types of fuel cells (Sections 7.2 to 7.6), is attracting much attention at the present time and several organizations are developing such SOFCs. Fabrication costs using tape-casting, slurry-sintering, screen-printing, and plasma-spraying are considerably lower than the CVD and ECD methods used for the fabrication of tubular solid oxide fuel cells. Individual layers of the electrode, electrolyte, and interconnection are thinner in the planar SOFCs than in the tubular SOFCs, reducing the ohmic and mass transport losses. Also, the bipolar construction allows for electron transfer from the anode of one cell to the cathode of the neighboring cell over the entire active cell area resulting in much lower ohmic overpotential losses in a cell stack than in a tubular cell bundle with the unipolar design.

Several modifications are being investigated for the planar design SOFCs. In the electrolyte-supported cells, the thickness of this layer is 50 to 150 μm. Because of this, the ohmic resistance is high and the SOFC will still have to operate at 1000 °C, as in the case of the tubular SOFC. To overcome this problem, investigations are in progress in several laboratories (e.g., Pacific Northwest, General Electric, Ceramatech, Z-tech, University of Utah) to develop planar SOFCs with the support being the anode or the cathode. In such a case, the electrolyte layer thickness can be reduced to 5–20 μm. This has a significant effect of lowering the ohmic overpotential in the electrolyte layer and consequently the cell operating temperature can be lowered to about 800 °C. Optimization of the composition and microstructure is essential to attain a high level of performance. A schematic of a planar SOFC is illustrated in Figure 9.51.[72] The material for the current collector or a stack series connection is a thin metallic plate containing the channels for the delivery of gases to the electrodes. Ferritic stainless steel is one of the typical materials used for the construction of the bipolar plate. To prevent the corrosion of the stainless steel, a layer of a stable metallic oxide is coated on the steel. Efficient sealing of the edges is vital in the case of planar SOFCs; this is a non-issue in the case of tubular SOFCs. The material for

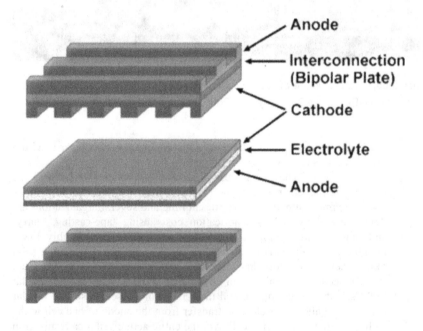

**Figure 9.51.** Schematic of planar SOFC. Reprinted from Reference 72, Copyright (2002), with permission from Elsevier.

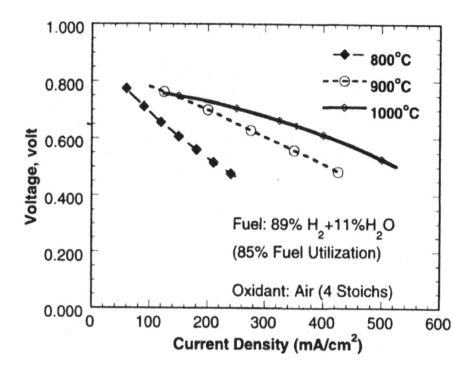

**Figure 9.52.** Cell potential vs. current density plots in Siemens-Westinghouse SOFC single cell at temperatures between 800 and 1000 °C. Reprinted from Reference 70, Copyright (2000), with permission from Elsevier.

the seal must be compatible with the other cell component materials. The seals must be gas-tight and electrically insulating to prevent crossover of reactant gases or short-circuiting of the cell. Glass ceramics are used for the seals. Two problems encountered with the sealing material are the brittle nature of the glass seals and chemical reaction of the glass seal material with the cell component materials. Modification of the compositions of the sealing materials are being investigated to minimize these problems.

Alternate materials for the cell components are being investigated for planar SOFCs. Presently, the best performance is with the same materials as used in tubular SOFCs. Some alternate materials for operation of planar SOFCs at a temperature of about 800 °C include: lanthanum strontium ferrite and praseodymium strontium manganite for the cathode;[73] and gadolinium-doped ceria, and lanthanum-strontium-gallium-magnesium oxide, for the electrolyte.[74]

9.7.2.2.  *Performance Characteristics.*  In this fuel cell, hydrogen and methane have been used as fuels. When methane is used as the fuel, it undergoes the steam-reforming reaction at the anode electrocatalyst to produce CO and $H_2$ and also the shift-conversion to produce $CO_2$ and $H_2$. The reactions are relatively fast because the $H_2$ produced is consumed in the anodic reaction. Oxygen and air have been used as the oxidant. The open circuit potential of the cell corresponds to the reversible potential because both the half cell reactions are quite reversible at this temperature, as in the case of the MCFCs. However, there is a Nernstian effect, with utilization of the reactants during cell operation, reducing the open circuit potential.

The operating temperature has a strong effect on the cell performance. In a study by Siemens-Westinghouse with the state-of-the-art single cells (150 cm long and 2.2 cm in diameter), with 89% $H_2$ and 11% $H_2O$ as the fuel feed and air as the oxidant, the cell potential vs. current density plots at temperatures of 800 to 1000 °C exhibited the behavior as shown in Figure 9.52. The fuel utilization was 85% and air was delivered at four times the stoichiometric rate. The slope of the linear section of the $E$ vs. $i$ plots decreases markedly with decreasing temperature. The reason for this is that the ionic conductivity of the YSZ electrolyte has the highest value at 1000 °C. As in the case of MCFCs, the $E$ vs. $i$ behavior is linear over the entire current density range because, at the high temperature, the exchange current density for the oxygen reduction is quite high such that the exponential term in the Tafel equation can be linearized (just as in the case of the electro-oxidation of hydrogen even at low temperatures). The linearization can also be done for the equation of the cell potential vs. current density in the mass transport region. It can be observed from Figure 9.52 that there is a departure from linearity at higher current densities in the $E$ vs. $i$ plot for the cell operating at 1000 °C. This is probably due to the increasing effects of mass transport limitations. The ohmic overpotential is not only in the electrolyte, but also from the ohmic resistance of the cathode and the interconnection (current collector). According to one study, the increase in cell potential ($\Delta E$) with temperature between 900 and 1000 °C may be expressed by the empirical equation[75]

$$\Delta E = 1.3\,(T_2 - T_1) \qquad\qquad (9.26)$$

where $E$ is in mV and $T$ is in °C. In this study, it was also shown that:

$$\Delta E = 0.008\,(T_2 - T_1)\,i \qquad\qquad (9.27)$$

where $i$ is the current density.

The effect of pressure on the cell potential vs. current density behavior has also been evaluated in the same tubular SOFCs. The results are depicted in Figure 9.53. Increasing the pressure increases the Nernst reversible potential but the

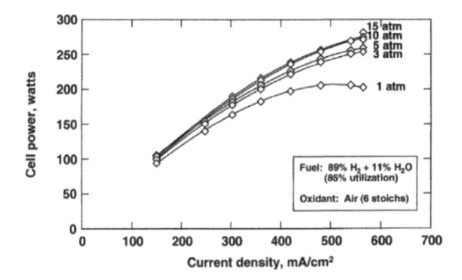

**Figure 9.53.** Effect of pressure on performance of Siemens-Westinghouse SOFC single cell. Reprinted from Reference 70, Copyright (2000), with permission from Elsevier.

more important effect is to enhance the kinetics of electro-reduction of oxygen. The pressure effect on the increase of cell potential ($\Delta E$) at 1000 °C can be expressed by:

$$\Delta E = 54 \log \frac{P_2}{P_1} \qquad (9.28)$$

where $P_2$ and $P_1$ are the higher and lower cell pressures, and $\Delta E$ is in mV.

Investigations have also been made to test the effects of impurities on the cell performance. The impurities tested were mostly those found in coal-gasified fuels. These are $H_2S$, HCl, and $NH_3$. The only impurity that caused a severe degradation in performance is $H_2S$, even at a level of 1 ppm! This is also the case in MCFCs. In both these types of fuel cells, nickel is the anode catalyst. As stated in Section 9.6, the mechanism of poisoning is by oxidation of $H_2S$ first to $SO_3$ and then the conversion of $SO_3$ to S, which strongly adsorbs on the active nickel sites. The effect is reversible because when the fuel is changed to pure hydrogen or a simulated reformate fuel with no $H_2S$, the SOFC regains its original performance.

Life-testing has been conducted extensively by Siemens-Westinghouse. In a paper by Singhal, it is stated that tests have been made up to about 25,000 h under a

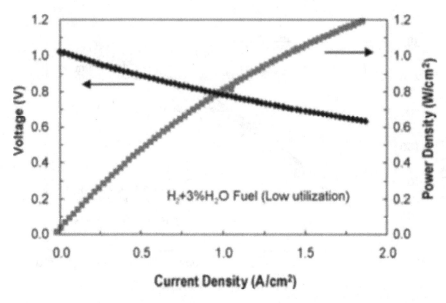

**Figure 9.54.** Plots of cell potential and power density vs. current density for planar SOFC. Reprinted from Reference 72, Copyright (2002), with permission from Elsevier.

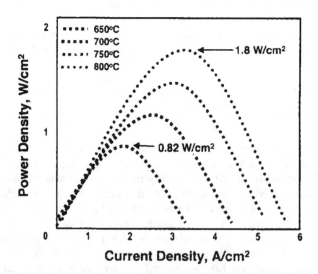

**Figure 9.55.** Effect of temperature on performance of General Electric Company's planar SOFC single cell. Reprinted from Reference 79, Copyright (2003), with permission from Elsevier.

variety of operating conditions.[76] The degradation rate of the cell potential was found to be about 0.1%/1000 h. If this is the case, the Siemens-Westinghouse SOFC has met the target of a degradation rate of 1 mV/1000 h in cell potential. This target is for 40,000 h of operation and cells have not yet been tested for this period of time.

The planar SOFC design has been attracting increased attention during the last decade because of the high level of performance achieved, nearly a factor of two higher than SOFCs with the tubular structure. Such an increase is possible due to ultra-thin layers for the anode, cathode, and electrolyte (see Section 9.7.2.1). A power density of as high as 1.8 W/cm$^2$ was attained by Kim et al.[77] This even greatly exceeds the highest power density achieved in PEMFCs. The reason for the dramatic increase is because the slope of the linear region $E$ vs. $i$ plot is only about 0.2 $\Omega$/cm$^2$. There is also a marked decrease in the oxygen overpotential at 800 °C in the SOFC as compared with a PEMFC operating at 80 °C. Thus, there is a parallel shift upwards in the $E$ vs. $i$ plot for the SOFC as compared with a PEMFC. Singhal and coworkers at Pacific Northwest National Laboratory have investigated the cell potential vs. current density behavior in planar SOFCs.[78] In a cell with a strontium-doped lanthanum ferrite cathode, a YSZ electrolyte, and a Ni/YSZ anode, the maximum power density was 1 W/cm$^2$ at 0.7 V (Figure 9.54). An even higher level of performance was reported by researchers at the University of Utah (Figure 9.56). Though there is no available data on the structure and composition of the electrodes and electrolyte, Minh at General Electric Company investigated the effect of temperature on the performance of a planar SOFC.[79] The results are depicted in Figure 9.55.

### 9.7.3. Cell Stacks: Design, Assembly, and Performance Characteristics

9.7.3.1. *Siemens-Westinghouse Tubular SOFCs.* A schematic design of a 24-cell stack is illustrated in Figure 9.56. The cells are assembled in series and parallel. The interconnection is used to connect the cells in series; there is a nickel felt strip between the anode of one cell and the interconnection of the next cell. The nickel felt strip is also located between the cathodes for the parallel connection. The nickel felt strips serve for electronic conduction and mechanical integrity permitting thermal expansion/contraction. In the state-of-the art stacks, the tube length is 150 cm and has a diameter of 2.2 cm. In a 200-kW stack there are 1152 cells (48 bundles of 24 cells). As stated in the Section 9.7.2.1, one advantage of the tubular design is that there is no need for gas-tight seals. The manner in which this is accomplished is illustrated in Figure 9.57. The air is delivered through the central alumina tube to the bottom of the cell and it flows upwards to the open end on the outside of the alumina tube. The fuel flows on the external side of the single cell, parallel to the air-flow. Thermal management in the tubular SOFC is relatively

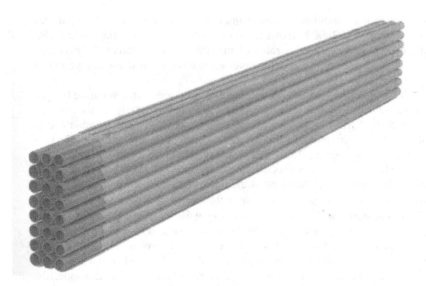

**Figure 9.56.** Schematic of a 24-cell stack (three-in-parallel by eight-in-series) in Siemens-Westinghouse tubular SOFC. Reprinted from Reference 79, Copyright (2003), with permission from Elsevier.

simple. The air is heated in the stack itself by the heat generated during fuel cell operation. The air flows at four times the stoichiometric rate. The fuel utilization is about 85%. The excess air and un-utilized fuel are combined and combusted and the temperature of the exit gases is 1000 °C. This heat is utilized for preheating the inlet air stream.

Stacks rated at 25, 100, and 200 kW have been constructed and their performances evaluated. The 25 and 100 kW stacks were operated at 1000 °C and 1-atm pressure. In order to enhance the power densities, the 200 kW stacks were operated at 3–5 atm. Even though the dimensions of the cells and number of cells in the stack were the same as in the 100 kW stack, the 200 kW system, hybridized with a gas turbine generating 50 kW, exhibited an energy conversion efficiency (from chemical energy of fuel to electrical energy) of 57%, based on the lower heating value of the fuel. The Siemens-Westinghouse stacks have undergone life-testing up to about 50,000 h. Cell stacks with power ratings of 25 and 100 kW have exhibited performance degradation of less than 0.1%/1000 h.

9.7.3.2. *Planar SOFCs.* There are several organizations worldwide engaged in the development of planar SOFC stacks. Basically there are two types of cell stacks that have been investigated. In one, the cells are electrolyte supported and the YSZ membrane has a thickness of about 100 μm. Because the thickness is quite high, the ohmic resistance in the cell is high (about 0.7 $\Omega/cm^2$) at 800 °C. The alternative is to

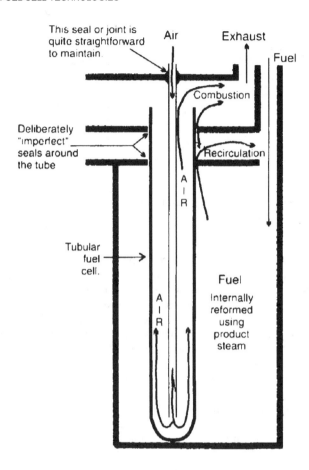

**Figure 9.57.** Flow pattern of reactant gases in Siemens-Westinghouse SOFC single cell. Reprinted from Reference 80, Copyright (1999), with permission from John Wiley & Sons, Ltd.

to have an electrode- (anode- or cathode-) supported design; the anode-supported design is preferred. Schematics of both the electrolyte-supported and anode-supported cells are illustrated in Figure 9.58. The thickness of the three component layers is also indicated in this figure. The thermal expansion coefficients are well matched for the three cell components. The anode-supported cell design is favored more than the cathode-supported one for the following reasons: the diffusion coefficient of the fuel gas is three to four times higher than that of air; the

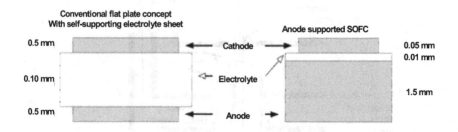

**Figure 9.58.** Schematics of electrolyte supported and anode supported planar SOFCs. Reprinted from Reference 54.

metallic plate exhibits better thermal shock resistance than LSM due to its higher conductivity and plasticity; and the porosity of the anode is optimized by reducing nickel oxide to nickel. A significant advantage of the planar SOFC stack is that metallic bipolar plates (ferritic stainless steel) is most commonly used for the construction of the bipolar plate. It may be quite advantageous to nickel-plate the stainless steel on the anode side to inhibit hydrogen embrittlement. The bipolar plates are quite thin. Flow channels are stamped on the bipolar plate. Using these thin structures, one could have 2 cells/cm.

To prevent oxidation on the anode side, a surface film of a metallic oxide is deposited. Potential candidates are oxides of aluminum, chromium, and titanium. It is essential that the surface film of the oxide does not inhibit electron transfer across the bipolar plate. A major challenge in the development of planar SOFCs has been the selection of the most appropriate sealing material. Glass ceramics (e.g., alumina glass) are currently used for the seals. An alternative to the glass sealing is compressive, non-bonding sealing. This has been evaluated at the Pacific Northwest National Laboratory. The leak rates were found to be minimal in such cells operating at 800 °C. The planar SOFC stack is at an infant stage of development as compared with the tubular SOFC stack. There is thus hardly any data in the published literature on the performance of cell stacks.

## 9.8.    COMPARATIVE ASSESSMENT OF SIX LEADING FUEL CELL TECHNOLOGIES

### 9.8.1.    Essential Characteristics: Single Cell Level

9.8.1.1. *General Comments.*    Table 9.10 presents the operating and performance characteristics of the six leading fuel cell technologies. By no means

**TABLE 9.10**
**Essential Characteristics of Six Leading Fuel Cell Technologies: Single Cell Level**

| | PEMFC | DMFC | AFC | PAFC | MCFC | SOFC |
|---|---|---|---|---|---|---|
| Fuel | $H_2$ (pure or reformed) | 1-2 M $CH_3OH$ | Pure $H_2$ (> 99.5%) | Reformed $H_2$ | $H_2$ and CO reformed fuel direct/indirect | $H_2$ and CO reformed fuel direct/indirect |
| Operating temperature (°C) | 50-90 projected up to 130 | 25-60 projected up to 145 | 50-250 under pressure at T > 100 | 180-200 | 650 | 800-1000 |
| Electrolyte | Perfluorosulfonic acid membrane, alternate membranes under investigation | Perfluorosulfonic acid membrane, alternate membranes under investigation | 8-12 M KOH molten KOH at about 200 °C | 85-100% $H_3PO_4$ 100% in UTC fuel cell PC 25 | 38% $Li_2CO_3$ /62% $K_2CO_3$ 50% $Li_2CO_3$ /50% $Na_2CO_3$ | 88% $ZrO_2$ /12% $Y_2O_3$ |
| Ionic conductor | $H^+$ | $H^+$ | $OH^-$ | $H^+$ | $CO_3^{2-}$ | $O^{2-}$ |
| Electrocatalyst loading (mg/cm$^2$) | Pt on C (~0.2-0.4) | Pt on C (~0.2-0.4) | Pt on C (0.4), Ni/LiNiO$_x$ (Pt-Pd and Pt-Au in NASA cell) | Anode: Pt (0.25) cathode: Pt-Co-Cr (0.4) | Ni/LiNiO$_x$ | Ni/Sr doped with LaMnO$_3$ (most advanced) |
| Efficiency (%) | 45-55 | 30-35 | 45-55 | 40-50 | 50-55 | 50-55 |
| Power density (W/cm$^2$) | 0.7 | 0.2 (maximum attained) | Terrestrial: 0.1 space: 0.8 | 0.2 | 0.12 | Tubular: 0.25-0.30 planar: 1.0-1.2 |
| Poisoning species | Anode: CO > 10ppm electrolyte: mono and divalent cations | Anode: CO, COH intermediate electrolyte: mono and divalent cations | $CO_2$ generating carbonates in electrolyte and solid carbonate in pores | CO > 1% $H_2S$ > 1 ppm | $H_2S$ > 0.5 ppm | $H_2S$ > 1 ppm |
| Lifetime/ degradation rate [(h)/(mV/1000h)] | 10,000/3-5 | 1000/10 | 10,000/5-10 | 40,000/1 | 10,000/2 | 10,000/0.7 |

does this table cover all essential characteristics. The reader can obtain some understanding of other aspects from the preceding Sections and from the published literature. In the following two Sections, brief remarks will be made on essential characteristics at single cell and stack levels.

9.8.1.2. *Fuel.* The choice of the fuel depends on whether it is readily available. Of the three primary fossil fuels (coal, petroleum, and natural gas), the preferred fuel is natural gas. The second is coal, in view of its relative worldwide abundance. Natural gas is a clean fuel with high hydrogen content. One advantage of the MCFCs and SOFCs is that natural gas can be directly delivered to the fuel cell after removal of poisoning species. In the case of the PAFC, natural gas will have to be processed via steam reforming and shift conversion. Thus, a PAFC system will involve an additional sub-system. The ideal fuel for PEMFCs and AFCs is pure hydrogen (> 99.5% purity). As in the case of the PAFCs, $H_2$ is produced by steam-reforming and shift-conversion. However, a preferential oxidizer or pressure swing absorption processor is required to reduce the CO level to less than 10-50 ppm. For AFCs, it is also necessary to remove the $CO_2$ completely by using scrubbers. Methanol is an ideal liquid fuel that is produced on a large scale by steam-reforming natural gas and combining $H_2$ and CO. Its energy density is half that of gasoline. Methanol is the most electroactive organic fuel and has been a "fuel cell researcher's dream" to directly use it in fuel cells.

9.8.1.3. *Operating Temperature.* A wide range of operating temperatures is covered by the six types of fuel cells. The low temperature fuel cells (T < 100°C) have the advantage of fast start-up time, which is required for transportation and portable power applications. However, since the efficiency of energy conversion (chemical to electric) is 40 to 50%, and the temperature differential between the fuel cell and the environment is small, methods for heat removal are problematic, particularly at high power densities (as for PEMFCs). Water is the ideal coolant for these systems. Also, the waste heat is of relatively low quality and finds hardly any applications. The intermediate temperature PAFC has the advantage that air could be used as a coolant, though there are systems that use a liquid coolant (ethylene glycol, water). The waste heat can be utilized for fuel processing and/or for space heating and hot water. There are significant increases in operating temperature when moving over to MCFCs and SOFCs. An advantage of the high temperature fuel cells is that such temperatures significantly increase the electrode kinetics of the half-cell reactions and thus reduce the activation overpotentials, particularly for the electro-reduction of oxygen. Another advantage is that CO is a fuel and not a poison; very probably CO is converted to $H_2$ plus $CO_2$ by the water gas shift-conversion reaction on the anode electrocatalyst and/or the catalyst for the fuel cell reaction. Mass transport limitations are also greatly reduced in high temperature fuel cells.

9.8.1.4. *Electrolyte and Ionic Conductor.* The low and intermediate temperature fuel cells predominantly use acid electrolytes, the exception being the alkaline fuel cell. Acids, such as sulfuric acid, perchloric acid and trifluoromethane sulfonic acid have been used in fuel cells developed during the 1960s and 1970s. But due to problems such as anion adsorption, instability, and the toxic nature of the acids, they were removed from the list of electrolytes. A breakthrough was the discovery of the perfluorosulfonic acid (PFSA), DuPont's Nafion, and this acid has been used in PEMFCs and DMFCs since the 1980s. This acid has a high strength (PFSAs are referred to as super acids) and is like a plastic. It is used as a thin layer (20-100 $\mu$m thick) in PEMFCs. The layer is thicker in DMFCs (about 175 $\mu$m) to minimize crossover of methanol. Protons have the highest mobility due to the Grotthus mechanism of proton transport (hopping of protons from one water molecule to another). The hydroxyl ion also has a relatively high conduction rate compared to other anions. Here again, proton hopping is involved.

For the intermediate temperature fuel cell (PAFC), 100% phosphoric acid is the electrolyte. Its conductivity is relatively low at temperatures below 150 °C. Above this temperature, this acid polymerizes to pyrophosphoric acid which has a higher ionization constant. But even so, the specific conductivity of this acid at 200 °C is lower than that of PFSA at 80 °C. For the AFC, KOH is the ideal electrolyte, but its concentration has to be quite high (8-12 M). This also causes problems of corrosion of cell components, *electrolyte weeping,* and $K_2CO_3$ formation by reaction with $CO_2$ present in the reactant streams.

MCFCs use a molten carbonate electrolyte, the state-of-the-art electrolyte being a eutectic mixture of $Li_2CO_3$ and $K_2CO_3$. Electrolytic conduction occurs via the carbonate ion. It has a considerably lower mobility than the $H^+$ or $OH^-$ ion. Furthermore, in a fuel cell, the ideal situation will be when the conducting ion is produced at one electrode, and then transported and consumed at the other electrode. In the case of the MCFC, the carbonate ion is produced at the cathode by reacting $CO_2$ with oxygen and electrons, and is then transported to the anode where it combines with $H_2$ to produce water, carbon dioxide, and electrons. The carbon dioxide is then shuttled back to the cathode. One problem with the $Li_2CO_3/K_2CO_3$ electrolyte was that it caused corrosion of the cathode and bipolar plate. This problem has been solved by choice of materials and operating procedure.

The SOFC is a two-phase system, whereas all others are three-phase. Thus, mass transport limitations of the reactant gases to the active sites are greatly reduced; diffusion of reactants is through a gaseous medium rather than through a liquid. Transport through the electrolyte occurs via the oxide ion. The electrolyte has a defect structure and the oxide ion transport occurs via the vacancies in it. The specific conductivity of the electrolyte is low. It is for this reason that very high operating temperatures are necessary (e.g., 1000 °C). High operating temperatures also cause problems: thermal compatibility of cell component materials, and inter-diffusion of ions. This is one of the reasons that the future direction of research and development is moving towards development of SOFCs for operation at lower temperatures (about 750–800 °C) with the same electrolyte composition for fabrication of ultra-thin electrolyte layers, or towards finding alternate electrolytes.

9.8.1.5. *Efficiency and Power Density.* The efficiencies for electrochemical energy conversion in all the types of fuel cells, except DMFCs, are nearly the same. But one must take into consideration that the low and intermediate temperature fuel cells use pure hydrogen. If one takes into account the efficiency for conversion of natural gas to hydrogen is about 70%, PEMFCs and AFCs will be less efficient, overall, than MCFCs and SOFCs. PAFCs are more efficient than PEMFCs and AFCs because the heat rejected from the fuel cell is at least of medium grade and can be utilized for fuel processing. The MCFCs and SOFCs have the distinct advantage of providing high-grade heat for the fuel processing reaction. Moreover, natural gas is directly fed to the fuel cell. In addition, there is sufficient high-grade heat for hybridizing the MCFC or SOFC with a gas turbine for additional generation of electricity and thus further increasing the efficiency for chemical to electric energy conversion.

The PEMFC has exhibited the highest power density, about three to six times higher than all other types of fuel cells. It is for this reason that scientists, engineers, and policy makers were encouraged to develop PEMFC power plants for electric vehicles. An added asset of a high power density fuel cell is that it can significantly lower capital costs for the production of electrochemical stacks for fuel cell power plants/power sources. Since the PAFCs, MCFCs, and SOFCs have relatively lower power densities and operate at intermediate and high temperatures, these are more appropriate for stationary applications. The current view is that the DMFCs will mostly be considered for the small portable power applications.

9.8.1.6. *Electrocatalysts / Loadings.* The state-of-the-art low and intermediate temperature fuel cells used Pt or Pt alloy electrocatalysts. By use of high surface area carbon supported nanosized particles of the electrocatalyst (to increase the surface area per gram of the electrocatalyst), the loadings have reached very low levels. Because of the high power density of the PEMFC, the cost of the electrocatalyst can be reduced to as low as $15/kW. In view of the considerably lower power densities of the DMFCs, AFCs, and PAFCs, this cost will increase by a factor of two or three. In the case of AFCs, there are good prospects of using non-noble metal electrocatalysts (e.g., heat-treated cobalt tetraphenyl porphyrin). The MCFCs and the SOFCs have the attractive feature of using non-noble metal and metal oxides (single or mixed) as electrocatalysts. However, these are mostly unsupported electrocatalysts prepared by tape-casting. Alternatives for SOFC fabrication are chemical vapor deposition and electrochemical vapor deposition methods. The particle sizes of the electrocatalysts are considerably higher (carbon-supported Pt electrocatalyst particles have a diameter in the range 2-4 nm, whereas nickel, nickel oxide, and the rare earth oxides probably have diameters in the range $0.1–1.0$ μm). Thus, the true active surface area of the fuel cell electrodes is considerably lower in MCFCs and SOFCs than in the low and intermediate temperature fuel cells, requiring a scale-up of electrode area. Another consideration is that an ionic conductor is required in the active layer of the SOFC anode to increase the three dimensional reaction zone. It is for this reason that the electrolyte $ZrO_2/Y_2O_3$ is also included in the active layer.

9.8.1.7.  *Poisoning Species.*  For the low and intermediate temperature fuel cells, the poisoning species are mainly CO and $H_2S$. The tolerance level of CO is less than 10 ppm in PEMFCs and AFCs; in the PAFC, it is 1%. To enhance the tolerance level of CO to about 50 ppm in a PEMFC, a Pt-Ru alloy is used instead of Pt as the anode electrocatalyst, but it is necessary to use higher noble metal loadings. CO and COH species are formed as intermediate species during electrooxidation of methanol. Thus, in this case too, a Pt-Ru electrocatalyst with a noble metal loading of 2–4 mg/cm$^2$ is necessary to remove the adsorbed intermediate by its oxidation with OH groups formed on Ru. $H_2S$ is an even deadlier poison than CO. Its removal by hydro-desulfurization, using ZnO, is vital before processing the primary fuel natural gas. CO is a reactant in MCFCs and SOFCs, but $H_2S$ is a poison because it forms nickel sulfide on the anode electrocatalyst. In MCFCs, there is the additional problem of $H_2S$ impurities causing the formation of alkaline sulfites and sulfates, which changes the composition of the electrolyte and hence its specific conductivity. The poisoning effects of CO and $H_2S$ are reversible because the adsorbed species on the anode can be removed by changing the fuel from reformed $H_2$ to pure $H_2$, but this procedure will be complex during operation of fuel cell power plants/power sources.

9.8.1.8.  *Lifetime / Degradation Rate.*  Lifetime studies have been extensively carried out in single cells for all types of fuel cells, except perhaps for the DMFCs. This is one of the most challenging problems for meeting the targets for power generation/cogeneration, transportation, and portable power applications. The competing technologies (see Chapter 11) have met these goals. The reasons for the difficulties of fuel cells in meeting these goals are two fold: corrosion and poisoning. Except for the SOFC, all the other types of fuel cells contain strong acid, alkaline, or molten carbonate liquid electrolytes, requiring corrosion resistant materials. Also, in the region of the cathode, the fuel cells are in an oxidizing environment, which enhances the corrosion rate. Very probably, the corrosion rate in the SOFC fuel cell is much less. Secondly, poisoning of the anode electrocatalyst by CO, $H_2S$, and other impurities further reduces lifetime. In the case of the DMFC, the problem is more severe because the poisoning species (CO, COH) are formed as intermediates during electro-oxidation of methanol.

It has been the consensus among government/private sponsors and fuel cell researchers and developers that the electrochemical cell stacks in fuel cell power plants will have to be replaced after 5 years of continuous operation for power generation/cogeneration, 3000 h over a 5-year period for transportation vehicles, and 1000 h of operation for portable power. The goals set by the US Department of Energy are that the degradation rate will be less than 1 mV/1000 h for power generation/cogeneration; a higher degradation rate, perhaps by a factor of 5, may be acceptable for transportation, and a still higher rate by another factor of 5 for portable power. The fuel cells that have met this target for the power generation/cogeneration application are the PAFC and SOFC. The MCFC degradation rate is approaching the targeted value.

## 9.8.2. Cell Stacks

9.8.2.1. *General Comments.* A few of the most important characteristics of fuel-cell stacks were chosen by the author and are presented in Table 9.11.

9.8.2.2. *Electrode Area.* One advantage of the PEMFC is that its power density can be three to five times higher than in all other types of fuel cells. Thus, for the same power density, the electrode area can be correspondingly decreased. But to achieve the same potential, the number of cells in the stack will have to be the same. Until the present time, the PAFC, MCFC, and SOFC have been developed for power generation/cogeneration applications. In the case of the PAFC and MCFC, the electrode area in the most advanced power plants is 8,000-10,000 cm$^2$, while for the SOFC it is 1000 cm$^2$. Fuel Cell Energy, in their most advanced power plants, have designed and constructed modules generating 250 to 275 kW. The most advanced Siemens Westinghouse power plant has 1152 cells, connected in series and parallel. Ballard Power Systems, Inc. designed and constructed a 250 kW power plant operating on natural gas with a fuel processor. The electrode area is 250 cm$^2$. The maximum electrode area in the AFC is 500 cm$^2$, whereas in the DMFC it is 150 cm$^2$.

9.8.2.3. *Bipolar Plate.* Graphitic carbon with a binder and fluid-flow channels is most commonly used material for the bipolar plate in low and intermediate temperature fuel cells. It is very probably the most expensive component in an electrochemical cell stack because of the high costs of fabrication of the plates and machining the fluid-flow channels; the cost of the materials is relatively low. To lower the costs for the bipolar plate, conductive plastics (graphite with polyethylene or polypropylene) were used by Alsthom/Exxon and ELENCO in AFCs. But the specific resistance of this material is high, about 1 $\Omega$/cm$^2$ (it has to be lower by a factor of 10, at least). In the NASA Space AFCs, the bipolar plate is gold-plated magnesium. Siemens, Adelphi, and some smaller companies have evaluated metallic bipolar plates (e.g., aluminum with a carbon layer and binder, or gold-plated stainless steel) for PEMFC stacks. The advantage of the metallic plates is that these can be made very thin (< 2 mm) and the flow-fields for the reactant gases be stamped. Thus, the power density (W/L) can be increased. Though the costs of these stamped metallic plates will be considerably lower than that of the carbon bipolar plates, their research and development is at an infant stage. The MCFC uses metal oxide coated stainless steel plates. One problem with coated plates is that they have to be pore free. Otherwise, electrolyte enters these pores and cause slow corrosion of the plates, which yield metallic impurities and leads to the formation of passivating layers. The state-of-the-art Siemens Westinghouse fuel cells use strips of magnesium-doped lanthanum chromite for interconnection (anode of one cell to the cathode of the next). The fabrication techniques are expensive (see Section 9.7). Less expensive techniques are being explored (e.g., tape-casting, plasma-spraying), particularly for planar SOFCs.

**TABLE 9.11**
**Essential Characteristics of Six Leading Fuel Cell Technologies: Stacks**

| | PEMFC | DMFC | AFC | PAFC | MCFC | SOFC |
|---|---|---|---|---|---|---|
| Electrode area (cm2) | 10-250 | 10-150 | 100-500 | Up to 10,000 | Fuel Cell Energy up to 10,000 | Siemens-Westinghouse Tubular, 1000 |
| Bipolar plate/interconnection | Graphitic carbon with binder | Graphitic carbon with binder | Terrestrial: graphitic carbon with binder Space: Au-plated Mg | Graphitic carbon with binder | Metal oxide coated stainless steel | Mg-doped lanthanum chromite |
| Rated Power (kW) | 0.01-250 | 0.01-1 | 1-100 | 40-200 | 100-250 | Tubular: 100 planar: up to 1 |
| Cogeneration | Electricity/space heating and hot water above 5 kW | None | None | Electricity/space heating and hot water | Electricity/high grade heat for chemical plant or for gas turbine hybrid system. | Electricity/high grate heat for gas turbine hybrid system |
| Efficiency (%) | 40 | 25 | Terrestrial: 40 space: 60 | 40 | 50-60 | 50-65 |
| Heat rejection coolant | Water | Water | Electrolyte circulation | Ethylene glycol or air | Air | Air |

**TABLE 9.11. Continuation**

| | PEMFC | DMFC | AFC | PAFC | MCFC | SOFC |
|---|---|---|---|---|---|---|
| Lifetime (h) | Present: 3000 projected: 40,000 | Present: 500 projected: 5000 | 5000–10,000 | 40,000 in UTC Fuel Cells PC25 | > 10,000 achieved by Fuel Cell Energy | > 20,000 |
| Poisoning species | Anode: $CO > 10$ppm electrolyte: mono and divalent cations | Anode: CO, COH intermediate electrolyte: mono and divalent cations | $CO_2$ generating carbonates in electrolyte and solid carbonate in pores | $CO > 1\%$ $H_2S > 1$ppm | $H_2S > 0.5$ppm | $H_2S > 1$ppm |
| Lifetime / degradation rate [(h)/(mV/1000h)] | 10,000/3–5 | 1000/10 | 10,000/5–10 | 40,000/1 | 10,000/2 | 10,000/0.7 |

9.8.2.4. *Rated Power.* The PEMFC power level has a wide range. The assembly of the stacks is also relatively simple, particularly with the proton exchange material also serving as a gasket material. During the last decade, the emphasis was on transportation applications and electrochemical cell stacks (10 to 50 kW) have been built by several companies (e.g., Ballard Power Systems Inc., UTC Fuel Cells, Plug Power, Toyota, Honda, Paul Scherrer Institute/Volkswagen). AFC stacks (7 to 50 kW) have also been assembled by Union Carbide and ELENCO for this application. There is considerable interest in developing PEMFC and DMFC cell stacks for low power applications (laptop computers, cellular phones, games, etc.). Thus, mini and micro fuel cells are being designed and assembled. PEMFC stacks with rated power levels of 5 to 50 kW are being developed for stand-by power and emergency power. PAFC, MCFC, and SOFC stacks are mainly considered for power generation/cogeneration applications and are rated in the multi-kilowatt to megawatt range. Planar SOFC stacks have been developed so far at the 1–5 kW level.

9.8.2.5. *Cogeneration.* The ideal fuel cells for cogeneration applications are the PAFC, MCFC, and SOFC systems. The most beneficial utility of the PAFC will be to generate electricity plus heat for space heating and hot water. The added advantage of MCFC and SOFC power plants will be that additional electricity can be generated by utilizing the high-grade waste heat in a gas turbine (a hybrid power plant). This significantly increases the efficiency for chemical to electrical energy conversion.

9.8.2.6. *Efficiency.* The efficiency for conversion from chemical energy to electrical energy in a fuel cell stack is reduced by about 5% from that in a single cell because of the energy requirements for auxiliaries (heat removal, pumps, blowers, compressors, etc.). MCFC and SOFC cell stacks have a higher demonstrated and projected efficiency because of their internal reforming and hybridization with gas turbines. The DMFC has the lowest efficiency because of activation overpotential losses at both the anode and cathode. The AFC in the NASA-space fuel cell has a high efficiency because it uses high loadings of electrocatalysts (about 4 mg/cm$^2$) to reduce activation overpotentials and pristine $H_2$ and $O_2$ as reactants.

9.8.2.7. *Coolant.* For the low temperature fuel cells (< 100 °C), water is the coolant fluid. In the case of the PAFC, both air and liquid coolants have been used successfully. In UTC Fuel Cells PC 25, a liquid coolant (water) is used. The MCFCs and SOFCs have a significant advantage in that the reactant air, which enters the fuel cell at three to four times the stoichiometric rate, serves as the coolant. Also endothermic-internal reforming provides significant cooling. The reason for the high efficiency for heat removal is the very high temperature differential between the cell stack and environment.

9.8.2.8.   *Lifetime.*  This topic was dealt with in Section 9.8.1.8. Table 9.11 shows the achieved and projected lifetimes. To make realistic estimates, it is necessary to conduct lifetime studies on a multitude of cell stacks for the desired periods of time. This is, however, not very practical because of the long periods of time required and the high costs of conducting these studies.

## 9.8.3.   Techno-Economic Challenges to Reach Era of Terrestrial Applications in the 21$^{st}$ Century

There is a great enthusiasm and momentum among the policy makers, researchers and fuel cell developers to accelerate the entry of fuel cell power plants/power sources into the commercial sector in the 21$^{st}$ century. Up to the mid 1990s, the bulk of the R&D efforts were to develop power plants for power generation/cogeneration applications. It was predicted that the PAFC, MCFC, and SOFC would be the first, second, and third generation power plants, respectively. Since the early 1990s, there was a great impetus for developing fuel cells for transportation applications because of the US Partnership for a New Generation of Vehicles Program involving the collaboration of the US Federal Government with the three major automobile companies: General Motors, Ford, and Chrysler. Daimler/Chrysler, followed by Toyota, Honda, Ford/Mazda, and Volkswagon, made the most progress. All these developments were with PEMFC power plants.

For both power generation/cogeneration and transportation applications, the capital costs and lifetimes have been the overriding factors in retarding the entry of fuel cells into the commercial sector. Since the late 1990s, the portable power application has been a springboard for development of PEMFCs and DMFCs (see Chapter 8) since higher capital costs ($1000–5000/kW) are sustainable.

Except for the PAFC, which reached the energy sector in the mid 1990s, there is no other type of fuel cell that has reached this stage. According to the authors, the most critical challenges for the six types of fuel cells reaching the era of terrestrial applications in the 21$^{st}$ century are as follows:

### 9.8.3.1.   *PEMFCs*

- Find alternate/modified proton conducting membranes, lower their cost by a factor of ten and operate at temperatures up to 120 to 130 °C.
- Further lower Pt loading to about 0.25 mg/cm$^2$ on both electrodes.
- Lower costs of production of bipolar plates and possibly replace carbon with metallic plates.
- Find alternatives to Pt-Ru to enhance CO tolerance level and reduce noble metal loading.
- Operate PEMFCs with no or under humidification of reactant gases.
- Investigate causes for performance degradation and find methods to reduce this behavior.
- Lower fabrication costs of cell stacks.

### 9.8.3.2. *DMFCs*

- Find alternative/modified membranes to minimize crossover rate from anode to cathode.
- Find better electrocatalysts than Pt/Ru to lower activation overpotential for electro-oxidation of methanol.
- Find alternatives to Pt cathode electrocatalyst, which will be more tolerant to $CH_3OH$ transported from anode.
- Lower fabrication costs of cell stacks.

### 9.8.3.3. *AFCs*

- Find methods to reduce $CO_2$ levels in fuel and air streams to zero.
- Develop fuel cells with immobilized electrolyte to minimize ohmic overpotential losses.
- Find more corrosion resistant materials for electrocatalyst supports.
- Replace Pt electrocatalysts with transition metal/metal oxide electrocatalysts.
- Develop more efficient methods to solve *electrolyte-weeping* problems.
- Lower fabrication costs of cell stacks.

### 9.8.3.4. *PAFCs*

- Mature technology developed with advances made in R&D from single cell to power plant level: may have reached the limit in terms of performance characteristics.
- Investigate potential for reducing anion adsorption and overpotential for electro-reduction of oxygen.
- Reduce manufacturing cost of cell stack at least by a factor of four.

### 9.8.3.5. *MCFCs*

- Find methods to enhance power density by at least a factor of two to three – predominantly by minimizing ohmic overpotential.
- Find more corrosion resistant materials for cathode electrocatalyst and bipolar plate, particularly for pressurized operation.
- Lower manufacturing cost.

### 9.8.3.6. *SOFCs*

- Find methods to reduce fabrication costs of cells and stacks.
- Advance technologies for development of planar SOFCs, which are promising for achieving power densities as high as PEMFCs.
- Find better sealant materials for cells and cell stacks with planar configuration.

## 9.9.    OTHER TYPES OF FUEL CELLS

### 9.9.1.  Carbonaceous Fuels

9.9.1.1.  *Higher Hydrocarbons.*  During the 1960s, saturated hydrocarbons from $C_1$ to about $C_{10}$ were investigated for direct utilization in fuel cells. The electrolyte was phosphoric acid and the operating temperature 150 °C. Platinum was the electrocatalyst and its loading was about 4 mg/cm$^2$. Methane showed the highest activity but the efficiency and the power density was very low. Fundamental studies were carried out to elucidate the mechanisms and kinetics of electro-oxidation of these fuels and it was concluded that the rate-determining step was the oxidation of Pt by water to PtOH species, according to the intermediate step,

$$Pt + H_2O \ \rightarrow \ PtOH + H^+ + e^-_0 \tag{9.29}$$

which was followed by the removal of the intermediates of the carbon species on the electrode at neighboring sites. A second reason for the ultra-low activity of the higher hydrocarbons ($C_2H_6$ and above) was that their electro-oxidation involves the cleavage of the C-C bond, which is highly energy intensive. Experimental studies were also conducted with the unsaturated hydrocarbons, ethylene, and acetylene. It was found that these fuels had slightly higher activities than ethane because of their higher formation rate of the adsorbed species on Pt.

There is revived interest in utilizing hydrocarbon fuels directly in fuel cells. Some studies were conducted with the hydrocarbons in a PEMFC but the efficiencies were quite low. However, in some recent work by Gorte and his coworkers at the University of Pennsylvania, the performances obtained in SOFCs with fuels such as n-butane, toluene, and a synthetic diesel were quite impressive.[81-83] The Ni-YSZ cermet used in the state-of-the-art tubular SOFCs showed low performance because of carbon formation on the nickel electrocatalyst. The extent of carbon formation could be minimized by using a high ratio of $H_2O$ to fuel in the anode feed (about 4:1). However, the electrocatalytic activity was quite low. But when the Ni-YSZ cermet was replaced by a Cu-CeO$_2$-YSZ cermet, the activity for electro-oxidation of the hydrocarbons was greatly enhanced. The results of the experiments in very small cells (electrode area 0.5 cm$^2$, 700 °C, 1 atm) with decane, toluene, and a synthetic diesel as the fuels in a very short-term lifetime study are shown in Figure 9.59. The water to fuel ratio was 1.5:1.0 to minimize carbide formation on the copper. It was interpreted by Gorte and co-workers that it is necessary to optimize the structure of the three phase boundary (electrode/electrolyte/gas) to facilitate the flow of reactants to the active sites through the porous structure, transport of oxide ions from the cathode, through the electrolyte to the active sites; and transport of electrons from the active layer to the current collector. A redox catalysis mechanism was proposed for the high electrocatalytic activity of the anode electrocatalyst; ceria partially reduced by the fuel is electronically conducting, and is regenerated by oxidation with the oxide ions

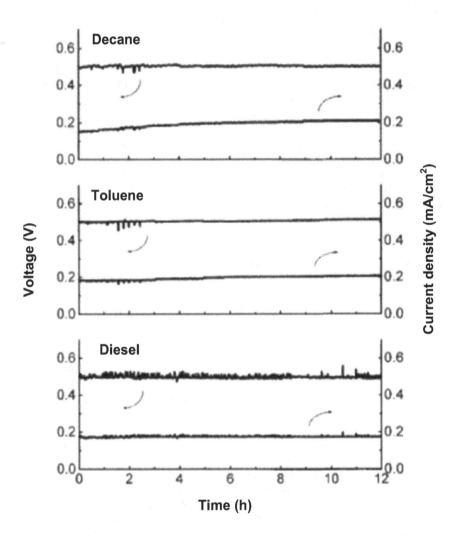

**Figure 9.59.** Cell potential and current density as a function of time for a planar SOFC with decane, toluene, and synthetic diesel as fuels; 700 °C, 1 atm. Reprinted from Reference 82, Copyright (2001), with permission from The Electrochemical Society, Inc.

transported from the cathode. It was also reported that the sulfur tolerance is higher with the $Cu$-$CeO_2$-YSZ anode than with the Ni-YSZ used currently in SOFCs. The performance degradation with the former electrocatalyst is due to the formation of $Ce_2O_2S_2$. Gorte and coworkers reported that the activity for electrooxidation of methane was low and they interpreted this result as being due to its inertness, even in heterogeneously catalyzed reactions.

9.9.1.2. *Direct Ethanol Fuel Cells (DMFC).* In recent times, there has been increasing interest in developing a direct ethanol fuel cell (DEFC), which is a potential competitor for DMFCs. Ethanol is a liquid fuel, like gasoline and methanol. Ethanol's specific energy and energy density (8.01 kWh/kg and 6.34 kWh/L) are higher than those for methanol (6.09 kWh/kg and 4.82 kWh/L). These values are more than half that for gasoline. Ethanol is a non-toxic fuel, unlike methanol, and is produced on a large scale, particularly in Brazil by fermentation of sugar cane. In the USA, the biomass fuel source material is corn. In Brazil, during the 1970s and 1980s, the production was greatly scaled-up because about 50% of the automobiles were powered by ethanol-fueled internal combustion engines. In the mid-western states of the USA, ethanol is used as an additive to gasoline (about 15% ethanol) to increase its octane number as well as its efficiency for fuel utilization.

It was stated in Section 9.3 that methanol is the most electro-active organic fuel ($CH_3OH$ is partially oxidized methane). So one may wonder why the higher partially oxidized hydrocarbons (like ethanol, propanol, ethylene glycol, glucose, etc.) are not as electro-active as methanol. The reason is that the complete oxidation of these fuels involves the breakage of C-C bonds, for which the energy requirements are high.

In some early work on DEFCs at temperatures below 100 °C, it was found that acetaldehyde and acetic acid were the main products. The recent work with DEFCs has utilized PEMFC technology. In work by Lamy et al.[84] and by Zhou et al.,[85] investigations were made on alternatives to the Pt-Ru electrocatalyst used in DMFC. Lamy et al. demonstrated that an 80% Pt-20% Sn alloy was a better electrocatalyst than Pt-Ru or Pt. Zhou et al. varied the composition of the Pt-Sn alloy (Pt-Sn/C, $Pt_3Sn_2$/C, and $Pt_2Sn$/C) and reported that $Pt_2Sn$ showed the best performance. In studies by Arico et al.[86] and by Yang et al.,[47] composite membranes (Nafion/$SiO_x$) were used. The operating temperature was also increased to 145 °C. Experiments were also carried out without humidification of oxygen or air. The electrocatalyst for ethanol oxidation was 50% Pt-50% Ru and the noble metal loading was 4 mg/cm$^2$. For the oxygen electrode, a Pt/C electrocatalyst was used with a noble metal loading of 4 mg/cm$^2$. Plots of the cell potential and of the power density vs. current density plot from the work of Yang et al. are shown in Figure 9.60. The operating temperature and pressure of oxygen were 145 °C and 3 atm. Gas chromatographic product analysis revealed that the Faradaic efficiency for conversion of ethanol to $CO_2$ was 95%; the side product was the partially oxidized acetaldehyde. There was also some unreacted ethanol. The performance of

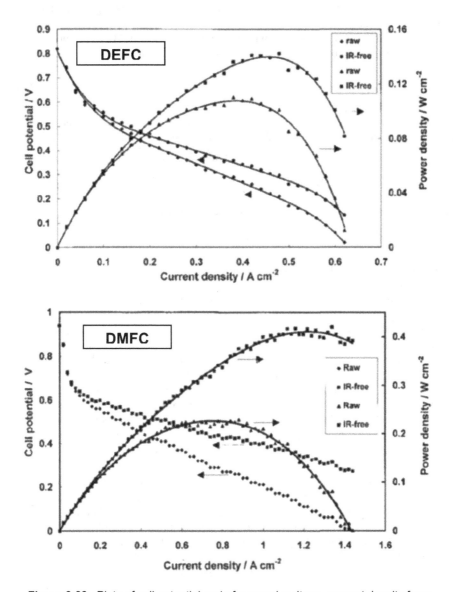

**Figure 9.60.** Plots of cell potential and of power density vs. current density for a DEFC with a Nafion/SiO₂ composite membrane; temperature 145°C, pressure 3 atm at cathode with no cathode humidification. Similar plots for DMFC also shown. Reprinted from Reference 86, Copyright (1998), with permission from The Electrochemical Society, Inc.

the DEFC was about half as good as that of the DMFC operating under the same conditions. Future research will focus on finding alternate electrocatalysts (one possibility is with ternary alloys of Pt with metals like Ru, Mo, or W) as promoters for the electro-oxidation of ethanol to occur via a redox mechanism.

## 9.9.2.  Nitrogeneous Fuels

9.9.2.1.  *Ammonia.*   Ammonia is produced on a large scale for the agricultural industry. It is a liquid fuel and has high hydrogen content (17.6%, c.f. gasoline 17.8%). The overall reaction in an ammonia/air fuel cell is:

$$NH_3 + 3O_2 \rightarrow 2N_2 + 6H_2O \tag{9.30}$$

The anodic reaction in the fuel cell, in an alkaline electrolyte (KOH), is:

$$NH_3 + 3OH^- \rightarrow \tfrac{1}{2} N_2 + 3H_2O + 3 e^-_o \tag{9.31}$$

The standard reversible potential for the cell is 1.23 V. An acid electrolyte cannot be used because ammonia is basic in character and will form a salt by reaction with the acid (e.g., formation of ammonium sulfate, if $H_2SO_4$ is used as the electrolyte). In some early work in the 1960s, Allis Chalmers Manufacturing Company, Electrochimica Corporation, and General Electric Company developed ammonia/air fuel cells. In the Allis Chalmers fuel cell, the strong KOH electrolyte was held in a porous diaphragm. The electrodes, a platinum electrocatalyst for the anode and a silver oxide electrocatalyst for the cathode, were pressed against the porous diaphragm. The open circuit potential was only 0.45 to 0.55 V, probably because of the high crossover rate of the ammonia from the anode to the cathode. Because of the relatively low efficiency of the fuel cell at temperatures in the range 30 to 80 °C, Electrochimica Corporation developed ammonia/air fuel cells for operation in the intermediate temperature range (190–300 °C). The electrolyte was molten KOH contained in a magnesium oxide matrix. Platinum catalyzed nickel electrodes were used. The performance of the fuel cell increased significantly with temperature. At 300 °C, the open circuit potential was 1.0 V. At a cell potential of 0.5 V, the current density was 40 mA/cm$^2$. Teflon was used as the sealant material for the cell. General Electric's ammonia/air fuel cell exhibited the best performance in the 1960s. Teflon-bonded platinum black electrodes (Pt loading 50 mg/cm$^2$) with a Niedrach-Alford structure were used. The electrolyte was 54% KOH. At 140 °C, the open circuit potential was 0.8 V and at a cell potential of 0.5 V, the current density was 500 mA/cm$^2$.

There has been a recent study by Wojcik et al.[87] on an ammonia/air fuel cell using a solid oxide electrolyte. The decomposition of ammonia was carried out on an iron-based catalyst, contained in a packed-bed at the front end of a tubular cell. Silver and platinum were tested as anode electrocatalysts. The electrolyte was yttria-stabilized zirconia. In another approach, platinum was used as the anode electrocatalyst. There was no iron-based catalyst for the decomposition of ammonia to hydrogen. Thus, the cell was a direct ammonia/air fuel cell. The operating temperatures were 800, 900, or 1000 °C. This fuel cell had about the same level of performance as a hydrogen-air fuel cell. The open circuit potential was 0.9 V. At a cell potential of 0.5 V, the current density was 300 mA/cm$^2$. The authors stated in their paper that the cell performance was high because of the high surface area of the platinum particles and the high porosity within the anode structure.

The advantages of using ammonia as a fuel cell are its low cost and ease of storage. The disadvantage is its toxicity, though small quantities can be detected by its pungent odor.

### 9.9.2.2. *Hydrazine.* Hydrazine, $N_2H_4$, is a fuel used in rockets by NASA. It is a liquid at low temperatures (melting point, 15 °C; boiling point, 113.5 °C) and is soluble in aqueous solutions. Its cost, particularly at high concentration, is high. In the 1960s, several companies (Chloride Electric Storage Company, Shell Research Company, Allis Chalmers Manufacturing Company, Electrochimica Corporation, and Monsanto Company) were engaged in the development and testing of hydrazine/air fuel cells. The overall reaction in the hydrazine/air fuel cell is:

$$N_2H_4 + O_2 \rightarrow N_2 + 2H_2O \tag{9.32}$$

while the anodic reaction is:

$$N_2H_4 + 4OH^- \rightarrow N_2 + 4H_2O + 4e^-_0 \tag{9.33}$$

The thermodynamic reversible potential is 1.56 V, while the observed open circuit potential (OCV) is about 1.16 V. There are two reasons for the OCV being considerably less than the reversible potential: some decomposition of $N_2H_4$ to its components; and crossover from the anode to the cathode, which depolarizes the oxygen electrode. The electro-oxidation of hydrazine is faster in basic than in acidic media. Platinum, palladium, and nickel boride have been evaluated as electro-catalysts for the electro-oxidation of hydrazine. The exchange current density on smooth noble metal electrodes for this reaction is as high as for the electro-oxidation of hydrogen ($10^{-3}$ A/cm$^2$).

Most noteworthy in the early work on hydrazine/air fuel cells was Allis Chalmers' development of a 3 kW power source to power a golf cart. The electrodes consisted of porous nickel sheets with palladium and silver as the electrocatalysts for the electro-oxidation of hydrogen and the electro-reduction of oxygen, respectively. The electrolyte was recirculated 25% KOH containing 3% hydrazine. The cell potential vs. current density plots were linear over the whole range, and the maximum power density of a cell was 250 mW/cm$^2$.

DuPont's Nafion 117 proton exchange membrane was used as an electrolyte by Yamada and coworkers in a direct hydrazine fuel cell.[88,89] An unsupported platinum black electrocatalyst was used for the anodic reaction; carbon black supported Pt (60 wt%) was the electrocatalyst for the oxygen reduction reaction. The electrocatalyst loadings were 2 mg/cm$^2$ for the anode and 3 mg/cm$^2$ for the cathode. The open circuit potential was 1.2 V, which is about 0.36 V less than the reversible potential. This loss was due to the crossover of hydrazine from the anode to the cathode. Alternate electrocatalaysts that were evaluated are Pd, Rh, and Ru, but Pt was shown to be the best. At a cell potential of 0.6 V, the current density was about 100 mA/cm$^2$. When rhodium or ruthenium was used as the electrocatalyst, there was a catalytic decomposition of hydrazine; the hydrogen, thus produced, also served as the anodic fuel. A carbon cloth with a thin carbon diffusion layer, on which the electrocatalyst layer was applied, was used to facilitate the diffusion of oxygen to the active layer. For the current collector, a sintered fiber matrix was used. The fuel was 10 wt% $N_2H_4$. The operating temperature was 80 °C.

### 9.9.3. Metal Fuels

9.9.3.1. *Zinc.* A brief description of zinc-air batteries (primary and secondary) was presented in Chapter 3. The reader may ask the question as to why we include this power source as a fuel cell. The reason is that if the fuel and the oxidant are delivered to the electrochemical cell externally, the cell is an electrochemical energy converter. Reference was made in Chapter 3 to such a device by circulating zinc particles contained in the electrolyte and feeding air into the cell for oxygen reduction, creating a fuel cell. In some recent work at Lawrence Berkley National Laboratory (LBNL), a packed bed of zinc powder was used as the anodic reactant. The electrolyte (8 N KOH) was circulated through the packed-bed tube and upward around the back of the current collector. The air electrode was external. After the cell was discharged, the electrolyte and residual particles were removed and replaced by a fresh zinc bed and electrolyte. A conceptual design was made for a 50 kW (peak power) battery. It was projected that the specific power and specific energy would be about 100 W/kg and 110 Wh/kg, respectively. It was proposed that the zinc oxide/potassium zincate produced during discharge could be used for the regeneration of zinc in an external reactor using electricity from the grid. The application envisaged for the 50 kW battery/fuel cell was electric vehicle propulsion.

Metallic Power in the USA has developed a zinc/air fuel cell. Zinc, in the form of pellets, is fed from a fuel tank to the anode compartment and air is delivered to the oxygen electrode, as in all other types of fuel cells. The electrolyte is 8 M KOH. Zinc oxide/potassium zincate formed during discharge is then transferred to a regenerator, where electricity from the grid is used to regenerate the zinc pellets. The projected efficiency for the regenerative system (electric energy generated by the fuel cell/electric energy utilized for the regeneration of zinc) was predicted to be about 50%, which is higher than for a regenerative hydrogen/air fuel cell (20 to 40%). These efforts are no longer pursued because of poor economics of this system.

9.9.3.2. *Aluminum.* Aluminum, like zinc, is a lightweight element and thus an aluminum/air battery/fuel cell could have a high specific energy (400 Wh/kg). These batteries have been developed using alkaline and neutral (saline) electrolytes. The caustic electrolyte is the preferred one because it has a higher specific conductivity and a higher solubility for the anodic reaction product:

$$Al + 3OH \rightarrow Al(OH)_3 + 3e^-_o \qquad (9.34)$$

$$Al(OH)_3 + KOH \rightarrow KAlO_2 + 2H_2O \qquad (9.35)$$

Aluminum also corrodes to some extent in alkaline electrolytes, producing hydrogen.

$$2Al + 6H_2O \rightarrow 2Al(OH)_3 + 3H_2 \qquad (9.36)$$

To reduce the corrosion rate, an alloy of aluminum is used, most commonly aluminum/gallium. Use of aluminum alloys also minimize oxide films formed on the surface during operation of the aluminum/air fuel cell. Sodium chloride (12 wt%) has also been used as an electrolyte. However, the performance of the fuel cell is much lower, 30–50 $mA/cm^2$.

Aluminum/air fuel cells were developed by Despic and coworkers at the Institute for Electrochemistry in Belgrade during the 1970-1980s. The intended application was as a power source for electric vehicles and it was projected that the range would be about 1000 km. In this system, aluminum plates were used for the anodes and these could be easily replaced after discharging about 80 to 90% of its capacity.

The other demonstrated and potential applications for aluminum/air fuel cells/batteries are ocean buoys, under-water vehicle propulsion, reserve batteries, and batteries for remote power.

### 9.9.4. Regenerative Fuel Cells

9.9.4.1. *Electrical: Hydrogen/Oxygen and Hydrogen/Halogen.* A regenerative fuel cell is similar to a secondary battery and thus may be considered as an energy storage device. As an example, a fraction of the electric energy generated by a thermal power plant during low power demands (evenings and nights) can be used for electrolysis of water, hydrochloric acid, or hydrobromic acid. The products can be used to generate electricity in fuel cells during peak power demands. Lead acid batteries are also used for this purpose by some electric utilities. The advantage of regenerative fuel cells over secondary batteries is that the fuel cell is a sub-system separate from the energy storage sub-system. Thus, the fuel cell sub-system can be reduced in size to be capable of using the available power during off-peak times for electrolysis and to generate the necessary power during peak demands. The fuel cell reactants produced during the electrolysis mode can be stored in other sub-systems. Energy storage in these sub-systems (e.g., $H_2$, $O_2$, $Cl_2$, $Br_2$) is relatively inexpensive compared to the capital costs of the electrolysis and fuel cell plants.

There are two types of $H_2/O_2$ regenerative fuel cells. In both these cases, the same electrochemical power plant served both functions: electrolysis and fuel cells. The overall efficiency for the electric $\rightarrow$ chemical $\rightarrow$ electric conversion is significantly higher for the higher temperature (1000 °C) regenerative SOFC system than for the lower temperature (80 °C) PEMFC system. The main reason for this is that the overpotentials for oxygen evolution during electrolysis and for electro-reduction of oxygen during fuel cell operation are high (about 400 mV) in the regenerative PEMFC. At the high operating temperature of the SOFC, these overpotentials are much lower. In the low temperature PEMFC system, the maximum reported efficiency is about 40%, while in the high temperature SOFC, it is 60–65%. Another design for the regenerative $H_2/O_2$ fuel cell is where the electrolysis and fuel cell power plants are separate units. The benefit of such separation is that the efficiencies can be enhanced by using different and more efficient electrocatalysts for oxygen evolution and oxygen reduction, but still using proton exchange membrane electrolytes; for example, high surface area mixed oxides of Ir-Ru and Rh for oxygen evolution and the conventional Pt on carbon for electro-reduction of oxygen.

In order to increase the efficiency for a low temperature regenerative proton exchange membrane fuel cell, researchers at Brookhaven National Laboratory investigated $H_2/Cl_2$ and $H_2/Br_2$ systems.[90] The reason for this approach was that the electrochemical reactions of the halogens, i.e.,

$$Cl_2 + 2e^-_o \rightarrow 2Cl^- \tag{9.37}$$

and

$$Br_2 + 2e^-_o \rightarrow 2Br^- \tag{9.38}$$

are highly reversible, unlike in the case of oxygen reduction/evolution. Thus in the $H_2/Cl_2$ and $H_2/Br_2$ cells, the open circuit potentials correspond to the reversible potential, 1.35 V and 1.0 V, respectively, under standard conditions. The main efficiency loss was found to be due to ohmic overpotential in PEMFCs. The efficiency of the regenerative halogen systems was found to be as high as in advanced secondary batteries. A detailed comparative engineering analysis was made for both these systems. It was concluded that even though the $H_2/Cl_2$ system has a higher cell potential during fuel cell operations than the $H_2/Br_2$ system, from an applications point of view the $H_2/Br_2$ system is preferred because $Br_2$ formed during electrolysis is in liquid form and can be stored easily in compact containers; in the case of $Cl_2$, the methods considered for storage were either as a compressed gas or as a solid chlorine hydrate. Interest in developing regenerative $H_2/Cl_2$ and $H_2/Br_2$ faded away because of the highly corrosive nature of HCl, HBr, $Cl_2$ and $Br_2$.

Another problem, which was not too severe, is that both the halogens are soluble in water (much more so than $H_2$ or $O_2$) and thus, there was some crossover of these reactants during the fuel cell operation from the cathode to the anode causing coulombic efficiency losses.

There was a recent study of the $H_2/Cl_2$ system by Thomassen et al. for cogeneration of electricity and HCl.[91] Instead of the perfluorosulfonic acid membrane, the HCl electrolyte was contained in a polyether-ethyl ketone (PEEK) separator. A commercially available Pt catalyzed electrode (0.5 mg $Pt/cm^2$) supported on carbon was used for the hydrogen ionization reaction and a high surface area Rh/C (1.0 mg $Rh/cm^2$) was used for the reduction of chlorine. The cell operating temperature was 50 °C. The open circuit potential corresponded to the reversible potential. The plots of cell potential ($E$) vs. current density ($i$) were linear over the entire range, confirming that the loss in energy conversion efficiency is due to ohmic overpotential. The effect of the concentration of electrolyte on the E vs i plot was also investigated in the range 0.01–5 M. An increase in concentration significantly decreased the ohmic overpotential losses. At 50 °C and using 3 M HCl, the power density was 0.5 W at a cell potential of 0.5 V. In this work, too, it was found that the corrosive nature of the electrolyte on the anode and cathode electrocatalysts decreased the cell performance as a function of time (this study was conducted for 120 h).

### 9.9.4.2.   *Solar: Photovoltatic Power Plants for the Decomposition of $H_2$ and HBr.*   In the Section 9.9.4.1, the electric power needed for the decomposition of water or of the hydrogen halide was generated by a thermal power plant. Another electric power generator considered for water electrolysis is a photovoltaic power plant. The benefit of using solar energy is that its cost is zero. There were several organizations developing and demonstrating the total $H_2/O_2$ system—i.e., use of photovoltaic plants to generate electricity, use this electric energy for the decomposition of water, store the hydrogen and oxygen as compressed gases, and use these as reactants in fuel cells for generation of electricity at night and on cloudy days. Most noteworthy of such development/demonstration is the complete set up for a 7 kW regenerative system at Humboldt State University (HSU) in California. An alkaline electrolysis plant was installed by Teledyne Energy Systems at HSU and the fuel cell power plant was designed and developed by the researchers at HSU. The disadvantages of this system are the low efficiency (about 10%) of photovoltaic generators using doped, amorphous silicon for the solar cells; large land requirements need for the array of solar cells because solar insolation is about 100–200 $mW/cm^2$; and the high capital costs of the photovotaic power (about $10,000/kW).

An alternate renewable energy source for this application is wind energy (see Chapter 11). Wind generators are steadily increasing the level of power generation in the USA, European countries (particularly Denmark), and the far-eastern countries. The most efficient and reliable wind generators are from Denmark. The capital cost of this type of power plant is steadily decreasing, currently $4,000/kW. Furthermore, the efficiency for energy conversion in a wind generator is as high as

65%. Each generator can produce power at a level of about 125 kW. These generators are also being located offshore for environmental reasons.

## 9.9.5.  Thermal

The essential principle of a thermally regenerative fuel cell is to decompose the products of a fuel cell reaction into its reactants in an external reactor and flow these reactants into the fuel cell. The thermal energy source for the decomposition reaction could be waste heat from a thermal power plant, nuclear or solar energy. The efficiency for energy conversion for such a system is Carnot limited. In the 1960s, the Allison Division of General Motors developed a potassium-mercury fuel cell.[92] The electrolyte was a molten salt mixture of the hydroxide, bromide, and iodide of potassium. The half-cell reactions in the cell are:

$$\text{Anode:} \qquad\qquad K \rightarrow K^+ + e^-_0 \qquad\qquad\qquad (9.39)$$

$$\text{Cathode:} \qquad\qquad K^+ + e^-_0 + xHg \rightarrow KHg_x \qquad\qquad (9.40)$$

A temperature of 630 °C was used for the regeneration process in a boiler. The cell potential current density was linear over the entire range because both electrode reactions are very fast and thus activation overpotential losses are negligible. At a cell potential of 0.4 V, the current density was 100 mA/cm$^2$. The efficiency of the fuel cell was 70%. It was proposed that the K/Hg fuel cell system could be used for space power applications using nuclear energy as the primary energy source.

## 9.9.6.  Radiochemical

The principle of this method is to use nuclear energy to decompose the product of a fuel cell reaction. Thus, the overall process in the radiochemically regenerative fuel cell occurs via nuclear $\rightarrow$ chemical $\rightarrow$ electrical energy. The efficiency of a radiochemically regenerative fuel cell is the product of two efficiencies: efficiency of conversion of nuclear energy into chemical energy of the reactants of the fuel cell and the fuel cell efficiency. The former is equivalent to the radiation yield, which is defined as the number of molecules decomposed per 100 eV of radiation energy.

An energy conversion system that has been investigated is the $H_2/O_2$ regenerative fuel cell. Water decomposition was affected by using $\alpha$ radiation.

$$H_2O + \alpha \rightarrow H^\bullet + OH^\bullet \qquad\qquad\qquad (9.41)$$

These radicals dimerize to produce $H_2$ and $H_2O_2$. A possible detrimental reaction is the recombination of the radicals to produce water. The radiation yield depends on the specific ionization of the radiation source and on the pH of the

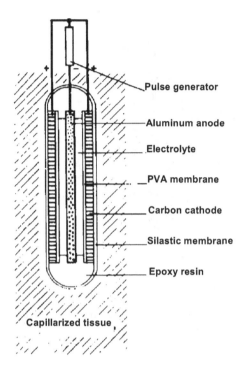

**Figure 9.61.** Schematic of Siemens biogalvanic cell. Reprinted from Reference 93, Copyright (1975), with permission from Plenum Press.

solution. The overall efficiency of the radiochemically regenerative fuel cell was 1% in both acid and alkaline media.

### 9.9.7. Biochemical

9.9.7.1. *Biogalvanic Cells.* There has been great interest in developing implantable power sources. These can be of two types: biogalvanic cells and biofuel cells. In the biogalvanic cell, a sacrificial anode is used, e.g., aluminum with an oxygen cathode. The source of oxygen is from the blood in living systems. Thus, this electrochemical cell is an aluminum-oxygen battery. Such a power source was developed by Siemens in Germany. A schematic of its design is illustrated in Figure 9.61. The anode is located between two cathodes to form a chamber

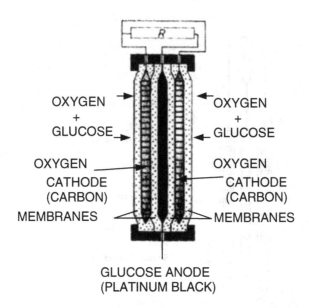

OXYGEN →
+
GLUCOSE →

OXYGEN ←
+
GLUCOSE

OXYGEN
CATHODE
(CARBON)

OXYGEN
CATHODE
(CARBON)

MEMBRANES

MEMBRANES

GLUCOSE ANODE
(PLATINUM BLACK)

**Figure 9.62.** Schematic of Siemens biofuel cell. Reprinted from Reference 93, Copyright (1975), with permission from Plenum Press

cell. The chamber space is filled with a physiological solution, which serves as the electrolyte. The compartments on the exterior side of the anode retain the anodic reaction product (aluminum oxide). The interior and exterior surfaces of the cathode were covered with hydrophilic and hydrophobic membranes, respectively, to make electrolyte contact with the electrodes and to prevent direct contact of body fluids with the electrode. The reactant oxygen is transported to the cathode via the gaseous phase. In the experimental studies, the biogalvanic cells were implanted in animals and the performances measured for a period of one year. The cell produced an electric power of 0.6 µW at 0.6 V. It was predicted that the lifetime of such a cell would be about 15 years.

9.9.7.2. *Biofuel Cells.* Siemens was also engaged in the development of a biofuel cell. The reactants were glucose and oxygen from blood in living systems. A schematic of such a cell is shown in Figure 9.62. The anode is sandwiched between two selective and porous cathodes. Thin hydrophilic membranes separate the electrodes to prevent short-circuiting. Platinum black served as the anode electrocatalyst. The electrocatalyst for oxygen reduction was high surface area carbon. The outer surface of the cathode was coated with a body-compatible hydrophilic membrane. The entire single cell was packaged completely. In the cell the glucose was possibly oxidized only partially to gluconolactone. Under physiological conditions, an *in vitro* cell produced a power of 100-150 μw at 0.5-0.6 V. The power level is quite low because of the low activity of the electrodes for electro-oxidation of glucose. *In vivo* experiments were not carried out. The power levels, generated in both biogalvanic and biofuel cells are adequate for pacemakers. To power artificial hearts, a power level of about 10 W is necessary.

A novel approach is being used by Heller et al. at the University of Texas[94] and by Palmore et al. at Brown University[95] to develop miniature biofuel cells, using glucose and oxygen as reactants. According to Heller, a battery of the same size can only produce about one-tenth of the power as the biofuel cell. A schematic of such type of biofuel cell is depicted in Figure 9.63. Enzymes are used as bioelectrocatalysts: glucose oxidase for the oxidation of glucose to gluconolactone, and laccase for electro-reduction of oxygen. The protons released during the electro-oxidation of glucose are transported through the electrolyte contained in a thin plastic membrane. The enzymes are tethered to the carbon fiber electrodes for optimum contact with the current collectors. Osmium containing polymers are used

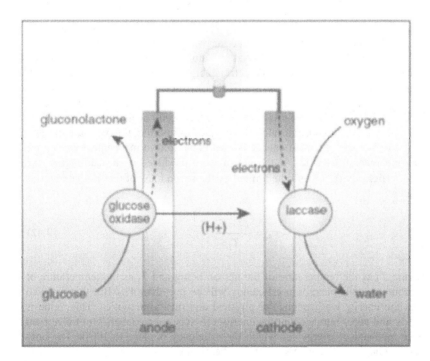

**Figure 9.63.** Schematic of biofuel cell proposed by Chen et al. Reprinted from Reference 94, Copyright (2002), with permission from AAAS.

for the fabrication of the tethers. In the work of Chen et al.,[94] the electrolyte was acidic, but enzymes are most electrocatalytic at neutral pH. Palmore et al. have used molecular biological techniques in order that the laccase enzyme could be at least half as active in acidic pH as in a physiological medium (neutral pH). The authors project that by using such an approach, body fluid can be used to extract power from a biofuel cell implanted in living systems.

In one of the investigations of Heller et al., the anode contained 35 wt% of glucose oxidase, 60 wt% of the polycationic redox polymer (2-2'-bipyridine complex of $Os^{2+}/Os^{3+}$), and the cathode electrocatalyst was bilirubin oxidase. The biofuel cell was operated at 37 °C; at a current density of 1 mA/cm$^2$; the cell potential was 0.45 V; the power output was 0.45 mW/cm$^2$. The authors proposed that such types of biofuel cells could power autonomous sensor-transmitters in animals and plants. In an article published in Science by Service,[96] it was stated that other applications could include implantable sensors to monitor glucose levels in diabetics and other chemicals that cause onset of heart disease and cancer. According to Service, a technical challenge is to reduce the problem caused by constituents of blood fluid, which adsorb on the enzyme electrocatalysts and decrease their activities for the fuel cell reactions.

### 9.9.7.3. *The Predominantly Electrochemical Nature of Biological Power Producing Reactions.*

This Section was mainly extracted from a short note by Bockris and Srinivasan, published in Nature.[97] One of the most fascinating types of fuel cells is the chemical to electric energy converter in living systems. It has a high efficiency (about 35%), no performance degradation, and longest lifetime, over 70 years in human beings. One can make a simple energy metabolism calculation as follows: the food intake by an adult is 3,000 kcal/mole per day. If the energy is produced at a constant rate over the 24-hour period, this will amount to a power of 160 W. The chemical energy is then converted to mechanical energy for physical and mental activities. If the power were produced by a heat engine, the maximum efficiency ($\varepsilon_m$) would be limited by the second law of thermodynamics as expressed by:

$$\varepsilon_m = \frac{T_2 - T_1}{T_2} \qquad (9.42)$$

Taking $T_2$ as the temperature of the human being and $T_1$ as the temperature of the environment, our maximum efficiency will be less than 4%. But this is much less than the energy we generate for physical and mental activities. This value is about 35% and takes into account the energy efficiency for conversion of electrical energy to mechanical energy for which the efficiency is close to 100%. The only alternate mechanism for conversion of the chemical energy to mechanical energy is via an electrochemical pathway, i.e., chemical to electrical to mechanical energy. Fuel cells in living systems consume carbohydrates as the fuel and oxygen as the cathodic reactant, just as in the case of practically all other types of fuel cells dealt with in this Chapter. Glucose is oxidized to gluconolactone and possibly a fraction to $CO_2$, while oxygen is reduced to water and small amounts of hydrogen peroxide. As mentioned in Section 9.9.7.2, these biofuel cell reactions require enzyme catalysts; these are generally gucose oxidase for the anodic reaction and cytochrome C (containing a redox center with $Fe^{2+}/Fe^{3+}$) for the cathodic reaction. Other reactions that contribute to the energy conversion at the anode are adenesine triphosphate to adenesine diphosphate and $NADH \rightarrow NAD + H^+ + e^-$ (NAD is nicotinamide adenine dinucleotide). These reactions also involve transport of electrons, protons, and ions (e.g., sodium pump: $Na^+$ transfer from the exterior to the interior of the biological cell and $K^+$ transfer in the opposite direction). The electrical energy will then have to be converted to mechanical energy. In a fuel cell system, this is done with electric motors. In living systems, this is affected by muscle function, contraction, and expansion.

There is an interesting correlation in performance degradation in traditional fuel cells (for power generation/cogeneration, transportation, and portable power) and fuel cells in living systems. Performance degradation is similar to a failing heart. The causes in fuel cells are loss of active sites in electrodes, corrosion, and mass transport limitations. In the heart, the causes are loss of active sites after a heart attack, leading to arrhythmia problems (ventricular tachycardia and fribrillation), fluid accumulation, causing mass transport problems, and increased resistance to blood flow in arteries. A comparative examination of the cell potential vs. current density plots for a degrading fuel cell and a pulse rate vs. stress and load for a normal and damaged heart are very similar.

## Cited References

1.  P. Costamagna and S. Srinivasan, *J. Power Sources*. **102**, 242 and 253 (1999).
2.  M. Wakizoe, O. A. Velev, and S. Srinivasan, *Electrochim. Acta,* **40,** 335 (1995).
3.  I. D. Raistrick, US Patent 4,876,115 (1989).
4.  S. Srinivasan, E. A. Ticianelli, C. R. Derouiu, and A. Redondo, *J. Power Sources*. **22**, 359 and 389 (1988).
5.  E. A. Ticianelli, C. R. Derouin, A. Redondo, and S. Srinivasan, *J. Electranal. Chem.* **251**, 275 (1988).
6.  E. A. Ticianelli, C. R. Derouin, A. Redondo, and S. Srinivasan, *J. Electrochem. Soc.* **135**, 2209 (1998).
7.  S. Srinivasan, D. J. Manko, H. Koch, M. A. Grayatzellali, and A. J. Appleby, *J. Power Sources* **29**, 367 (1990).
8.  S. Srinivasan, O.A. Velev, A. Parthasavathy, D. J. Manko, and A.J. Appleby, *J. Power Sources* **36**, 299 (1991).
9.  Z. Poltarzewski, P. Staiti, V. Alderucci, W. Wieczorek, and N. Giordano, *J. Electrochem. Soc.* **139**, 761 (1992).
10. H. P. Dhar, US Patent 5,318,863 (1994).

11. K. T. Adjemian, S. Srinivasan, J. Benziger, and A. Bocarsly, *J. Power Soucres* **109,** 356 (2002).

12. C. Yang, P. Costamagna, S. Srinivasan, J. Benziger, and A. Bocarsly, *J. Power Sources* **103**, 1 (2001).

13. M.-K. Song, Y.-T. Kim, J.-S. Hwong, J. M. Fentor, H. R. Kunz, and H.-W. Rhoe, *Electrochemical Society Meeting Abstracts*, Paris, France, April 27–May 02, 2001, Abstract No. 1217.

14. K. T. Adjemian, Investigation of High Temperature Operation of Proton Exchange Membrane Fuel Cells, Ph.D. Thesis, Princeton University, NJ, 2002.

15. D. Blom, J. Xie, and L. Allard, *Fuel Cell Seminar Abstracts*, Miami Beach, Florida, Courtesy Associates, Washington, D.C., 2003, p. 239.

16. C. Yang, Composite Nafion/Zirconium Phosphate Fuel Cell Membranes: Operation of Elevated Temperatures and Reduced Relative Humidity, Ph.D. thesis, Princeton University, NJ, 2003.

17. A. Parthasarathy, C. R. Martin, and S. Srinivasan, *J. Electrochem. Soc.* **138**, 916 (1991).

18. Z. Ogumi, Z. Takehara, and S. Yoshizawa, *J. Electrochem. Soc.* **131**, 769 (1984).

19. K. T. Adjemian, S. Srinivasan, J. Benziger, and A. Bocarsly, *J. Power Sources* **109**, 356 (2002).

20. S. L. Chen, L. Krishnan, S. Srinivasan, J. Benziger, and A.B. Bocarsly, *J. Membrane Sci.* **243,** 327-33 (2004).

21. S. Hivano, J. Kim, and S. Srinivasan. *Electrochimica Acta* **42**, 1587 (1993).

22. S. Mukherjee and S. Srinivasan, *J. Electoanal. Chem.* **357**, 201 (1993).

23. (a) Y. W. Rho, O.A. Velev, S. Srinivasan, and Y.T. Kho. *J. Electrochem. Soc.* **141**, 2084 (1994). (b) Y. W. Rho, O. A. Velev, S. Srinivasan, and Y. T. Kho *J. Electrochem. Soc.* **141**, 2089 (1994).

24. J. Kim, S. M. Lee, S. Srinivasan, and C. E. Chamberlin.*J. Electrochem. Soc.* **142,** 2670 (1995).

25. J. Yu, B. Yi, D. Xing, F. Liu, Z. Chao, Y. Fu, and H. Zhang. *Phys. Chem. Chem. Phys.* **5**, 611 (2003).

26. M. Watanabe, H. Uchida, and M. Emovi. *J. Electrochem. Soc.* **145**, 1137 (1998).

27. S. Gottesfeld and T. A. Zawodzinski, R. C. Alkive, H. Gevischev, in: *Advances in Electrochemisty and Electrochemical Engineering*, edited by R. C. Alkire, H. Gerischer, D. M. Kolb, and C. W. Zobias (Wiley-VCH, Weinheim, Germany, 1994), Vol. 5, 195.

28. C. Vickers, *Fuel Cell Seminar Abstracts*, Orlando, Florida (Courtesy Associates, Washington, D.C. 1996).

29. M. Wakizoe, H. Murata, and H. Takei, *Fuel Cell Seminar Abstracts*, Palm Springs, California (Courtesy Associates, Washington, D. C., 1998), No. 16-19.

30. A. B. LaConti (private communication), 1977.

31. T. Fuller, J. Meyers, M. Steinbegler, and M. Perry. *Electrochemical Society Meeting Abstracts*, Paris, France. April 27-May 02, 2003, Abstract No. 1236.

32. F. N. Büchi and S. Srinivasan, *J. Electrochem Soc.* **144**, 2767 (1997).

33. K. Haraldsson, T. Markel, and K. Wipke, in *Fuel Cell Science, Engineering and Technology*, edited by R. K. Shah and S.G. Kandlikar, The American Society of Mechanical Engineers, New York, NY, 2003.

34. K. M. Oseen-Senda, J. Panchid, M. Feidl, and O. Lottin, in *Fuel Cell Science, Engineering and Technology*, edited by R. K. Shah and S. G. Kandlikar, The American Society of Chemical Engineers, New York, NY, 2003, Abstract No. 455.

35. C. Y. Chow, B. Wozniczka, J. K. K. Chan, US Patent # 5804326 [Ballard Power Inc. (1998)].

36. T. O. Pavela, *Ann. Acad. Sci. Fennicae.* A II, 59 (1954).

37. E. W. Justi and A. W. Winsel, Brit. Patent, 821,688 (1955).

38. J. E. Wynn, *Proc. Ann. Power Sources Conf.* **14**, 52 (1960).

39. H. F. Hunger. *Proc. Ann. Power Sources Conf.* **14**, 55 (1960).

40. J. N. Murray and P.G. Grimes, in *Fuel Cells,* American Institute of Chemical Engineers, New York, NY, 1963, p. 57.

41. R. W. Glazebrook, *J. Power Sources.* **7**, 215 (1982).

42. K. Tamura, T. Tsukui, T. Kamo and T. Kudo, *Hitachi Hyoron.* **66**, 49 (1984).

43. S. R. Narayanan, T. I. Valdez, and N. Rohatgi, in *Handbook of Fuel Cells*, edited by W. Vielstich, A. Lamm, and H.A. Gasteiger (John Wiley and Sons, West Sussex, England, 2003), Vol. 4, pp. 894 and 1131.

44. C. Lamy, J. Leger, and S. Srinivasan, *Modern Aspects of Electrochemistry,* edited by J. O'M. Bockris, B. E. Conway, and R. E. White (Kluwer Academic/Plenum Publishers, New York, NY, 2001) No. 34, p. 68.

45. C. E. Heath. *Proc. Ann. Power Sources Conf.* **18**, 22 (1964).

46. X. M. Ren, M. S. Wilson, and S. Gottesfeld, *J. Electrochem Soc.* **143**, L12-L15 (1996).

47. C. Yang, S. Srinivasan, A. S. Arico, P. Creti, V. Baglio, and V. Antonucci, *Electrochemical and Solid State Letters* **4**, A31-A34 (2001).

48. R. Dillon, S. Srinivasan , A. S. Aricò, and V. Antonucci, *J. Power Sources* **127**, 112 (2004).

49. A. S. Arico, P. Creti, P. L. Antonucci, and V. Antonucci., *Electrochemical and Solid State Letters* **1**, 66 (1998).

50. R. Doshi, V. L. Richards, J. D. Carter, X. Wang, and M. Krumpelt, *J. Electrochem Soc.* **146**, 1273 (1999).

51. D. Buttin, M. Dupont, M. Straumann, R. Gille, J. C. Dubois, R. Ornelas, G. P. Fleba, E. Ramunni, V. Antonucci, V. Arico, P. Creti, E. Modica, M. Pham-Thi, and J. P. Ganne, *J. Appl. Electrochem.* **31**, 275 (2001).

52. K. Kordesch and G. Simader, *Fuel Cells and Their Applications* (VCH, Weinheim, New York, NY, 1996), p. 56-7.

53. E. R. Gonzalez, K. L. Hsueh, and S. Srinivasan, *J. Electrochem. Soc.* **130**, 1 (1983).

54. National Energy Technology Laboratory, , Office of Fossil Energy *Fuel Cell Handbook*, Sixth Edition (U.S. Department of Energy, Morgantown, West Virginia, 2002).

55. S. Srinivasan, B. B. Dave, K. A. Murugesamoorthi, A. Parthasarathy, and A. J. Appleby, in *Fuel Cell Systems,* edited by L. J. M. J. Blomen and M. N. Mugerwa (Plenum Press, New York, NY, 1993), p. 37.

56. A. J. Appleby, *Corrosion* **43**, 398 (1987).

57. J. R. Selman, in *Fuel Cell Systems*, edited by L. J. M. J. Blomen and M. N. Mugerwa (Plenum Press, New York, NY, 1993) p. 345.

58. G. H. Broers, J. A. A. Ketelaar, *Ind. Eng. Chem.* **52**, 303 (1960).

59. B. S. Baker, *J. Electrochem. Soc.* **127**, C340 (1980).

60. C. Y. Yuh, J. R. Selman, *J. Electrochem. Soc.* **138**, 3649 (1991).

61. A. Baker, S. Gionfriddo, A. Leonida, H. Maru, P. Patel, Internal Reforming Natural Gas Fueled Carbonate Fuel Cell Stack, in Final Report prepared by Energy Research Corporation for the Gas Research Institute, Chicago, IL, under Contract No. 5081-244-0545, March, 1984.

62. R. J. Remick, Effects of Hydrogen Sulfide on Molten Carbonate Fuel Cells, DOE/MC/20212-2039, Report, 168, pp., 1986.

63. H. Ghezel-Ayagh, A. J. Leo, H. Maru, and M. Farooque, in *Fuel Cell Science, Engineering and Technology*, edited by R. K. Shah and S. G. Kandlikar (The American Society of Mechanical Engineers, New York, NY, 2003), p. 24.

64. K. A. Murugesamoorthi, S. Srinivasan, and A. J. Appleby, in *Fuel Cell Systems*, edited by L. J. M. J. Blomen and M. N. Mugerwa (Plenum Press, New York, NY, 1993), p. 465.

65. J. Weissbaert, R. Ruka, *J. Electrochem. Soc.* **109**, C87-88 (1962).

66. A. O. Isenberg, in *Electrode Materials on Processes for Energy Conversion and Storage*, edited by J. D. E. McIntyre, S. Srinivasan, and F. G. Will (The Electrochemical Society, Inc., Pennington, NJ, 1977), p. 682

67. W. Fischer, F. J. Rohr, U.S. Patent #3,718,506 (Feb. 27, 1973).

68. K. M. Myles, and C. C. McPheeters, *J. Power Sources* **29**, 311-319 (1990).

69. A. C. Khandkar, S. Elangovan, and J. J. Hartvigsen, *Ceramic Transactions* **65**, 263-277 (1996).

70. S. C. Singhal, *Solid State Ionics* **135**, 305 (2000).

71. A. O. Isenberg, W. A. Pabst, and D. H. Archer, *J. Electrochem. Soc.* **117**, C197 (1970).

72. S. C. Singhal, *Solid State Ionics* **152-153**, 405 (2002).

73. G. Ch. Kostogloudis, N. Vasilakos, and Ch. Fitkos, *J. European Ceramic Soc.* **17**, 1513 (1997).

74. L. T. Nguyen, K. Kobayashi, Z. Cai, I. Takahashi, K. Yasumoto, M. Dokiya, S. Wang, and T. Kato, in: *Proceedings of the Seventh International Symposium on Solid Oxide Fuel Cells,* edited by H. Yokokawa and S. C. Singhal (The Electrochemical Society, Pennington, NJ, 2001), p. 1042, PV 2001-16.

75. S. W. Nam, S. A. Hong, and I. Y. Seo, in *Proceedings of the Fifth International Symposium on Solid Oxide Fuel Cells,* edited by U. Stimming, S. C. Singhal, H. Tagawa, and W. Lehnert (The Electrochemical Society, Pennington, NJ, 1997), (SOFC – V), PV 97-40, p. 293.

76. S. C. Singhal, Recent Progress in Tubular Solid Oxide Fuel Cell Technology, in *Proceedings of the Fifth International Symposium on Solid Oxide Fuel Cells*

edited by U. Stimming, S. C. Singhal, H. Tagawa, and W. Lehnert (The Electrochemical Society, Pennington, NJ, 1997), (SOFC – V), PV 97-40.

77. J. W. Kim, A. Virkar, K.-Z. Fung, K. Mehta, and S. C. Singhal, *J. Electrochem. Soc.* **146,** 69 (1999).

78. S. P. Simner, J. W. Stevenson, K. D. Meinhardt, N. L. Canfield, in *Proceedings of the Seventh International Symposium on Solid Oxide Fuel Cells*, edited by H. Yokokawa, S. C. Singhal (The Electrochemical Society, Pennington, NJ, 2001), PV 2001-16, p. 1051.

79. K. Kendall, N. Q. Minh, and S. C. Singhal, in *High Temperature Solid Oxide Fuel Cells: Fundamentals, Design and Applications*, edited by S.C. Singhal and K. Kendall (Elsevier, Oxford, UK, 2003), p. 197.

80. J. Laramie and A. Dicks, *Fuel Cell Systems Explained* (John Wiley & Sons, Ltd., Chichester, 1999).

81. S. Park, R. Craciun, J. M. Vohs, and R. J. Gorte, *J. Electrochem. Soc.* **146,** 3603 (1999).

82. H. Kim, S. Park, J. M. Vohs, and R. J. Gorte, *J. Electrochem. Soc.* **148,** A693 (2001).

83. R. J. Gorte and J. M. Vohs, *J. Catalysis* **216,** 477 (2003).

84. C. Lamy, E. M. Belgsir, and J. M. Leger, *J. Applied Electrochem* **31,** 799 (2001).

85. W. J. Zhou, Z. H. Zhou, S. Q. Song, W. Z. Li, G. Q. Sun, P. Tsiakaras, and Q. Xin, *Applied Catalysis B: Environmental* **46,** 273 (2003).

86. A. S. Arico, P. Creti, P. L. Antonucci, and V. Antonucci, *Electrochemical and Solid State Letters* **1,** 66 (1998).

87. A. Wojcik, H. Middleton, I. Damopoulos, and J. Van Herle, *J. Power Sources* **118,** 342 (2003).

88. K. Yamada, K. Asazawa, K. Yasuda, T. Ioroi, H. Tanaka, Y. Miyazaki, and T. Kobayashi, *J. Power Sources* **236,** 236 (2003).

89. K. Yamada, K. Yasuda, H. Tanaka, Y. Miyazaki, and T. Kobayashi, *J. Power Sources* **122,** 132 (2003).

90. D. T. Chin, R. S. Yeo, J. McBreen, and S. Srinivasan, *J. Electrochem. Soc.* **126,** 713 (1979).

91. M. Thomassen, B. Borresen, G. Hagen, and R. Tunold, *J. Appl. Electrochem.* **33,** 9 (2003).

92. B. Agruss, *Proc. Ann. Power Sources Conf.* **17,** 100 (1963).

93. S. Srinivasan, G. L. Cahen, G. E. Stoner, in *Electrochemistry: The Past Thirty and the Next Thirty Years*, edited by H. Bloom and F. Gutmann (Plenum Press, New York, 1975), p. 57.

94. T. Chen, S. C. Barton, G. Binyamin, Z. Q. Gao, Y. C. Zhang YC, H. H. Kim, and A. Heller, *J. Am. Chem. Soc.* **123,** 8630 (2001).

95. G. T. R. Palmore and H. H. Kim, *J. Electroanal. Chem,* **464,** 110 (1999).

96. R. F. Service, *Science* 296, 1223 (2002).

97. J. O'M. Bockris and S. Srinivasan, *Nature* **215,** 197 (1967).

## CHAPTER 10

# APPLICATIONS AND ECONOMICS OF FUEL-CELL POWER PLANTS/POWER SOURCES

## 10.1. SCOPE OF CHAPTER

The fuel cell is the most efficient, environmentally friendly, and elegant energy conversion device. One may ask the question: "Why has it taken more than 165 years, since its invention by Sir William Grove, to reach the era of applications and commercialization?" Other energy conversion technologies have reached this stage in considerably shorter times with billions of dollars invested in their development and commercialization. The advances in these technologies have decelerated the progress in fuel cell technology. During the last century, NASA's space program, the Energy Crisis in 1973, the USA Partnership for the Next Generation of Vehicles (1993-2000), along with environmental legislations in the USA, several European countries, and Japan have stimulated the development of fuel cell power plants/power sources for power generation/cogeneration, transportation, and portable power applications. Significant progress has been made in the six leading fuel cell technologies since the early 1960s. This progress led to the first application of fuel cell as power sources for NASA and Russian Space Vehicles. The overriding factors slowing the entry of fuel cell power plants and power sources into the terrestrial energy sector are the capital cost, short lifetimes of the electrochemical cell stacks, and choice of fuel and complexities of fuel-processing. On the positive side, the hope is that multi-billion dollar investments in the research and development programs of the leading fuel cell technologies since the early 1960s

---

This chapter was written by S. Srinivasan and E. Miller.

until the end of the 20th century will pave the way for fuel cell power plants/power sources to enter the energy sector for commercial applications during the 21st century.

The bulk of the investments in fuel-cell research, development, and demonstration in the latter half of the 20th century focused on power generation/cogeneration primarily for enhancing the conversion efficiencies of primary fossil energy sources (natural gas, petroleum, coal) into electricity. This approach was logical in respect to energy conservation and import reduction of these resources. Electric power generation accounts for about one third of the energy use in the USA. The second highest energy consumer, at about an equal level, is the transportation vehicles. Energy conservation and legislations requiring significant reductions in the level of environmental pollutants ($H_2S$, $NO_x$, $CO_x$) has encouraged the development of fuel cell power sources. In the last two decades, momentum for the development of portable power sources for commercial and defense applications has been increasing.

This sequence of applications is addressed in the following Sections of this Chapter. The final Section is on economics and technical challenges to enter the commercial energy sector in the 21st century. The economic projections are estimates because fuel cell power plants/power sources are not at a stage of mass production, and it is still uncertain how capital costs can be significantly reduced to the levels of competing technologies. Only the most significant technologies will be presented here. For a more complete coverage of this topic, the reader is referred to the recently published Handbook of Fuel Cells edited by Vielstich, et al. (See Suggested Reading).

## 10.2.  POWER GENERATION/COGENERATION

### 10.2.1.  High Power Level (100 kW to MWs)

10.2.1.1. *Fuel-Cell Energy MCFC and MCFC/GT Hybrid Power Plants.* Since the mid 1990s, Fuel Cell Energy (FCE), formerly Energy Research Corporation, has been the world leader for the production of MCFC power plants in the power level range from 250 kW up to 2 MW. The company invented the concept of the *direct* fuel cell (DFC), in which the reforming of the fuel (e.g., natural gas) occurs within the anode chamber using a catalyst on the backside of the electrode, with further reforming of the natural gas and CO on the anode electrocatalyst. The DFC has a significant advantage over the *indirect* fuel cell because in the latter, the fuel processor and electrochemical cell stacks are separate sub-systems. In the former, the waste heat generated by the exothermic fuel cell reaction is directly and efficiently transferred to the endothermic reforming reaction. Furthermore, the efficiency for the conversion of chemical energy of the fuel to electrical energy is enhanced because the product of the fuel processing is constantly consumed by the anodic reaction.

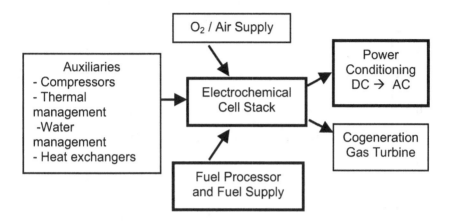

**Figure 10.1.** Primary (bolded) and secondary sub-systems in a fuel cell power plant for power generation/cogeneration.

A complete MCFC power plant consists of three main sub-systems: fuel processor, electrochemical cell stack, and power conditioning (Figure 10.1). Auxiliaries include pumps, blowers, and the equipment for waste heat removal. FCE has optimized the design of the power plant to maximize performance, lifetime, and thermal management. Prior to entry into the fuel cell, the fuel is processed to remove impurities such as $H_2S$, particulates, and in some cases, to convert the fuel into one that is compatible with the fuel cell (e.g., coal gasification and biogas production). The active electrode area in the electrochemical cell is 8000 $cm^2$, and there are 350 to 400 cells in a stack. The cells are assembled with external manifolds for the inlet and outlet of the gasses and form the building blocks of a DFC stack. The stack, then packaged in an insulated vessel, is the 275 kW module. Each module generates 250 kW of electric power. FCE, located in Danbury, Connecticut (USA), has the facilities for building and testing product lines including the DFC 300, DFC 1500, and DFC 3000 with present power ratings of 250KW, 1 MW, and 2 MW, respectively, at efficiencies ranging from 45 to 57%, for combined heat and power generation (electricity, steam/hot water, and absorption chilling). For a cogeneration power plant, the overall conversion efficiency of chemical into electrical and thermal energy can be as high as 80%. FCE is also developing MCFC/gas-turbine hybrid power plants to increase the efficiency of electric power generation to approximately 65%. It is interesting to compare the efficiency characteristics of DFCs with other types of fuel cells, as well as thermal energy conversion systems, as a function of power rating (Figure 10.2). The DFC fuels are capable of operating on a variety of fuels (natural gas, biogas, and coal-derived gas).

Some characteristics for the DFC 300, DFC 1500, and DFC 3000 systems are illustrated in Figure 10.3. DFC power plants emit considerably lower levels of $NO_x$

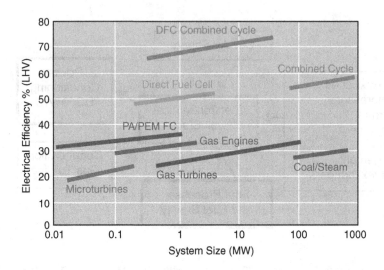

**Figure 10.2.** Comparison of efficiency of the fuel-cell energy with that of other types of fuel-cell power plants and with thermal power plants as a function of power rating. Reprinted from *Fuel Cell Energy: An Industry Leader*, Fuel Cell Energy Inc. April 2003 brochure, http://www.fce.com/downloads/profile_sheet_04_24_03.pdf, Copyright (2003) with permission from Fuel Cell Energy, Inc. All rights reserved.

| DFC300 | DFC1500 | DFC3000 |
|---|---|---|
| A self-contained commercial-grade power plant providing high quality, baseload electric power using natural gas. | These power plants consist of power generation and balance of plant skids capable of providing high quality, baseload electric power using natural gas. | |
| Output: 250 kW | Output: 1000 kW | Output: 2000 kW |
| Footprint: 10.5' w x 28' l | Footprint: 42' w x 39' l | Foot print: 42' w x 57' l |
| Efficiency: 47% | Efficiency: 49% | Efficiency: 50% |

**Figure 10.3.** Characteristics of fuel-cell energy's DFC300, DFC1500, and DFC3000 power plants. Reprinted from *Fuel Cell Energy: An Industry Leader*, Fuel Cell Energy Inc. April 2003 brochure, http://www.fce.com/downloads/profile_sheet_04_24_03.pdf, Copyright (2003) with permission from Fuel Cell Energy, Inc. All rights reserved.

and $SO_2$ as compared with the same from thermal power plants. It is also worth noting that the emission level of $CO_2$, a greenhouse gas causing global warming, would be considerably lower from DFC power plants than from thermal power plants.

In respect to applications of DFCs, FCE is focusing on customers with concerns about energy conversion efficiency, reduction in levels of environmental pollutants, reliability, and uninterrupted power. The last concern is becoming most essential because of power blackouts lasting from a period of few hours to a few days in several states in 2003. Energy policy makers and electric utilities are expressing greater interest in dispersed power generation levels in the 50 to 100 MW range to overcome the problems caused by damage to long distance power transmission lines. DFC power plants in the range of power levels from 250 kW to about 2 MW are installed in and planned for hospitals, schools, universities, hotels, and other commercial/industrial facilities throughout some European countries, Japan, and the USA. For this purpose, FCE has alliance agreements with several organizations. A brief summary of FCE's delivery, assembly, and operation of the DFC plants follows:

(a) FCE installed a 2 MW utility scale power plant at Santa Clara, California, Municipal Electric Utility, using natural gas as fuel in April 1996. This is the largest MCFC power plant tested to date. The peak power attained was 1.93 MW ac; the emission level of $NO_x$ was 2 ppm, while the $SO_2$ level was undetectable. The efficiency for electric power generation was 46%. The testing was terminated in March 1997.

(b) Subsequently, MTU, an affiliate of the Daimler-Chrysler group in Germany, demonstrated a 250-kW plant with cells from FCE.

(c) A 250-kW DFC grid-connected power plant in the FCE facility at Danbury, Connecticut, USA, was operated successfully from February 1999 for 11,600 h; operation was then terminated for a post-test analysis. A total electric energy of 1.9 MWh was generated. The degradation rate in performance was 0.3%/1000 h, which was within that projected.

(d) DFC/Gas-Turbine Hybrid power plant, using a Capstone (250 kW) Turbine Corporation, modified Model 330 microturbine was evaluated in the year 2002. The heat generated in the DFC 300 provided the thermal energy for the microturbine. This proof of concept demonstration was found useful for the design and development of a 40 MW DFC/GT power plant, projected to have an efficiency of 75% for electric power generation.

(e) In September 2003, FCE installed a 250 kW DFC power plant in Los Angeles and linked it to the power-grid system. This fuel cell uses biogas generated from a sewage treatment plant. The DFC consumes 50% less fuel than a conventional power plant, with zero emission level of pollutants. In another alliance, FCE and Caterpillar, Inc., will site a DFC 300 power plant in southern California. These two companies are also in collaboration to build and install power plants in the power range 250 kW to 3 MW.

(f)  FCE and its US distribution partner have installed two 250 kW power plants in New Jersey, USA, one at the Sheraton Parsipanny Hotel and the other at the Sheraton Hotel, Raritan. These power plants will provide about 25% of the electrical and thermal power requirements. The installation was in October 2003.

(g)  In October 1993, FCE installed a DFC, operating on coal-mine methane at the AEP Ohio Coal LLC Rosewall site in Hopedale, Ohio. According to Dr. Hans Maru, Chief Technology Officer at FCE, this demonstration will have many advantages: (i) the fuel cell will generate electricity using coal derived methane as fuel and (ii) methane, a greenhouse gas, emission will be eliminated.

(h)  In an international joint venture, Marubeni in Japan and FCE, will deliver two 250 kW DFC power plants to the First Energy Service Company in the city of Ina, Nagano Prefecture in Japan. The DFC will operate on liquefied natural gas (LNG) and  supply both electric and thermal power.

(i)  Finally, the first installation of a 250 kW DFC in an educational institution was in the Environmental Science Center of Yale University in New Haven, Connecticut, USA. This unit will provide the center with 25% of the required electric and thermal power. Plans are also underway to install such power plants in other universities, Grand Valley State University in Michigan and Ocean County College in New Jersey, USA.

It is also worthy of mention that FCE has recently obtained a contract from the US Department of Navy to develop 500 kW DFCs, operating on diesel fuel, for power requirements aboard US Naval ships and facilities.

### 10.2.1.2. Ansaldo Ricerche (ARI, Italy), 100 kW MCFC Power Plant.

ARI's project, sponsored by the European Union Program on Fuel Cells, involves collaboration with Babcock Wilcox Espanola (Spain) and ENEC Ricerche (Italy). The heart of the 100 kW power plant (Series 500) consists of two electrochemical modules, with the needed auxiliaries. ARI conceptualizes each stack with the auxiliaries as a compact unit (Figure 10.4). Each unit is located inside a pressurized vessel. The operating pressure is 3.5 bar in this design. The goals are to  have good thermal management, locate key components in a single block, reduce gas volumes, minimize differential pressure between internal/external sides of components, and reduce pressure gradients between anode/cathode, anode/vessel, and cathode/vessel. The intended fuels for delivery to the MCFC are natural gas, coal gas, landfill gas, or biogas. Most of the work has been with natural gas. A schematic of the Series 500 100 kW system, including Balance of Plant (BOP), is depicted in Figure 10.5.

To date, two plants have been installed and performance evaluated in ENEL near Milan, Italy and an "Iberdrola" site in Guadix, near Madrid, Spain. The performance goals have been met with respect to electricity and thermal

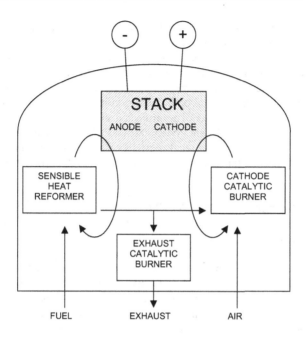

**Figure 10.4.** Ansaldo Recerche's design of an MCFC stack with auxiliaries.

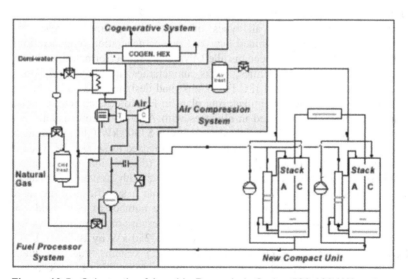

**Figure 10.5.** Schematic of Ansaldo Recerche's Series 500 100-kW system including BOP.

power generation. The Series-500 power plants will serve as building blocks for larger size power plants, delivering power up to a few megawatts. Such power plants will take into consideration integration of some of the BOP components to reduce footprint area and costs. Efforts will also be made to reduce overall size, by placing the hot components inside a vessel, reduce the number of components, simplify maintenance procedures, and optimize a skid-mounted layout for reducing installation time, footprint, and costs. The projected efficiency for electric power generation is about 45%. In the present system, 80% of the gas exhaust heat content was recovered. It is very possible that Ansaldo will also develop a hybrid MCFC/Gas-Turbine system for additional electric power generation, as well as to provide thermal power.

10.2.1.3. *IHI (Japan) 100-kW MCFC Power Plants.* From the early 1970s to the mid-nineties, four companies in Japan were engaged in research and development of MCFC power plants: Ishikawajima-Harima Heavy Industries Co. (IHI), Hitachi, Mitsubishi Electric Corporation, and Toshiba Corporation. Recently, it appears that IHI is the only company that has continued such activities. A design from Chuba Electric Company to build a 300 kW MCFC power plant, integrated with biogas derived from garbage was also developed. In this subsystem design, a stack consisting of 250 cells, with an electrode area of 1 $m^2$, was assembled and installed in the Kawogoe Power Generation Laboratory of the MCFC Research Association in December, 2002. The performance in a 10-cell block revealed that the open circuit potential was practically the same in all cells ($\sim 0.95$ V), and the cell potential under a 50% load was 0.75 V.

10.2.1.4. *UTC-Fuel Cells 200-kW PAFC, PC 25, Power Plant.* This power plant was the first of all types of fuel cell systems to reach the commercialization stage for combined heat and power generation (cogeneration). Since 1996, over 250 units have been installed, evaluated for performance, and grid-connected in several electric utilities, banks, universities, hospitals, and hotels. UTC Fuel Cells (formerly named IFC Fuel Cells) had designed and built a billion dollar, completely automated, manufacturing plant in Hartford, Connecticut, USA. UTC Fuel Cells was also involved in alliances with Ansaldo Ricerche, Italy and Toshiba Corporation, Japan. The initial sales price was $3000/kW (still government subsidized) but during the last two to three years it increased to $4000/kW. Though the performance of the power plant met its goals (degradation rate of about 1 mV/1000h), the cost was too high to be competitive with conventional energy conversion systems (Chapter 11). It was projected that a significantly increased volume of production would lower the costs but the number of units sold was insufficient over the seven-year period. Furthermore, because of the major advances made with the PEMFC power plant technology (up to 250 kW by Ballard), UTC-Fuel Cells has terminated the production of the PC 25 systems. Plans are under consideration to convert the fuel-cell production facility to building PEMFC power levels at about the same power level.

In order for PAFC systems to be competitive with PEMFC, MCFC, and SOFC power plants in respect to performance and capital costs, further research and technology development work will be necessary. Although work is not ongoing, the UTC-Fuel Cells PC-25 system can serve as a model for power generation/cogeneration. A brief description of the design and assembly of this system and its performance characteristics is presented in this Section. A schematic of the power plant is depicted in Figure 10.6. The system operates at about 250 °C and produces 200 kW of DC power, which is converted into AC power using a very advanced solid-state power conditioner. The system utilizes an efficient heat-transfer from the electrochemical cell stack, where the exothermic fuel cell reaction occurs, to the fuel processor for the endothermic fuel cell reaction. Since the overall efficiency for the power plant is about 40 to 45%, there is still sufficient waste heat, which can be captured for space heating and hot water. The mode of operation of the system is illustrated in Figure 10.7. The specifications and operational performance characteristics of the system are presented in Table 10.1.

**TABLE 10.1**
**Some Specification and Operational Performance Characteristics of UTC Fuel Cells 250-kW PC-25 Power Plant[a]**

| | |
|---|---|
| Rated electrical capacity | 200 kW / 235 kVA |
| Voltage and frequency | 480/227 V, 60 Hz, 3 phases<br>400/230 V, 50 Hz, 3 phases |
| Fuel consumption | Natural gas: 991 liter/min |
| Efficiency (LHV basis) | 87% total: 37% electrical, 50% thermal |
| Emissions | < 2 ppmv CO, < 1ppmv $NO_x$, negligible $SO_x$<br>(on 15% $O_2$ dry basis) |
| Thermal energy available:<br>   standard:<br>   high heat options: | 2,638 W @ 60 °C<br>131,900 W @ 60 °C<br>131,900 W @ 120 °C |
| Sound Profile | Conversational level (60 dBA at 10 meters) acceptable for indoor installation |
| Modular power | Flexible to meet redundancy requirements as well as future growth in power requirements |
| Flexible siting options | Indoor or outdoor installation<br>Small foot print |
| Power module | Dimensions 3 x 3 x 5.5 meters<br>Weight 18,000 kg |
| Cooling module | Dimensions 1.2 x 4.2 x 1.2 meters<br>Weight 770 kg |

[a]http://www.utcfuelcells.com/commercial/features.shtm

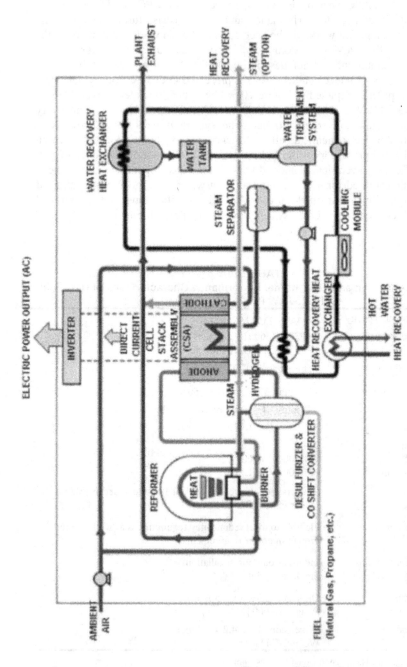

**Figure 10.6.** Schematic of UTC Fuel-Cell PC25 200-kW power plant. Reproduced from Fuel-Cell Brochure, with permission from Toshiba International Fuel Cells Corporation, http://www.tic.toshiba.com.au/website/newtic/data/fc.pdf

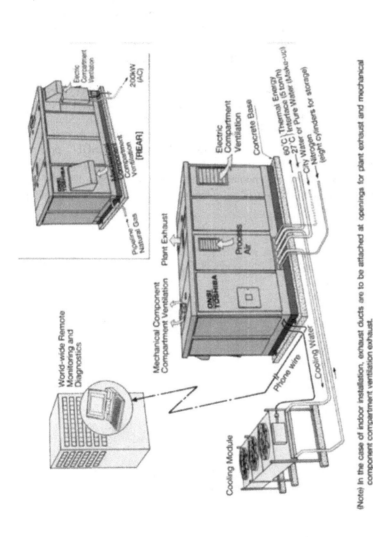

**Figure 10.7.** Mode of operation of a UTC-Fuel Cell's PC25 200-kW power plant. Reproduced from Fuel-Cell Brochure with permission from Toshiba International Fuel Cells Corporation. http://www.tic.toshiba.com.au/website/newtic/data/fc.pdf.

(Note) In the case of indoor installation, exhaust ducts are to be attached at openings for plant exhaust and mechanical component compartment ventilation exhaust.

The projected applications for this system by UTC Fuel Cell are for cogeneration in hospitals, restaurants, residential and commercial buildings, hotels, hospitals, swimming pools, and stand-by or emergency power in telecommunication and computer centers. The PC25s have been installed and operated worldwide in several of these applications (see UTC Fuel Cells website and references in suggested reading list). The major hurdles for a successful entry into the commercial sector are the high capital cost and relatively lower level of performance as compared to the MCFC, SOFC, and PEMFC power plants.

10.2.1.5.*Siemens-Westinghouse Tubular SOFC Power Plant (100-500 kW)*. Westinghouse Electric Corporation in Pittsburgh, Pennsylvania, USA has been the world leader in tubular SOFC technology since the 1960s. Siemens, Germany, recently merged with Westinghouse for this project, and this merger has been of great benefit in advancing the technology. A number of SOFC power generation/cogeneration systems have been built and sold to organizations in the USA, Netherlands, and Germany. A schematic, representing the process diagram of a Siemens-Westinghouse SOFC power plant, generating electricity and high-grade heat is illustrated in Figure 10.8.

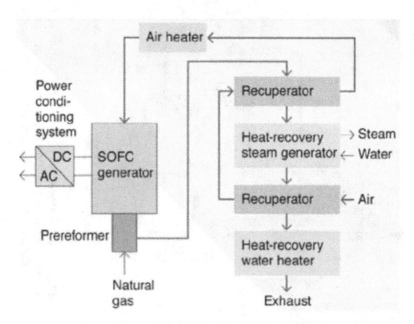

**Figure 10.8.** Process diagram of a Siemens-Westinghouse SOFC power plant for combined heat and power generation (CHP). Reprinted from *Operation Principle*, http:/www.powergeneration.siemens.com/en/fuel cells/ sofc/operation/index.cfm, Copyright with permission from Siemens-Westinghouse Power Corp.

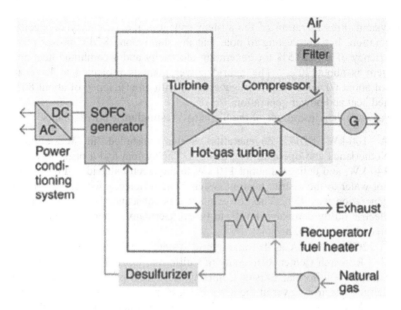

**Figure 10.9.** Process diagram of a Siemens-Westinghouse hybrid SOFC/gas turbine power plant. Reprinted from *SOFC/Gas Turbine Hybrid Diagram*, http:/www.powergeneration.siemens.com/en/fuelcells/ hybrids/performance/index.cfm, Copyright with permission from Siemens-Westinghouse Power Corp.

The fuel, desulfurized natural gas, is delivered into an ejector and then into the SOFC generator. The ejector also serves another function: by using a partial vacuum, it pulls out the anodic products from a re-circulating plenum near the top of the stack. Water, one of the products, is needed for the internal reforming reaction. The fuel mixture then enters the pre-reformer, containing a catalyst for the steam-reforming reaction, and is converted into higher hydrocarbons, methane, hydrogen, and carbon monoxide. The reformed fuel mixture, along with steam then enters the plenum, where it is distributed to the outer surface of the cells, flowing upwards, where the anodic reaction occurs. The cathodic reactant, preheated in a recuperator is delivered to the individual single cell tubes internally in the stack, via an air-manifold. The fuel cell reaction occurs in each cell in the stack and the operating conditions are maintained for a fuel utilization of 85% and an oxygen utilization of about 50%. The temperature in the stack is around 1000 °C. The unspent fuel (15%) and the air are mixed for combustion to preheat the reactant air and heat exchanged to generate steam and high-grade heat for thermal power generation.

Siemens-Westinghouse has also developed a SOFC/Gas-Turbine hybrid for enhancing the efficiency for electricity generation. A schematic of the process diagram for such a generator is shown in Figure 10.9. It is a highly integrated compact system. Pressurization of the system enhances the efficiency of electric power generation. It is interesting to note that the stand-alone SOFC power plant has an efficiency of about 45% for generating electricity and a combined heat and power system is about 70%. The hybrid power plant is projected to have an efficiency of about 60% for electricity generation, with an efficiency of about 80% for combined heat and power generation.

There are two demonstrations of the Siemens-Westinghouse power plants:

- A 100-kW SOFC cogeneration system installed in Westerwoof, Netherlands and operated for 16,600 h. The system had a peak power of 140 kW, and delivered about 110 kW to the electrical grid and 65 kW of hot water to the district heating system. The efficiency for electric power generation was 48%. The power plant was subsequently relocated at an electric utility company, RWE, in Essen, Germany, where it operated for an additional 3700 h.

- A 220-kW SOFC/Gas Turbine hybrid system installed in the Irvine Fuel Cell Research Center, University of California, USA and operated by the Southern California, Edison Company. This is a pressurized system. It delivers electric power at the level 200 kW from the SOFC generator and 20 kW from the microturbine generator. The overall efficiency for electric power generation is about 55%. It is projected that this can be increased to 65 to 70%. The efficiency for combined heat and power generation of this unit is about 80%.

**10.2.1.6.** *Ballard 250-kW PEMFC Power Plant.* The major activity at Ballard Power Systems, Inc. is focused on PEMFC power plants for electric vehicle propulsion. But since 1994, Ballard has also been engaged in the development of combined heat and power (CHP) generation systems. Ballard first developed 10- and 30 kW systems, which led to a 250 kW system; the latter was installed in Cinergy Technology, Inc. in 1999 and field-tested for 4 years. Ballard is in alliance with Ebara in Japan and Alsthom in Europe. The second system installed in Ebara had an efficiency of 34% for electric power generation at a level of 210 kW and an efficiency of 42% for heat generation.

## 10.2.2. Low to Medium Power Generation (5 to 100 kW)

**10.2.2.1.** *Plug Power 5-kW Power Plant.* Plug Power has been actively involved in the development of two types of 5 kW systems: the Gen Sys System for continuous combined heat and power generation of residential applications and the Gen Core System for telecommunications, cable-broadband, and uninterruptible applications. The former includes as its sub-systems the electrochemical cell stack, energy storage sub-system for the fuel and the power conditioning equipment, and the latter has an additional reformer for the processing of natural gas (lpg is also used as fuel). Some technical specifications and operating characteristics are listed in Table 10.2. The hydrogen fuel in the anode chamber is dead-ended; the anode chamber is periodically purged to remove part of product water (otherwise the electrode will get flooded), and inert gases, mostly $N_2$ and small amounts of $CO_2$. Auxiliaries for thermal management are also included in the total system. Three types of power plants have been developed and demonstrated: (i) Grid Parallel CHP system, (ii) Grid Parallel System with Grid Standby Capability, and (iii) Grid Independent CHP System.

Since 2002, ten Gen Sys Systems have been installed and operated in residential and operational facilities at a US Army Research Center. These systems operated for 80,000 total hours and delivered 215 MWh of electrical energy. Plug Power has also been working with the Long Island Power Authority (LIPA) in the installation and performance evaluation of seventy-five systems at LIPA's sub-station in West Babylon, Long Island, New York, USA. These systems are connected to the power grid. Plug Power and LIPA plan to install 45 more systems in Long Island in the near future. Plug Power projects that the 5 kW systems will be commercialized by the year 2005 at an initial cost of $15,000 per unit. It is probable that with mass production, the cost could be reduced to $1000/kW, which can make the PEMFC systems competitive with diesel generators and microturbines.

*10.2.2.2. General Electric 5-kW Planar SOFC Power Plant.* Minh and co-workers at General Electric Company in the USA have been actively involved in the development of low-power level SOFC power plants for cogeneration applications. The sub-systems consist of the electrochemical cell stack, fuel-processor auxiliaries for fuel delivery thermal management, electronic controls, and regulators. Techno-economic analysis for the 5 kW system showed the efficiency of

**TABLE 10.2**
**Some Technical and Operating Characteristics of Plug-Power PEMFC Power Plant[a]**

| | |
|---|---|
| Continuous power rating | 5 kWe (9 kWth) |
| Power output | 2.5–5 kWe (3–9 kWth) |
| Power quality | IEEE 519 |
| Voltage and frequency | 120/240 V, 60 Hz |
| Fuel | Natural Gas |
| Emissions | < 1 ppm $NO_x$, < 1 ppm SO |
| Sound profile | Conversational level (60 dBA at 10 meters) acceptable for indoor installation |
| Operating: | |
|    temperature | −18 °C through 40 °C |
|    elevation | 0 m through 1,800 m |
|    installation | Outdoor |
|    electrical connection | Grid parallel |
| Physical characteristics | Dimensions: 2.1 x 0.8 x 1.7 meters |

[a]http://www.plugpower.com/documents/GenSys5cMP5withstandby.pdf

**TABLE 10.3**
**Performance Characteristics of Daimler-Chrysler Vehicles Operating with Ballard Fuel Cell[a]**

| Vehicle's name | Year Shown | Engine Type | Cell Size | Range | Max. speed | Fuel Type |
|---|---|---|---|---|---|---|
| NECAR 1 180 van | 1993 | 12-fuel cell stacks | 50 kW/PEM | 81mi 130km | 56 mph 90 km/h | hydrogen 4,300 psi |
| NECAR 2 V-class | 1995 | Fuel cell | 50 kW/PEM | 155mi 250km | 68 mph 110 km/h | hydrogen 3,600 psi |
| NECAR 3 A-class | 1997 | 2-fuel cell stacks | 50 kW/PEM | 250mi 400km | 75 mph 120 km/h | methanol 40 liters |
| NECAR 4 A-class | 1999 | Fuel cell | 70 kW/PEM | 280mi 450km | 90 mph 145 km/h | liquid hydrogen |
| Jeep Commander 2 | 2000 | FC/battery hybrid | 90 kW/PEM | 120mi 193km | N/A | methanol |
| NEBUS | 1997 | Fuel cell | 150 kW/PEM | 155mi 250km | N/A | hydrogen 4,200 psi |

[a]W. Vielstich, A. Lamm, H. Gasteiger (Eds.), *Handbook of Fuel Cells* (John Wiley & Sons, London, 2003), Vol. 4.

electric-power generation to be about 40% and when mass produced, the power plant will cost about $400/kW. Plans are underway to develop a SOFC/Microturbine electric power generator and the forecast is that this system will have an efficiency of 65% for electric power generation. It is probable that the efficiency for CHP generation will be increased by 20%. The electrochemical cell stack in the hybrid system will operate at 800 °C.

## 10.3.  TRANSPORTATION

### 10.3.1. Daimler-Chrysler PEMFC Powered Vehicles

The Daimler Chrysler Company has been engaged in the development and demonstration of fuel-cell powered automobiles and buses. In the Daimler Chrysler Vehicles, the electric power is generated by the Ballard PEMFC power plants. To date, there have been six demonstration vehicles: four automobiles, one bus, and one light/duty vehicle performance. Parameters of these vehicles are presented in Table 10.3. In all vehicles, except NECAR 3, the fuel carried on board was hydrogen, mostly as compressed gas and, in one case, liquid hydrogen (LH$_2$). It was concluded that use of LH$_2$ powered vehicles would be highly energy consuming, complex, and expensive. Methanol was used in NECAR 3, which was reformed to hydrogen on board the two seat compact vehicle. A methanol-powered PEMFC power plant was also tested in the Jeep Commander 2, a sports utility vehicle (SUV). The NEBUS' capacity is 60 passengers. This type of bus has been operated and its performance evaluated in Hamburg and Berlin, Germany, Mexico City, Mexico and in Australia. Several buses, with further advancement in the propulsion systems, will be tested in European countries in the near future. It is also projected that Daimler-Chrysler's automobiles will appear on the scene in public roads in the year 2005.

## 10.3.2. General Motors/Opel PEMFC Powered Vehicles

In the 1960's, General Motors (GM) was the first company in the world to demonstrate a fuel-cell powered vehicle. The vehicle was a Corvair Van with an alkaline fuel-cell power plant developed by the Union Carbide Corporation. Hydrogen and oxygen were cryogenically stored in the vehicle. There were 32 cells in the cell stack. The designed power level was 32 kW and the peak power attained was 60 kW. The driving range of the vehicle was 240 km and the maximum speed was 110 km/h. The power plant was highly excessive in weight and volume: it practically occupied the entire space in the van! GM resumed its fuel cell activities in 1996 when they received a contract from the US Department of Energy for the design, development, and demonstration of a 50 kW PEMFC power plant with a methanol fuel-processor. This project was further stimulated by the USA Partnership for Next Generation of Vehicles Program. The projected applications for these power plants, as well as those of higher or lower rated power levels, included automobiles, buses, trucks, recreational vehicles, motorcycles, golf carts, and forklift trucks. They are also being considered for stationary applications. The designed power levels range from 15 to 120 kW. During the past seven years, significant increases of power densities and specific power have been made. At the present time, the values of these parameters for the electrochemical cell stack are 1.75 kW/ 1 and 1.25 kW/kg—about seven times higher values than when the program started.

The start-up time for a 70 kW system was 5 minutes. The fuel cell power plants have been operated with pure hydrogen and hydrogen produced by reforming methanol and gasoline. GM is working in collaboration with Giner, Inc., to develop high-pressure water electrolyzers. GM has also acquired 24% of the Company Hydrogenics and 15% of General Hydrogen, both in Canada, for the production of hydrogen and to set up an infrastructure for the transmission, distribution, and storage of hydrogen. In recent times, GM has demonstrated three types of PEMFC powered vehicles:

- Zafira-Opel compact van with an on-board methanol reformer /50 kW PEMFC power plant. The highest speed of the vehicle was 120 km/h and the accelerating time from zero to 100 km/h was 20 s.
- Cheverolet S-10 Pickup with an onboard reformer and 25 kW power plant. The maximum speed attained was approximately 120 km/h.
- Hy-Wire proof of concept vehicle with an onboard 94 kW PEMFC. The highest speed of the vehicle was 160 km/h.

### 10.3.3. Toyota PEMFC/NiMH$_x$ Powered Vehicle

Since 1996, Toyota Motor Corporation, Japan, has been actively engaged in the development and demonstration of a Fuel-Cell-Battery-Hybrid-Vehicles power. The fuel cell power is provided by a PEMFC power plant, while the Ni/MH$_x$ battery is used to capture the kinetic energy during braking at traffic lights and motion downhill. The battery power is then consumed during start-up and acceleration. The battery power is also essential for cold-start up. The energy storage device is a 5 kWh nickel/metal hydride battery. The overall efficiency of the battery for discharging/charging is about 65%. The main components of the power train are illustrated in Figure 10.10.

Toyota initiated a lease program of a fuel-cell hybrid vehicle in the year 2003. This vehicle is a mid-size Highlander SUV. Its design and performance characteristics, along with some other Toyota vehicles, are listed in Table 10.4. The gaseous hydrogen is compressed in tanks at a pressure of 250 atm. The hydrogen content is 7% by weight. The University of California, Davis, USA has recently entered a 30-month lease for two of Toyota's hybrid electric vehicles. Researchers at the Institute for Transportation Systems, in this University, are engaged in the performance evaluation of these vehicles under simulated driving conditions, both urban and highway drive cycles. The cost to lease each vehicle is $80,000/m. It is a highly sophisticated vehicle with excellently controlled electronics. Toyota has also installed hydrogen generation/storage and delivery systems in California, USA. An electrolysis plant furnished by Stuart Electrolyzer in Canada generates the hydrogen.

Toyota's view is that hydrogen will be the fuel of the fuel cell hybrid vehicles in the near to intermediate term. However, this company is also engaged in R&D activities to develop a fuel-processor for methanol and gasoline on board the electric vehicle.

**TABLE 10.4**
**Characteristics for the Toyota Fuel Cell Hybrid Vehicles**[a]

| Vehicle | Dimensions (m) | Cell size (kW) | Range (km) | Max. speed (km/h) | Seating capacity (people) | Secondary power | Fuel storage |
|---|---|---|---|---|---|---|---|
| Highlander SUV | 4.6x1.8x1.7 | 90 | 300 | 150 | 5 | NiMH | Metal hydride |
| FCHV-4 | 4.7x1.8x1.7 | 90 | 250 | 150 | 5 | N/A | Hydrogen gas |
| FCHV-5 | Differs from FCHV-4 only in component layout and design | | | | | | |
| BUS-1 | 10.5x2.5x3.4 | 80 kWx2 | 300 | 80 | 63 | NiMH | 3,500-psi H$_2$ |

[a] Compiled from Toyota web site http://www.toyota.com/

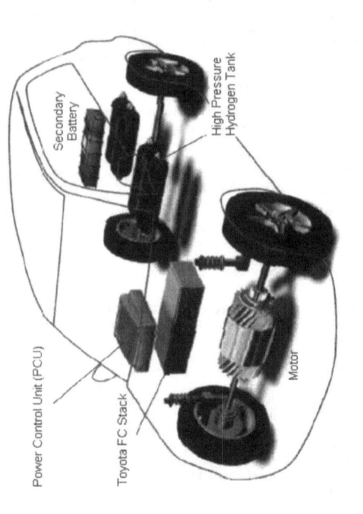

**Figure 10.10.** Main components of the Toyota PEMFC/NiMH$_x$ Hybrid. Reprinted from *Fuel Cell Driven Car*, http://www.toyota.co.jp/en/special/pdf/specialreport_10.pdf, Copyright with permission from Toyota Motor Corporation.

### 10.3.4. Honda PEMFC/Carbon Ultracapacitor Hybrid Vehicle

The main difference between the Toyota and Honda Vehicles is that, in the former, a Ni/MH$_x$ battery is the hybrid-power source, while in the latter, it is an ultracapacitor. Both are electrochemical energy storage devices (see Chapter 3). The ultracapacitor in the Honda electric vehicle contains high surface-area carbon electrodes in an acid electrolyte. The advantage of this ultracapacitor over a Ni/MH$_x$ battery is the higher specific power of the former. On the other hand, the Ni/MH$_x$ battery has a considerably higher energy density, which means that for the same mass, the kWh capacity for the battery is considerably higher. Ultracapacitors have very short response times. Since 1999, Honda has demonstrated four types of fuel-cell/battery hybrid vehicles in the USA and/or Japan:

- FCX-VI with hydrogen stored as metal hydride,
- FCX-V2 with methanol as fuel, subjected to autothermal reforming on board the vehicle,
- FCX-V3 with compressed hydrogen, and
- FCX-V4, which is an advanced version of FCX-V3 in terms of increased power output and mass of hydrogen stored.

A 60 kW Ballard fuel-cell power plant was used in these vehicles. All these vehicles have undergone road tests and performed satisfactorily. The FCX-V4 had a top speed of 140 km/h and a driving range 300 km. Hydrogen was stored at a pressure of about 300 atm in two tanks (130 liters). The FCX-V3 vehicle was driven on city and highway roads in Sacramento, California for more than 10,000 km.

### 10.3.5. Paul Scherrer Institute/Volkswagen AG/FEV Motorrentechnick GMB-1/Montena Components PEMFC Ultracapacitor Powered Automobile

The above mentioned organizations in Switzerland and Germany were actively involved in the development and testing of PEMFC power plant and installing it in a Volkswagen automobile, which was subjected to performance evaluation, even in the mountainous roads of Switzerland. This vehicle performed very well and the test-drive data agreed well with the computer simulations. It is worthwhile to provide a short description, which covers the spectrum from PEMFC power plant development to installation and testing of the Volkswagen vehicle.

There were six electrochemical cell stacks, each generating about 8 kW in the PEMFC power plant. Each stack contained 125 cells and the active area of the electrodes was 200 cm$^2$ per single cell. A novel approach was used for the production of the bipolar plates. Because of the expensive processing of milling the reactant flow channels in graphite blocks, a graphite-polymer compound was used for the bipolar plate. A molding process was used at 200 °C for introduction of the flow fields for delivery of the reactant gases to the electrodes in the single cells.

**Figure 10.11.** Six-array PEMFC stack, including manifolding plates in Paul Scherrer Institute/Volkswagen AG/FEV Motorventrechriih GMB-1/Monteun Components. Reprinted from Büchi, Ruge, Dietrich, Geiger, Hottinger, Marmy, Panozzo, Scherer, Rodatz, Tsukada, Roth, PSI Scientific Report 2001, V, 93-94 (2002). Picture reproduced by permission of Paul Scherrer Institute.

To provide cooling channels in the middle of the bipolar plates, two half plates with the cooling channels in between were glued together. The thickness of such a bipolar plate was 3.1 mm and its weight was 130 g. The pitch of a repetitive unit in the cell stack was 3.25 mm. Figure 10.11 illustrates the six-stack array, including the manifolding plates. Its total weight was 18.5 kg. In the power plant, the stacks are connected in series for entry of reactant gases and for product removal, while they are connected in parallel electrically. Thus, the power plant potential, during operation, is about 240 V. The specific power and power density of the stack are 330 W/kg and 450 W/l. The configuration of the power train is depicted in Figure 10.12. It includes a 6 kW double-layer capacitor. It is necessary to have a supercapacitor with an energy content of 300 to 400 W to deliver a peak power of 60 kW. As in the case of the $NiMH_x$ battery in the Toyota hybrid power plant, the electrochemical capacitor captures the energy during braking and delivers the power needed for start-up and acceleration. With the regenerative braking, the fuel consumption of the vehicle was reduced by 15%. The fuel used was pure hydrogen. The power plant

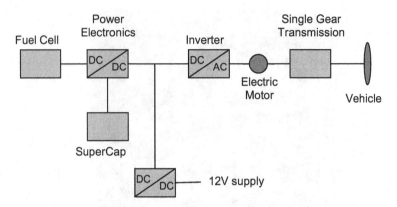

**Figure 10.12.** Layout of components in Paul Scherrer Institute/Volkswagen AG/FEV Motortechnik GMB/Monteua Hybrid Electric Vehicle. Reprinted from Dietrich, Büchi, Rodatz, Tsukada, Garcia, Wollenberg, Bärtschi, PSI Scientific Report 2002, V, 100-101 (2003). Picture reproduced by permission of Paul Scherrer Institute.

operated at 70 °C. The open circuit voltage was 350 V, and the voltage span during operation of the vehicle varied from 2 to 250 V. Figure 10.13 presents some interesting results showing the variation of power plant potential with current.

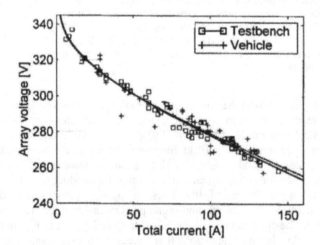

**Figure 10.13.** Variation of potential in Paul Scherrer Institute PEMFC power plant with current. Reprinted from Büchi, Rodatz, Tsukada, Dietrich, PSI Scientific Report 2002, V, 102-103 (2003). Picture reproduced by permission of Paul Scherrer Institute.

The agreement between the test bench and vehicle data is excellent. Test drives were conducted for more than 1000 km, including the mountainous roads in the Alps. Even at an altitude of 2000 m, the performance was very satisfactory.

### 10.3.6. Siemens PEMFC Propulsion System for Submarines

Siemens in Erlangen, Germany has been actively engaged in the development of fuel-cell power plants for submarine applications. For the first phase of the work in the 1970s, the type of fuel cell selected was the AFC system. Since the mid 1980s until recent times, there has been a transition to PEMFC systems. For this purpose, there was a transfer of the PEMFC technology from General Electric to Siemens. In the latest versions developed since 1992, the modules are designed for a power level of 30 to 50 kW. The fuel-cell power plants consist of eight modules (designed power level 270 kW, operating voltage 416–600 V). There are 16 cells in each module. The auxiliaries for the supply of gases, humidification, removal of product water (from both the anode and cathode sides), and removal of waste heat using a coolant are continued in each module. The response time for power generation was less than 20 ms. The gas purity levels were as high as 99.5%. The overall efficiency of the module for a delivered current of 800 A was about 60%. The performance degradation rate was 1 μV/h over a 1000-h testing period. The module weight is 650 kg and its volume 1.68 $m^3$. Siemens is projecting that the module power can be increased by a factor of four, which will increase the specific power and power density by a factor of three. Though this technology was developed for submarine application, it was also used to power buses and forklift trucks. In addition, it is being considered for stationary applications.

## 10.4. PORTABLE POWER

### 10.4.1. NASA's Space Shuttle AFC Power Plant

The AFC has been a power source in NASA's Apollo and space-shuttle vehicles. Some details of the two different AFC technologies were presented in Chapter 9. The current design, technology, and performance of the AFC power plant is considerably more advanced than in the original one. On each orbiter-space vehicle, there are three fuel cell power plants. Continuous power is delivered at a level of 2 to 12 kW with a peak power of 16 kW for 15 m. The voltage range for the operation is 28-32 V dc. The accessories include subsystems to maintain hydrogen and oxygen pressures, temperatures, and reactant flows. The reactants are stored as cryogenic liquids. The product water is used for drinking by the astronauts. The water must be articulately removed from the cells, as otherwise it will cause a flooding problem, which will result in an efficiency loss of the AFC power plant. The fuel cell stacks have to be replaced after about 2600 hours of operation mainly due to corrosion problems of the cell components including gaskets and separators. Also, it has been found that potassium carbonate forms by corrosion of the epoxies in the strong alkaline environment. New materials (e.g., polyether-ethyl ketone or PEEK) are being evaluated for insulator plates. The objective of advancing the AFC technology is to double the lifetime between cell stack replacements. At present, it costs about $3M to replace each cell stack. NASA has used this type of power plant in 113 space flights for about 90,000 operating hours. The performance and the reliability of the system are excellent. The orbiter's AFCs produced ten times more power than the Apollo's AFCs.

### 10.4.2. Power Sources for Commercial and Aerospace Applications: Low Power Level (100 W to 5 kW)

Since the late 1990s, there have been more than 800 companies involved in the design, development, and demonstration of fuel cells for power in range 100 W to 5 kW. Some characteristics of representative power sources are presented in Table 10.5.

What may be clear from this table is that the PEMFC is the preferred type of fuel cell for applications, such as residential power, stand-by power, emergency power, and auxiliary power units (e.g., automobiles). In most cases, the fuel for the PEMFC is pure or reformate $H_2$. In the case of the residential power application, the PEMFC power plant generates electricity, heat for space heating, and hot water. The latter is still relatively low-grade heat. Another application being developed is the reusable launch vehicle. Teledyne Energy Systems has obtained financial support from NASA for the Phase 2 developmental project.

SOFCs are gaining momentum for most of the above applications. These SOFCs generally have a planar design and performance characteristics are very satisfactory. The SOFCs use natural gas as the fuel either directly or indirectly. Some of the details of the technology development aspects are described in Chapter 9.

There are also projects underway for the development of AFC power sources for the applications noted above. However, in view of the technical problems connected with corrosion of the cell components and carbonation of the electrolyte, the author is of the opinion that such types of fuel cells will not meet the goals of capital costs and lifetimes.

### 10.4.3. Power Sources for Commercial Applications: Ultra-Low Power Level (1 to 100W)

Another area of application, which is receiving much attention, is for fuel cells to provide the power for electronic devices (1 W to 100 W) like cell phones, laptop

**TABLE 10.5**
**A Short List of Demonstrated and Potential Application of Low-Power-Level Fuel Cells (0.1–5 kW)**

| Organization | Type of Cell / Fuel | Rated Power | Design Operational Characteristics | Demonstrated and Potential Applications |
|---|---|---|---|---|
| Plug Power | PEMFC / $CH_4$, $H_2$ | 5 kW | | Residential, standby, and emergency power |
| Ballard Systems, Inc. | PEMFC / $H_2$ | 500 W–1 kW | | |
| H-Power | PEMFC / $CH_4$, $C_3H_8$ | 4.5 kW | | Residential (H-power merged with plug power in 2002) |
| Delphi | SOFC / $CH_4$ | 5 kW | Two 15-cell stack in series, 42 V | Auxiliary-power unit |
| General Electric | SOFC / $CH_4$ | 5 kW | 800 °C, 4 atm | Residential |
| | | | SOFC and SOFC/GT hybrid | |
| Giner, Inc. | PEMFC / $CH_3OH$ | 150 W | | |
| Teledyne Energy Systems / NASA | PEMFC / $H_2$ | 5–7 kW | 60 °C, 1.3 atm, 82 cell stack | Aerospace: reusable launch vehicle |
| Ceramic Fuel Cells Ltd. | SOFC / $CH_4$ | 1–10 kW | 850 °C | Residential / Cogeneration |
| Viessman Werke | PEMFC / $CH_4$ | 2-kW electrical 3-kW thermal | 70 °C, 1 atm, 60 cell stack, 39 V | Residential / Cogeneration |

computers, video cameras, and toys. Many of these devices currently use rechargeable batteries (e.g., Ni/Cd, Ni/MH$_x$, Li ion, etc). The motivation for replacing these batteries with fuel cells is the short storage time—the energy stored in the battery lasts at the most 5 h—and the long charging time required for the batteries. This is not the case with fuel cells. There is also an economic advantage to targeting this low power market, as the consumer is willing to pay a higher cost to support a longer lifetime. However, batteries still maintain some advantages over fuel cells in terms of packaging the power source within the device. Another disadvantage of the fuel cell is that air is the cathodic reactant, which has to be fed into the fuel cell (air-breathing or air-blown). This is not easily possible for the aforementioned applications. A solution that has been proposed for this problem is to have a small external fuel-cell power source for charging of the battery. This too is not very practical.

The types of fuel cells, which have been developed for the ultra-low power level applications, are the PEMFC and DMFC. Table 10.6 presents a selected list of DMFCs, which have been developed and demonstrated for these applications. The technology development aspects of the DMFCs were briefly described in Chapter 9. There are several organizations (universities, national laboratories, and industries) actively involved in this area of science, engineering, and technology as seen in Table 10.6. These are still at an infant stage in respect to replacing batteries with mini and micro fuel cells. But because of the rapid pace of development, this may become the first large-scale application of fuel cells.

## 10.5. ECONOMIC PERSPECTIVES AND TECHNICAL CHALLENGES

### 10.5.1. Rationale for this Section and a Look-back at the Past 50 years

It has been stated in various Chapters in this book that the fuel cell is a 19$^{th}$ century invention, a 20$^{th}$ century technology development, and a 21$^{st}$ century power plant/power source for power generation/cogeneration, transportation, and portable power applications. It was the NASA and Russian Space Programs that greatly accelerated the technology development not only of the two types of fuel cells–PEMFC and AFC—which provided auxiliary power of the space vehicles, but also of the PAFCs, MCFCs, SOFCs, and DMFCs. The "energy crisis" in 1973 and the environmental legislations in the 1980s also provided a boost for technology development.

However, the big question is: Why has it taken so long for fuel cell power plants/power sources to enter the energy sector? The thermal-power plants/engines were also 19$^{th}$ century inventions. The time span for their entering the energy sector was relatively short, about 20 years for power generation and transportation. Some primary and secondary batteries were also relatively old inventions, dating back to perhaps the 17$^{th}$ century. Gas turbines were 20$^{th}$ century inventions but within about 20 to 30 years these evolved as power plants for power generation and cogeneration

**TABLE 10.6**
**A List of Demonstrated/Potential Applications at Ultra-Low Power Levels**
**(10–100 W) (All are DMFCs)**

| Organization | Rated Power | Design Operation Characteristics | Demonstrated and Potential Applications |
|---|---|---|---|
| Motorola / Los Alamos National Laboratory | 2.5 W | 4 cells connected externally in series; methanol delivered through cartridges; 35 mW/cm$^2$ | Cell phone |
| Energy Related Devices / Manhattan Scientific, Inc. | | Cell connected externally in series; 1.5 M methanol; 370 Wh/kg, 250 Wh/L | Charge-cell phone battery |
| Jet Propulsion Laboratory | | 6 cells in series; 25 mW/cm$^2$ | |
| Bell Aerospace / Los Alamos National Laboratory | 50 W | 7 W/kg, 6 W/L | Military |
| Smart Fuel Cell, Germany | 40 W | 21 W/kg, 20 W/L | Multi-purpose |
| MTI MicroFuel Cells / Dupont | | Twin DMFC cells; 10 cm$^2$ active area | Portable electronics |
| Frannhoffer Institute for Solar Energy Systems | 10 W | Miniaturized fuel-cell stack | Power camcorder |
| Mesoscopic Devices | 15 W | 100% methanol; 12 V, 1.25 A, 480 | Portable electronics |
| Samsung Advanced Technology Institute | 15 W | Max- 7.5 W at 3.8 V; 2 stacks for 15 W | Laptop computer in 12h operation |

and for transportation (airplanes). Microturbines are increasingly being developed for low to medium power applications. Advanced rechargeable batteries such as Ni/Cd, Ni/H$_2$, Ni/MH$_x$, and lithium ion batteries had relative short transition times from the research to the technology development and commercialization stage.

A quantitative analysis as to why fuel cells have not yet reached the stage of being a competitive power-plant/power-source for terrestrial applications will not be made here. An attempt is made in the following Sections to provide a qualitative assessment and present some perspectives of accelerating the entry of fuel cells into the energy sector.

### 10.5.2. Largest Impact for Fuel Cells Entering the Energy Sector: Energy Conservation and Significantly Lowering Environmental Pollution

Of all the energy conversion technologies, the fuel cell and fuel cell hybrid with a gas turbine or battery have the best prospects for achieving high efficiencies. Power generation/cogeneration and the transportation applications consume about two thirds of the energy requirements in the USA. If the efficiency of an electrochemical energy conversion system is twice the efficiency of a thermal power plant power source, the energy consumption could be reduced by about 50%.

Environmental pollution and global-warming problems are mostly due to emissions of $CO_x$, $H_2S$, $NO_x$, and particulates from thermal power plants. Emissions of such pollutants from fuel-cell power/power sources are minimal or non-existent. Noise levels from these systems are also minimal.

### 10.5.3. Finding Niche Markets for the Entry of Fuel Cells in the Energy Sector

This has been one of the most challenging problems. In the power generation/cogeneration area, fuel cells cannot compete with large thermal systems (1000 MW and higher). Therefore, the goal has been to develop fuel-cell power plants at intermediate loads for dispersed power generation and for low-power levels for residential and commercial buildings. However, one major problem is that in the developed countries the thermal power plants are well established and have already captured the market. There are also excellent electric grid networks for their transmission over very long distances. The question then is whether fuel-cell power plants can provide the electricity for developing countries. However, are the economic situations in these countries satisfactory for introducing the new technology?

Transportation is another area where there is a major hurdle for fuel-cell power plants displacing IC and diesel engines. These are well established technologies used for more than 100 years in transportation vehicles.

There is an increasing interest in developing fuel cells, power-plants/power-sources for portable applications. However, this sector is a minor energy consumer and there are alternative power sources that satisfy such needs.

### 10.5.4. Cost Target for the Three Main Applications and the Need for a Realistic Cost-Benefit Analysis

In Chapter 11, tables include the present costs and/or projected costs for all types of energy-conversion power-plants/power-sources. The costs of fuel-cell systems are extremely high (by a factor in the range of 10 to 100 times) for the power generation/cogeneration, transportation, and portable power application. However, it is striking that the projected costs, with automation of technologies and large-scale projection, are about the same as for the state-of-the-art energy conversion technologies. The fuel-cell developers, who consider their detailed cost analysis to be proprietary, mostly make these cost estimates.

A detailed analysis has to take into account cost of construction of plants for the production of the power plants/power sources, the materials, all the components, and operations for large-scale production. On the financial side, one will have to take into consideration the discounted cash value of loans/bonds for the capital investments and the target rate of return of the investment. A GM technologist stated that when a plant was first constructed/assembled for the production of the Saturn automobile, the capital cost was about one billion dollars. Also, GM projected that for the company to attain a reasonable net profit, the volume of production should be about 300,000 cars per year for at least for 5 years.

## 10.5.5. Importance of Cost Reduction of Component Materials and Fabrication

The fuel cell technology is still at an infant stage in comparison with the other energy conversion technologies. Making capital cost estimates is difficult because many of the manufacturing techniques are still not fully automated. However, some projections of the costs can be made if there is a detailed inventory of all the components in the electrochemical cell stack and their material and manufacturing costs. There are some very expensive component materials, as for example, the current cost of the proton conducting membrane is $800/m$^2$ and its cost in a PEMFC will be approximately $200/kW. The projected cost for a fuel-cell power plant should be about $50/kW in order to be competitive with the conventional power plants. Other high-cost items are the platinum-catalyzed electrodes and bipolar plates, particularly for the low- and intermediate-temperature fuel cells. The noble metal loading is relatively low at the present time, about 0.5 mg/cm$^2$ on both electrodes in PEMFC. But the cost of production of electrodes is still high (total cost about $50 to $100/kW). In the case of the bipolar plates for the low and intermediate temperature fuel cells, graphitic carbon is the most commonly used material. Its cost is quite low but the cost of fabrication of the plates is extremely high, because of the need for machining the flow channels. One alternative is to use graphite/polymer composites. The flow fields can then be made using a molding technique. One disadvantage may be that the ohmic resistance will be higher than the state-of-the-art graphitic plate with a small amount of a resin (furfural), which serves as a binder. A second alternative is to use thin metallic plates with thin gold surface films or films of a graphite/polymer composite. The flow channels can then be stamped at relatively low cost.

In the case of the high temperature fuel cells (MCFCs and SOFCs), the costs of the component materials are very low but the fabrication costs are high. This is the case with the tubular SOFC for which extrusion, chemical-vapor deposition, and electrochemical vapor deposition techniques are used for the electrolyte, anode, cathode, and interconnection. The manufacturing costs of MCFCs can be reduced by automation of manufacturing techniques and volume of production.

Automation of techniques and finding niche markets for large-scale production are vital for a significant reduction of capital costs to be competitive with alternate-energy conversion technologies.

## 10.5.6. Performance Degradation and Operational Costs

The expected lifetime of fuel cells varies with the intended application. Lifetimes are the longest for continuous power generation/cogeneration (40,000 hours), somewhat lower for the transportation applications (3000 hours over a 5-year period), and low for portable power applications (about 1 month to 1 year). Because of the problems with stability of component materials in the electrolyte environment, there is degradation in the performance of an electrochemical cell stack. Other balance of plant components (BOP) are readily available at low cost and can be in operation for considerably long times. In the case of the electrochemical cell stacks, they will have to be periodically replaced (say 5 years for power generation/cogeneration and transportation). The cost of the electrochemical cell stack is approximately one third of the whole power plant. This will have to be taken into account in ascertaining the operational costs. Ultimately what matters is the cost of electric power generated (cents/kWh) for these two applications, i.e., capital cost, taking into consideration an amortization factor and the operational costs.

## 10.5.7. Technical and Economical Challenges

The technical challenges have been summarized in Chapter 9, in respect to fuel cell stacks. Another challenge is on the choice of fuels, fuel processor, fuel storage, and fuel transmission and distribution. This subject is dealt with in Chapter 8. For the near future, primary fuels are natural gas, coal, and oil. Hydrogen, a secondary fuel, derived from the primary fuels is the ideal fuel for the low and intermediate temperature fuel cells (PEMFC, AFC, and PAFC). The challenges faced with when hydrogen is the fuel are with respect to production, storage, and transmission and distribution, particularly for the transportation application. According to some estimates, the capital cost will be about a trillion dollars to replace natural gas and petroleum refineries, and transmission and distribution networks in the USA. Methanol will be a better fuel from this point of view, the capital cost being about one billion dollars. Exotic methods have been proposed for hydrogen storage but still the best method for hydrogen storage is as a compressed gas. Storing hydrogen in lightweight materials and at high pressures can significantly increase the specific energy and energy density. A lightweight reliable material is a carbon composite.

In summary, for fuel cells to be competitive with the competing energy conversion technologies it is vital to:

- find niche markets, particularly in the power generation/cogeneration and transportation sectors;
- reduce capital costs by a factor of 10 to 100;
- enhance the lifetime, with minimal degradation of performance; and
- ascertain whether fuel cells can compete with the advanced rechargeable batteries, $Ni/MH_x$ and Li ion, for the portable power application at low and ultra-low power.

## Suggested Reading

1. A. Avadikyan, P. Cohendet, J. Héraud (Eds.), *The Economic Dynamics of Fuel Cell Technologies* (Springer, New York, NY, 2003).
2. W. Vielstich, A. Lamm, H. Gasteiger (Eds.), *Handbook of Fuel Cells* (John Wiley & Sons, London, 2003), Vols. 1–4.
3. Bent Sorensen, *Hydrogen and Fuel Cells: Emerging Technologies and Applications* (*The Sustainable World Series*) (Academic Press, Amsterdam, 2005).

# CHAPTER 11

# *COMPETING TECHNOLOGIES*

## 11.1   SCOPE OF CHAPTER

### 11.1.1. Technology Comparison

Energy provides essential human needs such as heating and cooling; electricity to our homes, factories, and businesses; propulsion for transportation; and the ability to operate a wide range of portable electronic devices. There are a number of different energy technologies that satisfy such needs, and one of the key questions any new energy technology is not whether it can meet these needs, but whether it can meet them better than existing technologies. How one defines the better technology is open to some debate, but this Chapter will compare various options on the basis of a range of factors including engineering and system aspects, suitability for specific applications, cost, efficiency, and environmental impact. The chapter will focus on the applications that fuel cells are best suited for—electricity generation, distributed and remote power, transportation, and portable power—and on the wide range of energy conversion technologies that will compete with fuel cells in these specific applications. The end of the chapter gives technical and economic comparisons of fuel cells and the competing technologies.

### 11.1.2. Power Generation and Transportation

Power generation and transportation are two energy sectors that, at the present time, rely almost exclusively on the combustion of fossil fuels in thermal power systems (heat engines). These sectors account for approximately 65% of primary

This chapter was written by S. Srinivasan and C. Yang.

energy use in the United States. The steam heat engines have been in operation since the middle of the 18$^{th}$ century, while the first internal combustion engine was developed in the 1860s. Current heat engines achieve efficiencies of the order of 20–40%. The use of fossil fuels leads to a number of problems including regional air pollutants that affect health and air quality and $CO_2$ emission that contributes to global climate change. Since the world's demand for energy is rising, especially in these sectors, it is vital to investigate the different energy technologies currently deployed, as well as advanced energy technologies. In order to mitigate some of the negative effects of energy use, many have proposed using fuel cell power plants for electricity generation and transportation. An important question for fuel cells is how well they can compete with the conventional and advanced systems, such as high efficiency combined cycle gas turbine power plants, for power generation and advanced spark ignition and diesel engines for transportation. Efficiencies for energy conversion and emission of environmental pollutants are two of the most important criteria in determining the choice of the power plant for such applications.

### 11.1.3. Environmental Considerations

Energy conversion has a major impact on the environment. There is an increasing understanding of the social costs of power plant emissions, in terms of health and respiratory impacts, heavy metal toxins, acid rain, atmospheric damage, and climate change. Environmental regulations mainly in the form of emission standards are increasingly important constraints on all power plants, especially combustion power plants. The increasing costs of improving the least efficient and most polluting fuels and power plants will open a window of opportunity for cleaner, alternative technologies. Additionally, policies such as taxes on carbon or emissions trading may encourage the use of environmentally friendly technologies.

### 11.1.4. Portable Power Applications

In the 1950s and 1960s, NASA first used fuel cells, powered by hydrogen and oxygen, to power early space flights and provide potable water for astronauts. This was the first case where fuel cells displaced batteries. Other important applications include power for cellular phones, portable laptop computers, and auxiliary power for a wide range of applications. Portable power applications may require high energy and power densities. These engineering characteristics and the economics are two major considerations when choosing among various battery technologies and fuel cells. In general, considerations of efficiency and emissions are not as crucial, although issues related to the disposal of toxic materials are a concern.

## 11.2.  POWER GENERATION AND COGENERATION

### 11.2.1.  Electricity Generation in a Global Energy Context

Electricity generation is one of the major uses of primary energy in the world. It accounts for approximately 40% of the total energy consumed in the USA; transportation, residential and commercial non-electrical heating, and industrial processes account for the balance of energy consumption. Because of its importance and widespread use, there are great incentives for examining the technologies currently deployed for utility scale electricity generation. Fuel cells are being championed as a higher efficiency and environmentally cleaner means of electricity generation. In order for fuel cells to be competitive with the current technologies as well as advanced fossil fuel, renewable, and nuclear technologies, they must compete on a number of key issues including efficiency, environmental attributes, lifetime, and cost. In order to make an assessment of the potential for fuel cell penetration into this area, the following Sections will deal with current and advanced technologies, and specifically, principles of operation, system efficiency, costs, and environmental impacts.

### 11.2.2.  Fossil-Fuel Based Thermal Power Plants

11.2.2.1.  *Operating Principles: Steam Cycle.*  Fossil fuel-based steam power plants account for a high percentage of electricity generation in the world, as well as in the USA (50–60%). Including nuclear power plants, steam cycle power plants account for 75% of total electricity generation in the USA. The steam plant is a subset of the vapor power plant. Figure 11.1 shows an idealized version of the vapor-power-cycle plant (Rankine cycle).

The vapor power plant consists of the following key components:

- *Pump.* The energy, $W_p$, is used to compress the liquid working fluid to high pressure.
- *Boiler.* The heat, $Q_{in}$, is added from a heat source to the working fluid to vaporize it completely.
- *Turbine.* The vaporized working fluid is expanded (and cooled) back to the condenser pressure and this energy, $W_t$, is extracted from the working fluid.
- *Condenser.* The heat, $Q_{out}$, is extracted from the liquid-vapor mixture to return it to the saturated liquid state.
- *Generator.* Converts the mechanical shaft work from the turbine into electrical energy.

In most cases, the working fluid is water. Liquid water is compressed to high pressures and then heated at a constant pressure in the boiler to generate saturated steam. The turbine cools and extracts pressure from the steam, partially condensing

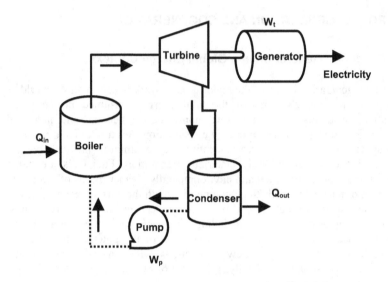

**Figure 11.1** Rankine cycle system.

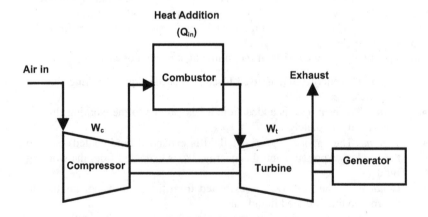

**Figure 11.2** Ideal Brayton Cycle system.

the steam to a liquid-vapor mixture (defined by the steam quality), and heat is rejected to the environment as the mixture is condensed completely back to a liquid state.

**11.2.2.2. Operating Principles: Gas Turbine Cycle.** Gas turbines are increasingly popular energy conversion devices and are used extensively in electricity generation, as well as for propulsion in most passenger and military aircraft. Gas turbines account for about 10% of all utility scale power generation in the USA. The simplest gas turbines (Brayton cycle) consist of the following components (Figure 11.2).

- *Compressor.* The energy, $W_c$, is used to compress inlet air to a high pressure.
- *Combustor.* Heat, $Q_{in}$, is added to raise the temperature and energy content of air, which is already at a high temperature and pressure.
- *Turbine.* It is used to extract work, $W_t$, in the form of shaft work and is used to power both the compressor and an 'electrical generator. The turbine extracts kinetic energy from the high pressure and temperature gas. The air is exhausted to the environment.
- *Generator.* It converts the output shaft work into electricity.

**11.2.2.3. Thermal-Power-Plant Efficiency.** The net work ($W_{net}$) output from the two heat engines described above is equal to the turbine work ($W_t$) minus the pump (or compressor) work ($W_p$):

$$W_{net} = W_t - W_p \qquad (11.1)$$

and the efficiency ($\eta_t$) is given by:

$$\eta_t = \frac{W_{net}}{Q_{in}} \qquad (11.2)$$

where $Q_{in}$ is the heat input.

The maximum amount of work ($W_{net}$) available (based upon the first law of thermodynamics) is:

$$W_{max} = Q_{in} - Q_{out} \qquad (11.3)$$

where $Q_{out}$ is the waste-heat output.

This results in a maximum thermal efficiency ($\eta_{max}$), which is expressed by:

$$\eta_{max} = \frac{Q_{in} - Q_{out}}{Q_{in}} = 1 - \frac{Q_{out}}{Q_{in}} = 1 - \frac{T_c}{T_h} \qquad (11.4)$$

The final expression in terms of temperatures—the temperature of the heat source (the boiler), $T_h$, and the temperature of the heat sink (the condenser), $T_c$, is made on the assumption of reversible heat transfer and represents the maximum (Carnot) efficiency for a heat engine operating between two temperatures. It reveals that the greater the temperature of heat addition ($T_h$ in the boiler) or the lower the temperature of heat rejection ($T_c$ in the condenser), the greater is the efficiency. In a steam plant, the working fluid is water. This ideal cycle can be modified in order to improve overall performance. Superheating the vapor beyond saturation, for example, raises the temperature at which heat is added and increases the efficiency. Reheating is another process that can increase the vapor-power cycle efficiency. Another approach is to utilize several turbines and reheat the working fluid in the boiler between neighboring turbines.

The key action in the steam cycle is the extraction of enthalpy from the high-pressure and high-temperature steam. The turbine inlet consists of nozzles and diaphragms to direct the steam flow into a high-speed jet, driven by the expansion of the steam from the inlet to the exhaust pressure. The turbine blades transfer kinetic energy to the shaft of the turbine. The rotational turbine energy is used to drive an electrical generator. In a gas turbine, the working fluid is a mixture of air, fuel, and combustion products.

### 11.2.2.4. Combustion and Environmental Concerns.

Fossil-fuel combustion is the predominant method of energy conversion in the world ($\sim$ 85%). One motivation for moving toward renewable energy technologies and fuel cells is the potential for mitigation of the negative environmental and health effects of traditional combustion technologies. Two major environmental concerns associated with fossil fuel combustion are the emission of carbon dioxide and of local air pollutants.

(a)  *$CO_2$ emissions.* Carbon dioxide, $CO_2$, is a harmful pollutant because of its potentially significant contribution to global warming and disruptive climate change. Anthropogenic $CO_2$ emission is currently around 22 Gt of $CO_2$ per year. Such a large flux into the atmosphere disrupts the natural carbon balance and increases the atmospheric concentration—currently about 380 ppm, compared to a pre-industrial concentration of 280 ppm. $CO_2$, like other greenhouse gases, is an effective absorber in the infrared (especially in the 13–18 µm range) and traps heat in the atmosphere. The Intergovernmental Panel of Climate Change (IPCC) estimates that a doubling of $CO_2$ concentration over pre-industrial levels could lead to a warming of 2–5 °C by the end of this century. Fossil fuel combustion is the most common source of greenhouse gases (accounting for 98% of $CO_2$ emissions and 82% of greenhouse gases based upon warming potential). An important strategy in reducing $CO_2$ emissions is the decarbonization of energy conversion. Figure 11.3 demonstrates the effects of lowering the carbon content (kg-$CO_2$/kWh generated) of fuels.

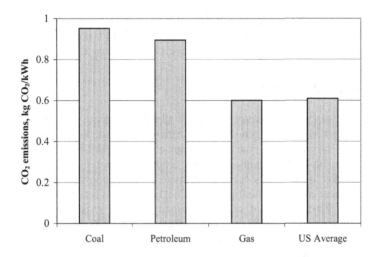

**Figure 11.3** Relative carbon intensity of electricity generation.

Even though coal is the major fuel source for electricity generation in the USA, the $CO_2$ emission level is about 0.6 kg/kWh (similar to that of natural gas), because of the additional use of lower and zero-carbon energy sources (natural gas, nuclear, and hydroelectric). The need to reduce carbon-emission is beginning to be recognized and could affect the usage of high carbon fuels such as coal and petroleum. Another means for reducing $CO_2$ emissions besides changing carbon intensity is to reduce the energy intensity, i.e., the amount of energy it takes to produce a given unit of GDP. Electric utility power generation accounts for 41% of total USA $CO_2$ emissions (2.2 Gt $CO_2$/y for electricity generation out of a total release of 5.4 Gt $CO_2$/y).

(b)  *Local and regional air pollutants.* Combustion of fossil fuels produces other pollutants that can lead to environmental problems such as respiratory disease, urban air pollution, and acidification of lakes and rivers. These effects are more localized than the problem of carbon emissions. Oxides of nitrogen ($NO_x$) are commonly released when burning fossil fuels because the high flame temperatures oxidize nitrogen from the atmosphere. $NO_x$ leads to ozone pollution and photochemical smog. Sulfur dioxide ($SO_2$) is another fairly common pollutant that is released during energy conversion. It is not as ubiquitous as $NO_x$ because sulfur needs to be present in the fuel to produce $SO_2$. Sulfur is present in coal (the commonly used fuel) and also in diesel fuel. $SO_2$ reacts with water in the atmosphere to produce acid rain, which can be devastating to biological systems (forests and lakes). It can also form sulfate aerosols, which affect visibility. Other pollutants will be dealt with in more detail when connected to specific energy conversion technologies.

## 11.2.3. Coal-Based Steam Power Cycles

11.2.3.1. *Operating Principles: System Configurations.* The previous Section summarizes the major operating characteristics of fossil fuel based steam power plants. The most common type of steam power plant in the USA uses coal or other solid fuel (~ 70%). The main differences among the several types of steam power plants are due to the type of fuel and equipment used to convert the chemical energy of fuel to heat energy for boiling water. Coal combustion is accomplished by several methods. First, the solid coal is crushed to form small particles. In pulverized coal burners, the coal is ground into particles about 40 μm in size. This powder is entrained by a stream of air into the combustion chamber. The powder is ignited and the combustion reaction is sustained by volatilization of the coal and subsequent burning. In the boiler, steam is created by pumping water through tubes that are heated mostly by radiation from burning particles. Current steam generators can achieve steam pressures and temperatures up to 300 atm and 600 °C. Another method for coal or other solid fuel combustion is in fluidized beds. The solid fuel and a bed material contained in a vessel are fluidized when air flows upward at sufficient velocity. Combustion temperatures in fluidized beds are significantly lower than those in conventional boilers.

11.2.3.2. *Thermal Efficiency Considerations.* The efficiency of a steam cycle deviates from the Carnot efficiency and the ideal Rankine cycle efficiency because of turbine and pump inefficiencies and heat and pressure losses. Variations on the simple Rankine cycle can improve thermal efficiency. Reheating will split the turbine expansion into two phases with reheating between the two turbines to raise the average temperature. Some steam from the turbine is used to preheat water entering the steam generator. These modifications are found in modern steam plants to increase efficiency and ensure high turbine steam quality. The average efficiency of pulverized coal power plants is approximately 34%, while modern plants capable of achieving higher steam pressures and temperatures can reach efficiencies of 40–43%.

11.2.3.3. *Emissions and Control.* Traditional coal-fired power plants produce a significant amount of pollution. Coal is the dirtiest and most carbon intensive of all common fuels. As discussed in Section 11.2.2.4., some of the most important pollutants are $CO_2$, $SO_2$, and $NO_x$. Coal has the highest carbon content per unit energy of all fossil fuels and is one of the most significant sources for anthropogenic greenhouse gas emissions. Beyond its potential contribution to significant climate change, coal has much more acutely harmful pollutants with well-documented and quantified impacts—i.e., $SO_2$ and particulates. $SO_2$ has a high solubility and is easily absorbed by the human respiratory tract, leading to constriction of airways and worsening of asthma symptoms at relatively low concentrations. The effects of fine sulfate aerosols and other particulates are strongly dependent on particle size; they also lead to respiratory problems and increased mortality. $NO_x$ emissions also contribute to tropospheric ozone formation and can increase respiratory problems.

The main benefits of fluidized bed-combustion-steam generators are that both $NO_x$ and $SO_2$ raw emissions are significantly lower, reducing the requirements for post-combustion exhaust cleanup. Limestone ($CaCO_3$), added to the fluidized bed, reacts with sulfur to form particulate $CaSO_4$, which can then be removed by particulate filters. $NO_x$ emission is reduced by lower combustion temperatures.

### 11.2.3.4. *System Considerations.*

The focus of this Section is on coal-fed steam plants, but steam generators can also operate on oil and gas. Nuclear power-steam plants are discussed in Section 11.2.5. The average sized steam plant in the USA generates 300 MW of electricity, with a range from 50 to over 1100 MW. Small turbines are found in combined cycle applications where a steam cycle is used as a bottoming cycle for lower temperature heat. Another coal-based power generation system is the integrated gasifier combined cycle (IGCC) system. This system has significantly reduced emissions and has high efficiencies for energy conversion. In this system, solid coal is converted into a clean gaseous fuel, a syngas composed mainly of CO and $H_2$. The gasifier breaks down the coal and partially oxidizes it in air or pure $O_2$. The sulfur and other impurities can be removed from the gas stream before combustion. The syngas is combusted to power a gas turbine cycle, which is then followed by a heat recovery steam generator (HRSG) that drives a steam cycle. Details of the gas turbine-combined cycle system are presented in the next Section. The IGCC system can achieve efficiencies around 45%, and projections of efficiencies for advanced systems are over 50%.

## 11.2.4. Natural-Gas Turbine Power Systems

### 11.2.4.1. *General Comments on Gas Turbine Systems.*

The basics of the Brayton cycle were described in Section 11.2.2.2. According to the US Department of Energy, natural gas turbines are expected to make up more than 80% of the new power generating capacity in the US over the next decade. Major reasons for this shift from traditional steam plants to gas turbine plants are the availability of natural gas, low installed capital cost, high efficiency, fast start-up, and good load following characteristics.

### 11.2.4.2. *System Configurations for Turbine Systems.*

Gas turbines are frequently found in both simple and combined cycle configurations. The simple cycle does not use the rejected exhaust heat from the turbine. These systems are inexpensive, simple to install, and typically used for reserve or peak power requirements (usually less that 2000 hr/yr operation). Combined cycle systems use the high-temperature exhaust heat of the gas turbine to drive a heat recovery steam generator (HRSG), which is the boiler in a steam cycle. The coupling of these two power cycles increases plant efficiency. These systems are more complex to design and install and are typically used for baseload (continuous operation) power. Gas turbine sizes range from less than 1 MW up to about 300 MW.

**11.2.4.3.** *Efficiency Considerations.* As described in Section 11.2.2.3, the Carnot efficiency of a heat engine is determined by the ratio of inlet and exhaust temperatures. In a gas turbine, these temperatures are related to the pressures before and after the turbine. Defining a compressor ratio, $r_c$, as the ratio of pressures that the compressor and turbine operate between, the isentropic efficiency of the simple Brayton cycle is given by:

$$\eta_{max} = 1 - \frac{T_c}{T_h} = 1 - \frac{1}{r_c^{(\gamma-1)/\gamma}} \qquad (11.5)$$

where $\gamma$ is the ratio of specific heats ($C_p/C_v$) for the working gas (air).

In a real situation with non-isentropic compressor and turbine efficiencies, the thermal efficiency will not increase indefinitely; instead, it reaches a maximum and then declines with further increases in the compressor ratio. The theoretical efficiency will also increase with the turbine inlet temperature. However, this temperature cannot increase indefinitely because of temperature limitations of the turbine materials. Research on advanced turbine materials and blade cooling raise prospects of working at higher temperatures and achieving higher efficiency.

Operation of a gas turbine with some modifications can increase cycle efficiency or power output. These include:

- reducing compressor work: cooling the inlet gas or having several compressor stages and cooling in between results in a denser, more easily compressed gas and a higher mass flow;
- recycling waste-heat: recuperating the turbines preheat-compressed gas to the combustor via a counterflow heat-exchanger coupled to the exhaust gases;
- increasing average temperature: including multiple turbines and reheating between turbines can increase the average temperature without increasing the turbine inlet temperature; and
- increasing mass-flow: injecting steam can lower combustion temperatures and increase the mass flow for an increased power output.

These cycle variations can be combined into various configurations depending upon the specific application and economics involved with each individual plant. A simple gas turbine cycle exhausts at high temperature, thereby wasting useful heat and limiting efficiency to around 27–30%. Utilizing this heat to drive the bottoming steam cycle in a combined cycle system can increase efficiencies to above 50%.

11.2.4.4. *Emissions and Control.* The primary pollutant from natural gas-fired turbine power plants is $NO_x$. Since most natural-gas resources are low in nitrogen content, this $NO_x$ formation occurs in the high temperature reaction between oxygen atoms and atmospheric nitrogen in the combustion zone. Since $NO_x$ formation is strongly temperature-dependent, the primary control for $NO_x$ formation is achieved by reducing temperatures by steam or water injection, lean combustion operation, or multi-stage combustors. Selective catalytic reduction (SCR) is used to lower post-combustion $NO_x$ emissions by combining vaporized ammonia to the exhaust gases and passing over a catalyst bed. Other pollutants such as carbon monoxide (CO) and other incomplete combustion products can be catalytically oxidized. $SO_2$ emissions are generally low because natural gas does not contain much sulfur. As shown in Figure 11.3, natural gas, which is mainly methane ($CH_4$), has a lower carbon content per unit heat energy than coal, thus yielding a lower $CO_2$ emission than electricity derived from coal. However, pipeline and other natural gas leaks can have significant effects on atmospheric radiative forcing since methane is a very powerful greenhouse gas with a higher global warming potential than $CO_2$ (~21x stronger than $CO_2$). Methane can also be involved in the formation of ozone and photochemical smog.

11.2.4.5. *Economic and System Considerations.* Overall annual electricity cost will depend on capital equipment cost, fuel costs, operations and maintenance costs, and the number of hours of operation. Given the higher fuel costs of natural gas vs. coal and the higher capital cost of steam cycles vs. gas turbines, natural gas turbines will be more economical for load following while steam power systems will be more economical for baseload power systems. Due to combined cycle plants having higher capital costs, but increased efficiencies (and thus lower fuel costs), the same result applies. In addition, natural gas is subject to greater fluctuations in price than other fuels.

## 11.2.5. Nuclear Based Thermal Power Plants

11.2.5.1. *General Comments on Nuclear-Power Systems.* Nuclear power was once touted as a future source of limitless and near-zero cost of electricity. Problems with long reactor construction times, budget overruns, higher operating costs, public perceptions of safety, and nuclear waste storage have retarded nuclear power development. Even so, nuclear power accounts for close to 20% of electricity generation capacity worldwide and nuclear power is gaining renewed attention as a reliable and well-established technology for providing $CO_2$-free electricity.

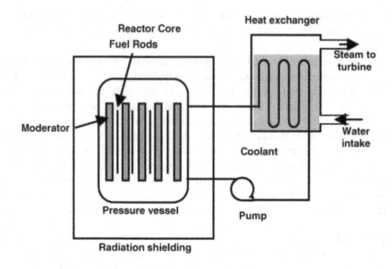

**Figure 11.4** Schematic of a pressurized water reactor.

**11.2.5.2. *Operating Principles and System Configurations.*** Most nuclear power plants operate on the steam cycle described in Section 11.2.2.1, although new designs are being developed which operate on a gas turbine cycle. The heat for the boiler is provided by fission, the splitting of a heavy atom into multiple fission products and heat. The majority of nuclear power plants use enriched $^{235}_{92}U$ as the fissionable material for heat and steam generation rather than the combustion of a fossil fuel (coal or natural gas). Isotopes of uranium (such as $^{235}_{92}U$ ) and plutonium are able to undergo a fission reaction when bombarded by neutrons. The following is the average result of the many possible fission reactions of uranium 235:

$$1\,n + {}^{235}_{92}U \rightarrow F_1 + F_2 + 2.47\,n + 203\,\text{MeV} \tag{11.6}$$

Any given fission reaction will produce an integral number of neutrons and a given amount of kinetic energy of the product fragments, $F_1$ and $F_2$. This kinetic energy, quickly dissipated by interactions with the surroundings, produces heat (203 MeV) and neutrons (an average number of 2.47n per $^{235}_{92}U$ atom). A chain reaction can be sustained if more than one of the released neutrons is absorbed by another fissionable material. The likelihood of a neutron being absorbed is increased by using moderators (such as water) to slow the speed of the neutrons. Currently, there are four major configurations of nuclear power plants that are commonly used:

(a) *Pressurized water reactor (PWR)*. Figure 11.4 shows a PWR. This is the most common type of reactor (57% of reactors worldwide). Water is used as both the moderator and coolant. The coolant water is in a separate loop from the working fluid of the steam cycle to prevent any radiation-contamination of the steam cycle equipment. Commercial PWR power plants range from 600–1200 MW and can contain over 50,000 uranium fuel rods in the core.

(b) *Boiling water reactor (BWR)*. This is the simplest type of nuclear reactor. BWRs account for around 21% of reactors worldwide. Instead of having two separate loops—one for the coolant water and another for the turbine water—they are combined to one loop. This is essentially a steam plant with a nuclear reactor-based steam generator. BWRs require a larger pressure vessel volume and radioactivity leaking from the fuel rods to the coolant water can be transferred to the turbines. Installed plant sizes range from 600–1400 MWe.

(c) *Gas cooled reactor (GCR)*. It uses $CO_2$ rather than water as the coolant. A variant, called the *high-temperature-gas cooled reactor* (HTGCR) operates at a fairly high temperature, which allows for superheating and reheating modifications of the basic steam cycle. Advanced designs of each of these types of reactors improve reactor safety and power density, simplify the plant design, and reduce construction costs. Additionally, there are new designs that operate with gas turbines rather than steam turbines.

(d) *Pebble-bed reactor*. This advanced design uses small tennis ball sized fuel pebbles instead of fuel rods. Helium gas is used as a coolant and is expanded through a gas turbine, which is able to operate at much higher temperatures, increasing efficiency.

*11.2.5.3. Efficiency Considerations.* Most nuclear plants operate on the Rankine cycle and are thus akin to fossil fuel steam power plants. The heat addition temperature, in part, determines the thermal efficiency of the cycle, and the efficiencies of PWR and BWR nuclear power plants are limited because of the low steam generation temperatures (~ 300 °C) needed to maintain a moderate reactor coolant pressure (70–200 atm). The thermal efficiency is thus around 32%. Another aspect of efficiency is fuel burn up, which indicates the extent to which heat is extracted from the uranium. The fissionable products in a fuel rod are not completely reacted before the fuel is replaced because of problems with fuel damage, production of gaseous fission products, and corrosion problems. As a result, some heat energy is left unutilized and lowers the ultimate fuel energy conversion efficiency.

*11.2.5.4. Environmental and Public Concerns.* There are significant environmental and safety concerns surrounding the use of nuclear power to generate electricity. Its proponents argue that it is a well-established carbon and air pollution free means of generating electricity. Nuclear power accounts for the largest non-fossil fuel based energy conversion in the world. With the large uranium reserves in the world and the possibility of developing fast breeder reactors, nuclear power is attracting more proponents, especially because of the increasing realization that $CO_2$ emissions need to be curtailed, and the environmental and health impacts of air pollution. Its opponents cite the main drawbacks of nuclear power—reactor safety and nuclear waste. Accidents at Three Mile Island in the USA and Chernobyl in Ukraine highlight the dangers of catastrophic accidents in a large nuclear power plant, including core meltdown and a large release of radiation. In the Chernobyl accident, 31 people died at the plant and several thousand deaths from cancer are attributed to the radiation release. Spent fuel from power plants is highly radioactive and needs to be quarantined from human contact for several thousand years. Finding a suitable storage facility has not been without controversy—both scientific and political. With respect to nuclear weapons proliferation, only a small quantity of fissionable material (~ 10 kg) is required to build a nuclear weapon and most power plants produce several hundred kilograms per year.

## 11.2.6. Hydroelectric Power Plants

11.2.6.1. *Operating Principles.* Hydropower is a clean and abundant energy source that relies on the hydrologic cycle. Water evaporates from the oceans and lakes and returns as precipitation. Some of this precipitation flows through rivers, lakes and reservoirs and the energy from this flowing water can be controlled and harnessed to provide mechanical energy and ultimately electricity. Hydropower was first developed during the Roman Empire by channeling water onto a water wheel to generate mechanical power for irrigation, milling, and other operations. In the middle of the 19th century, the hydraulic turbine replaced the water wheel. Hydroelectric power plants contribute about 20% of total worldwide electricity generation.

Large-scale hydroelectric power plants are typically sited where flowing water can be stored in reservoirs and controlled. The potential for useful and economical energy conversion depends on location, landscape and topography, and stream flow characteristics. The stream characteristics in turn, depend critically on the rainfall pattern and the watershed area (the area of land that contributes to flow in a given stream or river). The essential components of a utility-scale hydroelectric power plant are (Figure 11.5):

- a *reservoir/dam* that stores the water and blocks the downstream end of the reservoir;
- a *spillway* (or flood discharge structure) which controls the water level and the flow of water;
- *tunnels and penstock*, which transport the water to the turbines;
- a *turbine and generator* to convert the potential energy of the stored water into shaft work and finally electricity; and

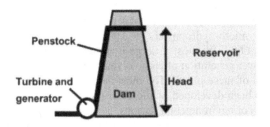

**Figure 11.5** Essential components of a utility-scale hydroelectric power plant.

- a *surge tank*, located between the tunnel and penstock or between hydraulic machine and tail water tunnel to dampen fluctuations in water pressure and level.

Hydraulic turbines are designed to use both the potential and kinetic energies of the water to convert it to mechanical energy.

**11.2.6.2. Performance Characteristics.** Hydroelectric power generation involves the conversion of the potential energy of water, with a mass, m, and height, h, to kinetic energy and then electric energy. Thus, assuming a 100% energy conversion and mass flow rate $\dot{m}$, the electrical power generated ($P$) is given by:

$$P = \dot{m}\,gh \qquad (11.7)$$

where $g$ is the gravitational constant.

Including the turbine ($\eta_T$) and generator ($\eta_G$) efficiencies, the power is:

$$P = \eta_T \eta_G\,\dot{m}\,gh \qquad (11.8)$$

The installed capacity is the designed maximum power from the turbines. In principle, there should be very few efficiency losses during power generation because turbine and generator efficiencies can be quite high. Large hydroelectric plants can have total efficiencies up to 90% but for small hydropower plants efficiencies could vary from 40–70%. Since precipitation patterns can be intermittent, the presence of a reservoir can mitigate some of the seasonal fluctuations in stream flow and allow continuous power generation.

11.2.6.3. *System Aspects and Economics*. Hydroelectric plants have been built in many sizes, from large hydro (> 100 MW) to small hydro (1–30 MW), mini- (100–1000 kW) and micro-hydropower (< 100kW). In 1997, the worldwide installed capacity was 660 GW, most in large hydropower plants (> 96%). Large-scale hydroelectric power generation often involves high transmission costs because of the remote location of these plants. The most attractive sites for hydropower in the USA have already been developed, while other potential sites are less attractive because of higher costs or environmental reasons.

An interesting variation on hydroelectric power, i.e., pumped storage, (which accounts for about 14% of all US hydropower) uses two reservoirs at different heights to store water during off-peak times and generate hydropower during peaking hours, when electricity generation is most expensive. In this system, the cheap, excess electricity is used to pump water from the lower reservoir into the upper reservoir, and this stored water can be used to drive the turbines when the electricity is needed. It appears unlikely that pumped storage will increase significantly because of the lack of suitable sites near large power plants.

The construction of hydroelectric power plants is capital intensive—capital costs vary widely, but generally are between \$1000 and \$3000/kW. However, since there are no fuel costs, the cost of electricity can be relatively low, i.e., \$0.02 to 0.08/kWh from large plants and \$0.03 to 0.10/kWh from smaller plants. Though this is a mature technology with a long history of implementation, advances can still improve dam structures and materials, turbines, generators, sub-stations, transmission lines, and environmental mitigation technologies, reducing costs and environmental impacts.

11.2.6.4. *Environmental Benefits and Concerns*. Hydroelectric power generation is apparently attractive because it is a clean, efficient, non-polluting, renewable energy system. In fact, it is the most advanced and mature renewable energy technology. Once constructed, these systems release almost no pollutants and require no fuel on resource use. Dams and the resulting reservoirs have other benefits including flood protection, stream-flow control, and recreation opportunities. However, there is an increasing awareness of a host of ecological and environmental consequences of building dams and creating large reservoirs. These major impacts may be summarized as follows:

- *Dam construction impacts.* Building large-scale hydropower plants involves construction of large dams, excavating underground channels, and the construction of large structures to house generators.
- *Ecosystems.* Large scale changes in water conditions like temperature and salinity can affect aquatic biodiversity, disrupt spawning runs, flood large land areas, and reduce the sediment load of a river.
- *Human displacement.* The construction of the Three Gorges Dam is expected to displace several million families and flood whole cities. The estimated cost is \$25,000/family.

## 11.2.7. Photovoltaic Power Generation

11.2.7.1. *Background and Operating Principles.* Currently, photovoltaic cells account for a tiny fraction (about 0.03%) of electricity generation in the USA, but solar panel manufacturing increased by about 20% annually during the 1990s. Worldwide annual production is approximately 400 MW/year and cumulative capacity is approaching 2 GW. Like the fuel cell, the photovoltaic or solar cell is a direct energy converter; in the latter case, visible light is converted directly into electricity.

Photovoltaic energy conversion occurs by exposing various configurations of a diode p-n junction to light; a semiconductor with a positively doped p-layer contains mobile positive charges or holes in contact with a negatively doped n-layer, which contains mobile electrons. A diode electric potential is attained when the electrons of the n-type move to fill the holes of the p-type and the p-type region becomes negative and the n-type region positive. When a photon with sufficient energy collides with an electron, it creates a free electron and a hole. When this process occurs near a p-n junction, the field will push the hole to the p-type side and the electron to the n-type side. If the two regions are connected through an external load, a current flows though it, allowing the electrons to reunite with the holes. This current flow across the cell electric field potential generates electrical power.

A typical device is made by diffusing a p-type material (e.g., B) into a heavily doped n-type (e.g., As into Si) wafer to form a p-film layer, a few $\mu$m thick. Electrical contacts to the p and n types are made by electroplating a metal (e.g., gold or copper) on each region. To achieve the desired DC power levels, photovoltaic cells can be connected in series and parallel; AC power can then be generated by using an inverter.

There are two types of solar cells–crystalline and amorphous. The former is made from silicon wafers, sliced from rods or blocks, and can either be monocrystalline or polycrystalline. Amorphous or thin film cells are made by depositing a thin layer ($\sim$ 1$\mu$m) of silicon on a substrate (glass or plastic). Silicon solar cells make up 95% of solar cells and are the most developed and commercialized types of photovoltaic cells. Other types of solar cells being developed use doped semiconductor materials like GaAs, CdTe, and CuInSe$_2$. Organic semiconductor cells are in the research stage and dye-sensitized TiO$_2$ cells are gaining momentum.

11.2.7.2. *Performance Characteristics.* Theoretically, with monochromatic light as the energy source, photovoltaic cells can achieve 100% efficiency. However, due to the radiation spectrum of sunlight, photons with less than the band gap energy cannot create electron-hole pairs, but instead create heat. Photons having energy greater than the band energy can generate electrons and holes; but because these photons will have more energy than the mean free energy of the carriers, only the energy corresponding to the band gap generates electricity while the excess energy is dissipated as heat. The theoretical maximum efficiency for crystalline Si cells exposed to sunlight is approximately 28%. Other inefficiencies are due to optical shading and cell and electrical contact resistance. Monocrystalline silicon can achieve about 24% efficiency in the laboratory while manufactured solar modules achieve around 15% efficiency. Polycrystalline Si solar cells can also achieve an efficiency of about 14% in real world manufactured solar modules. Amorphous silicon solar cells have much lower efficiencies—about 13% in the laboratory and 6% in manufactured cells.

   11.2.7.3. *Systems Aspects and Economics.*   Solar insulation can exceed 1 kW/m$^2$ on a flat surface on a clear, sunny day. However, the insulation will vary throughout the day with the sun's position, as well as the angle of incidence. Under the best circumstances, it is possible to average over 300 W/m$^2$ over the course of the day, which, given a 15% efficient photovoltaic system, translates to an average power output of 45 watts per m$^2$. It is therefore necessary to cover large areas with solar cells to generate significant power levels. Tracking systems can allow the panels to follow the sun and consequently capture slightly more incident radiation. Because photovoltaic cells only operate when sunlight is available, providing electricity during the night or during cloudy days requires another energy supply, as illustrated in Figure 11.6. In a small-scale hybrid system, excess electrical energy generated by the photovoltaic system is stored by charging a secondary battery. In grid-connected systems, the grid itself can be used as a backup. Excess solar electricity generation can be fed to the grid (via an inverter) and when demand exceeds generation, electricity can be obtained from the grid. In the USA, most states have some type of net or dual metering laws for residential customers to

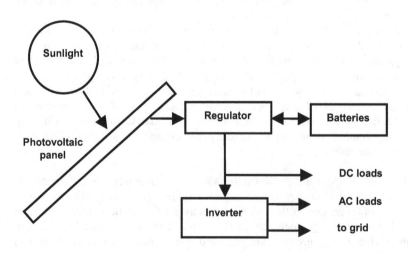

**Figure 11.6** Schematic of photovoltaic solar-power system.

facilitate the two-way flow. Despite the intermittency of solar energy, one benefit is that electricity is produced precisely when it is needed—air-conditioning peak demand corresponds with hot, sunny, summer afternoons.

Typical crystalline solar cell module costs are about $3.50/watt. The capital cost for the entire system depends on the type of photovoltaic system used. For example, costs can increase significantly if battery energy storage is added. The cost of electricity for renewable energy systems depends strongly on the capital cost of the energy conversion device, since there are no fuel costs and very low operation and maintenance costs. Thus, electricity costs for solar are currently quite high ($0.20–0.50/kWh) because of the high capital cost of the photovoltaic and energy storage systems and the low capacity factor (< 25%).

### 11.2.7.4. *Environmental Benefits and Concerns.*

Along with wind energy, photovoltaic systems are one of the most environmentally acceptable energy sources. It is a renewable energy technology that generates electricity without the use of fuel or other resources, and without emissions, such as $CO_2$. There are some fossil energy uses (amounting to as much as 10% of total expected energy generation) and pollutant emissions ($CO_2$, lead, CFCs, and chlorinated products) during the production of solar cells. One of the greatest disadvantages of photovoltaic systems is the large area of land required for electric power generation. Solar energy is a diffuse energy resource: 3 to 10 $km^2$ is required for a 100 MW plant, with the total amount of electricity generated per year being 180 GWh/y compared to an equivalent annual energy production from a 30 MW thermal power plant and a land area of about 0.01 $km^2$. However, incorporation of solar photovoltaic systems into rooftops and other building materials can essentially eliminate the land requirements and provide distributed generation at the point of demand.

## 11.2.8. Solar-Thermal Systems

### 11.2.8.1. *Background and Operating Principles.*

A solar-thermal power plant uses solar radiation to produce high temperature heat that drives a heat engine cycle (as described earlier in this Section) to produce electrical energy. Solar thermal power plants use reflectors and collectors to concentrate the sunlight onto a small area, which significantly enhances the temperature. The heat is collected by a working fluid and is coupled to a heat exchanger to drive a vapor power cycle. Solar thermal power plants have been confined to desert areas, with a high fraction of sunny, cloudless days. The installed, worldwide capacity is small, approximately 400 MW, and the amount of electric energy generated annually is about 1 TWh. Solar-thermal systems can also produce hot water and steam for industrial applications.

Several configurations of the thermal power plant have been developed. In each of these systems, there are four basic sub-systems: collector, receiver, transport and storage, and power conversion (Figure 11.7). The collectors or concentrators can

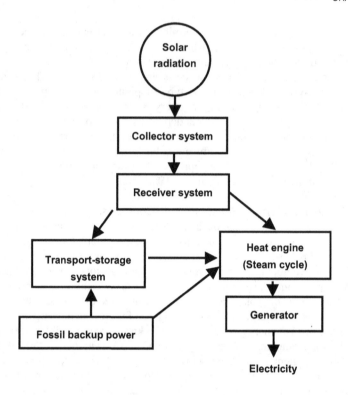

**Figure 11.7** Schematic of solar-thermal power plant.

take sunlight from a larger area to focus it onto the receiver. The receiver is the system where the solar radiation is converted into heat and carried away by the working fluid, the heat transport and storage system. Finally, the power conversion system is a conventional Rankine vapor power cycle. The different configurations which have been developed are:

- *Parabolic-trough system.* This system uses a parabolic mirror with a receiver pipe at the focus of the parabola and can reach temperatures up to 350 °C.
- *Parabolic-dish system.* It uses a tracking dish reflector to concentrate the sunlight onto the receiver, again at the focal point, and can reach temperatures up to 1000 °C.
- *Central-receiver system.* This system uses sun tracking mirrors (heliostats) to reflect solar energy to a receiver at the top of a tower. Because of the large collector area, the central receiver configuration can generate temperatures up to 1500 °C.

In each of these systems, the working fluid is transported to the power conversion system to generate electricity. Often, the solar heat is backed up or supplemented by a fossil fuel boiler.

**11.2.8.2. *Performance Characteristics.*** The conversion efficiency from solar energy to electricity may be expressed by the equations:

$$\eta = \frac{Net\ Electrical\ Output}{Solar\ Input\ to\ Collector} \tag{11.9}$$

$$\eta = \eta_c \eta_r \eta_t \eta_e \tag{11.10}$$

where $\eta_c$ is the efficiency of the collector system that depends on its reflectivity; $\eta_r$ is the efficiency of the receiver, the ratio of the energy absorbed by the working fluid to the energy incident on the receiver; $\eta_t$ is the efficiency of the transport system; and $\eta_e$ is the steam-cycle electrical-conversion efficiency determined by the temperature of the working fluid. Another important aspect is the capacity factor, which is typically around 20% for solar thermal plants. Solar insulation is relatively diffuse. Even with concentrating reflectors, the overall efficiencies of solar-thermal electrical power plants are about 15%, which is considerably less than the efficiencies of conventional thermal plants using fossil fuels.

**11.2.8.3. *Systems Aspects, Applications, and Economics.*** The Solar Two power plant, operated by Southern California Edison in the USA, can supply, on demand, power generation because thermal energy storage in a molten salt enables it to provide continuous power. The rated capacity of this plant is 10 MW. Solar thermal systems can also be coupled with conventional thermal boilers using fossil fuels to provide a continuous supply of electricity. Projections from the World Energy Assessment are that total solar thermal power generation will increase to 2 GW by 2020.

Projections for the cost of trough-based power plants are $3000–3500/kW, while the estimated cost for central receiver systems is in the range $4700–5000/kW. It is projected that in the long term these costs will decrease by 30 to 50%. The levelized energy cost is expected to be $0.14 to $0.18/kWh in the near term and could be reduced to $0.04 to 0.06/kWh in the long term, although this cost requires a credit of $25–40/ton for reduced $CO_2$ emission.

**11.2.8.4. *Environmental Benefits and Concerns.*** Similar to wind and photovoltaic solar cells, solar thermal power plants do not generate any pollutants during electricity generation. They rely on solar radiation that does not require any resource extraction. However, the land requirements can be extensive; a 20-MW plant will use about 1 km$^2$ of land area. Typical locations for installations are in the southwestern USA and the Middle East. Desert environments can be quite fragile, and the installation of large dispersed power plants can impact the ecology of these areas.

## 11.2.9. Wind Energy Systems

*11.2.9.1. Background and Operating Principles.*  Solar radiation is the primary source of atmospheric convection and consequently of wind energy. The uneven heating of the earth's surface leads to pressure differences and these convection currents. Using wind turbines, energy in the wind can be converted to rotating mechanical energy and coupled to generators, where it is converted to electrical energy. The exploitation of wind energy to do mechanical work, such as turning millstones or pumping water, dates back well over 2000 years. Significant development on modern wind turbines around 1980 has led to rapid growth and advances in wind energy systems. Currently, wind accounts for about 0.2% of electricity generation in the US, and there is an installed capacity of about 2 GW, most of it in California. Significant cost reductions and technology improvements in the last decade have spurred development of wind energy "farms" globally, and a significant growth rate of installed capacity (30–40% annually over the last few years in Europe and the United States).

Wind turbines extract power from the wind by taking advantage of aerodynamic lift as the wind passes over the wing-like turbine blades. Higher wind velocities on the curved side of the blade will lead to a lower pressure, and the pressure differential generates the necessary force, perpendicular to the flow, to turn the rotor. The power density (W/m$^2$) of the rotor area of a wind stream, $P$, is related to the cube of the wind velocity:

$$P = \frac{\rho v^3}{2}$$
(11.11)

where $\rho$ is the air density and v the velocity of the wind. The theoretical maximum power, $P_{max}$, which can be extracted from the wind is only about 60% of the total wind power density:

$$P_{max} = 0.592P$$
(11.12)

It is impossible to extract all the energy out of the wind, since the air leaving the turbine must have some velocity (i.e., kinetic energy).

As illustrated in Figure 11.8, a wind energy system is composed of the following components:

- *rotor/blades,* which extract kinetic energy from the wind to turn the turbine shaft;
- *generator* to convert the shaft work to electricity (AC or DC);

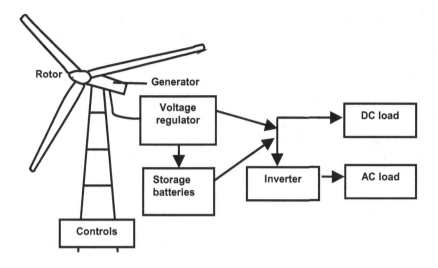

**Figure 11.8.** Schematic of a small hybrid wind energy system.

- *tower* to support the blades and generator at a suitable elevation; and
- *power converter/controller* to ensure the acceptable electrical output.

Advanced systems can also have a *yaw controller* to maintain optimal orientation of the rotor into the wind. There are two types of wind turbines: horizontal-axis and vertical-axis systems. The vertical axis wind turbine (such as the Darieus rotor) has the advantage that the drive train, generator, and controls are at the ground level for easier design, installation, operation, and maintenance. However, wind velocities increase with the height of the wind tower and higher power is more easily achieved with the horizontal-axis systems.

**11.2.9.2. *Performance Characteristics.*** The energy output of a turbine depends on the wind speed distribution and surrounding terrain. Because of the cubic relationship between power and wind speed, it is difficult to estimate the average power expected from a wind turbine using the average wind speed; knowledge of the details of wind speed distribution is necessary for this estimation. Wind resources are categorized by the average wind speed; a class 6 or 7 resource is a very windy location and has average wind speeds of 18–27 mph, which yields a wind power density of 0.5–2 kW/m$^2$ at a height of 50 m. On average, it is possible to capture only about 20–30% of the rated capacity (capacity factor), because of the intermittency of wind resources. The efficiency of a turbine is also optimized for a specific design wind speed; peak efficiency can reach 40–50%. The efficiency decreases with deviations from this design speed. High wind speeds can cause significant damage to the rotor, turbine, and generator; under such conditions, the excess energy must be dissipated by braking or by changing blade/rotor orientation. This "cut-out" wind speed is typically around 55 mph.

11.2.9.3. *Systems Aspects and Economics.* Wind turbines are found in many sizes. Some DC wind turbines, with rotor diameters of less than 1 m are capable of electric power generation up to 500 W. Some of the largest wind turbines have rotor diameters of over 70 m and are able to generate over 2 MW. Large wind turbines are typically grid connected and are often combined with transmission systems to maintain constant shaft speed to generate 60 Hz AC electricity. Small wind-energy systems offer several possible configurations depending upon the application:

- *hybrid systems*, in which wind turbines are coupled with another power source such as a generator or energy storage (Figure 11.8); and
- *grid-connected systems* which supply power to a house or cabin but use the grid for storage and backup power.

Since wind energy is typically intermittent, coupling the turbine with an energy storage system or the grid is required for constant and continuous power. The intermittency of wind power is important when considering large-scale wind power generation. Because wind power has a low capacity, increases in wind power generation to a significant fraction of total electricity generation would have to be coupled with breakthroughs in low-cost energy storage or long-distance transmission.

The economics of wind systems depends significantly on the installed cost of the wind turbine, as well as on the nature of the wind resource. Because there is no fuel cost and operational and maintenance costs are fairly low, the price of electricity is mainly determined by the amortized cost of the turbine divided by the annual electricity output. The capital costs will vary somewhat depending upon location and installation costs, while the annual electricity production can vary significantly with location. Typical turbine costs are around $700–1000/kW, excluding costs of other system components or installation. Electricity costs are expected to be below $0.04/kWh in the windiest locations (class 7 resources) and between $0.04–0.07/kWh for more moderate wind resources.

11.2.9.4. *Environmental Benefits and Concerns.* Wind turbines operate very cleanly, requiring no fuel and little maintenance, and their construction does not require any special materials or processes. They produce no pollutant emissions or greenhouse gases, require no fuel or other resources for operation, and use a truly renewable resource. As a result, they are touted as an important and economical means to reduce the environmental impacts of electricity generation. However, the main concerns that are raised have to do with the following:

- aesthetics and visual disruption of scenic landscapes and coastlines;
- sound produced by the wind turbine; and

- effects on bird migration and collisions of birds with the towers or rotor blades.

Research efforts into larger wind turbines yield slower, more silent rotor blades, which produce more power and thus require fewer towers. Other strategies include siting windfarms far offshore to minimize any visual impact, noise, and effects on birds. Given the significant environmental benefits of wind power, technical improvements and further studies should minimize these concerns.

## 11.2.10. Geothermal Energy Systems

*11.2.10.1. Background and Operating Principles.* The molten core of the earth and radioactive decay of materials in the earth give rise to abundant thermal energy, which moves to the earth's surface. This energy is quite diffuse (on average $0.1 \ W/m^2$) but can also be found concentrated in specific locations around the globe near tectonic plate boundaries and hot-spots. In these locations, hot springs have been used for bathing and washing for thousands of years. At the beginning of the $20^{th}$ century, geothermal energy systems were developed for space heating, industrial process heat, and electric power generation. Currently, the geothermal power generating capacity in the USA is 2.8 GW (around 0.3%) and is 8 GW world-wide. Some have estimated that a significantly higher capacity can be developed for electric power generation: 25–50 GW in the USA and over 100 GW in the rest of the world. Steam and hot water reservoirs make up only a small part of the useable geothermal resource. Magma and hot dry rock are also useful sources of thermal energy. The basis for electricity generation is the steam cycle, using the geothermal energy to supply steam (i.e., the boiler). The steam is expanded through a turbine driving an electrical generator. In addition to electricity generation, excess steam or hot water can be used for other purposes, such as process heat and space heat.

*11.2.10.2. Performance Characteristics and System Configurations.* There are three main types of geothermal-steam-power plants, depending upon the source of heat available:

- *Dry-steam plants* use steam piped directly from below the surface to drive a turbine, and the condensed water is returned via an injection well;
- *Flash-steam plants* pump pressurized high temperature water to the surface, where it is injected into a tank in which it quickly vaporizes because of the pressure drop. The energy of its expansion is then captured by the turbine; and
- *Binary-cycle power plants* use lower temperature water (< 200 °C), to vaporize another lower-boiling-point working fluid (in separate loops), which in turn drives a vapor-power cycle. Moderate-temperature geothermal heat is the most common source and binary cycle systems are being built to exploit them. Hot, dry rock is a more available geothermal

source than hot water or steam, and can potentially be exploited by crushing the rock to improve heat transfer and then injecting water into these areas.

Geothermal-electric-power systems use a vapor-power Rankine cycle. For most geothermal systems, steam is delivered at temperatures around 200 °C, and the efficiency is in the range of 5–20%, which is lower than for fossil fuel-fired thermal power plants. The efficiency for the direct use of thermal energy is between 50 and 70%.

**11.2.10.3.** *Systems Aspects and Economics.* The potential sites for geothermal electric power production are limited to specific areas near plate boundaries and volcanic areas. Typically, geothermal power plants have very high capacity factors (~ 95%) because of their simplicity and constant heat supply. One problem is that geothermal reservoirs may be finite and as heat is extracted, the quality of heat declines and the resource may become worthless. The capital cost of geothermal power plants is in the range of $800–3000/kW and the cost of electricity will depend on the quality of the geothermal resource, but some of the best sites can achieve a cost of $0.05 to 0.10/kWh. The viability of cogeneration from geothermal power plants is site-specific, because steam or hot water from geothermal resources cannot be economically transported over long distances.

**11.2.10.4.** *Environment Emissions Concerns.* Geothermal energy conversion systems are another example of a clean, renewable resource. They require no fuel supply and typically require less space to operate than a comparably rated coal power plant. They can reduce greenhouse gas (GHG) emissions significantly, but they are not zero. Dissolved gases in geothermal fluids are mostly $N_2$ and $CO_2$, with some $H_2S$ and smaller amounts of $NH_3$ and radon, which can be released to the atmosphere. Binary cycle systems are typically closed loop systems and the amount of dissolved gas released can be minimized. There are also small amounts of Hg and B found in the water. Most of these impurities are reinjected into drill holes. The $H_2S$ is removed by hydrodesulfurization. The amount of $CO_2$ released during geothermal power plant operation depends on the geothermal fluid properties but can be expected to be about one tenth the amount released from a fossil fuel plant (500–1000 g/kWh).

## 11.2.11. Ocean Energy Systems

**11.2.11.1.** *General Comments on Types of Systems.* Like solar radiation, the ocean is another widely available and distributed energy source. The energy in the ocean is in part kinetic energy in the form of tides, currents, and waves, and part thermal energy from the sun stored as heat. The quantity of energy stored in the ocean is enormous but can be quite diffuse; in some cases, it is about the same as from solar radiation and, as a result, it is difficult to harness. Marine energy systems can be classified into four main groups:

- Tidal-barrage energy
- Wave energy
- Tidal/marine currents
- Ocean thermal-energy conversion.

The following Sections briefly summarize some of the interesting aspects of these technologies.

**11.2.11.2.  *Tidal-Barrage Energy Systems.*** This type of energy system is equivalent to a low head hydrosystem and makes use of the rise and fall of the tides. A dam is built to separate the open water from a reservoir and the changing tides will drive water through the turbines. Electricity can be generated by water flowing either way. Also, the system can be used for pumped storage for cheap off-peak electricity. A 240-MW electric-generation power plant was built in France in the 1960s. There have been plans to design and build GW size plants in the Bay of Fundy in Canada and in the Severn estuary in Great Britain, but the costs are expected to be extremely high. Though such a system does not emit environmental pollutants, the effects on a tidal estuary and on the complex and often fragile ecosystem are poorly understood.

**11.2.11.3.  *Wave Energy Systems.*** Waves are a source of kinetic energy, which can be converted via a generator to electrical energy. Technologies have been developed to exploit wave energy offshore, as well as along the shoreline, but most demonstration units have been built along the shoreline because of the relative simplicity. There are several types of shoreline wave energy units but each relies on the periodic nature of waves. The oscillating water column device uses the vertical motion of the wave to compress an air column and drive a turbine. A tapered channel device uses wave energy to move water into an elevated reservoir, where it can be used to drive a turbine. These systems can be used to drive electrical generators.

**11.2.11.4.  *Tidal Marine Currents.*** Tidal power energy from ocean currents can be converted to electrical energy using large submerged turbines. Since the motion of sea water is rather slow in most places, the amount of power/energy, which can be extracted, is small; this is due to the cubic relationship between velocity and power. Grid connected systems are being investigated. The turbine designs are similar to those for wind turbines. Marine-current turbines should operate near to the surface, where the velocity is highest.

**11.2.11.5.  *Ocean Thermal- Energy Converters (OTEC).*** OTEC systems are heat engines that rely on the temperature difference of the ocean at different depths to drive a Rankine (steam/vapor) cycle. Warm seawater is heat exchanged with a low-boiling temperature fluid such as ammonia. There are several challenges in developing large systems:

- the need to circulate large volumes of sea water through pipes,
- the small temperature change between source and sink, limiting the theoretical efficiency for thermal to electric energy conversion which at best is of the order of 7% in a practical system and less than half of this value, and
- the need to lower the cost of transmission of electricity to the land.

## 11.3.  SMALL-SCALE REMOTE AND DISTRIBUTED POWER

### 11.3.1. Background

This category comprises generators with outputs ranging from a kilowatt to several megawatts. In very remote locations, access to the electric power grid may not be cost-effective or even possible. Remote power refers to small-scale stationary applications that rely on electricity generated from these devices as their primary supply. Examples of these applications are telecommunications relay stations, remote industrial operations (e.g., mining, oil drilling/exploration), water pumping, remote village power, and exotic systems for powering satellites and space probes. Distributed power refers to any small scale power generation that can be sited closer to the end-user than a typical large scale power plant and may be connected directly to the end user or to the utility distribution system. Specific benefits of distributed generation include high reliability standby and premium power, peak-shaving, baseload generation, and cogeneration. Its use may delay the need to upgrade transmission and distribution systems.

### 11.3.2. Small-Scale Generation Technologies

11.3.2.1. *General Comments.*   Distributed and remote power technologies include many small-scale versions of technologies described in Section 10.2. These include gas-fired microturbines, solar photovoltaic (PV) systems, wind turbines, and engine generators. The remote power technologies may also include energy storage (i.e., batteries) for the intermittent renewable sources in order to provide continuous power. Power generation technologies that require fuels (i.e., diesel or natural gas) may not be suitable for some remote locations because of the cost and logistics of delivery.

Power customers such as hospitals, computer servers, financial institutions, research laboratories, and factories may need high-reliability power because of the critical nature of their business and the high cost of power disruptions and outages. Siting a small-scale plant nearby or operating their own generator (either as primary power or backup) often makes sense for backup systems for manufacturing plants. For applications that also need heat, turbine, engine, or fuel cell waste heat can often be a useful co-product along of the electricity.

**11.3.2.2. Reciprocating Engines.** Internal-combustion engine generators (e.g., spark ignition and diesel) are the most widely used and mature distributed power generation technologies. The details of their operation are described in the next Section on transportation power plants. These engines are coupled to a generator to provide electricity. They can operate on natural gas, gasoline, diesel fuel, and even landfill/digester gas. Depending upon the size and power output of these engines, the efficiency can range from 25–45%, the higher efficiencies being from larger diesel engines. Emission controls are necessary to reduce the $NO_x$ and CO emissions. Some systems are capable of cogeneration.

**11.3.2.3. Microturbines.** Microturbines are smaller versions of traditional gas-fired turbines (described in Section 11.2.2.2) ranging from tens to hundreds of kilowatts and are very close to commercialization with extensive demonstration and field testing. Typical microturbines use radial flow, have a single stage, and include recuperation to recover some waste heat and boost turbine inlet temperatures. Efficiencies for these microturbines can be as high as 25–30%. They can operate on natural gas, propane, hydrogen, or diesel. Cogeneration is possible with the production of hot water (50–80 °C) and overall efficiencies can be as high as 85%. In general, microturbines are generally compact and light weight, with high reliability and low ($NO_x$) emissions. Efficiencies without cogeneration, however, are not as high as larger gas turbines or combined-cycle systems.

**11.3.2.4. Renewable Sources.** Wind turbines may be used for remote and distributed power applications if the users are located in windy regions. Another renewable energy system suitable for remote and distributed power is photovoltaic-solar power. These technologies are described in previous Sections. They will require energy storage systems to provide continuous power. Typical capacity factors are around 20–25% for photovoltaics and 20–40% for wind systems. However, these systems, which require no fuel and little maintenance, can be sited in appropriate locations that are fairly inaccessible if the resources are available. Small-scale hydropower is another renewable energy source available for remote and distributed power. Systems can range from less than one kilowatt to many megawatts. Smaller systems do not require damming or even significantly change of the flow of rivers. Small systems, with low head, include low dams or weirs to channel water or simply "run of the river". Because these systems may not be able to store significant amounts of water, their electric output may fluctuate with seasonal changes in river flow.

For outer-space applications, solar power is the only renewable source available, which is extremely important because of the difficulty and excessive cost of transporting fuel into orbit for long-term missions. Many satellites and the international space station (ISS) use photovoltaic (PV) solar panels.

11.3.2.5. *Nuclear Sources.*   Nuclear-power sources are used in some military and space applications where high energy density is a requirement for long durations and fuel cannot be resupplied. Nuclear fuels have over one hundred times greater energy density than chemical fuels. Nuclear powered submarines use the Rankine cycle to operate the ship propellers as well as to generate electricity. The use of nuclear power is beneficial for submarines where there is inadequate air for combustion. Other uses of nuclear power include nuclear heat sources to power thermoelectric generators. Thermoelectric power generation is based on the Seebeck effect, which causes a gradient of electric potential when the junctions of two dissimilar conducting or semiconducting materials (A and B) are maintained at different temperatures.

A thermoelectric generator is a heat engine and uses the temperature difference to generate electricity. Thus, the maximum efficiency is given by the Carnot efficiency:

$$\eta = \frac{T_1 - T_2}{T_1} \qquad (11.13)$$

where $T_1$ and $T_2$ are the temperatures at the hot and cold junctions. Thermoelectric generators have no moving parts. Radioisotope-thermoelectric generators (RTG) rely on the radioactive decay of plutonium as their heat source and have flown on a number of NASA space probe missions including Cassini, Voyager, Galileo, Mars Viking, Mars Soujourner, Ulysses, and Pioneer.

## 11.4.   TRANSPORTATION-AUTOMOTIVE POWER PLANTS

### 11.4.1. Perspectives

Transportation is another large energy consumer in the world. In the USA, transportation accounts for 28% of total primary energy use (automobile use constitutes about 23%.) The transportation sector also accounts for about one third of $CO_2$ emissions in the USA. The vast majority of cars and trucks are powered by the combustion of a hydrocarbon fuel in an internal combustion engine (either diesel or spark ignition). The transportation sector is interesting to investigate because it is a significant user of energy and because of its significant environmental impact. Fuel cells are, therefore, undergoing intensive research and development for use as power plants in transportation vehicles. Advanced vehicles and technologies such as electric vehicles and hybrid electric vehicles are also discussed in Sections 11.4.5. and 11.4.6.

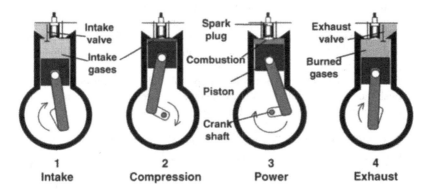

**Figure 11.9** Sequence of steps in a four-stroke spark-ignition cycle.

## 11.4.2. Internal-Combusion Engine

*11.4.2.1. Design Types.* Invented in the latter half of the $19^{th}$ century, internal combustion engines have been developed and refined for well over a hundred years. Designs are still being improved and the performance (power, efficiency, and emissions) can still be increased. The vast majority of automobiles are powered by a reciprocating (piston) internal combustion engine. They can be classified as operating under either an Otto cycle or a diesel cycle. The typical spark ignition engine, which is installed in the majority of passenger vehicles in the USA, is a homogeneous charge spark ignition engine. The air and fuel mixture are mixed before entering the combustion chamber, where ignition is initiated by a spark. Larger vehicles such as buses and trucks use Diesel engines because of their higher efficiency. These are stratified-charge compression-ignition (CI) engines, where air enters the combustion chamber and is compressed. Fuel is injected at the appropriate time and does not have a chance to fully mix with the inlet air before ignition is initiated by compression.

*11.4.2.2. Principles of Operation.* Most automotive engines are examples of four-stroke engines. Four stroke engines follow the sequence of steps in Figure 11.9. These may be briefly described as follows:

(a)  *Intake stroke.* The inlet valve is opened and the piston is moved down to draw in the inlet charge (step 4 → 1).

(b)  *Compression stroke.* The inlet valve is closed and the piston moves up to compress the inlet charge adiabatically (increasing pressure and temperature) (step 1 → 2).

(c) *Power stroke.* Combustion is initiated either by means of a spark or by the compression heating the mixture to the ignition temperature. The heat release raises the temperature and pressure applying a force on the piston and transferring kinetic energy to the crankshaft (step 2 → 3) ultimately powering the wheels.

(d) *Exhaust stroke.* The exhaust valve is opened and the piston moves up to push the burned gas mixture out of the cylinder (step 3 → 4). Figure 11.10 shows a pressure-volume trace for a four-stroke cycle.

11.4.2.3. *Work, Power, Efficiency, and Fuel Economy.* The net work of a given cycle will be equal to the pressure acting over the cylinder displacement:

$$dW = F \cdot dx = F / A \cdot A dx = P \cdot dV \tag{11.14}$$

During the power stroke, the high pressure burning gases will exert pressure on the piston transferring work to the piston and the crankshaft. During the compression stroke, the piston will be transferring work to adiabatic compression of the gases. Thus, the net work will be the difference between these two work values:

$$\text{Net work} = \text{Volume A} - \text{Volume B} \tag{11.15}$$

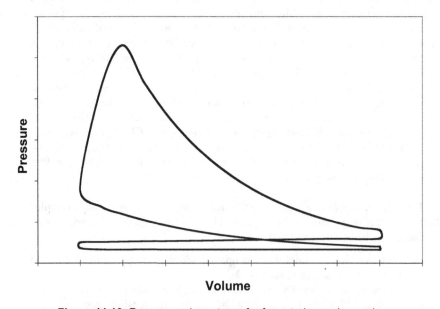

**Figure 11.10** Pressure-volume trace for four stroke engine cycle.

The power would depend on the net work per cycle and the engine speed (rev/min). The thermal efficiency of an engine ($\eta_t$) is determined by dividing the net output work ($W_{net}$) by the heat ($Q_{in}$) added in the fuel combustion:

$$\eta_t = \frac{W_{net}}{Q_{in}} \tag{11.16}$$

The overall net work, power output, and thermal efficiency of the engine will depend upon the specific operating conditions of the engine, such as level of exhaust gas recirculation, spark-timing, turbocharging, fuel-air ratio, throttle conditions, fuel octane value, and engine load and speed. The fuel economy of a vehicle (typically measured in miles/gallon(mpg) or liters/100 km) is dependent on a number of factors including engine efficiency (as mentioned above), mechanical and transmission efficiency, type of drive cycle, vehicle aerodynamics and weight, rolling resistance, etc. The average fuel economy of US vehicles in 2000 was 16.9 miles per gallon, which is equivalent to 14 l /100 km.

### 11.4.3. Internal-Combustion Engines: Spark-Ignition Engines

*11.4.3.1. Design and Principles of Operation.* The four-stroke spark ignition engine is a widely used example of the automotive combustion engine. Nearly all of the passenger cars in the US operate with a spark ignition engine. The fundamentals of operation, described in the last Section, will apply to both spark ignition and compression ignition engines.

The basic differences between a spark-ignition (SI) engine and other types of engines are based on the method of combustion initiation and of fuel-air mixing. In an SI engine, a spark plug is used to generate an electrical arc, which ignites the air-fuel mixture. This allows for precise control of the timing of the combustion event and can be altered under different circumstances. In addition, combustion in typical SI engines is initiated in a near stoichiometric premixed air-fuel mixture. The flame is initiated at the spark plug and propagates through the premixed gases. Heat release in combustion, which occurs fairly quickly, is aided by the turbulence of the intake gases; the pressure and temperature, attained during the power stroke, can increase rapidly before the piston expansion begins to extract energy and lower the pressure and temperature.

*11.4.3.2. Thermal Efficiency and Fuel Efficiency.* The theoretical efficiency of an ideal Otto cycle engine (assuming instantaneous heat release), which can approximate an SI engine operation, is based upon the compression ratio, $r_c$. The thermal efficiency ($\eta_t$) is given by:

$$\eta_t = 1 - \frac{1}{r_c^{\gamma-1}} \qquad\qquad (11.17)$$

where $\gamma$ is the ratio of specific heats of the working fluid at constant pressure and a constant volume. The overall compression ratio in SI engines is limited by knocking (unwanted compression-initiated ignition) in the unburned gas mixture. To limit knocking in high performance engines with increased compression ratios, high octane fuels are required. Typical compression ratios for a SI automobile engine are between 8 and 10, which thus limits the maximum theoretical thermal efficiency to around 50%. In practice, the efficiencies are considerably lower because:

- heat is not added instantaneously (and thus the maximum temperature and pressure are reduced from the ideal case),
- heat is lost to the environment, and
- energy dissipates via engine friction, which depends on engine speed.

Further, in order to meet emissions standards, exhaust gas recirculation (EGR) is utilized, which reduces the thermal efficiency. The best thermal efficiency achieved in the state of the art IC engines is about 35% without the use of EGR.

**11.4.3.3.** *Environmental Impacts: Emissions.* The most significant emissions from SI gasoline engines are CO, unburnt hydrocarbons, and $NO_x$ (mostly NO). The $NO_x$ contributes to tropospheric ozone production, which combined with hydrocarbon emissions produces photochemical smog. Exhaust gas recirculation (EGR) and varying the spark-timing decrease the maximum pressures and flame temperatures, and hence lower $NO_x$ levels, but reduce the thermal efficiency. The levels of these three pollutants are also controlled by a catalytic converter which reduces $NO_x$ and oxidizes CO and hydrocarbons. Current state-of-the-art engine and catalytic converter technologies are able to achieve very low emissions levels corresponding to the California emissions standards program LEV II (see Table 11.1). Other emissions include $SO_2$, due to small quantities of sulfur in gasoline.

**11.4.3.4.** *Typical Power Levels for Specific Applications.* Spark-ignition engines tend to be lighter and cheaper than diesel engines and are thus used for small to medium sized vehicles, typically in light duty or passenger applications. These range from motorcycles and scooters (from about 1–70 kW) to passenger cars and light trucks (20–200 kW). SI engines are also found in some small airplanes and helicopters. Gravimetric and volumetric power densities are approximately 0.15–0.5 kW/kg and 20–50 kW/l (displacement volume) respectively, at the rated maximum power. Power density may be increased with turbo-charging the engine, though the increased temperature and pressure could increase knocking, $NO_x$ emissions, and materials issues.

## TABLE 11.1
## Emissions Standards as Set by the State of California's LEV II program.

LEV II exhaust mass emission standards for new 2004 and subsequent model LEVs, ULEVs, and SULEVs in the passenger car, light-duty truck, and medium-duty vehicle classes

| Vehicle type | Durability vehicle basis (mi) | Vehicle emission category | NMOG (g/mi) | Carbon monoxide (g/mi) | Oxides of nitrogen (g/mi) | Formaldehyde (mg/mi) | Particulate from diesel vehicles (g/mi) |
|---|---|---|---|---|---|---|---|
| All PCs: | 50,000 | LEV | 0.075 | 3.4 | 0.05 | 15 | n/a |
| LDTs 8500 lb | | LEV, option 1 | 0.075 | 3.4 | 0.07 | 15 | n/a |
| GVW or less | | ULEV | 0.040 | 1.7 | 0.05 | 8 | n/a |
| Vehicles in this | 120,000 | LEV | 0.090 | 4.2 | 0.07 | 18 | 0.01 |
| category are | | LEV, option 1 | 0.090 | 4.2 | 0.10 | 18 | 0.01 |
| tested at their | | ULEV | 0.055 | 2.1 | 0.07 | 11 | 0.01 |
| loaded vehicle | | SULEV | 0.010 | 1.0 | 0.02 | 4 | 0.01 |
| weight | | | | | | | |
| | 150,000 | LEV | 0.090 | 4.2 | 0.07 | 18 | 0.01 |
| | (optional) | LEV, option 1 | 0.090 | 4.2 | 0.10 | 18 | 0.01 |
| | | ULEV | 0.055 | 2.1 | 0.07 | 11 | 0.01 |
| | | SULEV | 0.010 | 1.0 | 0.02 | 4 | 0.01 |

## 11.4.4  Internal Combustion Engines: Compression Ignition (Diesel)

While spark ignition engines are the power sources for the majority of passenger cars and trucks, diesel engines are widely used for heavy-duty commercial vehicles, such as large trucks and buses. In Europe, diesel fuel-powered passenger cars are also extensively manufactured due to their higher efficiency. At the present time, more than 30% of new cars sold in Europe have diesel engines. They are also used in large marine propulsion and railway locomotives.

*11.4.4.1. Applications and Principles of Operation.*  The fundamental operating principles of diesel engines are practically the same as those of SI engines. However, unlike the intake stroke of a SI engine, where a premixed fuel and air mixture enters the chamber, during the intake stroke of a diesel cycle, only air is drawn in. Fuel is injected later into the cylinder immediately before ignition. Another difference is that combustion is initiated by adiabatic compression of the fuel/air mixture. Diesel fuel needs the compression-ignition qualities (cetane number), which is not the case with high-octane fuels. Because the fuel is injected into the cylinder when the temperature and pressure are on the increase (i.e., during the compression stroke), the fuel is not completely mixed with air when combustion begins; the rate of heat release will be limited by the rate of mixing of fuel and air. The heat release occurs over a longer period of time during the power stroke and so the maximum pressure and temperature will not be as high as in a SI engine with the same compression ratio.

*11.4.4.2. Thermal Efficiency and Fuel Efficiency.*  Because the rate of heat release for a diesel engine is slower than that in a SI engine, its pressure peak will not be nearly as large as in the latter; this results in a pressure-volume trace different from that shown in Figure 11.10. As a result, the ideal diesel cycle approximates that of heat addition at constant pressure. The theoretical efficiency of an ideal diesel cycle is:

$$\eta_t = 1 - \frac{1}{r_c^{\gamma-1}}\left[\frac{\beta\gamma - 1}{\gamma(\beta - 1)}\right] \tag{9.18}$$

where $\beta$ is the ratio of volumes over which heat is added and pressure is constant. The term in the brackets indicates a term that is always larger than 1, so that the thermal efficiency is always lower than that of a SI engine for any given compression ratio. However, CI engines can operate at higher compression ratios because they rely on compression ignition to initiate combustion, and as a result, they are more efficient than SI engines. The compression ratio is limited to about 20–23 because of $NO_x$ and soot formation, which thus results in a maximum theoretical efficiency of about 55%. In practice, large diesel engines in trucks can achieve thermal efficiencies over 40%. Another contribution to the higher efficiency of diesel engines is the higher energy density of the fuel ($\sim$ 12% higher than gasoline on a volumetric basis).

*11.4.4.3. Environmental Impacts: Emissions.* Diesel engines are similar to gasoline spark-ignition engines in that some of the predominant pollutant-emissions are $NO_x$ and hydrocarbons. However, of equal or greater significance for urban air quality and health are the emissions of particulates (soot) and $SO_2$. Diesel fuels have higher sulfur content and their chemical composition leads to increased particulate formation. Diesels typically run lean (excess air) and as a result, CO emissions are not significant. The $NO_x$ formation rate can be reduced by altering the air to fuel ratio, fuel injection timing, exhaust gas recirculation (EGR), and compression ratio. Particulates and soot formation are dependent on the fuel composition, as well as on compression ratio. Particulate (PM) emissions can have significant impacts on public health and will be the limiting factor for the use of diesels in the future. Further, CI engines are relatively noisier than SI engines.

*11.4.4.4. Power Levels and Performance Factors.* Diesel engines are generally more expensive than SI engines but have longer lifetimes. Because of the higher compression ratios, diesels tend to be heavier and operate at lower speeds (and power). But since they operate more efficiently than SI engines, they are consequently used in applications where efficiency and the longer lifetime are important, including in heavy duty commercial applications like large hauling trucks, agricultural equipment, construction equipment, locomotives, and boats. Gravimetric and volumetric power densities are approximately 0.1–0.4 kW/kg and 18–26 kW/l (displacement volume), respectively, at rated maximum power.

### 4.4.5. Electric-Vehicle Power Plants

*4.4.5.1. General Comments.* Most people do not realize that many early automobiles were electric vehicles. They were competitive with internal combustion engine powered vehicles at the beginning of the 20th century. However, as both technologies developed, it became clear that the internal combustion engine vehicles would have a lower cost, and development of this type accelerated while development of electric vehicles declined. Then, in 1990 the California Air Resources Board adopted a requirement that 10% of all new cars offered for sale in 2003 and beyond must be zero emission vehicles (ZEVs). This reflected renewed interest in advanced batteries and electric vehicle technology. While this rule has since been modified, the mandate stimulated a great deal of work on battery-powered vehicles, since they were the only near-commercial vehicle technology. In recent years, research on electric vehicles has declined significantly as battery technology has not progressed rapidly enough to provide adequate vehicles range. However, many of the advances in a wide range of vehicle technologies, such as aerodynamics, lightweight materials, energy storage, electric motors, and electronic controls, will be useful for internal combustion engine, hybrid, and fuel cell vehicles.

4.4.5.2. *Principles of Operation.* An electric vehicle has a propulsion system that uses motors rather than a gasoline engine. Electrical energy stored on the vehicles, typically in batteries, is delivered to electric motors (via an electric controller) to turn the wheels. Generally, DC motors are more mature and will be used in low-cost and near-term electric vehicles, while AC motors are considered to have more potential benefits in terms of efficiency, reliability, and size.

Electric motors do not require a transmission system because their speeds and torque output are suitable for powering the wheels, although gears may be used to allow the motors to run at higher speed and thereby increase efficiency. Motors can be directly connected to individual wheels, potentially eliminating the need for a differential and improving handling and traction control. Electric motors have enough torque and can start from a stopped point while engines typically require gears to generate enough torque and need to idle at some low rpm, even under no load.

The control and power electronics are used to control the flow of power to the motors. Microprocessor-based control systems use a series of models that describe the driving situation, vehicle, and battery state to translate the driver's demands into the proper energy flow. Driver inputs can include steering, acceleration, and braking, while system control outputs can include motor torque and speed, gear ratio, regenerative braking, and battery charging.

Electric vehicles tend to be highly optimized, and significantly more efficient than one powered by an internal combustion engine. Advances to reduce the road load power (power to move a vehicle at a given speed and acceleration) include lighter weight, lower rolling resistance, and improved aerodynamics while the development of regenerative braking can reduce energy losses.

11.4.5.2. *System Issues.* A major area of research and development for electric vehicles is improvement of the energy-storage system: increasing the energy density and consequently the range between recharging, as well as decreasing the recharge time. The development of higher-capacity energy storage is a key challenge for developing vehicles that are convenient and acceptable to consumers. The ability of batteries to store only limited amounts of energy compared to liquid fuels is a major limitation on the electric vehicle. Battery development has been ongoing for many years. The US Advanced Battery Consortium (USABC) was founded in 1991, composed of the three major US auto manufacturers and the US Department of Energy. There was some progress in battery technology, but the research never led to large leaps in energy density. Other electricity sources, like fuel cells, have prospects to greatly improve the energy density and range by converting hydrogen and air directly to electricity.

Improving the electric-vehicle battery charging was another important area of research because of the typically long charging times. Convenience for the consumer is of primary concern as drivers are used to refueling their vehicles in a matter of minutes, whereas batteries are best charged slowly at a low current and would take several hours. Inductive coupling methods have been developed for rapid charging at high currents.

11.4.5.3. *Environmental Benefits.* Electric vehicles have numerous benefits in large urban areas with poor air quality. The total emissions associated with operating electric vehicles depend on the details of electricity generation (i.e., the type of power plant, fuel, and plant efficiency); however, it is expected that there is a significant reduction in total emissions compared to conventional vehicles. In addition, one of the main benefits is that emissions are shifted both spatially and temporally so that none of the emissions occur from the tailpipe. The generation of photochemical smog and ground level ozone is very location and time dependent so that these shifts can greatly reduce the impact of transportation on urban air quality.

## 11.4.6. Hybrid Powered Vehicles

11.4.6.1. *Unique Characteristics and Benefits.* Hybrid vehicles are automobiles that have two distinct power sources that complement one another to power the vehicle. Hybrids are being developed and utilized because of the main benefit of increased vehicle efficiency as well as the indirect benefit of reduced pollutant emissions ($CO_2$, $CO$, $NO_x$, and VOC). There are various hybrid configurations, but the most common one consists of a SI or CI engine and electrical battery storage as the power sources. Other possible energy storage devices are flywheels and ultra-capacitors. These energy storage devices are described in more detail in the portable power section.

11.4.6.2. *Fundamentals of System Operation.* The configuration of the IC engine-battery hybrid-power source with respect to how mechanical power is supplied to the wheels can be *parallel* and *series* (Figure 11.11.) In the parallel configuration, the SI or CI engine (as described in Sections 11.4.3 and 11.4.4) converts the combustion of fuel into the rotation of the crankshaft that powers the transmission or generates electricity via a generator. Electricity supplied by the battery can power an electric motor that is also connected to the transmission; thus, the two power sources provide power to the wheels. In a series hybrid, there is only one energy path. The engine is connected only to the generator. The series hybrid electric car thus essentially functions as one with an on-board internal combustion engine-powered recharging station. In both designs, the motor/generator stores excess energy in the battery during braking/decelerating or when the power generated by the engine is greater than needed. The battery is then capable of providing the extra power to the transmission for start-up and acceleration.

11.4.6.3. *Thermal Efficiency and Fuel Efficiencies.* Engine thermal efficiency and vehicle fuel economy are increased in hybrid vehicles, as compared

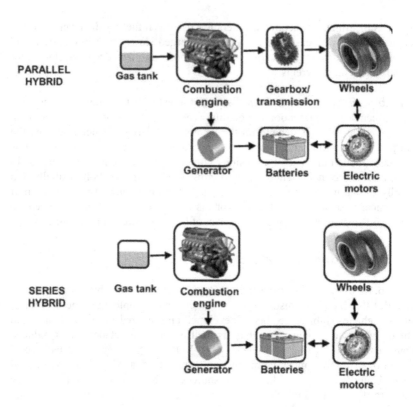

**Figure 11.11** Schematic of parallel and series gasoline-battery hybrid vehicles.

with only gasoline-engine powered vehicles. These gains can be attributed to several factors:

- *Smaller engine size.* Engines can be sized to provide much less power than the maximum required by the car. The hybrid engine will be more efficient for several reasons. Part load efficiencies are typically higher than the efficiency at rated power and the range of power outputs will be much smaller than for a traditional engine. Also, several hybrid designs turn off the gasoline engine when idling.
- *Regenerative braking.* It permits recapturing some of the energy that would otherwise be lost as heat in the brakes by using the electric motors to act as generators and using that energy to charge the energy storage devices.
- *Advanced vehicle design.* Lightweight materials, reduced-rolling resistance, and improved aerodynamics improve fuel economy. The road-

load equation describes the mechanical power requirements ($P_{road}$) for a vehicle under certain driving conditions:

$$P_{road} = C_R M_v g V + \frac{1}{2} \rho_a C_D A_v V^3 \tag{11.19}$$

where $C_R$ is the coefficient of rolling resistance; $M_v$ is the mass of vehicle; $g$ is the acceleration due to gravity; $\rho$ is the air density; $C_D$ is the drag coefficient; $A_v$ is the frontal area of the vehicle; and $V$ is the vehicle speed. Reductions in $C_R$, $M_v$, and $C_D$ will reduce power requirements to power a vehicle at a given velocity. The first two factors increase engine efficiency while the third factor increases fuel usage efficiency for a given engine thermal efficiency. For example, even with a slightly heavier total vehicle weight and similar vehicle aerodynamics, the Honda Civic hybrid model claims a 40% gain in fuel economy over the standard Civic [from 35 mpg (6.75l/100km) to around 50 mpg (9.65l/100km)] because of the inclusion of a smaller, more efficient engine.

Examining a typical drive cycle can help to understand the efficiency benefits of the hybrid vehicle. Figure 11.12 represents a plot of energy requirements as a function of power generated for a typical urban drive cycle of a small automobile. The significant fraction of the energy use occurs at a power below 18 kW. Yet there are occasions that the vehicle demands a power output beyond 40 kW. The hybrid allows the combustion engine (or any primary power source) to be smaller ($\sim$ 20 kW), and the batteries (or secondary energy storage device) can provide power for these infrequent transients.

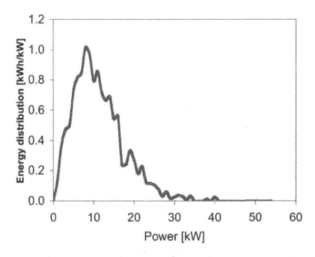

**Figure 11.12.** Typical driving-cycle power requirements.

11.4.6.4. *Emissions and Control.* Hybrid vehicles currently operate on four-stroke spark-ignition engines and have the same emissions characteristics and control processes as conventional vehicles. However, because of improved efficiency, the amount of fuel consumed and pollutants emitted will be reduced. The narrower range of engine operating speeds will also reduce emissions.

## 11.5 PORTABLE POWER

### 11.5.1. Background

Many different devices use portable power sources but the rapid growth in digital and wireless technologies is fueling the need for improvement and higher capacities. Cell phones, laptop computers, wireless organizers, and email devices are all drawing increasing amounts of power as they incorporate more powerful microprocessors, larger, full color screens, and wireless technologies to send and receive data almost anywhere. These high-tech devices, as well as other lower-tech devices, can draw as little as a few milliwatts in standby mode all the way up to 100 W. The most powerful devices can only operate at full power for a few hours before requiring an AC adapter or fully charged battery. Because of these short operating times and the growth of portable electronic devices, a great deal of research is being carried out to develop new power sources for this application, including advanced batteries and direct methanol fuel cells. Alternative energy and electricity storage devices are also discussed.

### 11.5.2. Batteries

Primary and secondary batteries make up the vast majority of energy storage devices used in portable power applications. They range from tiny batteries with capacities of 10 mWh to larger batteries with capacities of 25 kWh. The fundamental electrochemistry of batteries is described in more detail in Chapter 3.

Conventional primary batteries, such as alkaline $Zn/MnO_2$ batteries, are the most common and abundant. Other primary batteries include lithium batteries, which have approximately doubled gravimetric and volumetric energy density compared to the alkaline battery.

Secondary batteries, also called rechargeable batteries, can be reused over thousands of cycles. Conventional and advanced secondary batteries include lead-acid, Ni/Cd, Ni/MH, Li-ion, and zinc-air batteries. The lead-acid battery is the most widely used, lowest cost rechargeable system. It is most often used for automobile starting-lighting-ignition (SLI) systems. Nickel cadmium (Ni/Cd) batteries have higher energy density than lead-acid batteries. Nickel/metal hydride batteries (NiMH) are fairly common in a wide variety of applications including cell phones, laptop computers, cameras, and videocameras. The advanced Li-ion battery utilizes lithium because it has the highest electrochemical potential and currently provides the highest energy density. The zinc-air battery is a hybrid power source essentially combining a battery and a fuel cell. Because it uses an air/oxygen cathode, the zinc can take up a much greater portion of the total volume of the cell, providing a very high energy density. The performance characteristics (gravimetric and volumetric energy density, cost) of secondary batteries are generally poorer than those of primary batteries because of the complex design challenges involved in making a battery that can be recharged over hundreds or thousands of cycles.

## 11.5.3. Environmental Aspects

Because Cd is toxic, Ni-Cd batteries have for the most part been replaced by NiMH batteries, despite the poorer charge retention of the NiMH type. In general, the use of secondary batteries reduces the environmental impact associated with battery disposal.

## 11.5.4. Other Energy Storage Devices

11.5.4.1. *Flywheels.* Flywheels can store mechanical energy in the form of rotating kinetic energy. The benefit of any mechanical energy storage device is that the conversion between electrical energy and kinetic/mechanical energy can be very high using generators and motors. A flywheel is a disc shaped mass rotating about an axis. The kinetic energy stored in a flywheel is given by the equation:

$$E = \frac{1}{2} I \omega^2 \tag{11.20}$$

where $\omega$ is the angular speed (rev/sec) and $I$ is the moment of inertia, the resistance of a mass to angular motion. The further away from the axis the mass is distributed, the greater the moment of inertia and the greater energy storage is possible. In general, flywheels are shaped like bicycle tires with mass distributed around at a fixed distance from the axis. Energy is transferred to the disc by electric motors and extracted by running the motors as generators.

The advantages of flywheels compared to batteries are the very high round-trip efficiencies (electricity to kinetic to electricity), the potential for high power density, and a potentially long life. The disadvantages are the fairly low energy density and the hazard of uncontrolled release of energy upon failure.

11.5.4.2. *Compressed Air.* Large-scale systems exist where air is compressed and stored in underground caverns. Compression occurs when cheap, off-peak electricity is available, and electricity is generated during peak consumption periods by expanding the gas through a turbine-generator set. Some small systems may also be used as energy storage in vehicles or other applications.

Although small scale compressed air energy storage systems are not commercialized, there is some development going on as the secondary energy source for hybrid vehicles. The benefits of compressed air systems are high energy storage density and a potentially simple system with high efficiency. One disadvantage involves the heat that needs to be removed/added as the gas is compressed/expanded.

11.5.4.3. *Ultracapacitors.* Ultracapacitors (also called supercapacitors) can charge and discharge very rapidly, allowing very high-power operation for short times, or extended discharge times at low power. Ultracapacitors store energy electrochemically, although different types exist. Like simple fuel cells or batteries, they are composed of two electrodes and a separator/electrolyte. The electrodes are typically very high surface area porous carbon. Simple ultracapacitors store energy by charging the double layer. As charge is placed onto the electrode, ions migrate to the electrolyte/electrode interface, creating a double layer. In more complex ultracapacitors, some of the ions that enter the double layer are adsorbed on the surface of the electrode, and the charge may be transferred or merely intercalated into the electrode structure. Under these circumstances, the amount of energy/charge stored increases. The rate of charge and discharge may be reduced if charge transfer and ion intercalation reactions occur. Ultracapacitors store much less energy than batteries, but the energy can be released at significantly higher power. More details are given in Chapter 3.

## 11.6. TECHNO-ECONOMIC ASSESSMENTS: FUEL CELLS AND COMPETING TECHNOLOGIES

### 11.6.1. General Comments

This Section presents brief comparative assessments of fuel cells and the competing technologies discussed in the preceding Sections for the following applications:

- utility-scale power generation/cogeneration,
- small-scale remote and distributed power,
- transportation, and
- portable power.

Such comparisons can be rather challenging to make because many of the competitive technologies are in a significantly more mature state of development than fuel cells with well demonstrated value and wide-scale deployment. For these technologies, the operating parameters, system and environmental characteristics, and capital and operating costs are fairly well known. Fuel cells, as well as other alternative technologies, are in a rapid state of development, and much of the information necessary for detailed comparisons is not well known. For example, only a few fuel cells are currently in use, including the alkaline fuel cells developed for the US space program, phosphoric-acid fuel cells commercialized by UTC-Fuel Cells, and a number of demonstration units and prototypes. One final important point is that the competing technologies for fuel cells will, in 5–50 years, also have advanced.

Each of the applications listed above has specific requirements. The following Sections present tables of projected and actual values of important characteristics. The difficulty in comparing vastly different technologies, some of which are well established while others have yet to be proven, should be taken into account when viewing these tables. There are many criteria for judging technologies, and a balance of engineering and economic characteristics will determine how a utility, company, or individual chooses a power source for its specific projects.

## 11.6.2. Power Generation/Cogeneration

The most important characteristics of a power generation technology are reliability, efficiency, service life, capital and operating costs, and environmental impact. Fuel cells are being considered for utility scale power generation because of high efficiency and ultra-low levels of emission of environmental pollutants, and minimal health hazards. Additional benefits include operation with hydrogen, which is a flexible fuel made from any number of sources, including carbon-free renewables. A comparison of some of the engineering and economic parameters for fuel cells and other power generating technologies is made in Table 11.2.

Major competition for fuel cells over the coming decades will be natural gas and gas turbines in combined-cycle power plants. Coal is a very low cost fuel in much of the world and will also be used widely. Advanced "clean coal" power plants are being developed around the integrated gasifier combined cycle (IGCC), or gas turbines. Both IGCC and gas-turbine combined-cycle power plants have high efficiencies (40–60%) and low levels of environmental pollutants, perhaps comparable to fuel cells. Most importantly, their capital costs will likely be significantly lower ($500 to $1000/kW). The reduction of pollutants like $NO_x$, $SO_x$, and particulates can increase the costs of these technologies, but the most expensive environmental regulation would be a constraint on carbon dioxide emissions, especially for coal.

Fuel-cell power plants for this application are in the infant state of development and commercialization. Thermal power plants have been developed and commercialized during the past hundred years and well over several hundred-billion dollars have been spent for their development. On the other hand fuel cells for these applications have been researched and developed to a much lower extent, with investments of a few billion dollars. Thus, there is a need for automation of techniques for the manufacture of components, sub systems, and auxiliaries to lower the cost of fuel cells by at least a factor of four.

**TABLE 11.2**
**Techno-Economic Assessments of Fuel Cells vs. Competing Technologies for Power Generation/Cogeneration Applications**

| Energy-conversion system/fuel | Power level, MW | Thermal Efficiency, % | Lifetime, y | Environmental impact | Fraction of current US power generation, % | Capital Cost, $/kW | Electricity cost, ¢/kWh | Capacity factor |
|---|---|---|---|---|---|---|---|---|
| Fuel cells, PAFC | 0.2–10 | 35–45 | 5–20 | $CO_2$ | 0 | 1,500[a] | 6–10[a] | Up to 95% |
| Fuel cells, MCFC/ gas-turbine hybrid | 0.1–100 | 55–65 | 5–20 | $CO_2$ | 0 | 1,000[a] | 5–9[a] | Up to 95% |
| Fuel cell SOFC/ gas-turbine hybrid | 0.1–100 | 55–65 | 5–20 | $CO_2$, $NO_x$ | 0 | 1000[a] | 5–9[a] | Up to 95% |
| Coal-steam cycle | 10–1000 | 33–40 | > 20 | $SO_x$, $CO_2$, $NO_x$, PM | 50 | 1300–2000 | 2–5 | 60–90% |
| Advanced coal technology, IGCC | 10–1000 | 43–47 | > 20 | $CO_2$, $NO_x$ | 0 | 1500–2,000 | 6 | 75–90% |
| Gas turbine, NG | 0.03–1000 | 30–40 | > 20 | $CO_2$, $NO_x$ | 10 | 500–800 | 3–5 | Up to 95% |
| Combined cycle gas turbine, NG | 50–1000 | 45–60 | > 20 | $CO_2$, $NO_x$ | 2 | 500–1000 | 2–4 | Up to 95% |
| Microturbines | 0.01–0.5 | 15–30 | 5–10[b] | $CO_2$, $NO_x$ | 0 | 800–1500 | 6–10 | 80–95% |
| Nuclear | 500–1400 | 32 | > 20 | Radioactive waste | 20 | 1500–2500 | 1.5–5 | 70–90% |
| Hydroelectric | 0.1–2000 | 65–90 | > 40 | Ecosystem changes, fish | 13 | 1500–3500 | 1–4 | 40–50% |
| Wind turbines | 0.1–10 | 20–50 | ~ 20 | Visual, noise | 0.01 | 1000–3000 | 4–6 | 20–40% |
| Geothermal | 1–200 | 5–20[c] | > 20[c] | $SO_x$, $CO_2$ | 0.04 | 700–1500 | 5–8 | Up to 95% |
| Solar | 0.001–0.1 | 10–15 | 25 | -- | 0.001 | 2000–4000 | 10–40 | < 25% |

[a] Projections
[b] Before overhauls/rebuilding
[c] Depends on geothermal resource

Energy technologies that do not produce carbon emissions include nuclear power and those based on renewable sources (e.g., hydroelectric, geothermal, solar, and wind power). Electricity production could shift towards these sources if carbon dioxide emissions are regulated. Fuel cells potentially have zero carbon emissions if hydrogen is made from electrolysis of water using electric energy derived from these sources, but the cheapest hydrogen for many decades will likely come from natural gas and coal. Therefore, the production of hydrogen will produce $CO_2$. Nuclear power plants have had significant problems with public perceptions for the difficulty of safety and nuclear waste disposal. However, a number of new reactor designs are being developed which could significantly lower cost and improve safety and reliability. If public perception is altered and more emphasis is placed on carbon-free energy sources, nuclear power could continue to provide a significant fraction of our electricity. Hydroelectric and geothermal power plants are limited by available resources, and most of the best sites have already been developed, limiting the extent of future development. Wind turbines are starting to become economically competitive with more traditional electricity sources and rapid development of wind farms is occurring. Some forecasts call for wind power to provide 3–5% of electricity in the USA by 2020.

The benefits of fuel cells for electricity generation are the very high efficiencies achievable with high temperature fuel cells (SOFC and MCFC), especially when coupled with gas turbines in a combined cycle configuration. These systems are able to achieve efficiencies anywhere from 50 to 75%. In addition, the emissions of $NO_x$, particulates, $SO_x$, and other pollutants will be zero or near zero. However, the $CO_2$ emissions of fuel cell systems will depend on how the hydrogen is produced. Coal based hydrogen production would still produce $CO_2$ although this technology, coupled with fuel cell power plants, could reduce carbon emissions due to increased efficiency. However, even with very optimistic cost projections, it will still be a challenge for fuel cells to reach capital costs parity with the fossil fuel combustion based systems. While a great deal of engineering development is currently going into lower cost materials and manufacturing techniques for fuel cells, it is unclear if cost targets necessary to compete with gas turbines are likely. In addition, the hydrogen fuel for these fuel cells will also be higher in cost than other fuels (coal, natural gas), because hydrogen will be produced from these fuels. One of the other main concerns about fuel cells is the lifetime. To date the projected goals are that the electrochemical stacks will last for 40,000 hours with minimal degradation in performance; after this period, the cell stacks may need replacement. However, if the costs for refurbishment for cell stacks can be considerably reduced by automation of techniques and mass production, these costs could be brought down to the levels of maintenance costs of thermal power plants. Further, in the case of the fuel cell power plants, the balance of plant sub-systems could have a lifetime of over 20 years, comparable to those of thermal power plants. Because of the scale-up limitations, lifetimes and capital costs, fuel cell power plants are also being considered for peak-power demands by electric utility, while the large conventional thermal power plants provide the base-load power. For such a case, even though the capital and fuel costs of the fuel cell are relatively high for base-load power, they may not be excessive for peak power generation.

### 11.6.3. Small-Scale Remote and Distributed Power

Because remote and distributed power spans a wide range of engineering and economic requirements, it is difficult to define the most important factors. The goal of this Section is to address some of the possible factors that influence choices made for these applications. For small-scale remote and distributed applications, the important performance characteristics are similar to those for utility-scale electricity generation. In general, economic characteristics such as capital cost and operation, maintenance, and fuel costs and engineering characteristics such as efficiency, reliability, and capacity are key considerations for choosing power sources for these applications.

In general, higher unit costs are often tolerated in these applications because there are no acceptable lower cost alternatives. Providing grid electricity for remote terrestrial applications is generally very costly (involving the building of transmission lines), so that significantly higher capital and operating costs for stand-alone power generation systems are often tolerated. For remote military, marine and space applications, costs can be much higher and very specific engineering requirements can determine which technologies are chosen. By placing generating capacity near the end user, distributed generation often allows the utility to add additional generating capacity to the grid without upgrading substations and transmission lines. In addition, distributed generation also offers the possibility of high reliability standby power in the event of outages, peak power shaving, and cogeneration.

For many remote power applications, gasoline and diesel engine generators are used because they are proven technologies with low capital and fuel costs. In locations that have access to natural gas, propane, or landfill gas, microturbines are at an early stage of commercialization and appear to offer many similar characteristics to reciprocating engines, while also providing the possibility for cogeneration hot water.

Fuel cells are currently used for remote power applications on the Space Shuttle because of the benefits of using hydrogen fuel, which has a high gravimetric energy density. In addition, the product water is used as a potable water source for astronauts. But even for space applications, the energy source will depend upon the requirements of the mission and the resources available. For example, the International Space Station (ISS), when completed, will rely mainly on about 240 kW of solar modules and battery storage. For long duration space probes which do not receive enough solar radiation, NASA has used thermoelectric generators with the radioactive decay of nuclear materials as the heat source.

As with large-scale utility generation, the advantages of fuel cells are high efficiency, near-zero emissions of criteria pollutants, and minimal noise pollution. In remote applications where the difficulties are in generating any power at all, these reasons may be less compelling. In remote power applications, it may be impractical or uneconomical to provide fuel. In these cases, renewable energy sources, such as solar, wind, or small hydro often make sense. Which energy source is utilized will depend upon what resources are available in that location. In general, acceptable solar resources may be more widely distributed, but capital costs are higher than for wind power. The cost of electricity will depend strongly on the capacity factor and quality of the resource. For remote and off-grid homes and buildings or other remote applications, photovoltaic solar panel systems often make sense, especially coupled with battery energy storage, and perhaps an engine generator backup.

Depending upon the availability of fuel in these locations, fuel cells for remote applications may make sense. The use of waste heat in addition to electricity may also be beneficial for specific purposes. Propane or natural gas reformers can produce hydrogen for use in fuel cells. These reformers add to the capital cost of the fuel cell systems. The benefits of fuel cell systems include lower noise than internal combustion engines, significantly higher efficiency, and lower emissions. These smaller fuel cell systems could be phosphoric acid (PAFC) or proton exchange membrane fuel cells (PEMFC).

The benefits associated with distributed generation often permit higher costs and lower efficiency than for central station power. Premium power customers demand high quality and reliable electricity for critical loads such as in hospitals, banks, apartment buildings, factories, and computer server farms. The increase in digital technologies is also increasing the need for high quality electricity. Peak power is also becoming important as utilities begin to charge real-time prices for electricity. Technologies including microturbines and diesel engines are suitable for providing backup and peak power. Fuel cells are a possible alternative to these more conventional technologies.

## 11.6.4. Transportation

The most important performance characteristics of power sources for automotive applications are cost, energy conversion efficiency (%), power density (W/l), specific power (W/kg), energy density (kWh/l), and specific energy (Wh/kg), in addition to lifetime and impact on environment and health. Enthusiasm and interest for the development of fuel-cell powered vehicles was stimulated in the late 1970s with the program *Fuel Cells for Electric Vehicles*. However, with rapidly decreasing oil prices in the 1980s, the incentives for alternative energy programs decreased. The California zero-emission vehicle (ZEV) mandate has renewed interest in fuel-cell powered vehicles. At about the same time, several battery programs with the lead acid battery and the advanced nickel/metal hydride and lithium ion batteries as power sources for electric vehicles were stimulated. In this Section, a techno-economic assessment is made of these power sources for electric vehicles, in comparison with the conventional spark ignition and diesel engine powered vehicles, as well as with the battery powered and hybrid systems (IC

## TABLE 11.3
### Techno-Economic Assessments of Fuel Cells vs. Competing Technologies for the Transportation Applications

| Energy conversion system / fuel | Power Level, kW | Thermal efficiency, % | Specific power, kW/kg | Power density, kW/l | Vehicle range, km | Environmental impact | Capital cost, $/kW |
|---|---|---|---|---|---|---|---|
| Fuel Cells, PEMFC / $H_2$, methanol or gasoline onboard fuel-processor | 10–300 | 40–45 | 400–1000 | 600–2000 | 350–500 | $CO_2$ | 100[a] |
| Fuel Cells, PEMFC direct $H_2$ | 10–300 | 50–55 | 400–1000 | 600–2000 | 200–300 | $CO_2$ | 100[a] |
| SI engine / gasoline | 10–300 | 15–25 | >1000 | >1000 | 600 | $NO_x$, $CO_2$ HC, CO | 20–50 |
| Diesel engine | 10–200 | 30–35 | >1000 | >1000 | 800 | PM, $CO_2$ $NO_x$ | 20–50 |
| CIDI / battery hybrid | 50–100 | 45 | >1000 | >1000 | >800 | PM, $CO_2$ $NO_x$ | 50–80 |
| ICE-battery hybrid / gasoline | 10–100 | 40–50 | ~1000 | ~1000 | >800 | $NO_x$, HC, CO | 50–80[a] |
| Battery / lead acid, Ni-MH | 10–100 | 65 | 100–400 | 250–750 | 100–300 | $CO_2$ | >100 |

[a] projected costs

Note: Capital costs for fuel cell and battery power plants are projected

engine/battery), the latter incorporated into Toyota and Honda hybrid electric vehicles (see Table 11.3). Fuel-cell powered vehicles are much less mature compared to conventional vehicles, while hybrid vehicles have significant experience over several years and are maturing rapidly.

Current and advanced hybrid are the most compelling vehicle technology with respect to increasing fuel economy of vehicles, reducing emissions, and meeting cost goals. The US Partnership for a New Generation of Vehicles (PNGV) Program (1993 to 2001) selected three advanced hybrid vehicle technologies: the diesel engine (CIDI)/battery, the fuel-cell battery, and the gas turbine/battery. The main competition for fuel-cells in a hybrid vehicle is from the diesel engine, which provides high efficiency and fairly low costs. Table 11.3 shows that PEMFC power plants for automobiles and PAFC power plants for fleet vehicles may be able to compete with ICE or diesel engine powered vehicles with respect to performance characteristics. However, the major challenge is in meeting the fuel cell cost targets of less than \$50/kW. At the present time, fuel cell costs are off by a factor of about 100. In addition, the cost of hydrogen fuel will also be significantly higher than either gasoline or diesel fuels. If onboard reforming is to be utilized, the added capital cost of the reformer will also contribute significantly to overall cost. Cost is again the major consideration and the prospects for the dramatic cost reductions necessary to be competitive with internal combustion engines appear unlikely in the near term. The recommended methods for cost reduction are the same as those of power generation/cogeneration applications (see Section 11.6.2).

At the present time, hydrogen, stored as a compressed gas, appears to be the most feasible storage technology. The best cylinders made of composite materials for pressures up to 400 atm, yield a hydrogen content of about 6 wt%. This will amount to between 10–20% of the energy content in gasoline. Thus, the driving range of the vehicle will be limited. However, the higher efficiency of fuel will increase the driving range. Improving hydrogen storage and developing the infrastructure for producing, distributing, storing, and dispensing hydrogen to vehicles will be quite expensive. The challenges may be greater than those associated with reducing particulate and sulfur emissions from diesel engines and developing a higher energy-density battery. A major R&D effort has been proposed by the present USA administration to advance the technologies for hydrogen storage, transmission, and distribution for the transportation application.

$CO_2$ emissions are an important consideration in the future, and therefore, emissions from transportation need to be addressed because it is such a large contributor. While efficiency improvements can reduce emissions, both electricity and hydrogen can be made from carbon-free sources and greatly reduce carbon emissions from automobiles.

## 11.6.5. Portable Power

For portable power applications, the competitors for fuel cell power sources are secondary batteries—Ni/Cd, NiMH, and lithium ion batteries (Table 11.4). Because

**TABLE 11.4**
**Techno-Economic Assessments of Fuel Cells vs. Competing Technologies for Portable Power Applications**

| Portable-power system and fuel | Grav. Energy density (Wh/kg) | Vol. energy density (Wh/L) | Power density (W/kg) | Capital cost ($/kWh) |
|---|---|---|---|---|
| DMFC | > 1000 | 700–1000 | 100–200 | 10–50[a,b] |
| Lead-acid battery | 20–50 | 50–100 | 150–300 | 70 |
| NiCd battery | 40–60 | 75–150 | 150–200 | 300 |
| NiMH battery | 60–100 | 100–250 | 200–300 | 300–500 |
| Li-ion battery | 100–160 | 200–300 | 200–400 | 200–700 |
| Alkaline battery (primary) | 100–200 | 300–450 | 200–400 | < 50 |
| Flywheel | 50–400 | 200 | 200–400 | 200–500[a] |
| Ultracapacitor | 10 | 10 | 500–1000 | ~ 100[a] |

[a] projected costs
[b] $/kW

of the toxic nature of cadmium the NiMH batteries are fast replacing the Ni/Cd batteries for various applications. Also, since 1997, the lithium ion battery has made impressive headway into these markets, the main reasons being its higher specific energy and energy density and its lower self-discharge rate relative to the NiMH battery. Another advantage is the potential of 3 V, whereas the operating potentials for NiMH batteries and fuel cells are 1.1 and 0.7 V, respectively. The number of Li ion cells will therefore be less than in an NiMH battery or fuel cell of the same voltage. Again, the high cost of fuel cells is a hurdle that needs to be overcome, but the advantageous characteristics of fuel cells, such as using higher energy density fuel (compared to batteries), will allow them to be competitive with batteries for specific portable applications such as cell phones, laptop computers, wheel chairs, golf carts, etc. For these applications, the required power levels will range from a few watts to a few kilowatts.

Other energy storage technologies such as flywheels and ultracapacitors do not have high energy density and may not be suitable for many portable applications. But some specialized applications can be found where their specific benefits may be useful, especially since they can charge and discharge at very high power.

## 11.6.6. Conclusions

Fuel cells still have a long and challenging road ahead before they can achieve significant usage in the applications described in this chapter. The major barrier is the high capital cost of fuel-cell systems. A great deal of research and development is underway to make engineering improvements in order to bring down this cost.

Automation of techniques for manufacture and mass production are vital to realize this goal.

The development of advanced technologies for utility scale electric power generation is an important goal because of the potential to increase thermal efficiency of the power plant and greatly reduce the emission of environmental pollutants and $CO_2$. Fuel cells offer the possibility of achieving these goals, along with several other advanced combined cycle systems operating with natural gas and coal. The main challenges for fuel cells are to demonstrate reliability and reduce capital costs. The cost reductions that are necessary to compete economically with the lowest cost technologies available today are unlikely, but there will be growing niche applications as the costs are lowered, and environmental benefits are considered or environmental regulations will be strengthened.

The fuel cell application with the greatest amount of interest is transportation, but it also appears to be the most difficult in the near term. A number of barriers must be overcome before commercial fuel cell vehicles can be at competitive prices and are embraced by consumers. The development of advanced internal combustion engines coupled with hybrid vehicle technology can significantly reduce the environmental impact of transportation energy use and thus reduce the incentives for developing fuel cell vehicles.

It appears that fuel cells will be commercialized first for portable power applications. In part, this is due to the high cost of portable energy (batteries) compared to the cost of fuels like gasoline, methanol, or even hydrogen. Also, in terms of performance, fuel cells appear to be able to meet or exceed the advanced battery systems. Although from a total energy perspective, the use of fuel cells for portable applications will not be a significant fraction of our energy use, it will be an important first step in the growth of fuel cells in a range of applications that will appear throughout the modern energy intensive world.

The presumption that fuel cells are an advanced energy technology that will revolutionize the world cannot be discounted. However, it is unlikely that this will occur in the next decade or so. Too many challenges exist, both from within fuel cell research, development, and engineering as well as from among competing technologies. The benefits of fuel cells are, in large, partly due to their efficiency, few moving parts, low levels of environmental pollutants, and negligible noise pollution; on the contrary the costs tend to be significantly higher. In general, this pattern is not uncommon. Environmental safeguards tend to benefit everyone while leading to higher costs. How and why people make decisions about their energy choices are important and have important policy implications. Environmental and social benefits of fuel cells or any other technology need to be mandated, regulated, or in some way encouraged for promotion of fuel cells to enter the energy sector.

## Suggested Reading

1. J. A. Fay and D. S. Golomb, *Energy and Environment* (Oxford University Press, NY/Oxford, 2002), p.p. 10-18.

2. National Research Council, *Review of the Research Program of the Partnership for a New Generation of Vehicles, First to Seventh Reports* (National Academy Press, Washington, D. C., 1995-2001).

3. United Nations Development Program, *World Energy Assessment: Energy and the Challenge of Sustainability* (United Nations Department of Economic and Social Affairs, World Energy Council, New York, NY, 2000).

4. C. Yang and S. Schneider, Global Carbon Dioxide Emissions Scenarios: Sensitivity to Social and Technological Factors in Three Regions, *Mitigation and Adaptation Strategies for Global Change* **2** -4, 373-404 (1997).

5. W. W. Pulkrabeck, *Engineering Fundamentals of the Internal Combustion Engine* (Prentice Hall, New York, NY, 1997).

6. L. F. Dibal, P. G. Boston, K. L. Westra, and R. B. Erickson (Eds.), *Power Plant Engineering,* (Chapman & Hall, New York, NY, 1996).

7. H. P. Block, *A Practical Guide to Steam Turbine Technology* (McGraw-Hill Book Company, New York, NY, 1996).

8. R. A. Hinrichs, *Energy, Its Use and the Environment* (Saunders College Publishing Company, New York, NY, 1996).

9. T. B. Johansson, H. Kelly, A. K. N. Reddy, and R. H. Williams, *Renewable Energy-Sources for Fuels and Electricity* (Island Press, Washington, D.C., 1993).

10. R. Stone, *Introduction to Internal Combustion Engines* (Society of Automotive Engineers, Inc., New York, NY, 1992).

11. K. C. Weston, *Energy Conversion* (West Publishing Company, New York, NY, 1992).

12. W. Bartok and A. F. Sarofim, *Fossil Fuel Combustion – A Source Book* (John Wiley and Sons, Inc., New York, NY, 1991).

13. J. B. Heywood, *Internal Combustion Engine Fundamentals* (McGraw-Hill Book Company, New York, NY, 1988).

14. J. H. Seinfeld, *Atmospheric Chemistry and the Physics of Air Pollution* (John Wiley and Sons, Inc., New York, NY, 1986).

15. J. A. Barnard and J. N. Bradley, *Flame and Combustion* (Chapman and Hall, New York, NY, 1985), 2nd edition.

16. E. Logan Jr., *Turbomachinery, Basic Theory and Applications* (Marcel Dekker, Inc., New York, NY, 1981).

17. J. O'M. Bockris, *Energy Options* (Australia and New Zealand Book Company, Sydney, Australia, 1980).

18. R. L. Loftness, *Energy Handbook* (Van Nostrand Reinhold Company, New York, NY, 1978).

19. R. Fine, D. Hart, J. Umanetz, and B. McCallum, *New Energy Sources for Today – The Renewable Energy Handbook* (Tutor Press, Toronto, Canada, 1978).

20. H. Thirring, *Energy for Man – From Windmills to Nuclear Power* (Harper and Row/Indiana University Press, New York, NY, 1958).

21. P. W. Beck, Nuclear Energy in the 21st century: Examination of a Contentious Subject, *Ann. Rev. Energy Environ.* **24**, 113 (1999).

22. M. S. Kazimi and N. Todreas, Nuclear Power Performance: Challenges and Opportunities, *Ann. Rev. Energy Environ.* **24**, 139 (1999).
23. S. Srinivasan, R. Mosdale, P. Stevens, and C. Yang, Fuel cells: Reaching the Era of Clean and Efficient Power Generation in the Twenty-First Century, *Ann. Rev. Energy Environ.* **24**, 81 (1999).
24. R. Howes and A. Fainberg (Eds.), *The Energy Sourcebook—A Guide to Technology, Resources and Policy* (AIP Press, New York, NY, 1991).

## PROBLEMS

1. List 5 differences between diesel engines and SI engines and their implications on their use for specific applications.

2. Explain the differences between engine thermal efficiency and automobile fuel economy.

3. Assuming all other parameters to be constant, how would the power to overcome aerodynamic resistance and rolling resistance change between a speed of 55 mph and 75 mph?

4. What is the difference between the net power output of a thermal cycle and the power output of a turbine?

5. Make a graph of how Carnot efficiency changes with temperature (from 0–2000 °C), and compare it with a graph of how the theoretical fuel cell efficiency (based upon the Nernst equation) changes with temperature.

6. Describe three factors that limit the efficiency of internal combustion engines (automotive and stationary).

7. How would the drive train of a fuel cell hybrid differ from an internal-combustion-engine hybrid?

8. If Carnot efficiencies improve with temperature, what limits the highest temperatures at which power plant technologies can operate?

9. Describe the environmental impacts of each of the energy technologies described in this chapter.

# CHAPTER 12

# CONCLUSIONS AND PROGNOSIS

Even though the concept of the fuel cell was discovered by William Grove in 1839, there was no successful attempt to engineer a fuel cell until the work of Francis Thomas Bacon that started in the early 1930s and culminated in the demonstration of a 5-kW fuel cell in 1959. This high temperature, high pressure alkaline fuel cell (AFC) had the capability of current densities of 1000 mA/cm$^2$ at 0.8 V, a performance comparable to the best fuel cells produced today. Pratt and Whitney adapted the Bacon design to produce AFCs for the NASA Apollo and shuttle programs. This development resulted in an enormous amount of activity in fuel cells in the U.S. in the late fifties and early sixties. Two famous projects from that time were the fuel cell tractor at Allis-Chalmers and the development of the solid polymer electrolyte fuel cell at General Electric. Because of trademark issues the latter has been called the PEMFC. These activities also resulted in the development of four other types of fuel cells: the PAFC, DMFC, MCFC, and SOFC. When it became obvious that carbonaceous fuels yielded very low current densities in low temperature fuel cells, interest in fuel cells rapidly waned by the end of the sixties.

Government funding for fuel cells was low during the seventies and eighties. Nevertheless, there was steady progress. The development of carbon-supported platinum catalysts reduced the platinum requirement for the PAFC by a factor of ten. United Technologies Corporation (UTC) delivered two 1 MW PAFCs, one to New York City and the other to Tokyo. During this time, International Fuel Cells, a subsidiary of UTC, developed 250-kW PAFC units for stationary applications. They have sold about 275 of these units. Westinghouse Corporation made considerable progress in developing the SOFC and Energy Research Corporation (now Fuel Cell Energy Inc.) advancing the technology of the MCFC.

In the last decade of the 20$^{th}$ century there was a renewed interest in fuel cells, including fuel cells for transportation and portable devices. The main drivers for

---

This chapter was written by A. Bocarsely and J. McBreen.

fuel cells for transportation are air pollution, global warming, and the heavy reliance of the US on imported oil. At the same time Ballard Power Corporation, a Canadian company, based in Vancouver, made great progress in improving the performance of the PEMFC. As a result, most of the world's automobile manufacturers have large development efforts on the PEMFC. However, several materials issues have to be resolved before polymer-electrolyte-membrane fuel cells (PEMFC) become commercial in automobiles.

Distributed stationary fuel cell power systems offer many advantages. They can be used for combined generation of heat and electric power (CHP). Fuel cells also are a highly reliable source of electricity. In the August 2003 New York City power outage, a UTC Power fuel cell kept the Central Park police station operating during the blackout. In May 2005, there was a similar blackout in Russia. Orgenergogaz, a subsidiary of Gazprom, the largest gas producing company in the world, never lost power at its control facilities because of its PureCell™ 200 system from UTC Power. This, again, demonstrated that independent fuel cell systems can provide uninterrupted power during widespread electrical grid failures.

Until now the PAFC has dominated on-site distributed power market. This market could easily be taken by the PEMFC if the costs come down and operation at 100 °C, or greater, is achieved.

The MCFC is also being considered for on-site power. Fuel Cell Energy, Inc. (FCE) has several sub megawatt plants operating in the U.S., Europe, and Japan. FCE is also developing an ultra-high efficiency hybrid system, a power plant designed to use the heat generated by the fuel cell to drive an unfired gas turbine for additional electricity generation. The overall efficiency of the system is expected to approach 75%. The MCFC needs a carbonaceous fuel. Part of the anode effluent has to be routed to the cathode to maintain an invariant carbonate electrolyte during cell operation.

Several SOFC technologies are also being considered for on-site generation. These include both tubular and flat plate designs, based on an oxide conducting yttria stabilized zirconia (YSZ) electrolyte. The sizes range from 5-10 kW, with operating temperatures in the 700–1000 °C range. Ceres Power Ltd. of the U.K. is developing an interesting low temperature SOFC. This is based on technology developed by the late Brian Steele and co-workers at Imperial College, London. The electrolyte is ceria gadolinia oxide (CGO). With this electrolyte, the cell can be operated in the 500-600°C range. This permits the use of stainless steel for the bi-polar plates and many of the-balance-of-plant components. The SOFC can be fueled by either hydrogen or carbonaceous fuels.

The introduction of a new energy infrastructure must be considered a revolution. However, it is not likely that we will wakeup tomorrow, next year or even in the next decade and find a transformed hydrogen economy with fuel cell powered automobiles and solar/fuel cell energy supplied to our neighborhoods. It is more likely that the introduction of fuel cells (hydrogen and other fuels) will first occur in specialized niche markets where the fuel cell offers a specific advantage that overcomes issues of efficiency, cost, and easy availability of the fuel supply. Residences that are off-grid, convenient applications such as an easily recharged long lasting direct methanol fuel cell for laptop computers, or load-leveling applications in conjunction with power generating facilities are possible entry-level applications. In contrast to the wholesale adoption of a *hydrogen or alternate energy economy*, a more realistic scenario can be constructed by considering two important features about hydrogen and how we use energy today: first, hydrogen is not a primary fuel and second, we currently have a mixed energy supply.

As we look at our current energy landscape, it is easy to see that we do not have a monolithic energy structure. Rather, we utilize a variety of energy sources and energy conversion devices depending on the exact application, pricing structure, length of time that the energy need be supplied, availability of the supply, and historical moment. More recently, environmental impact has started to become a player in this consideration, and at this moment in the first decade of the second millennium (2005), the political impact of using energy supplied by a second nation (energy security) is an ardent personal and political concern in the United States. Often these requirements are in conflict, and thus, we design compromise solutions. A good example of this type of interaction was the introduction of electrical in-floor heating as a housing option in the United States during the 1970's. With this introduction, homeowners had three energy options, electrical, gas, and oil. At its introduction, electrical heating was toted as efficient, cost effective, highly controllable, and clean to the interior environment. It had drawbacks related to high installation (capital) costs and high operating costs, since the cost of electricity from fossil sources typically costs more than the direct use of those sources. Yet, these limitations were expected to be overcome (in at least part of the population) by the efficiency and control advantages that electricity provides. Nonetheless, today in-floor electrical heating is a rarity. Its promoters did not foresee improvements in the efficiency of gas and oil heating, along with improved control of those systems. One fall out of this electrical heating technology is the use of zone heating by most modern heating systems independent of the energy source or energy conversion device. Today, we still see a very aggressive competition between oil and gas heating in those parts of the United States where both are available. In this case, the main driver is the relative cost of the two supplies or the cost of the energy conversion devices (furnaces). However, advertisers often point out the efficiency of conversion technology, impact on the environment of one or the other technology, and the anticipated availability (dependability) of the energy supply. Given this state of affairs it is likely that a mixture of energy sources and technologies will continue. Thus, the portent of a *hydrogen economy* should be modified to a situation where hydrogen and its associated infrastructures are added to the mix of potential fuel sources. Put simply, today we do not find a "one size fits all" approach to energy to be viable or desirable, so why should one believe that in the future there will be a single, all encompassing, energy source? The concept of a *pure* hydrogen economy with fuel cells associated with most end uses is unrealistic based on historical and present energy utilization.

It is equally important to understand that hydrogen is not a primary fuel in contrast to fossil fuels and nuclear fuels as well as renewable energy sources such as wind, geothermal, and solar. Hydrogen cannot be directly mined or harvested on the

Earth; it must be produced from a primary source such as the fuels and energy sources listed above. Thus, a direct comparison of a *hydrogen economy* to a *fossil-fuel economy* for example, is a comparison of apples to oranges. In fact, since our current dominant source of hydrogen is from natural gas processing, the relative advantages of hydrogen versus fossil fuels are not the important issues. Rather, given today's implemented technology a hydrogen fuel stream requires a *fossil-fuel economy*. While this may not be the case in the future, it will always be true that hydrogen must be produced from a primary energy source. Thus, in the pure *hydrogen economy* the typical source of hydrogen is from solar energy. Based on this analysis, the advantages and disadvantages of hydrogen as a fuel, should be weighed against other energy carriers, the most obvious being electricity.

In the context of this analysis one might reasonably envision an alternate energy economy in which fossil fuels along with renewable energy sources provide a mix of primary energy sources with hydrogen and electricity both utilized as secondary energy sources providing transportation and storage opportunities. In general (but not specific), in such a scheme it makes little sense to generate hydrogen from electricity since there is a significant energy loss in carrying out the conversion, and electricity is typically the desired end product. So under what conditions should one form hydrogen? The simple answer appears to be when either the infrastructure to bring electricity where it is needed is not available, or when it is inconvenient to use standard electrical supply and storage strategies. This analysis assumes that traditional *energy thinking* prevails in which environmental impact is not considered a determining factor. That is, traditionally both government incentives and free market forces have been relatively blind to degradation of the environment due to energy choices. Hydrogen and other renewable fuels become more attractive if this is not the case. To that end, the discussion of hydrogen as a fuel (or energy carrier) needs to be expanded. The initial primary concept behind the *hydrogen economy* is one of environmental compatibility. The key argument is that a hydrogen fuel used in conjunction with a fuel cell only produces water as a product; none of the environmental contaminates are associated with the conversion of fossil fuels to useful energy. This assumption needs to be examined more critically. This premise is only true if the hydrogen is produced in a renewable manner, for example by the electrolysis of water using solar, wind, or hydro- generated electricity. If instead, hydrogen is produced from hydrocarbon sources or by the electrolysis of water using electricity that is generated by burning hydrocarbon sources, then the production of the hydrogen generates both carbon dioxide (a key greenhouse gas) and potentially other gas phase pollutants. One might elect to mitigate this problem by using appropriate pollution abatement technologies, but this latter hydrogen cycle is not fundamentally environmentally benign. Given the current high cost of producing hydrogen from water by electrolysis, it is likely in the short term that hydrogen will continue to be produced directly from hydrocarbon sources, the major methodology for generating today's commercial hydrogen.

Since it is likely that in the near term hydrogen will not be produced renewably, one should also consider other potential fuels that might in part meet the environmental and energy needs. Many of these fuels could be considered hydrogen carriers within the context of a *hydrogen economy* and would include substances such as methanol, ethanol, sugars, and molecular hydrides. Of these fuels, only the direct methanol fuel cell is close to commercial viability today; however, all of the other substances have been considered for fuel cell applications including hydrides such as hydrazine and sodium borohydride. Depending on the fuel employed, one has to evaluate the environmental impact. Thus, for example a direct ethanol fuel cell would generate carbon dioxide as a product, thus greenhouse gas emission is not avoided. Nonetheless, production of hydrogen from fossil fuel sources has the same limitation. Further, if the ethanol is produced from fermentation of biomass, it can be argued that the process is *carbon neutral* since only previously fixed carbon dioxide is emitted.

Given this discussion and forecast, one can then consider the role of fuel cells in the near future and far future energy supply. As discussed in the pages of this text, the fuel cell does not represent a single device, but rather an electrochemical system that is flexible both with respect to the fuel employed (hydrogen vs. direct alcohol, etc.) and operating conditions (temperature, pressure, load, etc). Thus, the fuel cell meets the expected need for a mix of energy generating devices. Of course, within the concept of a diversity of energy sources and transformations, it must be immediately conceded that while the fuel cell offers new and useful opportunities, it will not be the only system of interest and thus, will compete with other alternatives. Therefore, one needs to ask: For which applications does the fuel cell represent the best available solution? In asking this question, one inevitably raises the issue of cost. However, the answer to this question is always open-ended since there is not a current fuel cell market upon which to base a prediction. Thus, there are always a number of "ifs" strung together before an answer can be provided. For the purposes of this discussion, we simply assume that the fuel cell represents an expensive (but not outrageously expensive) alternative. Therefore, in considering the near term, the benefit of convenience or the simple fact that there is no other alternative will be necessary to overcome the intrinsic cost of a new device with no predefined market. The claim of intrinsic cost is based on the assumptions that early generation commercial cells will require the use of a platinum-based electrocatalyst and an expensive electrolyte independent of the type of cell employed. Of the various hydrogen fuel cells that have been developed the PEMFC has received by far the most industrial R&D attention for the many reasons elucidated in this volume. However, interest in SOFC's have been growing.

Based on the criteria put forth here, it would appear that an early market entrance might be available for the direct methanol fuel cell. It's ability run low power demand small electronics such as laptop computers and cell phones for extended time periods along with ease of refueling (i.e., no need to electrically recharge) will provide these devices with extended capability and convenience that may negate higher costs. In addition, introduction of the fuel cell with a methanol fuel eliminates many of the roadblocks associated with the delivery and storage of hydrogen.

Hydrogen, however, may also have a role in the early fuel cell market. Although, there is currently great attention on hydrogen as a fuel in an electric car, this is not an obvious early market application because of the storage, volumetric energy density, and supply issues coupled with the present cost and reliability of fuel cells. However, because of hydrogen's excellent mass energy density and the efficiency of the hydrogen fuel cell, it may have a role to play in stationary fuel cell applications or in the powering of large vehicles where volumetric storage is less of an issue and the demonstration of functional fuel-cell technology holds market advantage from a public relations and advertising point of view. Thus, electricity generation load leveling, for example, may be an early beneficiary of hydrogen/air fuel cells. This stationary application clearly requires the conversion of electricity to another energy form that is storable. Hence, excess electricity could be used to split water, generating stored hydrogen. When an increase of electrical demand occurs a hydrogen fuel cell could then meet that demand. While load leveling is an important concern today, since electrical generation is continuous, independent of the demand cycle, it can be expected to be even more important as alternate energy sources are tapped. Many renewable energy sources such as wind and solar are by their very nature intermittent, thus load-leveling is not only desirable, but also enabling.

Similarly, large vehicles (buses, trucks, ships, etc.) that have available storage space for hydrogen and can be run as a fleet where they will receive the needed maintenance, are a possible implementation application for hydrogen fuel cell demonstration technology. In these cases, the advantage is not that other technologies fall short, per se, but that demonstration of a new environmentally clean technology that can eventually be based on a totally renewable energy source holds economic benefit. It is quite interesting to note that in this latter case, there is a fundamental chicken-versus-egg issue that only fleet operations can alleviate. Namely, if one builds a hydrogen fuel cell vehicle, how can it be fueled given the lack of a national hydrogen supply system? On the other hand, if one were to build a national hydrogen-supply system, who would use it, given the limited number of fuel cell vehicles anticipated to be available during the next decade? Thus, only a fleet type operation that provides hydrogen in a localized area for a specific set of vehicles can currently take advantage of hydrogen fuel cell technology.

Independent of ones prognostications about the implementation of a hydrogen economy or the introduction of consumer grade fuel cells, one conclusion is very clear. The fuel cell stacks presently available do not meet minimum criteria for broad-based introduction into society. Issues of reliability, cost, and efficiency still must be addressed with new materials, new engineering strategies, and an inclusive systems approach. Further, to date, little work has gone into manufacturing technology for fuel cells. While some effort has been focused on the manufacturing of hydrogen/air PEMFCs, facilities that can produce a large quantity of stacks with good quality assurance have not come on line yet. Perhaps the best example of a current manufacturing facility is United Technologies' fuel cell plant.

Researchers that can take a fresh look at the present technological issues and can step across the traditional boundaries of science and engineering while looking at the practical challenges faced by a consumer friendly fuel cell are needed. For PEM based systems, new polymer electrolytes or a decrease in the expense of sulfonated-perfluorocarbon membranes, lower cost, high efficiency catalysts, and improved engineering of the electrode-catalyst-electrolyte interface are mandatory. Long cell lifetime with limited maintenance must be demonstrated. This is a combined materials and systems engineering challenge. For SOFCs, one must add to this list of challenges the need to maintain system integrity through severe thermal cycling along the formation of high quality gas tight metal on ceramic seals.

Today, we find ourselves in a situation where electrochemist concern themselves with the current-voltage properties of the fuel cell along with the development of key material interfaces while materials scientists focus on new membrane materials, and engineers limit their prospects to stack design, transport issues and systems integration. Yet, as one develops an understanding of present fuel cell limitations, it becomes clear that the necessary solutions transcend all of these boundaries. This is the challenge put forth to the readers of this text.

# INDEX